Energy Systems in Electrical Engineering

Series Editor

Muhammad H. Rashid, Florida Polytechnic University, Lakeland, USA

Energy Systems in Electrical Engineering is a unique series that aims to capture advances in electrical energy technology as well as advances electronic devices and systems used to control and capture other sources of energy. Electric power generated from alternate energy sources is getting increasing attention and supports for new initiatives and developments in order to meet the increased energy demands around the world. The availability of computer–based advanced control techniques along with the advancement in the high-power processing capabilities is opening new doors of opportunity for the development, applications and management of energy and electric power. This series aims to serve as a conduit for dissemination of knowledge based on advances in theory, techniques, and applications in electric energy systems. The Series accepts research monographs, introductory and advanced textbooks, professional books, reference works, and select conference proceedings. Areas of interest include, electrical and electronic aspects, applications, and needs of the following key areas:

- Biomass and Wastes Energy
- Carbon Management
- Costs and Marketing
- Diagnostics and Protections
- Distributed Energy Systems
- Distribution System Control and Communication
- Electric Vehicles and Tractions Applications
- Electromechanical Energy Conversion
- Energy Conversion Systems
- Energy Costs and Monitoring
- Energy Economics
- Energy Efficiency
- Energy and Environment
- Energy Management, and Monitoring
- Energy Policy
- Energy Security
- Energy Storage and Transportation
- Energy Sustainability
- Fuel Cells
- Geothermal Energy
- Hydrogen, Methanol and Ethanol Energy
- Hydropower and Technology
- Intelligent Control of Power and Energy Systems
- Nuclear Energy and Technology
- Ocean Energy
- Power and Energy Conversions and Processing
- Power Electronics and Power Systems
- Renewable Energy Technologies
- Simulation and Modeling for Energy Systems
- Superconducting for Energy Applications
- Tidal Energy
- Transport Energy

The books of this series are reviewed in a single blind peer review process by the main editor and/or the advisory board.

Ersan Kabalci
Editor

Modern Control Methods for Power Electronics

Theories and Applications

Springer

Editor
Ersan Kabalci
Department of Electrical and Electronics Engineering
Nevsehir Haci Bektas Veli University
Nevsehir, Türkiye

ISSN 2199-8582 ISSN 2199-8590 (electronic)
Energy Systems in Electrical Engineering
ISBN 978-981-92-1973-5 ISBN 978-981-92-1974-2 (eBook)
https://doi.org/10.1007/978-981-92-1974-2

This Springer imprint is published by the registered company Springer Nature Singapore Pte Ltd.
The registered company address is: 152 Beach Road, #21-01/04 Gateway East, Singapore 189721, Singapore

Preface

Power electronics has become a basis of modern electrical and energy systems by enabling efficient conversion, control, and management of electrical power in applications ranging from renewable energy and electric vehicles to smart grids, industrial drives, and consumer electronics. As power electronic converters grow in complexity and performance requirements become more strict, conventional control strategies are often insufficient to address challenges such as nonlinear dynamics, parameter uncertainties, high switching frequencies, and real-time implementation constraints. This evolution has driven the rapid adoption of advanced, adaptive, and intelligent control methodologies in power electronics.

The book *Modern Control Methods for Power Electronics: Theories and Applications* is written to provide a comprehensive and structured presentation of recent control techniques applied to power electronic systems. It connects fundamental control theory with modern computational and data-driven approaches by offering both analytical accuracy and practical relevance. The book is organized to guide the reader from foundational concepts to advanced and emerging control strategies which make it suitable as both a textbook and a reference for research and industrial applications.

A brief overview of the contents of each chapter is provided as follows.

Chapter "Introduction to Advanced Power Converters and Control Methods" introduces modern power electronic converters and their role in energy and power systems. It discusses converter topologies, dynamic modeling approaches, control objectives, and performance metrics. The chapter sets the foundation for advanced control by highlighting the limitations of classical techniques and motivating the need for modern and intelligent control methods.

Chapter "Linear and Nonlinear Control for Power Converters" presents linear control techniques such as PID, state-space control, and frequency-domain methods that are followed by nonlinear control approaches designed to power electronic systems. The importance of techniques like feedback linearization, sliding mode control, and nonlinear modeling for controlling major nonlinearities, and fast dynamics is highlighted and discussed in detail.

Chapter "Digital Control of Power Electronic Systems" focuses on practical implementation, covers digital control fundamentals by including discretization methods, sampling effects, quantization, and delays. Digital controller design for power converters, microcontroller, and DSP implementation issues, and real-time constraints are examined to bridge theory with hardware realization.

Chapter "Lyapunov-Based Control Methods for Power Converters" introduces Lyapunov stability theory as a systematic tool for controller design and stability analysis. Energy-based and Lyapunov-based control strategies are developed for power converters that ensure global or asymptotic stability under nonlinear and time-varying conditions.

State estimation plays a critical role in modern control systems. Chapter "Extended Kalman Filter" focuses on the extended Kalman filter (EKF) for nonlinear power electronic systems. It explains theoretical formulation, implementation steps, and practical considerations with applications in state estimation, parameter identification, and sensor reduction.

Chapter "Fuzzy Logic-Based Control of Power Converters" introduces fuzzy logic control as a rule-based, model-free approach suitable for uncertain and nonlinear systems. Fuzzy controller design, membership functions, inference mechanisms, and tuning strategies are presented along with case studies demonstrating robustness and simplicity in power converter applications.

Artificial neural networks (ANNs) are explored as powerful tools for modeling, control, and estimation in Chapter "Artificial Neural Network-Based Control in Power Electronics". This chapter discusses ANN architectures, training methods, and real-time implementation, highlighting their ability to learn nonlinear behaviors and improve control performance in complex power electronic systems.

Chapter "Adaptive and Robust Control Techniques in Power Electronics" addresses parameter variations, load disturbances, and modeling uncertainties. Adaptive control strategies and robust control frameworks are presented, emphasizing their ability to maintain performance and stability under changing operating conditions and external disturbances.

Sensorless control techniques are discussed as cost-effective and reliable alternatives to sensor-based systems. Chapter "Sensorless Control of Power Converters" covers observer-based and estimation-based methods for eliminating physical sensors while maintaining accurate control and system performance.

Model predictive control is presented as an advanced control strategy capable of handling multivariable systems and constraints. Chapter "Model Predictive Control (MPC) in Power Electronics" explains MPC principles, cost function formulation, and real-time implementation challenges with applications to high-performance power converters and drives.

Chapter "Internal Model-Based Control for Grid-Connected Inverters" focuses on grid-connected inverter control using internal model principles. It discusses synchronization, harmonic compensation, power quality improvement, and compliance with grid codes that make it particularly relevant to renewable energy and smart grid applications.

Broader artificial intelligence (AI) techniques beyond neural networks are introduced in Chapter "Artificial Intelligence Control in Power Electronics". It covers intelligent decision-making, optimization-based control, and hybrid AI approaches, highlighting their potential to enhance autonomy, efficiency, and adaptability in power electronic systems.

As the final one, Chapter "Machine Learning Control in Power Electronics" explores emerging machine learning-based control methods including supervised, unsupervised, and reinforcement learning. It discusses data-driven modeling, learning-based control design, challenges in training and deployment, and future research directions in intelligent power electronics.

This book is intended for graduate students, researchers, and practicing engineers in power electronics, electrical engineering, and control systems. It can serve as a textbook for advanced courses, a research reference, and a practical guide for implementing modern control strategies.

The authors hope that this work will support learning, innovation, and further advancements in the rapidly evolving field of power electronics and control.

Nevsehir, Türkiye Ersan Kabalci

Contents

Introduction to Advanced Power Converters and Control Methods

Ersan Kabalci, Aydin Boyar, and Ayse Kocalmis Bilhan

Abstract This chapter presents a comprehensive overview of advanced power converters, covering DC–DC, AC–DC, DC–AC, and AC–AC conversion technologies that are critical in modern power electronic systems. It begins with DC–DC converters, detailing non-isolated and isolated topologies such as buck, boost, buck-boost, Ćuk, flyback, forward, push–pull, and dual-active bridge converters, including interleaved and soft-switching designs that improve efficiency and reduce electromagnetic interference (EMI). The AC–DC section addresses both uncontrolled and controlled rectifiers, single- and three-phase structures, and modern Power Factor Correction (PFC) techniques including bridgeless and totem-pole PFC. The DC–AC converter section discusses inverter topologies such as multilevel and modular inverters, grid-connected and standalone inverters, and their use in renewable energy and smart grids. The AC–AC section focuses on matrix converters, cycloconverters, variable frequency drives (VFDs), and solid-state transformers (SSTs) used in high-performance and high-frequency applications. This chapter serves as a foundational reference for researchers and engineers in the field of power electronics, highlighting the principles, performance criteria, and future directions of advanced converter technologies.

Keywords DC–DC · AC–DC · DC–AC · AC–AC · Converter

E. Kabalci (✉) · A. Boyar · A. K. Bilhan
Electrical and Electronics Engineering, Faculty of Engineering and Architecture, Nevsehir Haci Bektas Veli University, Nevsehir, Turkey
e-mail: kabalci@nevsehir.edu.tr

A. Boyar
e-mail: aydinboyar@nevsehir.edu.tr

A. K. Bilhan
e-mail: akbilhan@nevsehir.edu.tr

E. Kabalci (ed.), *Modern Control Methods for Power Electronics*, Energy Systems in Electrical Engineering, https://doi.org/10.1007/978-981-92-1974-2_1

Nomenclature

ASIC	Application-specific integrated circuit
DAB	Dual active bridge
DER	Distributed energy resource
DSP	Digital signal processor
EMI	Electromagnetic interference
ESS	Energy storage system
EV	Electric vehicle
GaN	Gallium nitride
HFT	High-frequency transformer
IGBT	Insulated-gate bipolar transistors
LRC	Load resonant converter
MEA	More-electric aircraft
MPC	Model predictive control
MPPT	Maximum power point tracking
NPC	Neutral-point clamped
PFC	Power factor correction
PRC	Parallel resonant converter
PWM	Pulse width modulation
QRC	Quasi-resonant converter
SCR	Silicon controller rectifier
SEPIC	Single-ended primary inductance converter
SiC	Silicon carbide
SMC	Sliding mode control
SPRC	Series-parallel resonant converter
SRC	Series resonant converter
SST	Solid-state transformer
V2G	Vehicle-to-grid
VFD	Variable frequency drive
VPP	Virtual power plant
ZCS	Zero current switching
ZCS-MR	Zero-current multi-resonant converter
ZVS	Zero voltage switching
ZVS-MR	Zero-voltage multi-resonant converter

1 Introduction

Power electronics has become an indispensable field in modern electrical and energy systems owing to its crucial role in efficient power conversion, energy management, and automation technologies. As the technological advancements force for higher efficiency, compact designs, low power consumption and smart energy solutions;

power converters have evolved to meet the increasing demand across various sectors, including renewable energy, electric vehicles (EVs), industrial automation, and smart grids. The development of electrical devices that can manage high power levels is fundamental to the evolution of power electronics. The restrictions and capabilities of power semiconductor devices have always played a significant role in shaping this field of research. The field developed in tandem with advancements in device technology until the late twentieth century, particularly prior to 1990. It often involved the development of devices to achieve the required power, voltage, and current ratings before new applications could be made available. Because of this dynamic, power electronics became a "device-driven" field, with a concentration on what the devices could accomplish rather than what the applications required (Rashid 2024a).

However, a significant paradigm change took place in the early 1990s. The focus shifted from the devices themselves to their applications as high-performance power semiconductors, such as metal–oxide–semiconductor field-effect transistors (MOSFETs), insulated-gate bipolar transistors (IGBTs), and later wide bandgap devices like silicon carbide (SiC) and gallium nitride (GaN), became more mature and commercially available. This change marked the start of the "application-driven" era. Engineers could now assume the availability of suitable switching devices for most practical power levels, allowing them to focus on optimizing systems for specific requirements such as efficiency, reliability, cost, size, and environmental footprint. This change was driven by two major developments. First, the widespread availability of modern semiconductor devices allowed researchers to develop systems customized to a wide range of power applications without having to wait for new device capabilities to appear. Second, the increased need for smaller, lighter, and more integrated electronic systems, powered by consumer electronics, electric vehicles, renewable energy systems, and the aerospace industry, pushed power electronics into a wider range of demanding applications. The drive toward downsizing and system integration continues to broaden the scope of power electronics, including it into devices where power conversion must be smooth and discrete.

Although semiconductor technology is still advancing, particularly with the use of wide bandgap materials, the industry is currently being driven by novel applications of these technologies. These modern materials have faster switching rates, higher thermal conductivity, and can function at higher voltages and temperatures, resulting in more compact and efficient power conversion devices. However, research and development are guided by system-level objectives, such as incorporating power converters into distributed renewable energy systems, high-efficiency data centers, or electric vehicles. Power electronic systems are fundamentally responsible for controlling the flow of electricity between sources and loads. Therefore, it is imperative that this energy management be carried out precisely, effectively, and consistently. The primary purpose of a power converter is to regulate the flow of energy from a source, like a solar panel, battery, or power grid, to a load, like an electronic device, motor, or light. Different parts of the converter cooperate to direct electrical energy along predetermined paths without needless loss or interference. Since the converter is directly connected to the source and load, any energy dissipated within it affects overall system efficiency. As a result, one of the fundamental principles of power electronics is the objective of lossless energy conversion. Power converters rely largely on switching components, which are typically semiconductor devices that turn on and off quickly to control the flow of electrical energy.

An ideal switch has no energy loss since it conducts current with no voltage drop when turned on, and it completely stops current regardless of voltage when turned off. This ideal behavior assures that the power dissipated, or the product of voltage and current across the switch, is always zero. Although practical devices differ from this ideal, contemporary semiconductor switches are getting closer to this performance. Mechanical switches, such as those seen in domestic lighting, demonstrate the concept of ideal switching, which has high reliability and little energy loss. These devices typically have a switching cycle count of over 100,000 and can function for decades. However, they are not suitable for high-frequency applications due to their mechanical nature. On the other hand, power electronic systems often require switching operations at tens or even hundreds of kilohertz to provide fast dynamic response and small passive components. This requirement is met by semiconductor switches, which operate quickly and without moving components, enabling power converters practical and efficient. In addition to switches, power converters use energy storage devices such as capacitors and inductors, which can temporarily store energy and release it when needed, thereby smoothing out voltage and current variations. Magnetic transformers are also widely used, particularly in applications that need electrical isolation or voltage transformation. These passive components, such as the switches, are essentially lossless and help the converter manage energy efficiently (Rashid 2024a; Erickson and Maksimović 2001). An important point is that power conversion is not just a passive procedure. It requires intelligent control. A control system analyzes a variety of factors from the source and load, as well as user or designer inputs, to determine the precise timing and sequencing of switching actions. This control is usually accomplished with low-power analogue and digital electronics, such as microcontrollers, digital signal processors (DSPs), or application-specific integrated circuits (ASICs). The system uses feedback and control algorithms to ensure that the power converter satisfies performance objectives such as voltage regulation, current limiting, power factor correction, and thermal management. A full power electronic system consists of the power converter circuit which consists of passive components and switches, the source, the load, and the control subsystem. Each component is crucial to ensure that the overall system runs smoothly, consistently, and safely. The principles of power electronics such as effective switching, intelligent control, and careful energy management are essential to the functioning of any system, whether it is generating electricity for an electric vehicle motor, powering a household appliance, transforming solar energy into grid-compatible electricity, or facilitating fast charging for mobile devices. Power electronics will only become more widespread as technology develops because of the demand for greater performance electronics, more intelligent, autonomous systems, and sustainable energy. This makes power electronics not only a fundamental technical field, but also a dynamic and continuously changing area of innovation.

This chapter provides a comprehensive introduction to advanced power electronics converters, control structure of power converter, and emerging advancement in power electronics controllers focusing on their importance in modern applications, historical evolution, and the key challenges and advancements shaping their future.

1.1 Importance of Power Electronics in Modern Applications

Power electronics is the foundation of modern energy conversion, regulation, and control systems, which are essential for facilitating functionality and efficiency in a growing number of industries. It enables the intelligent and smooth conversion of electrical energy, frequently between different voltage levels, frequencies, and forms (AC to DC or vice versa), ensuring optimal performance and integration in both conventional and modern technologies. Its significance extends far beyond simple power supply operations, enabling some of the most transformational changes in energy production, distribution, and consumption. As global energy systems continue to change in reaction to decentralization, digitization, and environmental concerns, power electronics will become more and more important in determining the infrastructure of the future.

In the shift to low-carbon energy sources, it allows for the flexible connectivity of intermittent renewable generation with traditional grids, ensuring power quality, synchronization, and real-time control. This feature is essential for managing the fluctuation of wind and solar energy, enabling grid operators to minimize outages and balance supply and demand. Power electronics operates as the technological interface that enables the substitution of clean, electric processes for fossil fuel-based equivalents in the electrification trends of transportation, manufacturing, and household applications. This encompasses not just electric propulsion and charging infrastructure, but also electric heating, smart appliances, and distributed energy management systems. Power electronics enhances system flexibility and security by reducing energy losses, reducing operational costs, and utilizing modern control techniques and high-efficiency topologies. Furthermore, developments in power electronics are directly related to the development of cyber-physical systems and the rapid adoption of digital technology. Smart homes, smart factories, and smart cities rely on networked systems that constantly monitor, analyze, and adapt to energy consumption. Real-time diagnostics, fault tolerance, predictive maintenance, and smooth energy consumption control are all made possible by embedded power electronics. Power electronics offers the responsive hardware backbone that allows artificial intelligence (AI), machine learning, and edge computing to interact with the physical world as these intelligent capabilities become increasingly integrated into energy systems. Power electronics is being pushed to operate in more severe conditions in high-frequency and high-power applications such as electric aircraft propulsion, solid-state circuit breakers, and ultrafast data transmission.

The use of wide bandgap semiconductors, such SiC and GaN, which provide better switching performance, thermal conductivity, and breakdown voltages than conventional silicon devices, is one of the advancements in materials science being driven by this. Furthermore, power electronics plays an important role in improving energy availability and equality. Decentralized renewable energy systems powered by wind or solar energy and regulated by power electronic interfaces deliver reliable and cost-effective electricity to rural populations in areas without stable grid infrastructure. Microgrids, which are supported by power electronics-based control and protection systems, provide a robust solution to grid instability and natural disasters, assuring continuous energy supply even in extreme conditions.

Power electronics remains an evolving subject of study in both research and academia, combining principles from electrical engineering, materials science, control theory, and computer science. The interdisciplinary nature of power electronics encourages constant innovation, resulting in advancements like wireless power transfer, modular multilevel converters, and enhanced pulse-width modulation methods. These technologies are not only improving existing applications, but also creating totally novel applications, such as implantable medical devices, enhanced robotics, and space exploration systems. In conclusion, power electronics is more than a support technology; it is a fundamental component of modern energy systems. Its ability to precisely regulate electrical energy enables crucial improvements in sustainability, automation, digitization, and mobility.

Power electronic converters, including AC–AC matrix converters, DC–DC converters, and DC–AC inverters, are essential components of the renewable energy cycle. In photovoltaic (PV) systems, DC–DC converters are used to maximize the power point of solar panels, ensuring that they maintain their peak performance under all environmental circumstances. This is accomplished using complex algorithms such as Maximum Power Point Tracking (MPPT), which continuously adjusts voltage and current levels to maximize energy gathering. Higher system efficiency and a lower cost per watt of generated power are closely correlated with greater energy capture. On the other hand, inverters play a crucial role in transforming the DC output from solar panels or battery storage systems into AC electricity that can be utilized by conventional appliances or fed into the utility grid.

In addition to simple conversion, inverters offer advanced features including harmonic filtering, voltage regulation, frequency synchronization with the grid, and islanding detection. Grid-tied systems provide bidirectional power flow, allowing consumers to become prosumers that consuming and supplying electricity to the grid based on generation and demand.

Wind turbines rely significantly on power electronics to convert energy and interface with the grid. Variable-speed wind turbines, which are more efficient than fixed-speed turbines, require power electronic converters to separate the mechanical rotor speed from the grid frequency. This allows for optimal energy harvest across a wide range of wind conditions. Full-scale converters in wind energy systems provide further control over reactive power, improve fault ride-through capabilities, and facilitate grid frequency regulation, making wind a more grid-compatible resource.

Energy storage systems (ESS), which include batteries, flywheels, and supercapacitors, are increasingly being used in combination with renewable energy sources to mitigate uncertainty and improve grid stability. Power electronics regulates the charge and discharge cycles of these storage devices, ensuring their safety, longevity, and optimal performance. Bidirectional converters deliver stored energy when generation is low, or demand is high and absorb excess power when there is a surplus. This dynamic balancing improves electricity flow and decreases dependency on peaking fossil-fuel generators. Advanced power conditioning devices, which frequently use a combination of various converter types, have a much broader role. Aside from MPPT, they provide real-time control over reactive power, phase balance, power factor correction, and voltage and frequency management. These capabilities contribute to the grid's integrity even as the proportion of renewable energy increases. Advanced power electronic interfaces are the only way to enable renewable energy sources to contribute to system stability, which has mandated by grid norms in many different countries. Furthermore, power electronics makes it possible for generation, storage, and consumption to be coordinated in real time with the rise of microgrids, virtual power plants (VPPs), and peer-to-peer energy trading systems. These new models of energy distribution would be impossible to implement without the precise and programmable capabilities of power electronic converters. Their adaptability to both grid-connected and island modes promotes resilience and autonomy in community-scale and remote renewable systems (Kabalci 2017, 2021a).

One of the most important applications of power electronics in this field is Variable Frequency Drives (VFDs), which control the speed and torque of AC motors through adjusting the frequency and amplitude of the supplied voltage. These drives use advanced power electronic inverters controlled by pulse-width modulation (PWM) techniques to provide finely adjusted power profiles to motors. VFDs help to reduce energy consumption during partial load operations, which are frequent in pump, fan, and conveyor applications. Motors frequently operate at constant speeds without VFDs and use mechanical throttling to regulate output, which is an energy-wasting technique. Power electronics eliminates inefficiencies by ensuring that motors only consume the amount of energy required for the activity at hand. More demanding applications use advanced inverter topologies including multilayer inverters and matrix converters to improve output waveform quality, reduce harmonic distortion, and increase voltage scalability. Multilevel inverters, for example, generate output voltages in levels that closely resemble sinusoidal waveforms, reducing stress on motor insulation and electromagnetic interference (EMI). This is especially useful in high-voltage or high-power situations, such as massive industrial motors or factory-wide renewable energy integration. Matrix converters, which eliminate the need for intermediary DC links, provide direct AC–AC conversion and are praised for their small design and regenerative braking capability. Power electronics plays an important role in system longevity and durability, in addition to performance and efficiency. Smooth motor start-up and shutdown procedures reduce mechanical shock, extending the life of the motors and associated machinery. Furthermore, regenerative braking technologies enable extra energy to be recovered during deceleration and utilized or sent back into the grid, improving overall energy efficiency (Rashid 2024a; Erickson and Maksimović 2001).

The integration of power electronics with digital control systems and industrial internet of things (IIoT) platforms has increased automation and motor drive system capabilities in recent years. Smart VFDs and power converters now have sensors and embedded controllers that relay operational data in real time. This connectivity provides predictive maintenance, in which advanced analytics detect wear patterns, temperature varieties, or forthcoming issues before they cause unnecessary downtime. These insights enable maintenance teams to act proactively, boosting plant reliability and lowering the costs related to emergency repairs or interruptions in production. Furthermore, integrated monitoring of energy functions provides precise information on power use at the device or system level. Facility managers can use this data to assess performance among equipment or locations, optimize production schedules, and adopt energy-saving initiatives. In industries where energy represents a large amount of running costs, this level of control can result in significant economic gains. Power electronics also facilitates automatic fault detection and protection procedures, hence improving operational safety. Modern drives and converters can detect overcurrent, overvoltage, and overheating problems in real time and initiate automated responses such as shutdowns or load reductions to minimize equipment damage. These built-in diagnostics help make industrial operations safer and more resilient, particularly in mission-critical areas (Kabalci 2019).

Smart grids and energy storage constitute a revolutionary horizon in modern power systems, with power electronics serving as a key enabler of flexibility, reliability, and sustainability. As electric grids grow from centralized, unidirectional systems to dynamic, decentralized networks with high levels of renewable integration, power electronics play an increasingly important role in controlling complexity and maintaining uninterrupted energy flow. In a smart grid context, distributed energy resources (DERs) such as rooftop solar panels, small-scale wind turbines, and residential or commercial battery systems provide variability and bidirectionality to the traditional grid paradigm. Unlike conventional power plants, which provide electricity in a controlled and continuous manner, DERs are intermittent and localized, necessitating rapid and sophisticated interfaces to ensure grid stability (Kabalci 2019; Kabalci and Kabalci 2018).

In addition to supporting current infrastructure, power electronics plays a crucial role in facilitating breakthroughs of the future. In aerospace and defence, for example, power electronic systems are critical in more-electric aircraft (MEA) projects that replace hydraulic and pneumatic systems with electrically driven alternatives to save weight and increase fuel efficiency. In medical equipment, clean and reliable electricity delivered by specialized converters is required for precise control of medical imaging systems, lasers, and surgical instruments. The fundamental ideas and elements of even newer technologies, such solid-state transformers and wireless power transfer, have their roots in power electronics.

Power electronics is essentially at the center of the worldwide energy revolution. Its capacity to manage, convert, and control electrical energy with high precision and little loss makes it critical for attaining sustainability objectives, improving energy security, and enabling smarter, more connected systems. Power electronics will continue to be a driving force in determining the future of energy and electrification across practically every industry, as companies demand greater efficiency, flexibility, and integration.

1.2 *Evolution of Power Converters*

Power conversion technology has evolved significantly over the last few decades, from simple rectifiers and thyristor-based controllers to high-frequency soft-switching converters and wide-bandgap semiconductor devices. The evolution can be divided into the following important phases:

- Early Rectifiers and Linear Regulators (1900s–1960s): Transformers, diode rectifiers, and linear voltage regulators were used for early power conversion, but they were large, inefficient, and produced a lot of heat. Early AC–DC conversion systems used tungsten arc rectifiers and vacuum tubes, which had low efficiency and a limited lifespan.
- Thyristor-Based Converters (1960s–1980s): The introduction of thyristors (SCRs) and triacs transformed power conversion, enabling phase-controlled rectifiers, cycloconverters, and early motor drive systems. Although commonly used, thyristor-based converters have large switching losses and long response times.
- High-Frequency Switching Converters (1980s–2000s): MOSFETs and IGBTs enabled the development of PWM switching converters, which significantly improved efficiency and reduced size. Bulky linear regulators were replaced by isolated flyback, DC–DC buck, and boost converters as conventional components in power supply architecture. Multilevel inverters were developed for high-voltage applications to improve waveform quality and reduce harmonic distortion.
- Wide-Bandgap Semiconductors and Digital Control (2000s–Present): The latest improvements include the use of SiC and GaN power devices, which provide improved efficiency, faster switching rates, and lower heat dissipation. Soft-computing and predictive control are being integrated into power converters to enable self-adaptive and fault-tolerant functioning. Smart grids and next-generation energy systems are being designed with bidirectional converters and solid-state transformers (SSTs).

Power converters have transformed from bulky, inefficient devices into compact, high-efficiency, and intelligent systems, revolutionizing energy conversion and control.

2 Fundamentals of Power Electronics Converters

Advanced power electronics converters are key components of modern energy systems, providing high efficiency, high power density, and intelligent power management solutions. These converters are essential for a variety of applications, including renewable energy systems, EVs, smart grids, industrial automation, data centers, and aircraft power systems. Fundamentally, power electronics converters change electrical energy depending on the needs of the application by changing its voltage level, current type (AC/DC), frequency, or phase. This transformation must be carried out with low losses, compact system design, and robust control mechanisms to ensure durability and adaptability in dynamic contexts.

This section presents brief information on the fundamental ideas of advanced power conversion, the classification of advanced converters, and the important performance measures that determine their efficiency, power density, and reliability in modern applications.

The goal of high-efficiency conversion of energy has become an essential component of modern power electronics. The industry has moved away from conventional silicon-based hard-switching technologies and toward a new paradigm in response to growing demands for small, high-performance systems spanning applications such integrated renewable energy, electric mobility, aerospace, and high-performance computing. This includes the use of wide-bandgap (WBG) semiconductors, advanced converter topologies, soft-switching techniques, and intelligent digital control schemes. Fundamental and advanced converter topologies, which control how energy is processed and transported within a system, are the foundation for these advancements. The majority of modern designs for non-isolated DC–DC converters are based on the buck, boost, and buck-boost topologies. The buck converter, which is frequently utilized in battery-powered systems and voltage management applications, efficiently decreases voltage. When driving high-voltage loads from lower-voltage sources, the boost converter is necessary since it increases the input voltage. The buck-boost converter adds flexibility to systems with varying input and output demands by offering output voltages both higher and lower than the input. Although these fundamental topologies are frequently employed, new non-isolated architectures like the Zeta converter, Single-Ended Primary Inductance Converter (SEPIC), and Ćuk converter have drawn interest due to their enhanced electromagnetic compatibility, continuous input and output currents, and built-in filtering features. These topologies are especially useful for applications like solar systems and automotive power distribution that demand low-ripple, high-efficiency power conversion (Saravanan et al. 2022; Bose 2007).

Power electronic systems are becoming more responsive and robust under dynamic working situations, in addition to being more efficient and compact, thanks to these developments. Soft switching techniques, including zero-voltage switching (ZVS) and zero-current switching (ZCS), have emerged as important equipment for reducing switching losses and increasing energy efficiency. Power devices in traditional hard-switching circuits suffer significant losses when they are turned on or off at high voltages and currents. On the contrary, soft switching minimizes energy loss and thermal stress by allowing transitions to take place while either voltage or current is zero. In addition to increasing efficiency, especially at high switching frequencies, this makes it possible to employ lighter and smaller passive parts, which is essential for reaching compact form factors and high-power densities.

Inverter topologies, which convert DC to AC, have also evolved significantly over time. AC motor drives and grid-connected applications continue to rely on the conventional full-bridge and half-bridge inverters. However, multilevel inverters like neutral-point clamped (NPC), flying capacitor, and cascaded H-bridge inverters are becoming more and more popular due to growing needs for higher voltage levels, better waveform quality, and fewer filtering requirements. These topologies are ideal for medium- and high-voltage industrial, renewable, and traction systems because they allow for higher voltage operation with less harmonic distortion and switching stress.

Additionally, improvements in resonant and quasi-resonant DC–DC converter technology have made it possible to build small, incredibly efficient devices that can soft-switch across a broad load range. Topologies like the LLC resonant converter are currently common in electric vehicle chargers and high-performance server power supplies owing to their high efficiency and low switching losses across a broad range of input and output conditions. Furthermore, topologies such as the dual-active bridge (DAB) provide galvanic isolation and bi-directional power flow, which makes them ideal for vehicle-to-grid (V2G) interfaces and battery energy storage systems.

These topology developments, which range from basic non-isolated converters to advanced multilevel inverters and resonant DC–DC topologies, serve as the foundation for modern high-efficiency power electronics. They offer a strong foundation for next-generation power systems when combined with wide-bandgap semiconductors and soft-switching strategies, and when improved by digital control algorithms and automated optimization. In addition to being more energy efficient, these systems also establish a new benchmark for intelligent and sustainable energy conversion since they can adapt dynamically, manage faults predictively, and optimize performance in real time (Hui and Chung 2007; Jovanovic and Irving 2016).

Advancements in high-efficiency power conversion technologies are not solely due to improved semiconductor materials or novel circuit topologies; they are also dependent on advanced control strategies that assure optimum efficiency, reliability, and adaptability under changing operating conditions. The function of control systems has changed from simple regulation to complicated, real-time decision-making processes that manage energy flow, guarantee safe operation, and optimize efficiency as power converters becoming faster, smaller, and more integrated. Control methods act as an essential connection between the hardware of converter and its functional objectives. While some applications still benefit from traditional analog control methods, digital controllers that provide more flexibility, accuracy, and programmability are gradually replacing them. Advanced features including adaptive tuning, nonlinear compensation, and smooth integration with system-level monitoring and communication protocols are made possible by digital control. These features are especially crucial in applications like grid-connected inverters, electric vehicles, and renewable energy interfaces where the power system needs to communicate with other smart devices or react dynamically to changes in demand and source. More advanced techniques including model predictive control (MPC), sliding mode control (SMC), and fuzzy logic control are being used in high-performance applications in addition to traditional feedback and feedforward procedures. These methods are ideal for power converters operating in nonlinear and time-varying settings as they provide rapid dynamic response and reliable performance under parameter uncertainty. For instance, MPC can use a system model to forecast future converter behavior and adjust switching actions that result in high efficiency and less component stress.

Artificial intelligence (AI)-based control techniques are starting to take center stage in the developing field of intelligent and self-governing energy systems. Machine learning algorithms such as neural networks and reinforcement learning are being developed in order to improve converter efficiency, forecast problems, and modify control policies in response to real-time operational data. These methods show significant potential in applications like microgrid controllers, hybrid ESSs, and battery management systems that involve complex multi-variable interactions. The operating features and design of the underlying power converter topologies have a significant impact on how well these control techniques work. The simplicity and adaptation of rapid, high-bandwidth control loops allow them suitable for non-isolated converters such as buck, boost, and buck-boost circuits and for the other types of power converters.

3 Types of Advanced Power Converters

Advanced power converters are designed to meet the evolving requirements of modern electrical systems by improving efficiency, power density, and reliability. These converters are categorized based on their input–output characteristics, functionality, and control strategies. This section provides a detailed breakdown of DC–DC, AC–DC, DC–AC, and AC–AC converters by highlighting their operating principles, topologies, and applications.

3.1 DC–DC Converters

DC–DC converters are widely used in battery-powered systems, renewable energy integration, EVs, and aerospace applications. They regulate voltage levels, enhance energy efficiency, and facilitate bidirectional power flow as needed. DC–DC converters have several key functions in power electronic systems. The efficient transformation of a DC input voltage (V_{DC}) into a desired DC output voltage (V_o) by enabling flexibility in power supply requirements for various devices and applications is comprised by DC converters. These converters are designed to maintain a stable output voltage, regardless of fluctuations in load conditions or in the input voltage. On the other hand, DC–DC converters minimize the AC ripple in the DC output to meet severe performance standards. This feature is essential for protecting sensitive electronic components that require a steady voltage supply to operate effectively. Many DC–DC converters provide electrical isolation between the input source and the load. This isolation protects the load from potential disturbances from the input and enhances safety by reducing the risk of damage. The DC–DC converters also achieve to protect both the power systems and the input sources from electromagnetic interference. The DC–DC converters are designed to meet various international and national safety standards. This compliance contributes to the reliability and endurance of the entire electrical system by ensuring safe operation within specified parameters (Saravanan et al. 2022; Hossain et al. 2018; Lakshmi and Hemamalini 2018; Rashid 2024b; Bose 2021a).

The earliest DC–DC converters were known as DC choppers, in which SCRs were used as the controlled switching devices. In contrast, modern DC–DC converters developed in recent years utilize advanced power switches such as IGBTs, MOSFETs, and other semiconductor devices. DC–DC converters are generally divided into two main categories: resonant DC–DC converters and hard-switched PWM DC–DC converters. Resonant converters employing soft-switching techniques have been continuously developed over the years. On the other hand, PWM converters offer several advantages including as low number of components, high efficiency, fixed-frequency operation, relatively simple control, commercial availability of integrated circuit controllers, and the ability to achieve high conversion ratios for both step-down and step-up applications. However, the rectangular voltage and current waveforms in PWM operation cause switching losses in semiconductor devices during turn-on and turn-off times, which limit the practical operating frequencies of PWM DC–DC converters to the several hundreds of kHz ranges. The PWM DC–DC converters are nonlinear variable-type systems that are mainly defined as isolated or non-isolated converters. The following parts of this section present the fundamental introduction of DC–DC converter topologies and analytical representations.

3.1.1 Non-isolated DC–DC Converters

Non-isolated DC–DC Converters are typically low-cost, high-efficiency power converter systems with direct electrical connections between input and output. These converters are favored, particularly in situations where input and output galvanic isolation is not necessary. They are commonly employed in battery systems, portable electronic devices, and automotive electronics due to their compact structure. Non-isolated DC–DC converters can be built in a variety of topologies including buck, boost, buck-boost, and Cuk converter. Each topology is designed to produce the desired output voltage within a given input voltage range. High switching frequencies and the flexibility to work with a limited number of passive components make these converters suitable for use in modern power electronics systems.

Buck (Step-Down) Converters

The output voltage of embedded systems and power management units is decreased to accommodate low-power loads. DC–DC buck converters are frequently used in portable electronic devices, including smartphones, tablets, and laptop computers. Furthermore, sensors, microcontrollers, and communication modules in automation systems commonly operate at low voltages. Thus, the input voltage needs to be filtered and decreased if it is not consistent or if it is received directly from the grid. At this stage, DC–DC buck converters manage voltage while also improving the

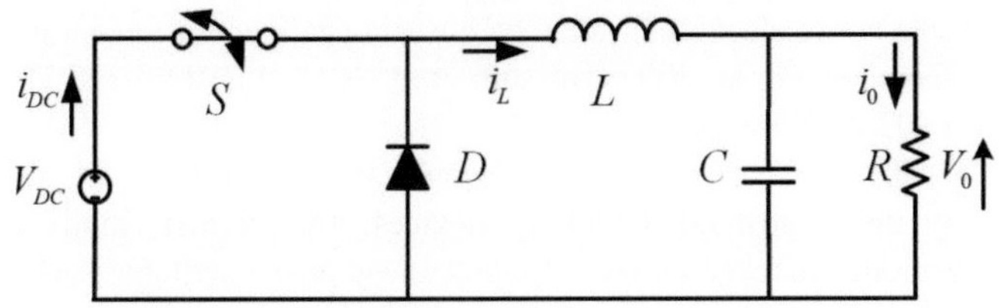

Fig. 1 Circuit diagram of the buck converter

overall safety and efficiency of the system through increased energy efficiency. The DC–DC buck converter topology, which is given in Fig. 1 consists of five fundamental elements: an inductor (*L*), a diode (*D*), a controlled switch (*S*), a capacitor (*C*), and the load. In this scheme, when the switch is closed (t_{on}), *D* is reverse polarized, and the current through the inductor (i_L) rises since the input voltage (V_{DC}) exceeds the output voltage. This phase comprises the transmission of energy from the source to the output because of the input voltage providing energy to the load and the output capacitor.

The voltage across the inductor reverses polarity when the switch is opened (t_{off}), causing the diode to become forward-bias and begin conducting. During this interval, the inductor current (i_L) flows toward the output and returns through the conducting diode (Bilhan and Kabalci 2024; Salimi et al. 2013).

The duty cycle (*d*) of the buck converter can be calculated as follows:

$$V_{L_{\text{on}}} t_{\text{on}} + V_{L_{\text{off}}} t_{\text{off}} = 0 \tag{1}$$

$$(V_{\text{DC}} - V_{\text{O}})dT + (-V_{\text{O}})(1 - d)T = 0 \tag{2}$$

$$V_{\text{O}} = dV_{\text{DC}} \tag{3}$$

$$d = \frac{V_{\text{O}}}{V_{\text{DC}}} \tag{4}$$

Boost (Step-Up) Converters

The output voltage is required to be increased for the load of embedded systems and power management devices. DC–DC boost converters are utilized in many different fields including portable devices, renewable energy sources, electric vehicles, industrial automation systems, and so on due to their compact structure, high efficiency, and stable output (Cheng et al. 2025). In renewable energy systems, boost converters are used for MPPT to ensure that the maximum amount of energy is harvested. Additionally, low irradiation levels reduce the voltage of PV panels, resulting in insufficient

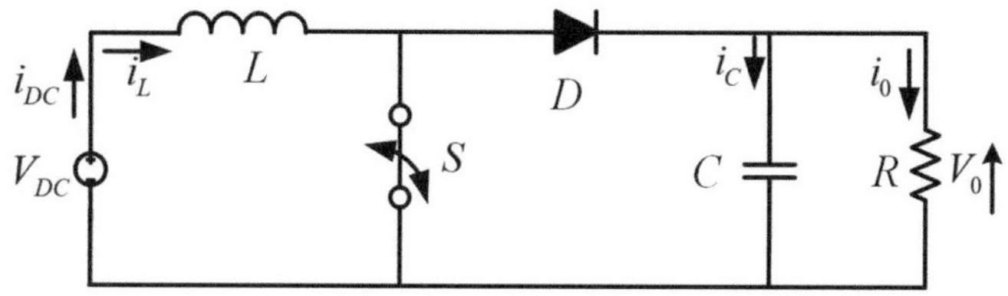

Fig. 2 Circuit diagram of the boost converter

output voltage to charge the batteries. In this situation, it should be guaranteed to have a steady output voltage and to use the boost converter circuit to supply the energy storage units. Another application field of boost converters can be described for EVs. Even if the voltage of an electric vehicle's battery may be reasonable, the driver circuits or other auxiliary systems may need a higher voltage to function. In this case, the boost converters increase the battery voltage to the required level and guarantee the system performs steadily. Furthermore, these circuits are widely utilized in hybrid automobiles to bridge the gap between low-voltage and high-voltage power electronics. The schematic diagram of the DC DC boost converter is illustrated in Fig. 2. During the interval in which the switch is closed the diode (D) is reverse biased and the inductor current (i_L) is increased. When the switch is opened in t_{off} period, the diode is forward biased, and the load is fed by the sum of the inductor and source voltages.

The voltage conversion rate (M) of the boost converter is calculated using the voltage change across the inductor and the duty cycle (d) as presented in the following equations.

$$V_{L_{\text{on}}} t_{\text{on}} + V_{L_{\text{off}}} t_{\text{off}} = 0 \tag{5}$$

$$V_{\text{DC}} dT + (V_{\text{DC}} - V_{\text{O}})(1 - d)T = 0 \tag{6}$$

$$V_{\text{O}}(1 - d)T = V_{\text{DC}} T$$

$$M = \frac{V_{\text{O}}}{V_{\text{DC}}} = \frac{1}{(1 - d)} \tag{7}$$

Buck-Boost Converters

The buck-boost converter is a robust and commonly used DC–DC power conversion topology that can provide an output voltage that is either higher or lower than the input voltage. This dual feature makes it especially helpful in systems where the input voltage varies significantly or unpredictably, and a controlled output

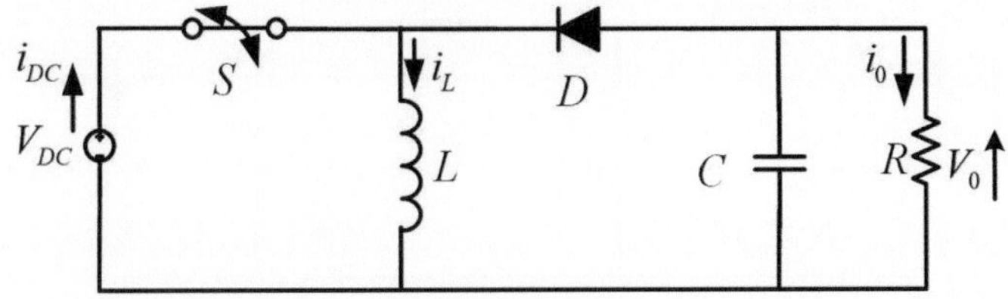

Fig. 3 Circuit diagram of the buck-boost converter

voltage is required. Applications include embedded electronics, electric vehicles, battery-powered gadgets, and renewable energy systems where steady performance is required despite input variations. The buck-boost converter combines the concepts of the buck (step-down) and boost (step-up) converters. When the duty cycle is properly set, it can decrease the voltage as a buck converter or increases as a boost converter. This dynamic control enables the buck-boost converter to maintain a consistent output voltage even when the input voltage decreases below or increases beyond the intended output level which making it suitable for power supplies that need to connect with sources such as batteries or photovoltaic panels (Mchaouar et al. 2025; Banaei et al. 2014). The schematic diagram of a buck-boost converter is given in Fig. 3. When the switch (*D*) is turned ON, a direct electrical path is formed between the input voltage source and the inductor. During this interval, the inductor is connected across the input supply that causes a linear increase in current through the inductor, and magnetic field of inductor stores energy. The diode becomes reverse-biased because the voltage on its cathode is higher than on the anode (connected to the output capacitor and load), preventing current from flowing into the output. As a result, no energy is delivered to the load during this time; the load is supplied solely by the output capacitor, which discharges slightly. This interval is critical for accumulating energy in the inductor, which will be transferred to the output in the next phase. When the switch is turned OFF, the current through the inductor cannot change instantly due to its fundamental characteristics. In response, the inductor reverses its polarity to keep the current flowing. During this time, the voltage polarity of inductor changes and the stored energy is released. The diode becomes forward-biased, which permits the inductor current to pass through the load and the output capacitor. The load is supplied and the capacitor charge is restored when energy is moved from the inductor to the output. A linear reduction in the inductor current occurs as the magnetic field decreases. During this phase, the buck-boost converter transfers the energy stored in the inductor during the ON interval to the load side. In this scheme, the inductor acts as a source for the load (Bahravar et al. 2012).

The voltage conversion rate (*M*) of the buck-boost converter can be obtained as follows:

$$V_{L_{\mathrm{on}}} t_{\mathrm{on}} + V_{L_{\mathrm{off}}} t_{\mathrm{off}} = 0 \tag{8}$$

$$V_{\mathrm{DC}} dT + V_{\mathrm{O}} (1 - d) T = 0 \tag{9}$$

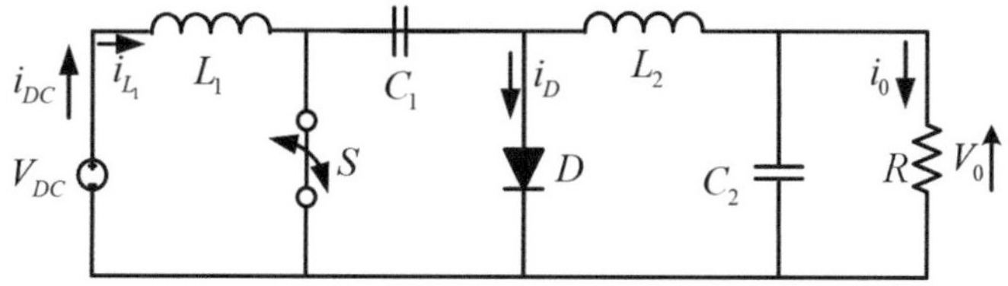

Fig. 4 Circuit diagram of the Ćuk converter

$$V_O(1-d)T = -V_{DC}dT \tag{10}$$

$$M = \frac{V_O}{V_{DC}} = -\frac{d}{(1-d)} \tag{11}$$

Ćuk Converters

The Ćuk converter is a DC–DC converter that delivers a controlled output voltage, which can be either higher or lower than the input voltage. The Ćuk converter, similar to the buck-boost converter, provides a non-pulsating current on both input and output sides. This results in reduced EMI and more efficiency that makes it appropriate for sensitive electrical systems, audio amplifiers, and battery-powered devices. The Ćuk converter is designed to overcome the disadvantages observed in the buck-boost converter in terms of polarity reversal. Therefore, an additional inductance (L_2) and capacitance (C_2) are introduced in the circuit diagram of the Ćuk converter. It is generally preferred in systems where sensitive loads are supplied, in applications where power factor is important, or in electronic circuits where it is desired to keep the voltage fluctuation at the output at a minimum level (Bilhan and Kabalci 2024). Ćuk converters are used in industrial control systems, power electronics laboratories, solar energy systems, and electric vehicle battery management systems. In the Ćuk converter, C_1 is energy transfer capacitor, while the C_2 is the filter capacitor. Inductor L_1 is located on the input side and filters the input current to ensure that it is continuous and smooth, while inductor L_2 is located on the output side and maintains the continuity of the output current. Figure 4 shows the basic topology of a DC–DC Ćuk converter.

The duty cycle for a Ćuk converter can be expressed as

$$V_{DC}dT = \left(V_{C_1} - V_{DC}\right)(1-d)T \tag{12}$$

$$V_{C_1} = \frac{1}{1-d}V_{DC} \tag{13}$$

$$(V_{C_1} + V_{C_2})dT = -V_{C_2}(1-d)T \tag{14}$$

$$V_{C_1}d = -V_{C_O} \tag{15}$$

$$V_{C_2} = -\frac{d}{1-d}V_{DC} \tag{16}$$

$$V_O = -V_{C_2} \tag{17}$$

$$d = \frac{V_O}{V_O - V_{DC}} \tag{18}$$

Interleaved Converters

Interleaving effectively reduces load current ripple in multiphase converters, leading to smaller, lighter, cheaper, and more efficient components compared to single-phase converters. It minimizes size and cost of the output capacitor while allowing for reduced inductor values (Biswas et al. 2019). Additionally, interleaving reduces component stress by enabling higher switching frequencies and further decreasing reactive components. Ideally, each phase is designed to manage an equal amount of power. However, variations in structure across the phases, which may arise from component tolerances, can create imbalances in the power distribution. When power flow is not equalized, overall efficiency decreases since the phases are optimized for a nominal power level. To address this issue, modular systems often employ current balancing algorithms to ensure equitable power sharing among the phases. Interleaved converters play an important role in meeting the high current requirements in battery charging systems and motor drives in electric vehicles. In addition, these converters are effectively used in renewable energy systems, especially in MPPT units of solar panels and in energy storage systems. In information technology hardware, these structures are preferred to ensure stability in low-voltage and high-current supplies of high-performance components such as CPUs and GPUs.

Single-phase DC–DC converters typically exhibit the highest efficiency at full load and the lowest efficiency when partially loaded. To achieve high efficiency across the entire loading range, multiphase interleaved DC–DC converters are introduced by adjusting the number of active phases in proportion to the output load power (Liang et al. 2019; Nguyen et al. 2021; Azer and Emadi 2020; Kabalci and Boyar 2019). The multiphase configuration of conventional buck, boost, and buck-boost converters can be enhanced using interleaved converter topology as illustrated in Fig. 5.

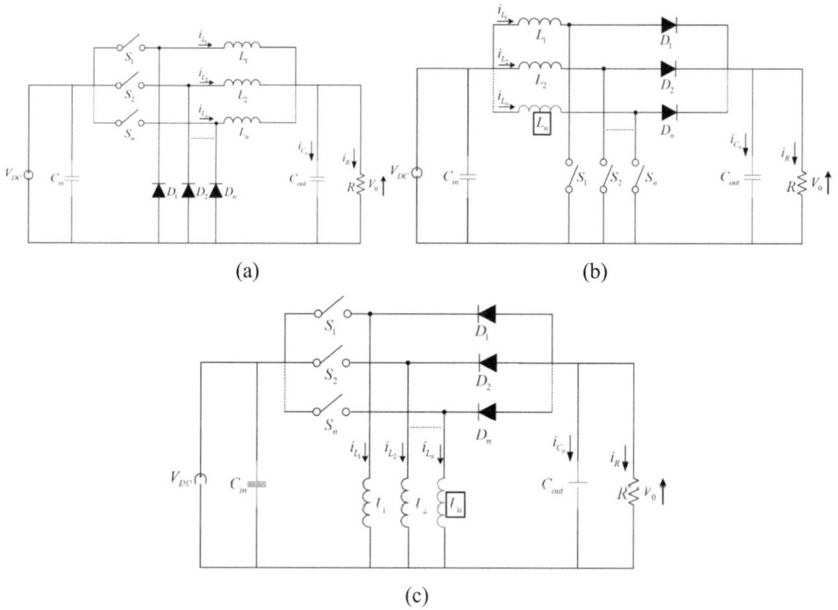

Fig. 5 Multi-phase interleaved converter topologies **a** interleaved buck converter, **b** interleaved boost converter, **c** interleaved buck-boost converter

3.1.2 Isolated DC–DC Converters

The isolated converters are important structures that provides electrical isolation between input and output in power electronic circuits. This isolation is usually achieved by a transformer and plays a critical role in system safety. Isolated converters are widely preferred to ensure user safety, especially in applications with high voltage input sources. In sensitive areas such as grid-connected power supplies, UPS systems, industrial automation solutions, and medical devices, isolation has become mandatory for both safety and performance. In addition, especially in renewable energy applications such as solar energy inverters, isolation is of great importance for both system protection and compatibility with the grid.

Flyback Converters

The flyback converter is a simple and cost-effective isolated converter topology that is frequently used in low-power applications. In terms of its operating principle, energy is stored in the magnetic field of a transformer during switching and transferred to the output when the switch is closed. The flyback structure offers advantages such as multiple outputs, low component count, and galvanic isolation. It is often used in adapters, chargers, and small power supplies (Boyar and Kabalci 2018; Zhu et al.

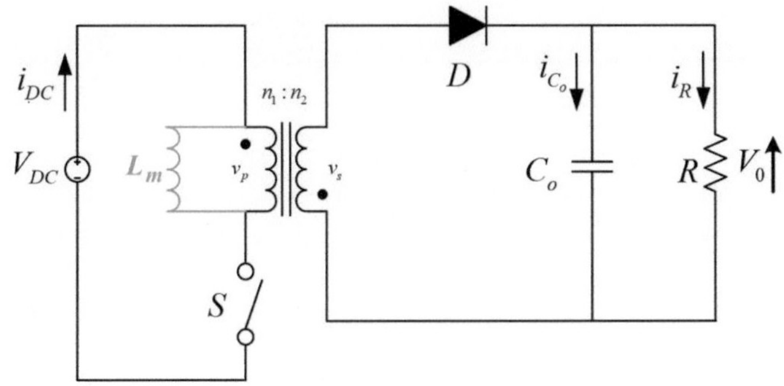

Fig. 6 The circuit diagram of flyback converter

2024; Ramos-Paja et al. 2022). The schematic diagram of the flyback converter is illustrated in Fig. 6.

The primary winding of the transformer is connected in series with the input V_{DC} source and the switch (S). Conversely, the secondary winding of the flyback transformer is connected in series with the diode (D) and the output circuit, which comprises the capacitor (C_o) and the resistor (R). The high-frequency transformer, which serves to isolate the source from the load within the converter, includes the magnetizing inductance (L_m) and is represented by an ideal transformer model with a turns ratio of n_1:n_2. The mutual inductance between the primary and secondary windings of a high-frequency (HF) transformer induces a polarity inversion between the primary voltage (v_p) and the secondary voltage (v_s). Although leakage inductances are not critical for the fundamental operation of the converter, they adversely affect switching transitions of semiconductors, particularly switches and diodes. Consequently, flyback converters commonly require the inclusion of snubber circuits to mitigate these effects.

The basic control approach and operation of a flyback converter are similar to those of a buck-boost converter. In a flyback converter, the magnetizing inductance replaces the inductor found in a buck-boost converter. When the switch is turned on, the primary current (i_p) increases, causing the primary winding to store energy. During this time, the diode connected to the secondary side remains open, preventing energy transfer from the input to the output. When the switch is turned off, a secondary voltage is induced, which is proportional to the transformation ratio multiplied by the output voltage. The mean voltage over inductors are zero at steady-state, the duty cycle (d) for a flyback converter can be expressed as

$$\frac{V_o}{V_{DC}} = \frac{n_2}{n_1} \cdot \frac{d}{1-d} \tag{19}$$

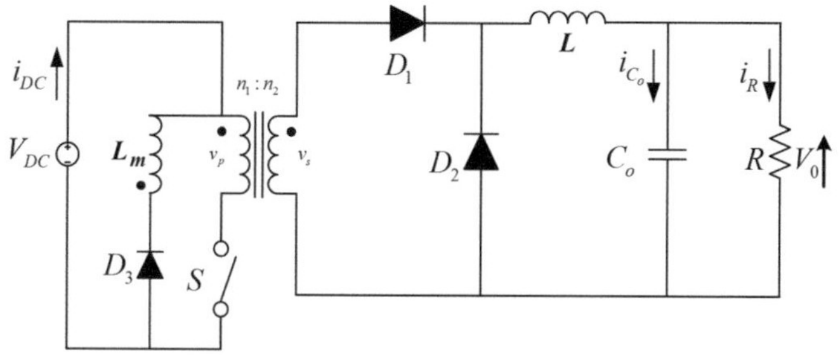

Fig. 7 Forward converter with core reset

Forward Converters

In comparison to the flyback converter topology, the forward converter offers several advantages including a lower peak output current and fewer issues related to leakage energy (Bilhan and Kabalci 2024; Park et al. 2023). It typically demonstrates higher energy efficiency and is preferred in applications that require moderate power outputs usually in the range of 100–200 Watts. However, the circuit configuration, particularly the output filtering stage, is more complex than that of the flyback converter. The forward converter scheme is given in Fig. 7. When the switch (S) is turned on, the supply voltage is applied to the primary winding of the transformer. Due to the transformer operation, a proportional voltage is induced on the secondary winding. This voltage forward biases diode D_1, thereby supplying power to the load.

A low-pass filter that is consisting of inductor L and capacitor C, smooths the DC voltage and reduces switching ripple. When the switch (S) is turned off, the currents in both the primary and secondary windings decrease to zero. However, diode (D) becomes forward biased, allowing the tertiary winding to conduct and provide a continuous path for the magnetizing current. As diode D conducts, the voltage across the tertiary winding is clamped to the input DC voltage, with the negative terminal located at the dotted end. Induced voltages arise simultaneously across both the main and secondary windings as a result of transformer operation (Tan 2002). The output voltage of converter can be controlled by adjusting the duty cycle. Typically, the duty cycle for the forward converter is kept lower than 50%. The voltage conversion rate of the forward converter can be calculated by using the voltage variation on the inductor and duty cycle (d).

$$\left(V_{\mathrm{DC}}\frac{n_2}{n_1} - V_{\mathrm{o}}\right)d = V_{\mathrm{o}} \cdot (1 - d) \tag{20}$$

$$M = \frac{V_{\mathrm{o}}}{V_{\mathrm{DC}}} = \frac{n_2}{n_1} \cdot d \tag{21}$$

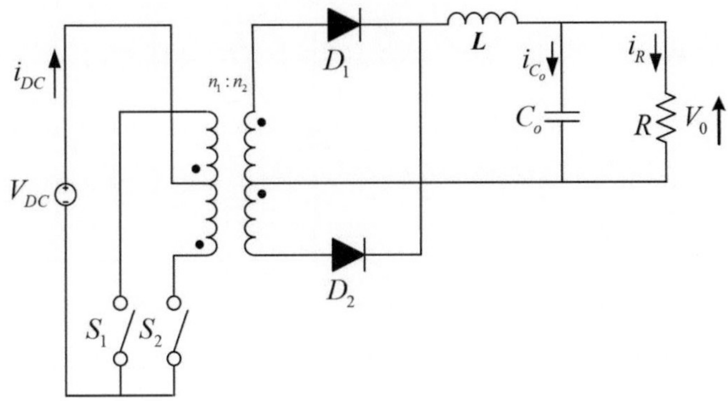

Fig. 8 Circuit diagram of the push–pull converter

Push–Pull Converter

The push–pull converter is a topology that is widely used, especially in medium and high power DC–DC conversion applications (Petit et al. 2012; Ivanovic and Knezic 2022). This converter type provides energy transfer from the primary winding alternately using two switches. Figure 8 illustrates the push–pull converter topology. One switch is in conduction while the other is in cut-off which allows the transformer to carry energy in both half cycles. The duty ratio must be smaller than 0.5. When switch S_1 is on, diode D_1 conducts and diode D_2 is off. Diode states are reversed when switch S_2 is on. The main advantage of push–pull converters is that they provide more efficient energy transfer by using the transformer in both positive and negative current directions. At the same time, their symmetrical structure provides lower EMI and the output voltage becomes more stable. These converters are frequently used in areas such as telecommunication systems, battery charging circuits, and renewable energy systems. The weakness of the push–pull structure is the need for precise timing between switches and the risk of saturation in the event of asymmetry in the transformer windings.

The conversion rate of the push pull converter can be calculated as given below.

$$M = \frac{V_o}{V_{DC}} = 2 \cdot \frac{n_2}{n_1} \cdot d \tag{22}$$

Half-Bridge and Full-Bridge Converters

Half-bridge and full-bridge converter topologies are presented in Fig. 9. The operating principle of the PWM-controlled half-bridge converter is similar to the operation of the push–pull converter. In comparison to the push–pull converter, the primary winding of the transformer in the half-bridge construction is fed more easily by

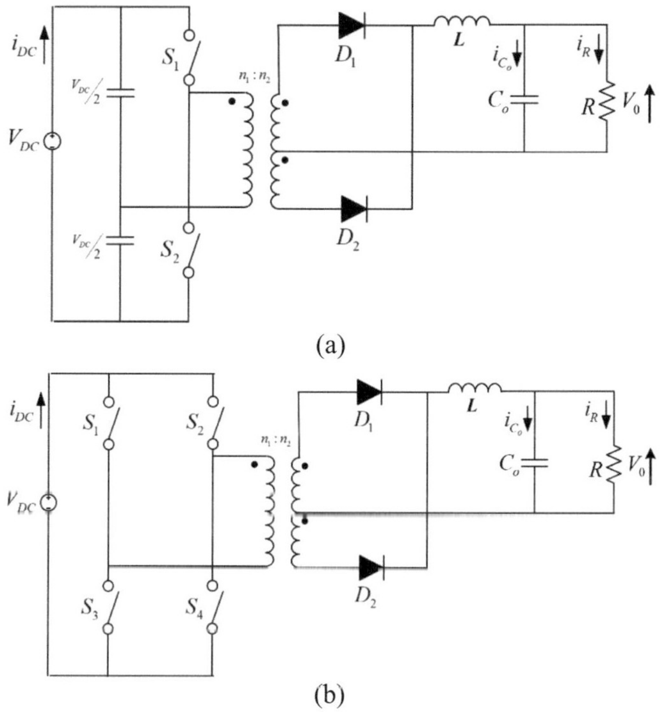

Fig. 9 Circuit diagram of **a** half bridge converter, **b** full bridge converter

employing two voltage-sharing input capacitors (Liu et al. 2013; Asl et al. 2018). However, it is commonly observed that two controlled switches are used in substitution of the input capacitors when comparing the PWM DC–DC full-bridge converter and the half-bridge construction. These switches are connected in pairs. When the S_1 and S_4 switches are turned on, V_{DC} voltage is applied to the primary winding of the transformer, and diode D_1 is turned on. Similarly, when the S_2 and S_3 switches are opened, $-V_{\mathrm{DC}}$ voltage is generated at the primary ends and in this case, diode D_2 becomes conductive. The duty ratio must be lower than 0.5.

The conversion ratio of the half bridge converter and full bridge converter can be calculated as, respectively.

$$M = \frac{V_{\mathrm{o}}}{V_{\mathrm{DC}}} = \frac{n_2}{n_1} \cdot d \tag{23}$$

$$M = \frac{V_{\mathrm{o}}}{V_{\mathrm{DC}}} = 2 \cdot \frac{n_2}{n_1} \cdot d \tag{24}$$

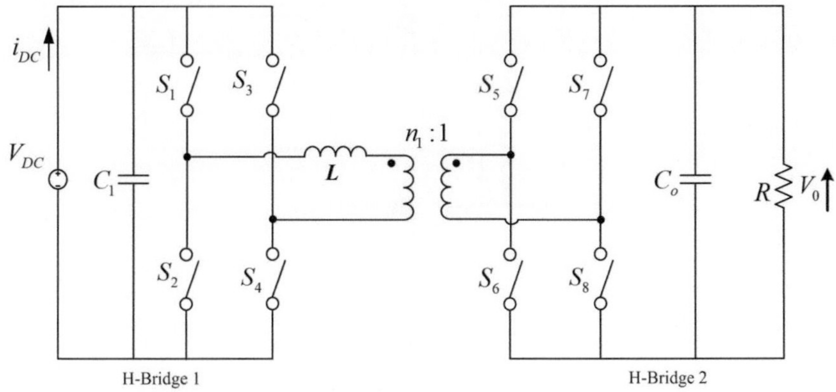

Fig. 10 Conventional dual-active bridge converter scheme

Dual-Active Bridge (DAB) Converters

DAB converters represent advanced DC–DC converter topologies that operate with remarkable efficiency and facilitate bidirectional power transfer while ensuring galvanic isolation between input and output (Shao et al. 2022; Li et al. 2023; Kheraluwala et al. 1992). A key feature of DAB converters is their design, which incorporates full-bridge inverter circuits on both the input and output ends. DAB converter topology is presented in Fig. 10. These full-bridge configurations enable the converter to efficiently manage electrical energy by alternating the direction of current flow, essential for bidirectional operation. The connection between the two full-bridge circuits is established through a high-frequency transformer (HFT), which not only increases the voltage levels but also provides electrical isolation between the input and output (Yadeo et al. 2021). The high-frequency operation of transformer minimizes losses and improves overall efficiency. Moreover, the active control of both sides allows for precise regulation of power flow, enhancing operational flexibility and responsiveness to varying load conditions. This control mechanism ensures optimal performance in applications such as energy storage systems, electric vehicles, and renewable energy integration that makes the DAB converter a crucial technology in modern power electronics.

Each full-bridge generates a square-wave AC voltage with a high frequency that is usually tens to hundreds of kHz. The phase shift between the primary and secondary bridge voltages determines the magnitude and direction of power transmission. A 0° phase shift causes not any net power transmission. A positive phase shift allows power to flow from the main to the secondary side while a negative phase shift reverses the power direction. The AC square-wave voltages generated on each side cause a voltage differential across the transformer. This voltage pushes current through the leakage inductance of transformer or an intentionally added inductor that allows for regulated energy transfer. The current form is commonly trapezoidal or roughly triangular with the amplitude determined by the phase shift and DC-link voltages.

3.1.3 Resonant Converters

The developments in power electronics over the last few decades have led to significant advances not only in the performance of power devices but also in new converter topologies and control techniques. The "hard switching" technique was used by conventional PWM converters, which were popular in the 1970s. During switching, the switch was subjected to both high voltage and high current. This resulted in substantial switching losses, overheating of the devices, and EMI issues. Although passive snubber circuits were employed to reduce switching losses, they limited switching frequency and decreased efficiency. Resonant converters, which created soft switching techniques including zero voltage switching (ZVS) and zero current switching (ZCS), have gained importance since the 1980s. These strategies significantly minimize losses by generating circumstances in which the voltage or current is near zero during switching moments. Since the 1990s, a new generation of soft-switching converters has emerged that maintains the typical PWM structure while using just the resonance in a regulated way during switching times. This has made it possible to operate at higher frequencies (500 kHz–several MHz), reduce EMI, and create more compact and efficient power converters. These converters are widely used in DC–DC, AC–DC and DC–AC applications (Hui and Chung 2007; Patin 2015).

Resonant DC–DC converters are modern power electronic circuits developed to reduce switching losses, increase efficiency and minimize EMI noises. The high switching losses and high electrical stresses encountered in traditional PWM converters have led to the emergence of new converter topologies based on resonance principles. Resonant converters enable soft switching by incorporating inductance (L) and capacitance (C) elements into a resonant circuit to provide zero voltage (ZV) or zero current (ZC) conditions at the time of switching. Figure 11 illustrates the voltage and current waveforms for turn-on and turn-off soft-switching scenarios as the semiconductor switch transitions between states using resonant circuits. In Fig. 12, two types of resonant-voltage switches and resonant current switches are shown.

In ZV resonant switching, the approach is centered around optimizing the timing of the switch operation by opening it when the voltage is zero. This is accomplished by incorporating a resonance capacitor (C_r) in parallel with the unidirectional switch (S). When the switch is conFig. d to conduct in only one direction, it allows the voltage across C_r to oscillate freely during both positive and negative cycles, facilitating full-wave operation. When an antiparallel diode is connected to the unidirectional switch, it clamps the resonant capacitor voltage to zero during the negative half-cycle. In contrast, the goal of ZC resonant switching is to achieve zero current conditions at the time that the switch turns off. To enable this, a resonant inductor (L_r) is connected in series with the unidirectional switch. S is a unidirectional switch, therefore only positive current can resonate, resulting in half-wave action. However, if a diode is connected in anti-parallel to the unidirectional switch, current can flow in both directions, causing the system to enter full-wave mode. In this situation, the current begins at zero when the switch is opened and gradually increases, subsequently

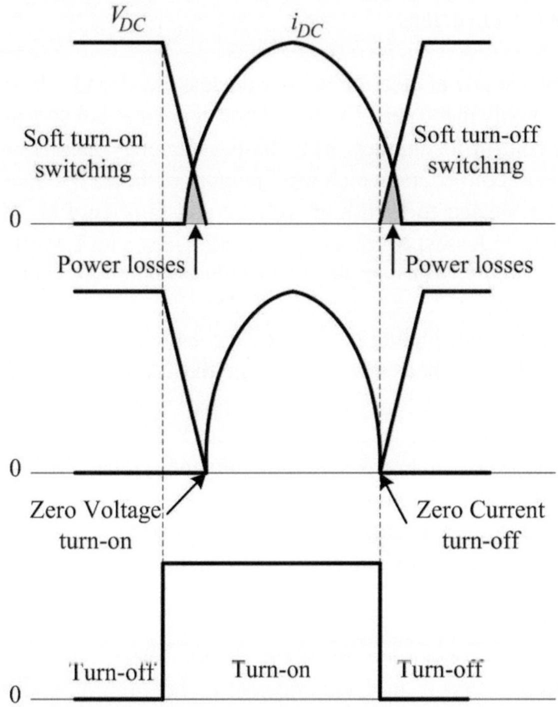

Fig. 11 Voltage and current waveforms of turn-on and turn-off switching conditions that use a resonant circuit

oscillating due to the resonance effect between the L_r and C_r components (Rashid et al. 2024).

Since resonant converters contain a wide variety of structures, different criteria are used in the literature to classify these converters. Among these criteria, the number of elements in the resonant tank, the circuit structure of the tank, and the operating modes are important. When DC–DC resonant converters are classified according to their operating modes, the general appearance of this classification is shown in Fig. 13 (Deshmukh et al. 2022; Mohan et al. 2002; Mudiyanselage et al. 2023; Yano et al. 2000).

Quasi-resonant Converters

Quasi-resonant converters (QRCs) represent a hybrid topology that integrates the characteristics of both resonant and PWM converters. The fundamental concept involves substituting the conventional power switch in PWM converters with a resonant switch. This modification allows a broad range of traditional converter topologies to be adapted into their resonant counterparts. In QRCs, the current and/

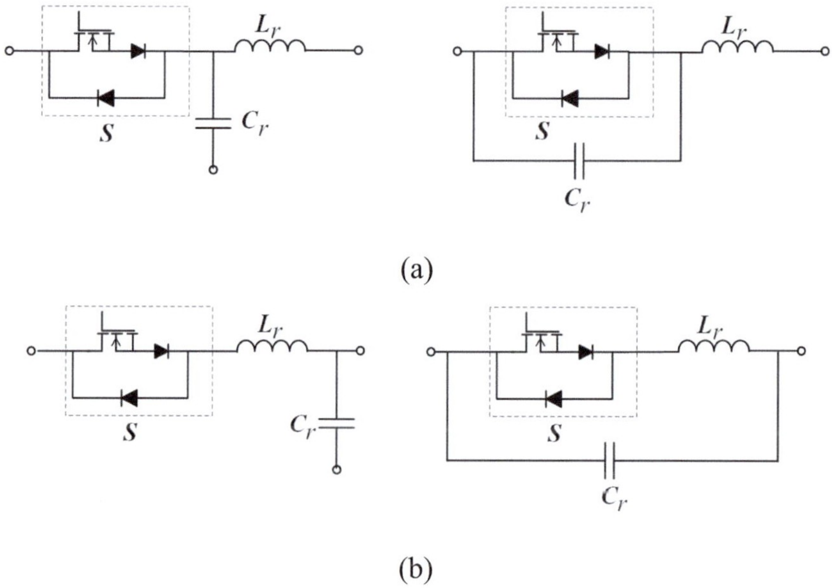

Fig. 12 Two types of **a** resonant-voltage switches (ZVS), **b** resonant-current switches (ZCS)

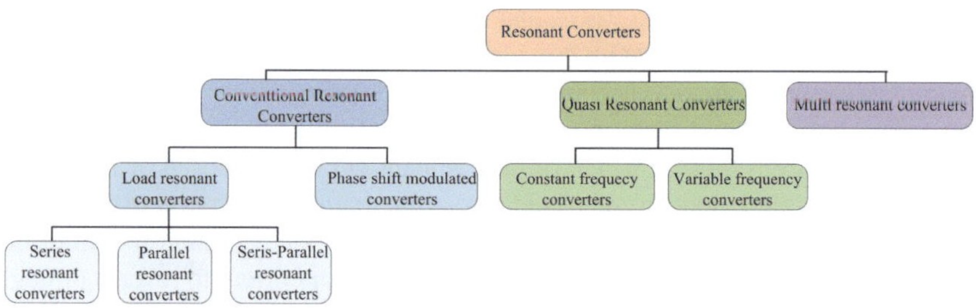

Fig. 13 The classification of resonant converters

or voltage waveforms of the switching device are intentionally shaped into quasi-sinusoidal forms, thereby enabling the realization of ZCS and/or ZVS. Furthermore, both ZCS-based QRCs and ZVS-based QRCs can operate in either half-wave or full-wave modes, offering flexibility in design and application (Hui and Chung 2007). In Fig. 14, half-wave and full-wave ZCS-Quasi-resonant buck converters are shown. Both circuits are formed by replacing the power switch in a conventional PWM buck converter with the ZC resonant switch.

In the half-wave ZCS-Quasi-resonant buck converter scheme, the converter topology is derived by replacing the conventional power switch with a PWM buck converter with a ZC resonant switch. The output filter inductor L is designed with a sufficiently high inductance to maintain an approximately constant current. Prior to

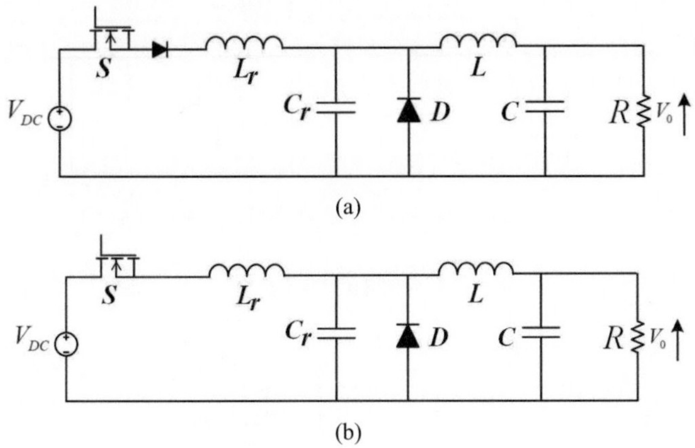

Fig. 14 ZCS-quasi-resonant buck converter. **a** Half-wave scheme, **b** Full wave scheme

the activation of the switch, the output current circulates through the freewheeling diode *D*, and the voltage across the resonant capacitor V_{Cr} is zero. When the switch is turned on under ZCS conditions, a quasi-sinusoidal current flows through the resonant elements L_{r} and C_{r}, the output filter, and the load. When the switch is turned off again with ZCS, it achieves soft commutation.

During and after the gate pulse, the resonant capacitor voltage V_{Cr} initially increases and then decreases, with its waveform shaped by the output current. Regulation of the output voltage is accomplished by modulating the switching frequency. The overall behavior and performance characteristics of the converter are primarily governed by the design parameters of the resonant circuit composed of L_{r} and C_{r}. When an antiparallel diode is connected across the main switch, the converter operates in full-wave mode. This operational mode is functionally similar to that of the half-wave configuration; however, it permits the inductor current to reverse through the antiparallel diode, thereby extending the resonant interval. Consequently, excess energy stored in the resonant circuit under light load conditions can be returned to the input voltage source, effectively reducing the sensitivity of the output voltage to variations in the load.

In ZVS-based Quasi-Resonant Converters (ZVS-QRCs), the resonant capacitor provides a zero voltage condition for turn-on and turn-off situations of the switch. In Fig. 15, half-wave and full wave ZVS-Quasi-resonant buck converters are illustrated. In the half wave ZVS-Quasi-resonant buck converter scheme, when the *S* is turned on, it conducts the output current, while the supply voltage V_{DC} reverse-biases the diode. Upon ZV turn-off condition of the switch, the output current is redirected through the resonant capacitor C_{r}. As the voltage across the resonant capacitor V_{Cr} rises to match the supply voltage V_{DC}, the *D* becomes forward-biased and conducts, initiating the resonant interval. When V_{Cr} subsequently drops to zero, the antiparallel diode across the switch begins to conduct, effectively short-circuiting the resonant capacitor and applying the input voltage directly across the resonant inductor L_{r}. The

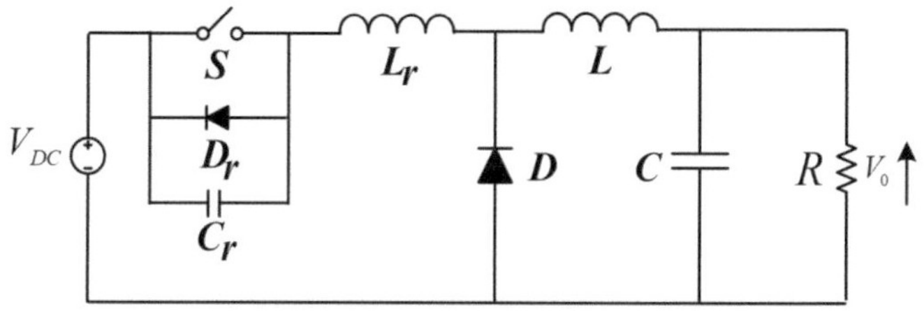

(a)

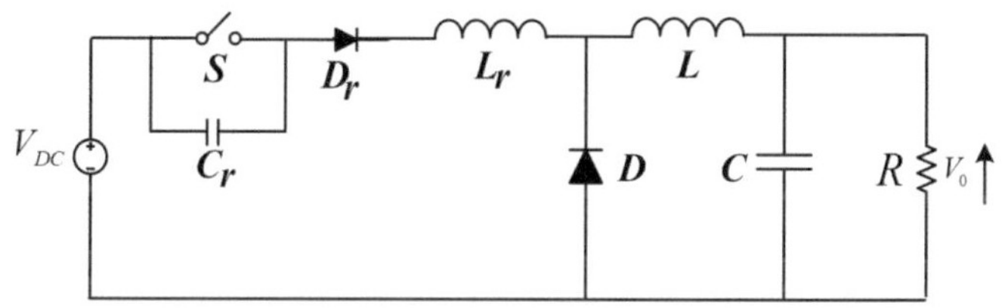

(b)

Fig. 15 ZVS-quasi-resonant buck converter. **a** Half-wave scheme, **b** Full wave scheme

inductor current I_{Lr}, then increases linearly until it reaches the level of the output current, at which point D turns off.

To ensure ZVS operation, the switch S must be triggered while the antiparallel diode is conducting. The operation of the full-wave mode closely resembles that of the half-wave mode; however, a key difference lies in the bidirectional nature of V_{Cr}, which can assume both positive and negative values.

Load Resonant Converters

Load resonant converters (LRCs) offer several advantages over traditional power converter systems. One of the key benefits is that the switches in these converters utilize soft switching, which eliminates losses and stresses associated with on–off switching. LRCs can also help reduce the size and weight of devices by operating at high frequencies, all while maintaining efficiency and avoiding additional stress on the system. Because of these features, LRCs are particularly favored in high-power applications. Structurally, LRCs are categorized into three main types: series resonant, parallel resonant, and series–parallel resonant. Series resonant converters (SRCs) have their load connected in series with the resonant circuit, which is made up of an inductor (L_r) and a capacitor (C_r). In contrast, parallel resonant converters (PRCs) have their load connected in parallel with the resonant capacitor (C_r). The Series–Parallel Resonant Converter (SPRC) combines the benefits of both SRCs and

PRCs by incorporating an additional capacitor or inductor into the resonant circuit. The half-bridge configurations of SRC, PRC, and SPRC are illustrated in Fig. 16.

SRC presents several notable advantages. Primarily, it effectively prevents transformer core saturation, as the series-connected capacitor inherently blocks any DC component. Moreover, the SRC exhibits high efficiency under light load conditions, which is attributed to the reduced device current and minimal conduction losses. Despite these benefits, the SRC also faces certain limitations. Specifically, regulating the output voltage becomes increasingly difficult at light load or no-load conditions. Additionally, the output DC filter capacitor is required to endure substantial ripple currents, posing challenges in systems characterized by low output voltage and high output current. Conversely, the PRC offers a distinct set of advantages. It demonstrates a high tolerance to short-circuit conditions, thereby rendering it suitable for applications demanding low output voltage and high output current. Nonetheless, the PRC is also subject to critical disadvantages. Chief among these is the elevated device current inherent to its operation. Furthermore, because the device current does not scale down with a decreasing load, the converter's efficiency substantially declines under light load conditions (Tarter 1993).

Multi-resonant Converters

Traditional ZCS and ZVS quasi-resonant converters (QRCs) are limited in that they optimize soft-switching conditions for either the active power switch or the output rectifier diode, but not both simultaneously. To address this limitation, the concept of the multi-resonant switch—an extension of the traditional resonant switch topology—has been introduced, enabling improved switching performance for multiple components within the converter. Multi-resonant converter circuits, such as zero-voltage multi-resonant converters (ZVS-MR) and zero-current multi-resonant converters (ZCS-MR), are designed to integrate important parasitic components into the resonant network. These components include the leakage inductance of the transformer, the output capacitance of the active switch, and the junction capacitance of the output diode. ZVS is typically chosen over ZCS in high-frequency DC–DC converter applications, especially when operating in half-wave mode. This desire stems from the fact that the parasitic capacitances of both the switch and the diode can be successfully used as integral components of the resonant circuit, enhancing switching performance and lowering losses. In Fig. 17, a buck ZVS-MR structure is illustrated.

Multi-stage Versus Single-Stage Converters

Single-Stage Converters

Single-stage DC–DC converters are advanced devices that combine all the essential power processing functions into a singular conversion stage. This innovative design

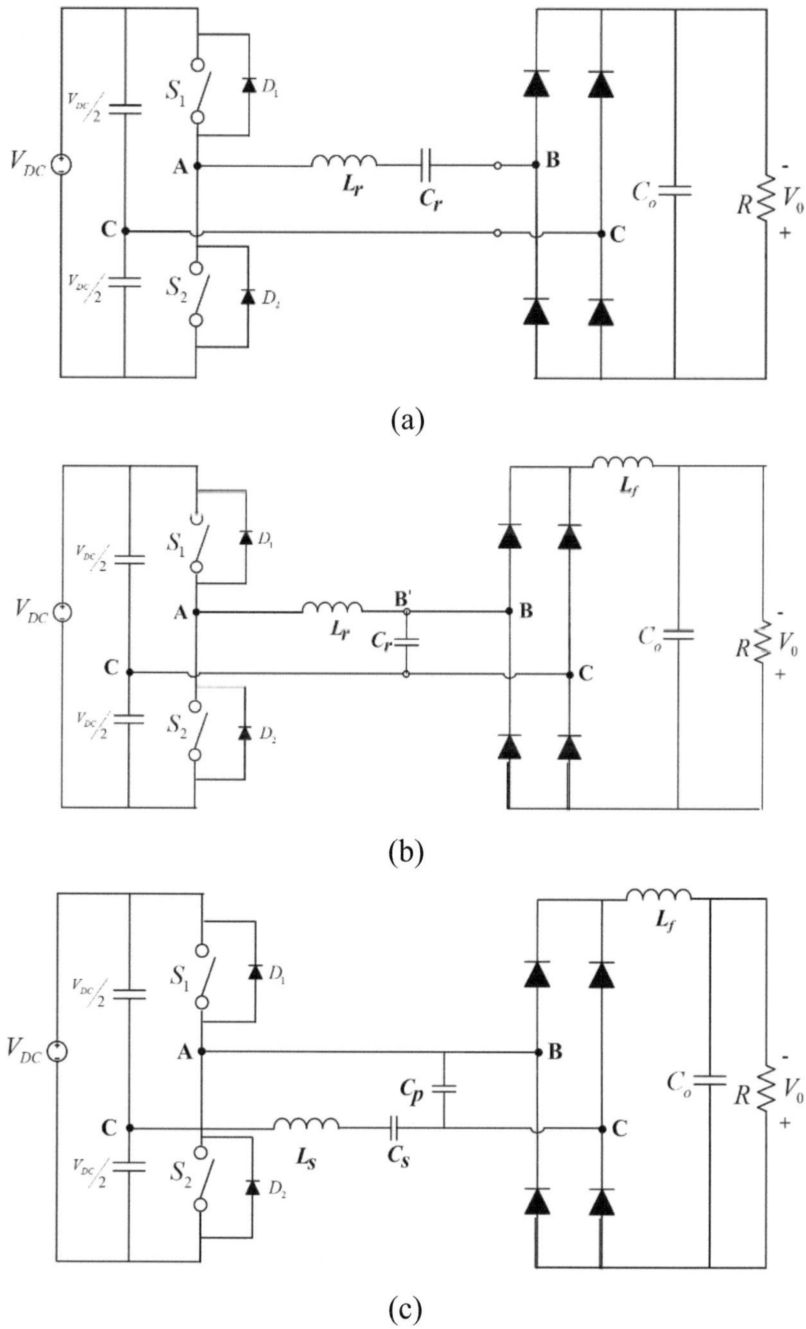

Fig. 16 The half-bridge configurations of SRC, PRC, and SPRC

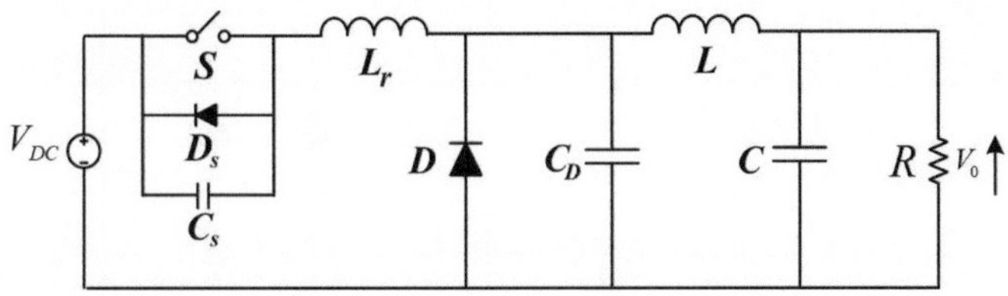

Fig. 17 A buck ZVS-MR circuit scheme

enables the management of critical tasks such as voltage regulation and current control, as well as additional functions such as galvanic isolation, which is required to protect sensitive components from high voltages, and power factor correction, that helps to improve power usage efficiency in applicable systems. This unified topology not only streamlines the circuitry but also significantly minimizes the overall number of components involved. As a result, the complexity of the control systems is greatly reduced, leading to advantages such as lower manufacturing costs, a more compact form factor, and enhanced power density. These characteristics make single-stage converters particularly suitable for low- to medium-power applications where space, simplicity, and cost-effectiveness are paramount. Applications range from portable consumer electronics, like smartphones and tablets, to LED drivers that power modern lighting solutions, and even compact renewable energy systems, such as solar power converters (Gürçam and Nuri Almalı 2025).

While single-stage converters offer numerous benefits, the integrated design can lead to certain compromises. For instance, the need to balance multiple functions within a single stage may limit the ability to optimize specific performance characteristics, such as achieving rapid transient response under sudden load changes or maximizing efficiency across varying operational conditions. Despite these design challenges, single-stage DC–DC converters remain a highly practical solution, especially in scenarios where ease of design, compact integration, and cost savings are of utmost importance.

Multi-stage Converters

Multi-stage DC–DC converters are advanced power conversion systems that divide the power processing tasks into multiple distinct stages, each designed to optimize a specific function for improved performance. In these designs, the initial stage typically focuses on voltage regulation or input conditioning, which may involve stepping the input voltage either up or down. This process ensures that the input voltage of the system is within the required range for subsequent stages. The following stages refine this input, providing precise voltage control, electrical isolation, or effective output filtering. This multi-tiered approach allows each stage to be carefully engineered for specific performance metrics, such as efficiency, dynamic response, and thermal management. Maintaining accuracy, scalability, and reliability is critical for overall system performance in high-power applications such advanced server power supplies, data centers, industrial automation systems, and electric cars. Efficient power management is essential in data centers since it has a direct effect on system reliability and operating costs. Multi-stage converters improve energy efficiency and modularity by enabling flexible power scaling based on changing load demands.

While multi-stage systems are more complex and expensive than single-stage converters, the benefits frequently outweigh the downsides. These converters are the favored choice for demanding power electronic applications because to their improved performance, EMI compliance, and effective thermal distribution. Using a multi-stage method, it is possible to design durable power systems that meet both operating efficiency and reliability requirements in demanding conditions (Abolghasemi et al. 2025).

3.2 AC–DC Converters

AC–DC converters, also known as rectifiers, are essential for converting grid AC voltage into stable DC voltage for power supplies, chargers, and renewable energy systems. Their efficiency and power quality are critical in minimizing losses and improving system performance. AC–DC converters can generally be classified into two main categories: uncontrolled rectifiers and controlled rectifiers. Uncontrolled rectifiers, which use a diode as a switch, provide a fixed DC voltage output and are typically used in applications where constant voltage is sufficient. In contrast, controlled rectifiers, which use a thyristor as a switch, offer a variable DC voltage output, making them particularly suitable for applications requiring adjustable power delivery, such as the control of DC motors. An overview of AC–DC converter (rectifier) classifications is shown in Fig. 18.

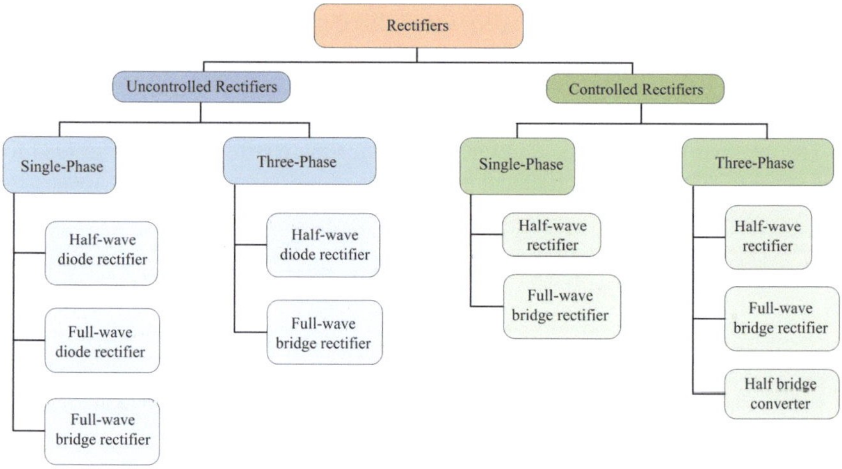

Fig. 18 The classification of AC–DC converter (rectifier)

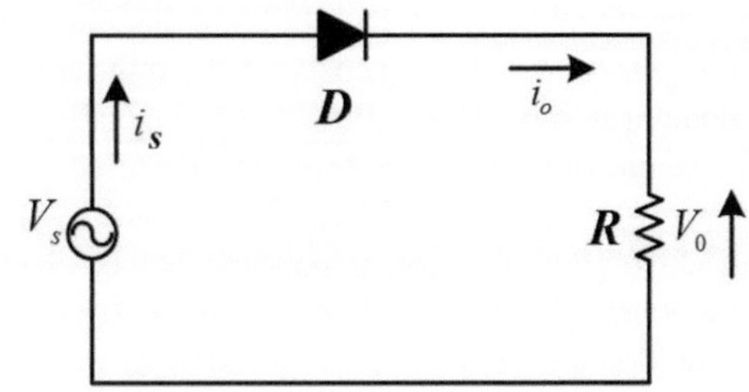

Fig. 19 The single-phase half-wave diode rectifier

3.2.1 Uncontrolled Rectifiers

Uncontrolled rectifiers are specialized circuits that convert alternating current to direct current using only diodes as switching components. Since diodes in these circuits only conduct when a positive voltage is applied, the output voltage is essentially dictated by the properties of the input signal and the configuration of the diodes. Uncontrolled rectifiers are widely used in power supply, battery charging systems, and other industrial applications because of their simple design, low cost, and dependable performance. These rectifiers can be used in both single-phase and three-phase circuits and are particularly useful in situations that require a constant output voltage.

Single-Phase Diode Rectifiers

Two main types of single-phase diode rectifiers exist for converting a single-phase AC source into DC voltage: the half-wave rectifier and the full-wave rectifier. The single-phase half-wave uncontrolled (diode) rectifier circuit is shown in Fig. 19. The circuit consists of an AC voltage source, a diode, and a resistive (R) load. The basic waveforms illustrating the operation of this circuit on a resistive load are given in Fig. 20. During the positive half-cycle of the AC voltage (V_s), diode D becomes forward-biased and conducts. Consequently, the load voltage (V_o) and current exhibit one pulse per cycle of the input voltage.

The average voltage value and the root mean square (rms) value of the load for half-wave rectifier can be calculated as follows:

$$V_{\mathrm{DC}} = \frac{1}{2\pi}\int_0^{\pi} V_m \sin\omega t d(\omega t) \quad V_{\mathrm{DC}} = \frac{V_{\mathrm{m}}}{\pi} = 0.318 V_{\mathrm{m}} \tag{25}$$

$$V_{\mathrm{DC}} = \sqrt{\frac{1}{2\pi}\int_0^{\pi} (V_{\mathrm{m}} \sin\omega t)^2 d(\omega t)} = \frac{V_{\mathrm{m}}}{2} \tag{26}$$

In Figs. 21 and 22, the single-phase full-wave diode rectifier circuit and its waveforms are illustrated. Diode D_1 conducts when the V_s voltage is positive and D_2 diode

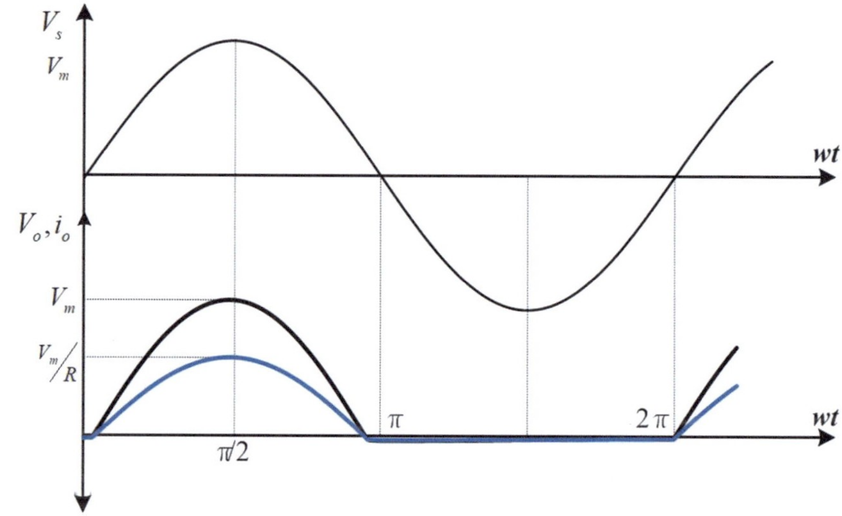

Fig. 20 The single-phase half-wave diode rectifier waveforms for a resistive load

Fig. 21 The single-phase full-wave diode rectifier

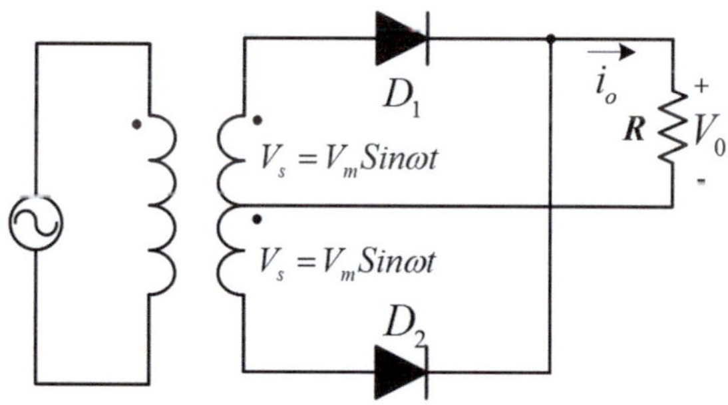

conducts when the V_s voltage is negative, resulting in a fully rectified voltage across the load. From the transformer's perspective, the DC currents of the two half-wave rectifiers are equal and opposite, so there is no DC component that would create a transformer core saturation problem. Compared to the half-wave rectifier shown in Fig. 20, it can be observed that the full-wave rectifier has twice the DC output voltage.

The single-phase full-wave diode bridge rectifier circuit scheme is shown in Fig. 23. During the positive half-cycle of the V_s voltage, the current flows to the load through D_1 and D_2 diodes and during the negative half-cycle of the V_s voltage, D_3 and D_4 conduct. The voltage and current waveforms of the single-phase full-wave diode bridge rectifier are shown in Fig. 24.

The average voltage value and the root mean square (rms) value of the load for full-wave rectifier can be calculated as follows:

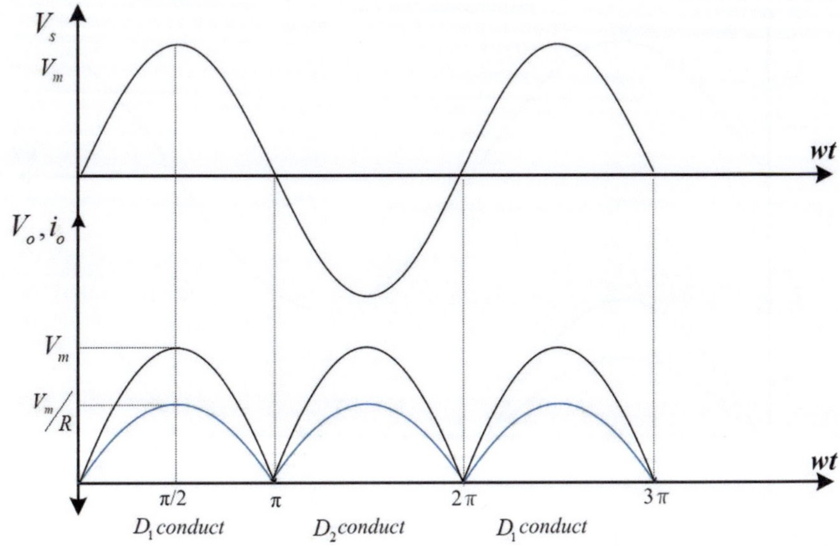

Fig. 22 The single-phase full-wave diode rectifier waveforms for a resistive load

Fig. 23 The single-phase bridge diode rectifier

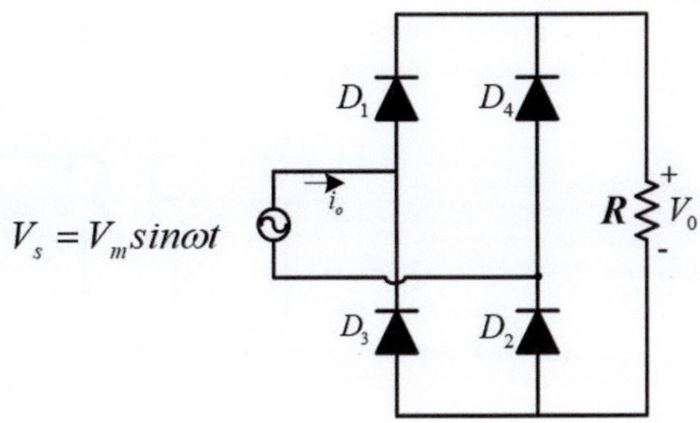

$$\overline{V}_{\text{o}} = \frac{2}{\pi} \cdot V_{\text{m}} = 0.637 V_{\text{m}} \quad (27)$$

$$V_{\text{rms}} = \frac{V_{\text{m}}}{\sqrt{2}} = 0.707 V_{\text{m}} \quad (28)$$

Three-Phase Diode Rectifiers

Three-phase uncontrolled (diode) rectifiers are widely used in high-power applications, especially in industrial motor drives, battery charging systems, DC power supplies, etc. They are generally divided into two main groups according to their

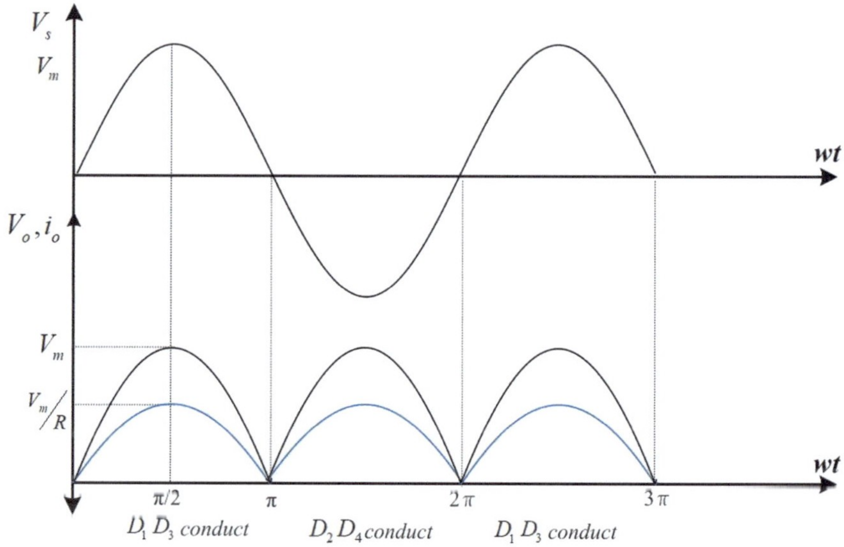

Fig. 24 The single-phase bridge diode rectifier waveforms for a resistive load

structures: half-wave rectifiers and full-wave rectifiers. Half-wave rectifiers use one diode for each phase to transmit only the positive alternation of each phase, while full-wave rectifiers are usually of a bridge type structure consisting of six diodes and provide a smoother and higher output voltage by using both alternations. Although these systems are preferred due to their simple structure, low cost, and reliability, they have limitations in terms of output voltage regulation and harmonic content due to their uncontrolled nature.

In Fig. 25, a three-phase half-wave diode rectifier is illustrated. In this circuit, a diode and an AC source are connected to each phase. Each diode only conducts during the positive voltage of the phase connected to it and allows current to pass through the load; in the negative voltage, the diode reversely polarizes and cuts off the conduction. Therefore, each phase contributes only during the positive voltage, and a fluctuating current flows through the load with a DC component. Since each phase has a 120° phase shift in the three-phase system, this structure results in less ripple in the output voltage compared to single-phase half-wave rectifiers and a smoother DC output is obtained. However, since there is still a significant ripple in the output, such circuits are generally preferred in applications where cost and filtering requirements are minimal. The source voltage and output voltage waveforms of the three-phase half-wave diode bridge rectifier are shown in Fig. 26.

The average voltage value and the root mean square (rms) value of the load for three-phase half-wave rectifier can be calculated as follows:

$$\overline{V}_{\mathrm{o}} = \sqrt{3}\frac{3}{2\pi}V_{\mathrm{m}} = 0.827V_{\mathrm{m}} \tag{29}$$

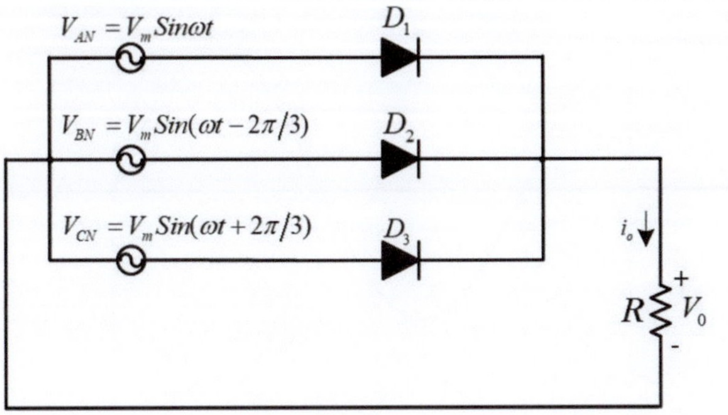

Fig. 25 The three-phase half-wave diode rectifier

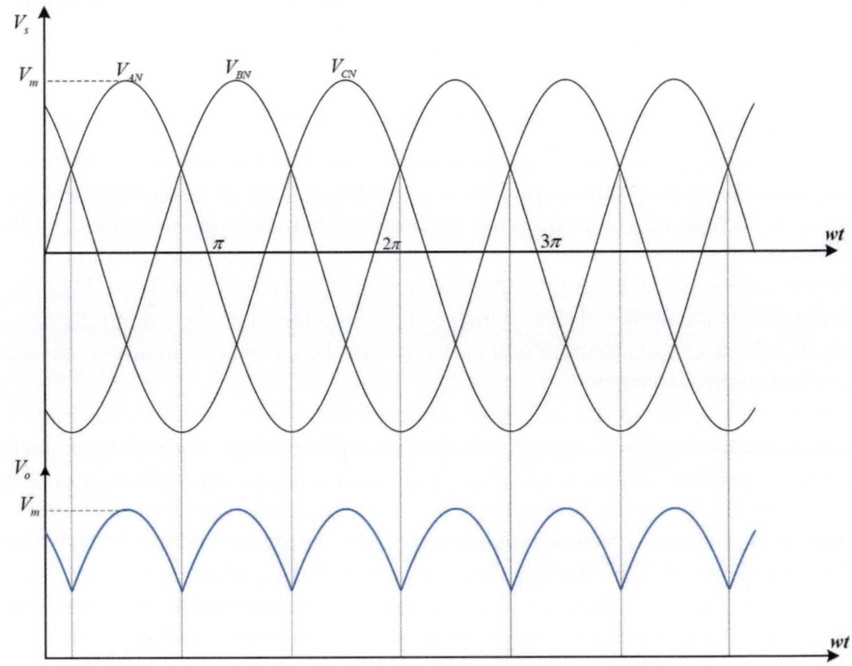

Fig. 26 The three-phase half-wave diode rectifier waveforms for a resistive load

$$V_{\text{rms}} = 0.814V_{\text{m}} \tag{30}$$

The three-phase full-wave diode rectifier is created using six diodes and is generally known as a "six-pulse" rectifier. The circuit scheme is illustrated in Fig. 27. The diodes are placed between the three-phase source and the load in a bridge connection;

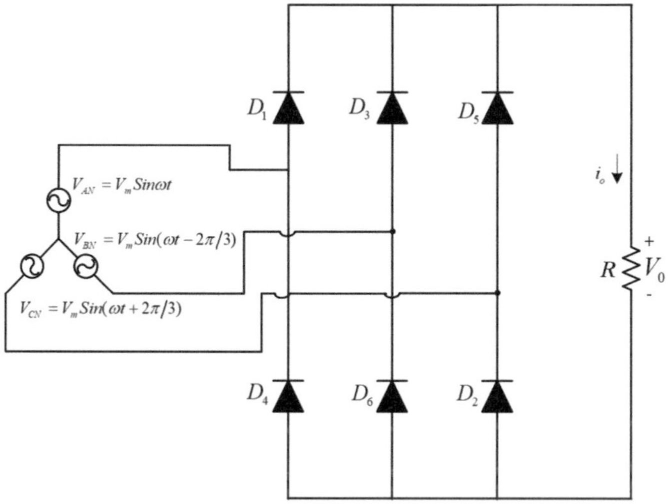

Fig. 27 The three-phase full-wave diode rectifier

two diodes are used for each phase, one conducting during the positive voltage of the phase and the other during the negative voltage. During the circuit's operation, only two diodes conduct at any time: one at the highest positive voltage of the source phases and the other at the lowest negative voltage. Thus, the instantaneous maximum and minimum voltage difference of the three-phase AC source is applied directly to the load, providing a highly efficient direct voltage output. Due to this structure, only small ripples occur in the output voltage, and the frequency of this ripple is six times the input phase frequency. This reduces the need for filtering and allows for more compact and cost-effective designs. Additionally, this type of rectifier contains fewer harmonics compared to single-phase rectifiers and offers more advantages regarding power factor. For these reasons, three-phase full-wave rectifiers are widely used in industrial power supplies, large DC motor drives, battery charging systems, and HVDC (high voltage direct current) transmission systems. The source voltage and output voltage waveforms of the three-phase full-wave diode bridge rectifier are shown in Fig. 28.

The average voltage value and the root mean square (rms) value of the load for three-phase half-wave rectifier can be calculated as follows:

$$\overline{V}_{\mathrm{o}} = \sqrt{3}\frac{3}{\pi}V_{\mathrm{m}} = 1.654V_{\mathrm{m}} \tag{31}$$

$$V_{\mathrm{rms}} = 1.655V_{\mathrm{m}} \tag{32}$$

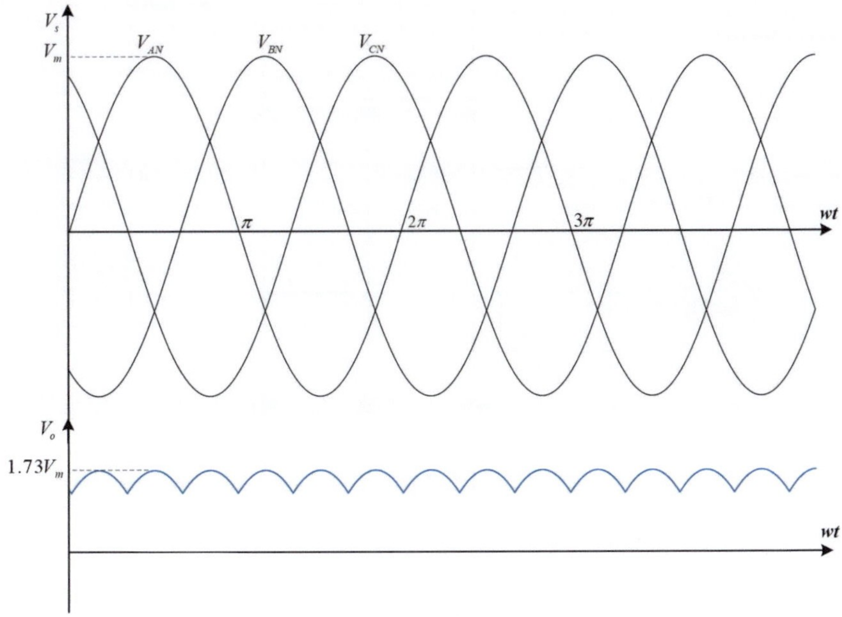

Fig. 28 The three-phase full-wave diode rectifier waveforms for a resistive load

Power Factor Correction (PFC) Techniques

Passive PFC

Passive PFC techniques are based on the use of passive circuit elements such as inductors and capacitors to improve the power factor and reduce harmonic currents returning to the grid. Such circuits are preferred in some low-power applications due to their low cost and structurally simple design. However, passive PFC systems are generally large in size and have limitations in terms of operating efficiency. In addition, since they are only effective in certain frequency ranges, they have limited performance under dynamic load conditions.

Active PFC

Active PFC methods correct the power factor using switching regulators and control circuits and significantly reduce total harmonic distortion (THD). These systems increase efficiency and comply with international power quality standards by making the current drawn from the grid in phase with the voltage and close to the sinusoidal form. Active PFC circuits are widely used in areas requiring high performance, especially LED drivers, telecom power supplies and industrial rectifiers. Although they are more complex and costly compared to passive methods, they are preferred due to the advantages they offer, such as compact structure and constant power factor over a wide input voltage range.

Bridgeless PFC and Totem-Pole PFC

Bridgeless PFC and Totem-Pole PFC structures aim to increase the overall efficiency of the system by reducing the conduction losses caused by traditional bridge rectifier elements. Since some of the bridge diodes are bypassed during the rectification process in Bridgeless PFC circuits, lower voltage drop and less heat loss occur. Totem-Pole PFC, on the other hand, offers high-frequency and high-efficiency solutions, especially when used with fast-switching semiconductors such as GaN (Gallium Nitride) or SiC (Silicon Carbide). These techniques provide great advantages in terms of energy efficiency and thermal management in high-wattage applications—especially in areas such as data centers, power-intensive industrial systems, and electric vehicle charging stations.

3.2.2 Controlled Rectifiers

Controlled rectifiers operate using a semiconductor switching element called Silicon Controlled Rectifier (most commonly known as a thyristor). The output voltage can be adjusted to the desired level by controlling the trigger angles of the thyristors. In this way, the desired voltage and current values on the load can be obtained. Controlled rectifiers not only perform the rectification process but also provide regulation of the energy flow. Controlled rectifiers are important in industrial applications as they increase energy efficiency and enable precise control of systems. Controlled rectifiers are preferred in applications where fixed voltage sources are insufficient, variable loads are present, or energy management is important.

Single-Phase Controlled Rectifiers

A single-phase half-wave rectifier uses a single thyristor to control the load voltage. The thyristor is turned on when the input voltage is positive and a firing current pulse (α) is applied to the thyristor gate terminal. The load voltage is controlled by delaying the firing pulse by an angle α. A simple circuit scheme of a single-phase half-wave rectifier is given in Fig. 29.

The source voltage, load voltage, and current waveforms of the single-phase half-wave controlled rectifier are shown in Fig. 30. Because of the resistive load used, the output current has the same load voltage waveform. The average voltage of the load can be calculated as follows:

$$\overline{V}_{\mathrm{o}} = \frac{V_{\mathrm{m}}}{2\pi}(1 + \mathrm{Cos}\alpha) \tag{33}$$

In Fig. 31a, b, the single-phase full-wave half-controlled and full-controlled bridge rectifier topologies are illustrated, respectively. As illustrated in Fig. 31, two thyristors and two diodes are used in a half-controlled bridge rectifier circuit, while four

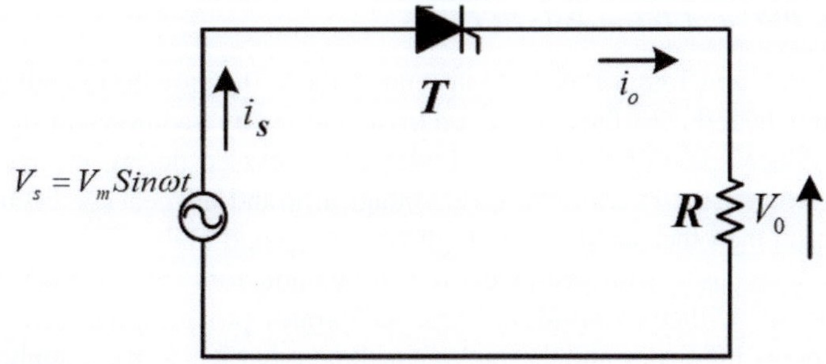

Fig. 29 The single-phase half-wave controlled rectifier

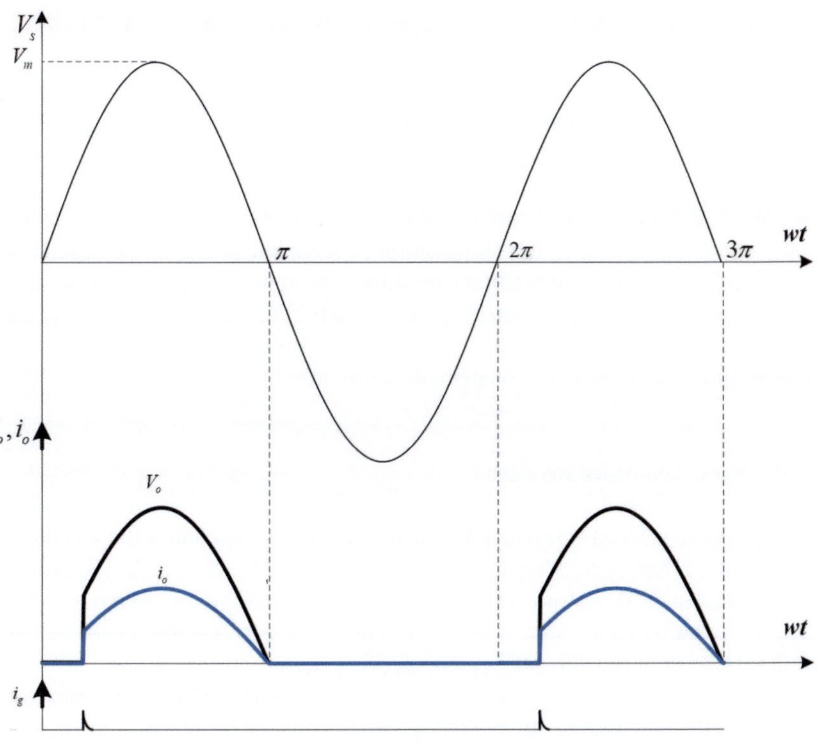

Fig. 30 The single-phase half-wave controlled rectifier waveforms for resistive load

thyristors are used in a full-controlled bridge rectifier circuit to regulate the average load voltage.

In the single-phase half-controlled bridge rectifier, thyristor T_1 must be fired on during the positive half-wave of the source voltage V_s, to allow the conduction of current. Alternatively, thyristor T_2 must be fired simultaneously during the negative half-wave of the source voltage. During the positive half-wave of the source voltage,

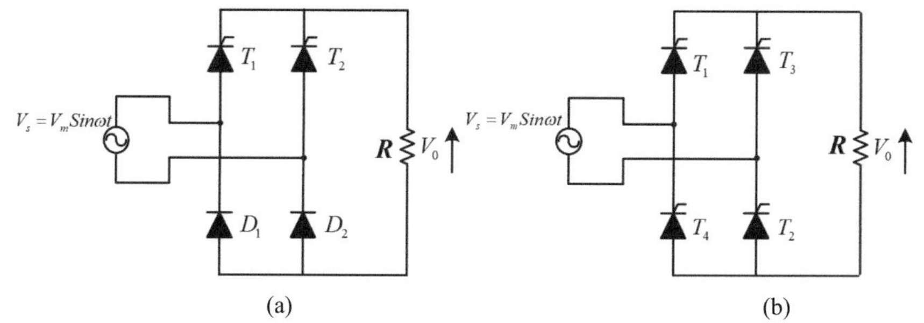

Fig. 31 The single-phase full-wave **a** half-controlled rectifier, **b** full-controlled rectifier

conduction occurs through thyristor T_1 and diode D_2. Conversely, during the negative half-wave of the source voltage, thyristor T_2 and diode D_2 conduct together, allowing current flow through the load. In the single-phase full-controlled bridge rectifier, thyristors T_1 and T_2 must be fired on simultaneously during the positive half-wave of the source voltage v_s, to allow the conduction of current. Alternatively, thyristors T_3 and T_4 must be fired simultaneously during the negative half-wave of the source voltage. In Fig. 32, waveforms for a full-controlled bridge rectifier source voltage, load voltage, load current, and firing points of thyristor are illustrated.

The full-controlled bridge rectifier average voltage of the load can be calculated as follows:

$$\overline{V}_o = \frac{2V_m}{\pi}\text{Cos}\alpha \tag{34}$$

Three-Phase Controlled Rectifiers

Three-phase controlled rectifiers consist of two structures: half-wave and full-wave. Both structures are connected to a three-phase AC source. In both half-controlled and full-controlled, thyristors are triggered with a certain firing angle (α) and only conduct at certain time intervals. This allows control of both the amplitude and the average value of the DC voltage obtained at the output. The thyristor will be in conduction when the anode–cathode voltage is positive and the firing current pulse i_G is applied to the gate terminal. As the firing angle is increased, the average output voltage decreases and after a certain angle, a reverse current direction occurs, and rectifiers can act as inverters. Figures 33 and 34 show the three-phase half-wave rectifier topology and its source voltage and load voltage waveforms for a resistive load.

The full-controlled half-wave bridge rectifier average voltage of the resistive load can be calculated as follows:

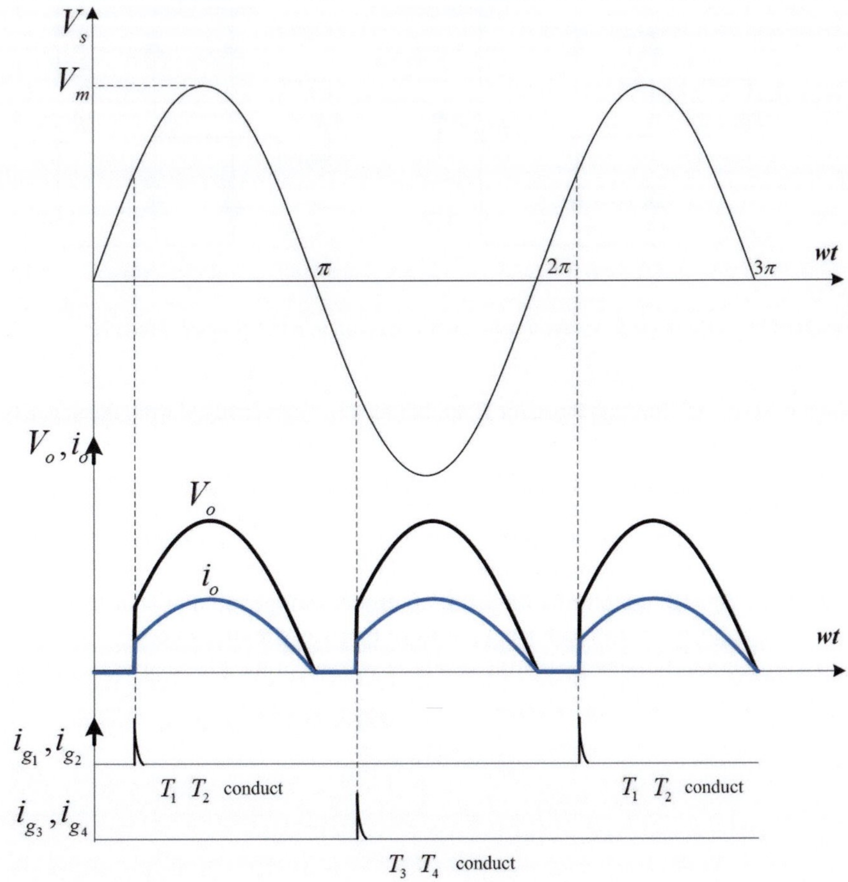

Fig. 32 The single-phase full-wave full-controlled rectifier waveforms

Fig. 33 Three-phase half-wave rectifier

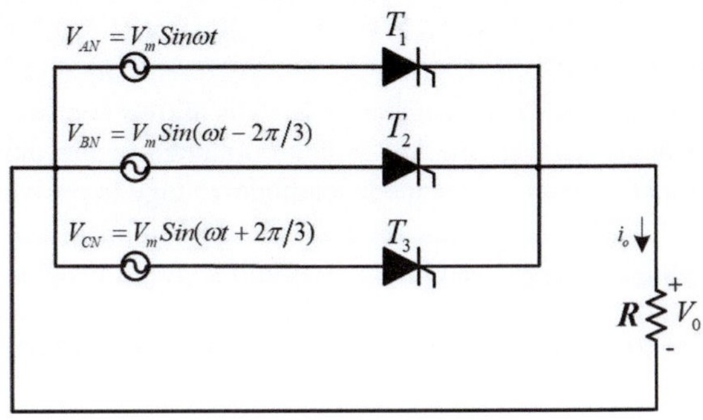

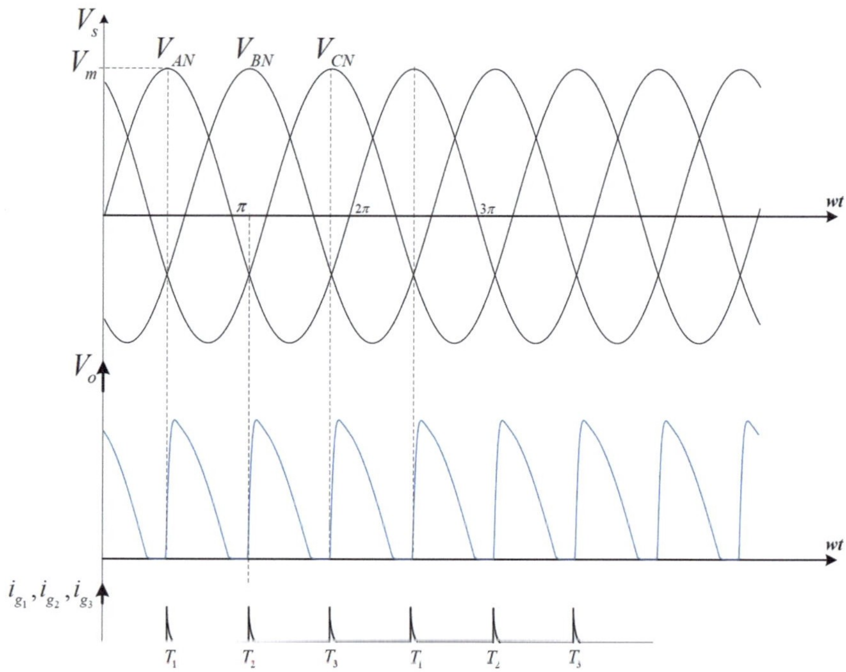

Fig. 34 Three-phase half-wave rectifier waveforms for resistive load

$$\overline{V}_{\mathrm{o}} = \frac{3\sqrt{3}}{2\pi} V_{\mathrm{m}} \mathrm{Cos}\alpha \tag{35}$$

In three-phase full-wave half-controlled rectifiers, both thyristors and diodes are used in a three-phase bridge structure, while full-controlled rectifiers consist only of thyristors. In the half-controlled circuit topology, as in the single-phase bridge, the upper elements are thyristors and the lower elements are diodes for ease of driving. In Fig. 35a, b, these two topologies are illustrated. A full-controlled three-phase bridge converter operates in rectification mode when the firing angle of each thyristor is less than 90°, and transitions to inverter mode when the angle exceeds 90°. In applications where inverter functionality is not required, the circuit can be simplified by substituting three of the controlled rectifiers with power diodes, thereby reducing system complexity without compromising rectification performance. This approach offers a cost-effective alternative, as diodes are significantly cheaper than thyristors and eliminate the need for additional firing angle control circuit. In Fig. 36, the source voltage and load voltage waveforms for a resistive load of three-phase controlled rectifier.

The three-phase full-controlled bridge rectifier average voltage of the resistive load can be calculated as follows:

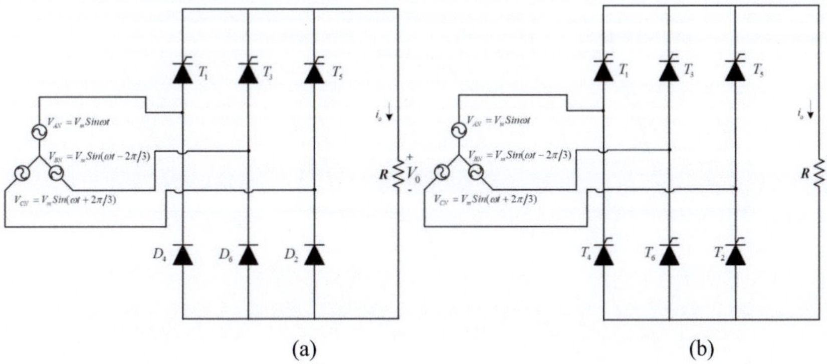

Fig. 35 Three-phase full-wave **a** half-controlled, **b** full-controlled rectifier

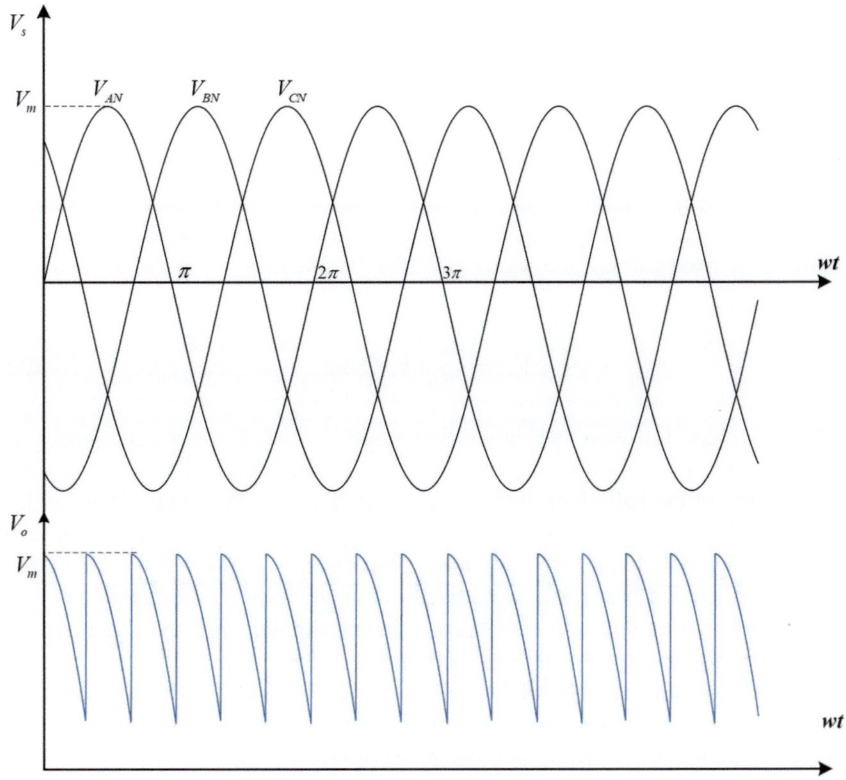

Fig. 36 Three-phase full-wave waveforms for resistive load

$$\overline{V}_o = \frac{3\sqrt{3}}{\pi} V_m \cdot \mathrm{Cos}\alpha \tag{36}$$

3.3 DC–AC Converters (Inverters)

Inverters are widely preferred in many areas such as uninterruptible power supplies, renewable energy sources, electric vehicles, motor drives. If the input voltage of a system feeding an AC load is DC, the conversion process to AC voltage is carried out by the inverter. With the increase in the tendency towards alternative energy sources due to the emergence of global warming problems, the efficiency and performance to be obtained from inverters have become important parameters. Inverters are divided into two main groups in terms of output voltage levels. These are; two-level and multilevel inverter topologies. The traditional two-level inverter circuit structure is shown in Fig. 37. When the switching component S_{x1} is conducting, the output voltage is $V_{DC}/2$, while when S_{x2} is conducting, it is $-V_{DC}/2$. Traditional two-level inverter topologies have disadvantages such as high switching losses and high voltage stress on the switching elements (Kabalci et al. 2023; Vijaya Sambhavi and Ramachandran 2023; Rana et al. 2019).

Multilevel inverter topologies have been suggested to increase the power level and improve the power factor over traditional two-level inverters. The development of three-level inverter topologies began with the Neutral Point Clamped (NPC) topology developed by Nabae et al. (1981). In this proposed topology, the third level was created with a neutral point using input capacitors and the developed circuit was called the NPC inverter topology. In recent years, multilevel inverter topologies that provide high power and low harmonic distortion have been widely utilized in

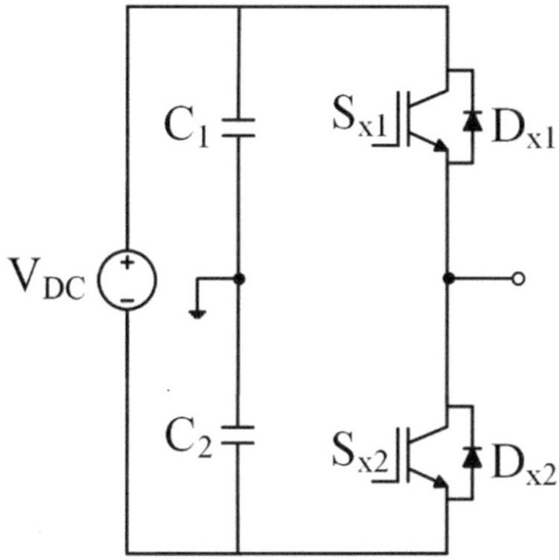

Fig. 37 Two level inverter structure

power electronics applications. Multilevel inverter topologies have advantages over traditional two-level inverter topologies such as lower harmonic distortion and lower *dv/dt* ratio, closer output waveform to sinusoid, high power quality, operation at low and high switching frequencies, and lower noise transfer of input current to the output. In addition, the increase in the number of components to be used along with the increase in the number of levels causes disadvantages such as increased cost and the need for more complex switching situations (Boyar 2024; Colak et al. 2011; Gupta et al. 2016; Kabalcı 2021b). Multilevel inverter topologies are examined in three main classes: neutral point clamped, capacitor clamped and cascade inverters (Kabalci et al. 2023).

3.3.1 Neutral Point Clamped (NPC) Multilevel Inverter

The NPC inverter topology, which is one of the multilevel inverter topologies, was proposed by Nabae et al. (1981). By using diodes in the neutral point clamped multilevel inverter topology, the voltage is clamped and the voltage stress on the switching components is reduced. Figure 38 illustrates the general structure of the three-level NPC inverter topology. There are four switching components and two diodes in the NPC inverter structure. D_{a5} and D_{a6} are called clamping diodes. The number of clamping diodes for the *m*-level NPC inverter is calculated with $(m - 1)(m - 2)$. In this inverter topology, there must be $(m - 1)$ auxiliary capacitors on the DC to make the phase voltage *m*-level. The NPC inverter topology is commonly utilized in industrial applications thanks to its simple structure and low cost (Kabalci et al. 2023; Nabae et al. 1981).

Power losses rise as a result of the unregulated current passing through the diodes in the NPC inverter topology. The switching states of the three-level NPC inverter are displayed in Table 1 where *x* represents phase *a*, *b*, or *c*. Three different states occur, namely P, O, and N, depending on the output voltage conditions. When S_{x1},

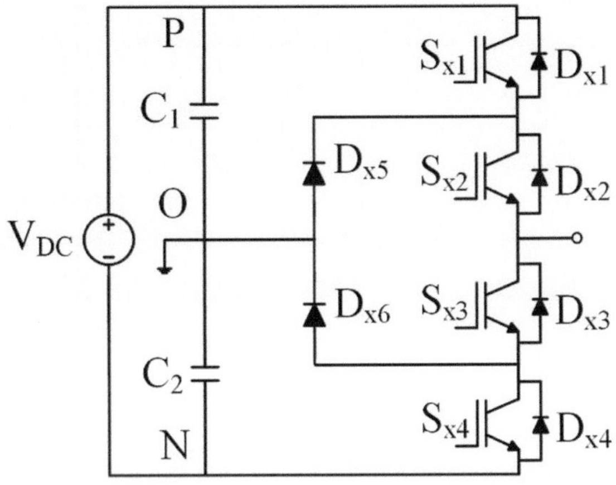

Fig. 38 NPC multilevel inverter structure

Table 1 NPC multilevel inverter switching states

S_{x1}	S_{x2}	S_{x3}	S_{x4}	Output	State
1	1	0	0	$V_{DC}/2$	P
0	1	1	0	0	O
0	0	1	1	$-V_{DC}/2$	N

S_{x2} are in conduction state and S_{x3}, S_{x4} are turned off, the output voltage is $V_{DC}/2$. When S_{x2}, S_{x3} are turned on and S_{x1}, S_{x4} switching components are turned off, the output voltage of the NPC inverter is zero. When S_{x1}, S_{x2} are turned off and S_{x3}, S_{x4} are turned on, $-V_{DC}/2$ is obtained from its output (Kabalci and Boyar 2018).

3.3.2 Flying Capacitor (FC) Multilevel Inverter

The flying capacitor inverter topology was proposed by Meynard, Foch, and Lavieville in the 1990s (Meynard and Foch 1992; Lavieville et al. 1997). The FC multilevel inverter topology was constructed by employing a capacitor instead of the diodes used in the NPC inverter (El-Hosainy et al. 2017). Figure 39 shows the FC multilevel inverter circuit structure. For the m level FC inverter, $(m-1)(m-2)/2$ flying capacitors, $2(m-1)$ switching elements and $(m-1)$ auxiliary capacitors are needed (Islam et al. 2014).

Table 2 displays the three level FC inverter topology's switching states. Three different states occur, P, O, and N, depending on the output voltage level. In P state, S_{x1}, S_{x2} are turned on and S_{x3}, S_{x4} are turned off and the output voltage is $V_{DC}/2$. When the S_{x2}, S_{x3} switching components are in conduction state and S_{x1} and S_{x4} are opened, the O state happens and the output voltage is 0. In the N state, S_{x3}, S_{x4}

Fig. 39 FC multilevel inverter structure

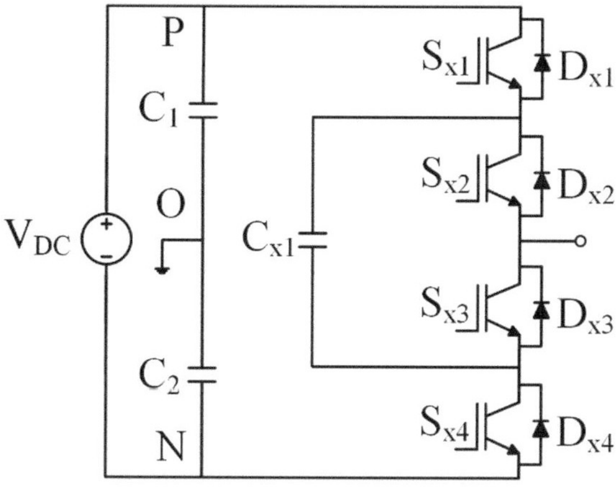

Table 2 FC multilevel inverter switching states

S_{x1}	S_{x2}	S_{x3}	S_{x4}	Output	State
1	1	0	0	$V_{DC}/2$	P
0	1	1	0	0	O
0	0	1	1	$-V_{DC}/2$	N

switching elements are turned on, S_{x1} and S_{x2} are opened. In this case, the output voltage is equal to $-V_{DC}/2$ (Vázquez and Vaquero 2024).

3.3.3 Cascade H-Bridge Multilevel Inverter

Unlike other topologies, the cascade H-bridge multilevel inverter topology does not have clamping diodes or capacitors. This topology is created by connecting H-bridge inverters in series, which have four switching elements in their structure. An independent DC source is required for each H-bridge inverter. In this topology, the number of switches in each phase to be used is determined by $2(m - 1)$ for an m-level inverter, while the number of independent sources is calculated by $(m - 1)/2$ (Rana et al. 2019; Kabalcı 2021). Figure 40 displays the three-phase five-level cascade H-bridge inverter topology. The DC sources need to be isolated from one another since the H-bridge inverter outputs are connected in series. It is recommended to use this topology with fuel cells or photovoltaic panels to obtain high level output voltage. The switching states of each H-bridge inverter are given in Table 3. If the voltage source is V_{DC}, the output voltage is $+V_{DC}$, 0, $-V_{DC}$ depending on the switching status. Three different voltage levels are obtained with each H-bridge inverter. It is easy to increase the output voltage level by connecting them in series. S_{x1} and S_{x3}, S_{x2} and S_{x4} switching elements are switched in reverse (Colak et al. 2011; Kabalcı 2021).

3.3.4 Active Neutral Point Clamped (ANPC) Multilevel Inverter

The use of multilevel inverters is rising with the developments in semiconductor material technology in DC–AC conversion systems requiring high power and efficiency. The ANPC inverter was proposed by Bruckner and Bemet in 2001 to eliminate the disadvantages of the NPC inverter topology (Bruckner and Bemet 2001). Instead of using the diodes found in the NPC inverter topology, switching components were used to create the ANPC inverter topology. The ANPC inverter structure has six switching components for each phase and two capacitors at the input to create the neutral point. The general structure of one phase of the ANPC inverter topology is displayed in Fig. 41. Compared to NPC, it features a lower total harmonic distortion (THD) ratio and two selectable current routes at the neutral point (Boyar et al. 2022; Boyar et al. 2023; Liu et al. 2022).

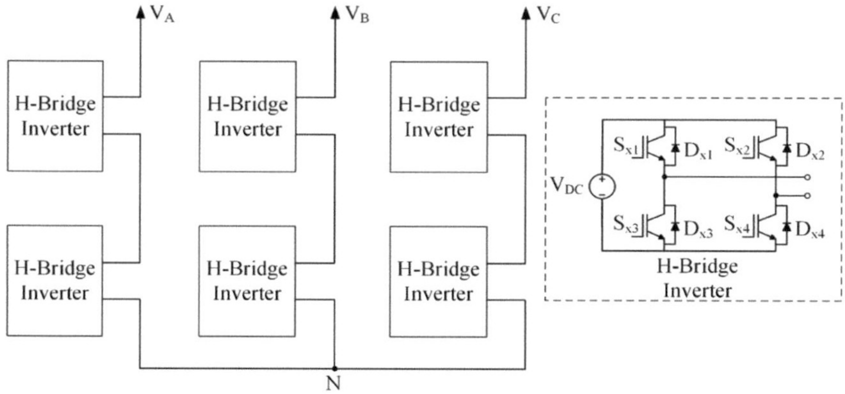

Fig. 40 The structure of three-phase five-level cascade H-bridge inverter topology

Table 3 H-bridge inverter switching states

S_{x1}	S_{x2}	S_{x3}	S_{x4}	Output
1	0	0	1	$+V_{DC}$
0	0	1	1	0
0	1	1	0	$-V_{DC}$

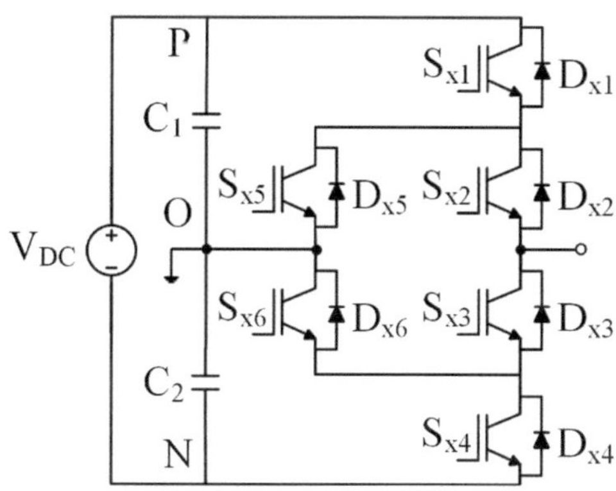

Fig. 41 The structure of ANPC inverter

The required switching states for the switching components in the ANPC inverter topology are given in Table 4. The ANPC topology contains six switching elements in each phase. Here, *x* stands for phases *a*, *b*, and *c*. If the input voltage of the ANPC inverter is V_{DC}, the output voltage produces three different voltage levels as $V_{DC}/2$, 0 and $-V_{DC}/2$ depending on the switching conditions. These are also defined as three special switching states, namely P positive, O zero, and N negative. In the ANPC switching method, there are four O switching states, denoted as OU1, OU2, OL1, and OL2. When the switching components S_{x2} and S_{x5} are closed, the output voltage of the ANPC is equal to the O state expressed by OU1 or OU2. The state of switch S_{x4} determines whether the output state is OU1 or OU2. If the switching elements S_{x3} and S_{x6} are closed and S_{x2}, S_{x4} and S_{x5} are opened, the output voltage will be in the O state, which represents the zero voltages of OL1 or OL2. The state of the S_{x1} switch determines whether the output state is OL1 or OL2. In the P state, the switching components S_{x1}, S_{x2}, and S_{x6} are closed, S_{x3}, S_{x4}, and S_{x5} are opened and the output voltage is equal to $V_{DC}/2$. To obtain the N state where the output voltage of the ANPC inverter is $-V_{DC}/2$, the S_{x3}, S_{x4}, and S_{x5} switching components must be closed and S_{x1}, S_{x2}, and S_{x6} must be opened (Hu et al. 2017; Xu et al. 2019; Kabalcı and Boyar 2021).

The transitions of the load current on the circuit components in the P–O case are displayed in Fig. 42. When switches S_{x1}, S_{x2}, and S_{x6} are on, the load current flows through switches S_{x1} and S_{x2}. When the S_{x1} is opened, the dead time event occurs. In this case, half of the current reaches the load via S_{x6} and D_{x3}, and the other half via S_{x2} and D_{x5}. When switches S_{x3} and S_{x5} are also conducting, the O state happens. In case O, the load current flows through two branches consisting of D_{x5}, S_{x2} and S_{x6}, D_{x3} (Boyar and Kabalci 2021; Feng et al. 2020).

In Fig. 43, the transition states of the load current in the N–O switching case are shown. The N state happens when switches S_{x3}, S_{x4}, and S_{x5} are turned on. In this case, the load current flows through S_{x3} and S_{x4}. Then, the dead time event happens when the S_{x4} switch is opened. While half of the current flows through S_{x2} and D_{x5},

Table 4 ANPC inverter switching states

S_{x1}	S_{x2}	S_{x3}	S_{x4}	S_{x5}	S_{x6}	Output	State
1	1	0	0	0	1	$V_{DC}/2$	P
0	1	0	1	1	0	OU1	O
0	1	0	0	1	0	OU2	
1	0	1	0	0	1	OL1	
0	0	1	0	0	1	OL2	
0	0	1	1	1	0	$-V_{DC}/2$	N

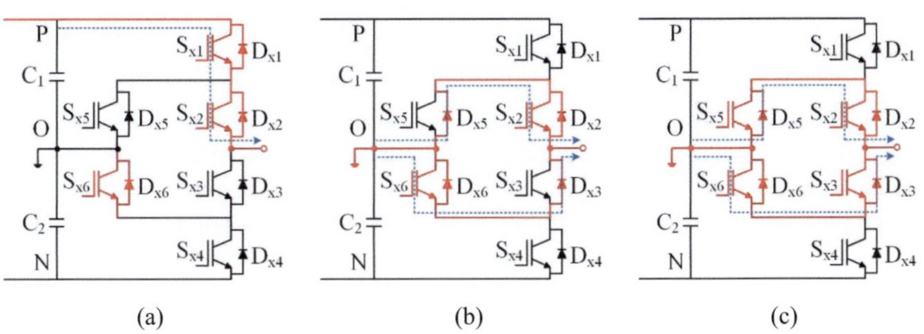

Fig. 42 ANPC inverter P–O switching state. **a** P **b** dead time **c** O

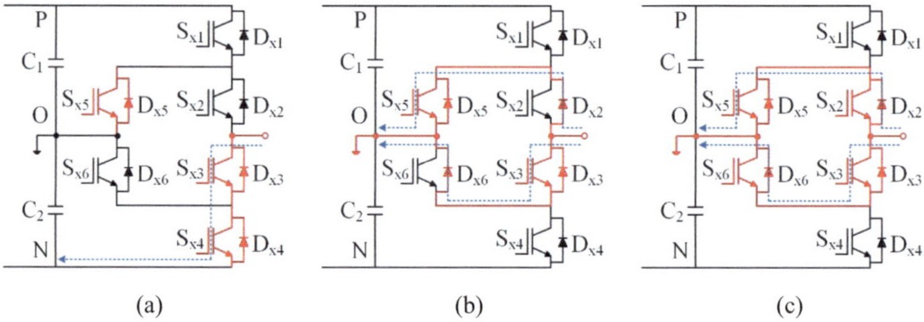

Fig. 43 ANPC inverter N–O switching state. **a** N **b** dead time **c** O

the other half flows through D_{x3} and S_{x6}. When S_{x2} and S_{x6} are in conduction state, the O state happens. The load current flows through two branches consisting of S_{x5}, D_{x2}, D_{x6}, and S_{x3} in this case (Boyar and Kabalci 2021; Feng et al. 2020).

3.3.5 T-Type Multilevel Inverter

T-type inverter topology is commonly utilized in medium and high-power applications such as solar energy systems, uninterruptible power supplies, motor drives due to its superiorities in harmonics and efficiency. T-type inverter topology can provide better quality power flow with THD at low switching frequencies compared to traditional inverters. It also provides lower voltage stress on switching elements and smaller filter sizes (Emon et al. 2021). The general structure of the three-level T-type multilevel inverter topology is displayed in Fig. 44. In the traditional two-level inverter topology structure, the DC-link midpoint is created, and two bidirectional switching elements are added. These two switches allow the current flow to be controlled by turning one or the other on or off. This topology offers an alternative to more complicated three-level topologies like ANPC. The T-Type inverter topology combines the benefits of three-level inverters, such reduced switching losses and better output voltage quality, with the advantages of two-level inverters, including low conduction losses, fewer components, and an easy-to-understand operating principle (Schweizer and Kolar 2013).

There are four switching components in the three-level T-type inverter topology. The switching states of a phase are given in Table 5. Depending on the switching states, three different states occur at the output: P, O, and N. Switching components S_{x1}, S_{x3} and S_{x2}, S_{x4} are switched inversely. When the switching components S_{x1} and S_{x2} are turned on and S_{x3} and S_{x4} are turned off, the P state happens and the output voltage is $V_{DC}/2$. When S_{x2} and S_{x3} are turned on and S_{x1} and S_{x4} are turned off, the O state occurs and the output voltage is obtained as 0 (Schweizer and Kolar 2013).

Figure 45 shows the P–O switching transition states of the T-type inverter topology for one phase. The P state occurs when the S_{x1} and S_{x2} are turned on. The load current flows through S_{x1} in the state P. Dead time state occurs when the S_{x1} is turned off. In

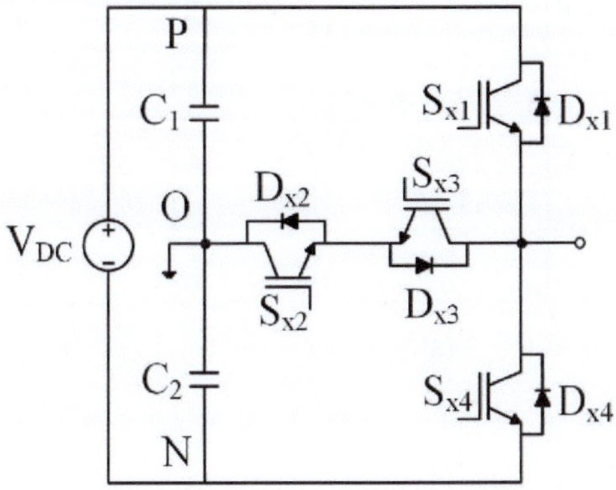

Fig. 44 The structure of T-type inverter topology

Table 5 T-type inverter switching states

S_{x1}	S_{x2}	S_{x3}	S_{x4}	Output	State
1	1	0	0	$V_{DC}/2$	P
0	1	1	0	0	O
0	0	1	1	$-V_{DC}/2$	N

this case, the load current flows to the load via S_{x2} and D_{x3}. In case O, S_{x2} and S_{x3} operate conducting mode and the load current flows through S_{x2} and S_{x3} (Schweizer and Kolar 2013; Wang et al. 2020).

The transition states of the load current in the N–O state are given in Fig. 46. When the S_{x3} and S_{x4} are in conducting mode, the N state arises and the load current flows through S_{x4}. In this case, the S_{x4} is turned off, dead time condition happens. The load current flows on D_{x2} and S_{x3}. In the O state, S_{x2} and S_{x3} are in conduction mode. In this case, the load current flows through S_{x2} and S_{x3} (Schweizer and Kolar 2013; Wang et al. 2020).

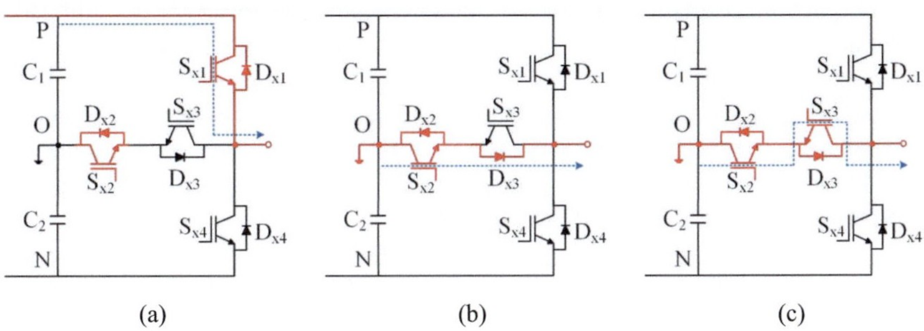

Fig. 45 T-type inverter P–O switching state. **a** P **b** dead time **c** O

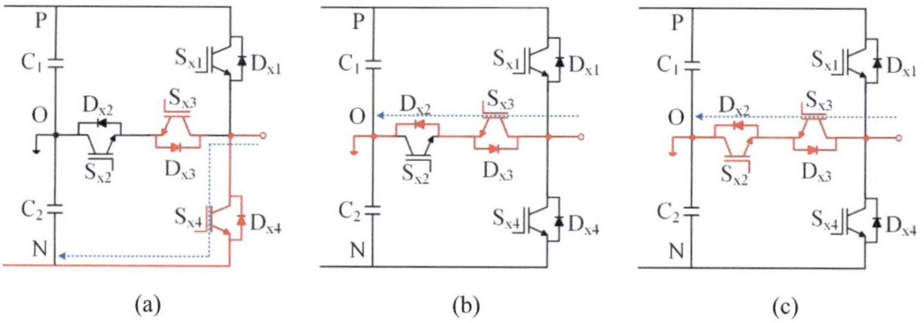

Fig. 46 T-type inverter N–O switching state. **a** N **b** dead time **c** O

3.4 AC–AC Converters

AC–AC power converters are crucial components in power electronics that enable the direct transformation of alternating current from one set of electrical characteristics—such as voltage amplitude, frequency, and phase—to another. Rather than passing through a DC link, these systems perform the conversion directly, depending on their architecture. Depending on the required power rating and application, both the input and output sides may be configured as single-phase or three-phase. In recent years, however, there has been growing interest in multiphase systems due to their improved performance in reducing harmonics, lowering phase currents, and offering enhanced redundancy and reliability—particularly in high-performance industrial environments.

Converters that modulate the RMS voltage level of a fixed-frequency AC supply are generally referred to as AC voltage controllers or regulators. These converters implement voltage regulation using either phase-angle control techniques with naturally commutated devices like SCRs and triacs, or by employing high-frequency forced-commutated switches such as IGBTs, GTOs, and IGCTs. The choice of switching technology depends largely on the target switching speed and power throughput. Meanwhile, cycloconverters represent another class of AC–AC systems that allow direct conversion between two frequencies without the use of intermediate DC stages. Traditionally built with line-commutated thyristors, their operation is limited to generating output frequencies that are a fraction of the input frequency. However, advances in high-speed, bidirectional semiconductor devices have enabled the development of matrix converters, which support independent control over output voltage and frequency, while maintaining near-sinusoidal waveform quality.

In terms of applications, AC voltage controllers are typically used in systems that require simple and effective voltage regulation, such as resistive heating loads, lighting dimming systems, transformer tap-changers, and soft starters for motors. Cycloconverters, due to their robust handling of high power and low-speed operation, are commonly found in specialized industrial sectors—such as rotary furnaces, large rolling mills, and propulsion systems in marine vessels. Matrix converters, although more complex, are becoming increasingly popular in applications that

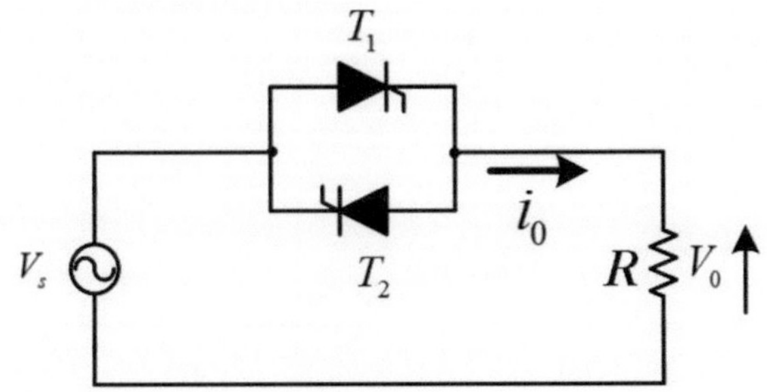

Fig. 47 Single-phase full-wave AC–AC controller

demand compact form factors, high efficiency, and precise dynamic control. This section explores the operating principles, circuit configurations, and control strategies of these AC–AC conversion technologies, while also highlighting their practical uses across various industries (Ali et al. 2024).

3.4.1 Single-Phase AC–AC Converters

Figure 47 shows the basic power circuit of a single-phase AC–AC voltage controller. This circuit structure consists of a pair of SCRs connected in reverse parallel (also known as antiparallel) between a single-phase AC supply and the load. This connection provides bidirectional full-wave symmetrical control. In addition to this connection type, there are also different connection types and usage cases of different elements for special applications. In lower power applications, this SCR pair connected in reverse parallel can be replaced with a triac as shown in Fig. 48a or a connection consisting of one SCR and four diodes can be made to reduce the device cost as shown in Fig. 48b. However, in this case, the device conduction loss increases. In addition, a single thyristor and a single diode circuit structure can be used as shown in Fig. 48c to provide unidirectional voltage control. However, this circuit structure is not very practical to use except for very low power heating loads since the DC component and harmonics increase. Figure 49 shows the waveforms of the source voltage, load voltage, and current of the single-phase AC–AC converter circuit, depending on the triggering locations of the thyristors. As can be seen from this figure, the T_1 thyristor is in conduction when the input voltage is positive, while the T_2 thyristor is in conduction when the output voltage is negative.

3.4.2 Three-Phase AC–AC Converters

Different circuit topologies can be used in three-phase phase-controlled AC–AC converter circuits depending on whether the load is star- or delta-connected. Figure 50 shows several circuit topologies suitable for this situation. In the star-connected topology of Fig. 50a, the SCRs are selected to carry line currents and withstand phase voltages, while in the delta-connected topology of Fig. 50b the capacity to

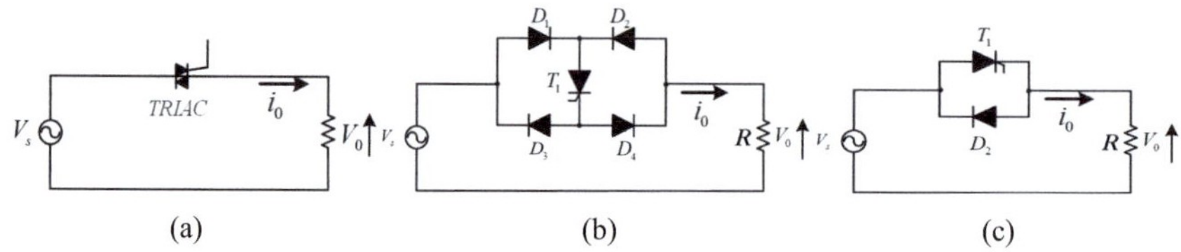

Fig. 48 Single-phase AC–AC controller alternative topologies **a** full-wave with triac, **b** full-wave with four diodes and one thyristor, **c** half-wave with one diode and one thyristor

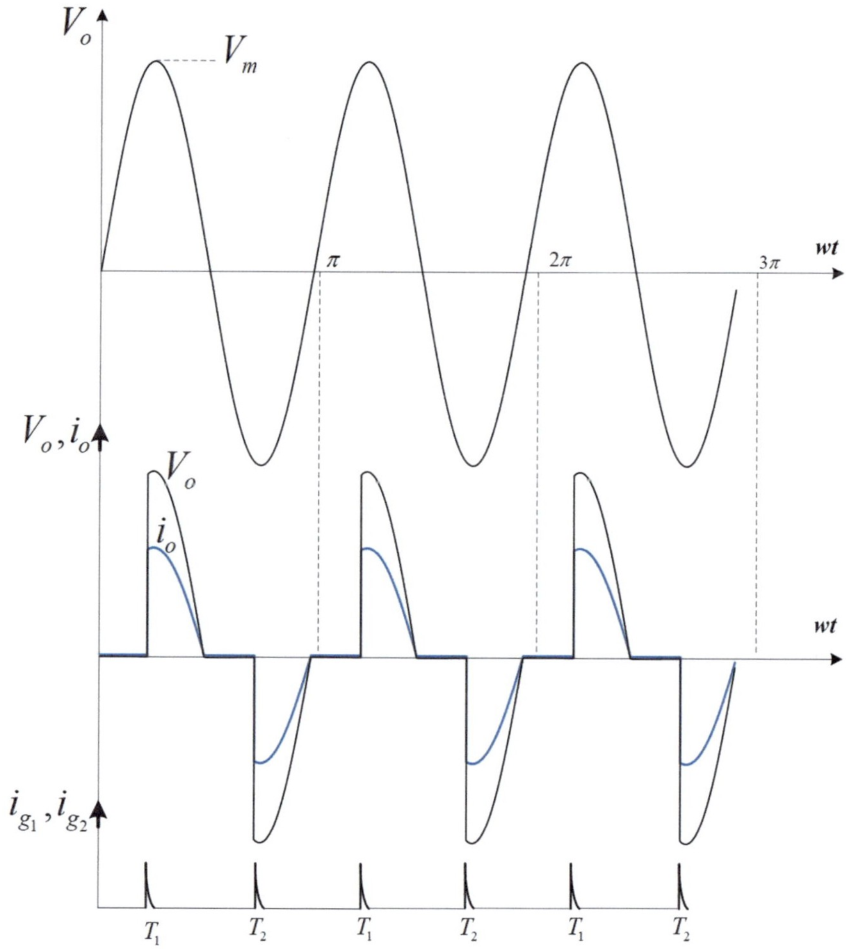

Fig. 49 Single phase full-wave AC–AC controller waveforms for resistive load

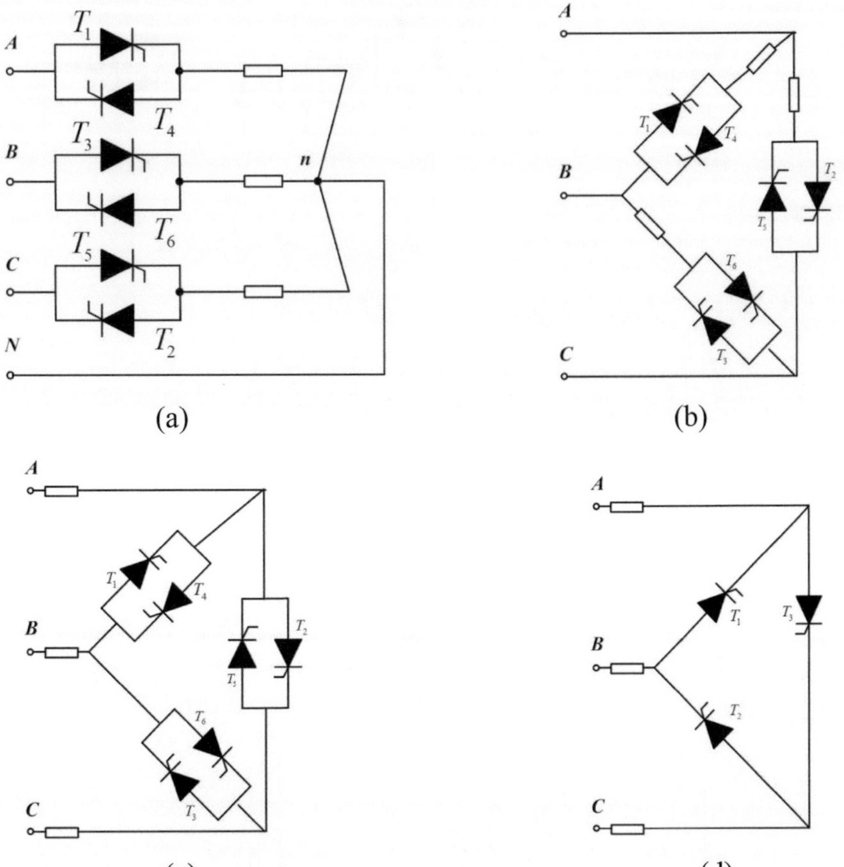

Fig. 50 Three-phase AC–AC converter topologies

carry phase currents and withstand line voltages must be used, while the power factor is slightly higher. The firing angle control range for these two circuits is 0°–180° for a resistive load. In the topology shown in Fig. 50c, where the controllers are connected in delta and the load is connected between the input voltage and the converter, current can flow between the two lines even if an SCR is conducting, so each SCR requires one firing pulse per cycle. In the topology shown in Fig. 50d, three SCRs are connected to the input voltage via the load in a delta-connected structure.

3.4.3 Matrix Converters

Matrix converters (MC) are new generation AC–AC power converters that use fully controlled bidirectional switches and can apply PWM (Pulse Width Modulation) based voltage control. Developed as an alternative to traditional rectifier-inverter or back-to-back topologies, this structure offers a single-stage solution that provides three-phase to three-phase conversion with only nine bidirectional switches (Wang et al. 2025; Zhang et al. 2024). While the bidirectional switch topologies are shown in Fig. 51a–d, the MC circuit topology is shown in Fig. 52. In the diode bridge bidirectional switch shown in Fig. 51a, the two-way flow of current is provided by the bridge diode. Its biggest advantage is that a single controlled element is used and a single control signal is required. Its disadvantage is that the number of elements is higher than other bidirectional switches. At the same time, since two diodes and one transistor are in conduction in both directions, the voltage drop on the elements during conduction is higher than the others. In this structure, only one isolated source is needed to drive the switch. It carries more current than other switches.

Figure 51b, c depicts two widely adopted topologies: the common emitter and common collector configurations, respectively. These arrangements facilitate smooth transition between bidirectional switches. Compared to the common emitter, the common collector configuration offers the benefit of requiring fewer isolated gate drive supplies. In Fig. 51d, an alternative bidirectional switch design is presented,

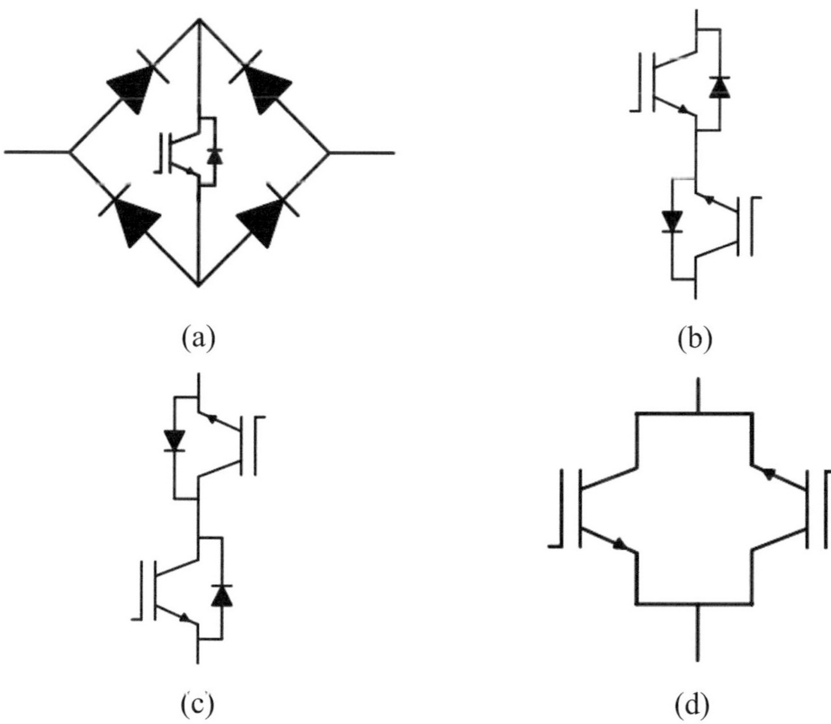

Fig. 51 Different bidirectional switch configurations **a** diode-bridge, **b** common emitter, **c** common collector, **d** reverse blocking

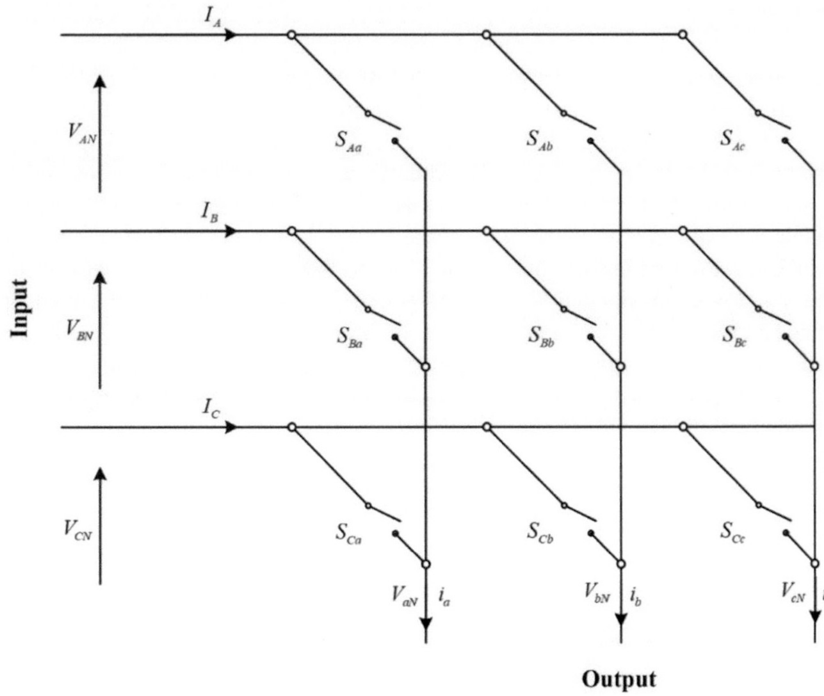

Fig. 52 Three-phase matrix converter scheme

utilizing two reverse-blocking IGBTs connected in antiparallel. This approach eliminates the need for external diodes, leading to lower conduction losses and enabling a more compact transistor design.

While bidirectional power flow between input and output is possible with MC as seen in Fig. 52, sinusoidal waveforms can be obtained on both input and output sides. In addition, the elimination of DC intermediate circuit reactive components allows the system to be designed more compactly. MCs are also important because they provide controllable input power factor independent of load current. In addition to the theoretical advantages of MCs, practical benefits such as better performance at high temperatures, lower voltage stress on switches and reduced maintenance requirements are also emphasized in the literature. Despite their numerous advantages, the large-scale deployment of matrix converters remains constrained by several factors that include the theoretical voltage transfer ratio limit of 0.866, the need for advanced control and switching algorithms, and the challenges related to the implementation and commercial availability of high-frequency fully controlled bidirectional semiconductor switches. In the system design, only certain switching combinations are allowed, and various modulation techniques (such as space vector modulation) are used to synthesize the desired voltages between the input and output phases through these combinations. While the performance of the system is tried to be optimized with these methods, the voltage transmission ratio can be increased in some cases by adding harmonic components (Wheeler et al. 2002; Raju and Mohan 2024).

In this configuration, V_{AN}, V_{BN}, and V_{CN} denote the three-phase input voltages, while V_{aN}, V_{bN}, and V_{cN} correspond to the three-phase output voltages. The nine bidirectional switches, S_{Aa}, S_{Ba}, S_{Ca}, S_{Ab}, S_{Bb}, S_{Cb}, S_{Ac}, S_{Bc}, and S_{Cc}, control the switching operation. Each output phase is formed by successively connecting the corresponding input phases to the output terminals within a defined switching interval. The switching period denoted as T_s is composed of the intervals T_{Aa}, T_{Ba}, and T_{Ca}. By activating switch S_{Aa} for the duration of T_{Aa}, the V_{AN} input phase is applied to the output. Similarly, closing switch S_{Ba} during T_{Ba} transfers the V_{BN} input phase, and engaging S_{Ca} during T_{Ca} delivers the V_{CN} input phase to the output. As a result, the V_{aN} output phase is synthesized through these time-sequenced connections. Figure 53 shows the V_{aN} output voltage for the RL load. Other phase outputs are obtained by similar switching.

The phase-to-phase connections made by matrix converters are controlled by various switching strategies in order to produce the desired voltages between the input and output phases. The most commonly used among these strategies are the Venturini algorithm, the modified Venturini method, and space vector modulation (SVM) techniques. SVM, in particular, draws attention with its high-quality output waveforms and improved voltage usage ratio. With these methods, the output frequency can be adjusted independently of the input frequency, while harmonic distortions can also be minimized (Ravindrababu et al. 2025).

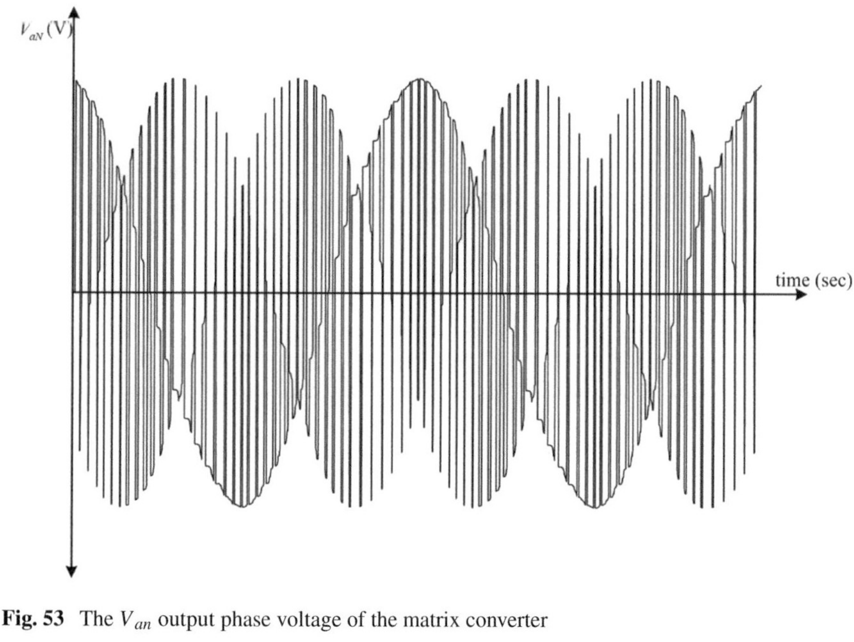

Fig. 53 The V_{an} output phase voltage of the matrix converter

From a hardware perspective, each switch in MC structures should generally be capable of conducting current in two directions and blocking voltage. For this purpose, reverse-blocking, self-controlled elements such as inverse parallel connected IGBTs or power MOSFETs are preferred. The most practical three-phase to three-phase MC topology uses a matrix structure consisting of nine bidirectional switches (BDS) that can connect each output phase to any input phase. Input filters in the system are necessary to suppress high-frequency fluctuations caused by switching; on the output side, the inductive character of the load plays a critical role in ensuring current continuity.

3.4.4 Cycloconverters

Cycloconverters are power electronic circuits that enable direct frequency conversion in alternating current (AC) systems (Bose 2021b). Their basic function is to convert the fixed frequency AC voltage at the input to AC voltage at a lower frequency. This conversion is performed without any intermediate direct current (DC) step. In this respect, cycloconverters are preferred especially in industrial applications where there are high power requirements. These converters typically use controlled semiconductor switching elements such as SCRs. These switches are triggered to generate sinusoidal voltage pulses from the input phases at specific timings. The output voltage obtained in this way is synthesized to include the desired low-frequency component. Cycloconverters use either bidirectional thyristors or antiparallel-connected thyristor pairs to accomplish this process.

Cycloconverter systems are generally classified as unidirectional for applications where current is required to flow in only one direction, or bidirectional for systems where current is required to flow in both directions, such as synchronous motor drives. Unidirectional structures can provide output voltage in only one polarity, while bidirectional structures provide output in both positive and negative polarities, allowing results closer to pure sinusoidal waveforms. In this way, less harmonic distortion occurs in the output power and smoother torque production is achieved in motor applications. In addition, cycloconverters are also categorized according to their inputs and outputs, such as single-phase to single-phase, single-phase to three-phase, three-phase to three-phase (Tahmid et al. 2025). In Fig. 54a, b, single-phase to single-phase cycloconverter and three-phase to three-phase cycle converter topologies are illustrated. Also, the output voltage for resistive load of single-to-single phase cycloconverter is given in Fig. 55.

Effective control strategies are essential for cycloconverters to accurately regulate the root mean square (RMS) output voltage and frequency. These strategies have a direct influence on the overall system efficiency and power quality, making the choice of control method a critical design consideration.

Among the most commonly used techniques, the phase control strategy stands out as the most basic and widely adopted. It functions by adjusting the firing angles of the thyristors to control the output parameters. Due to its simplicity and cost-effectiveness, it is suitable for many standard applications. However, this approach

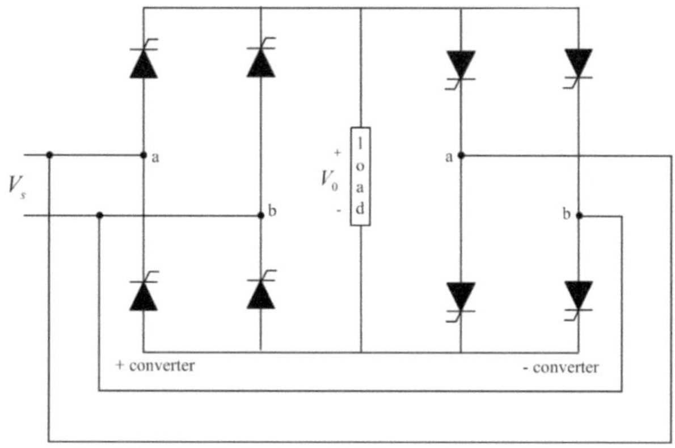

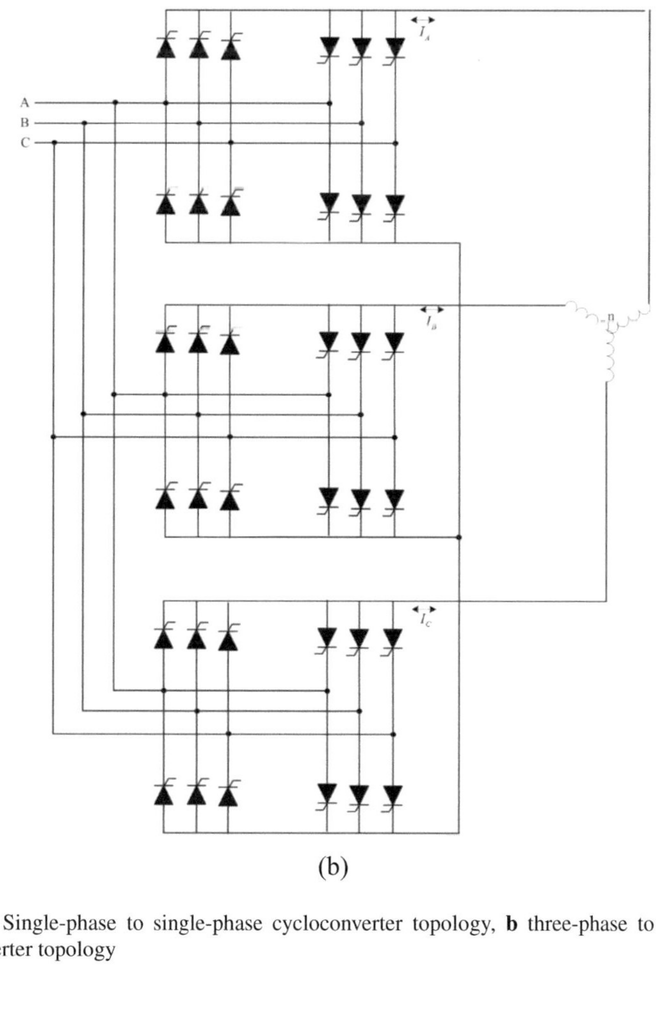

Fig. 54 **a** Single-phase to single-phase cycloconverter topology, **b** three-phase to three-phase cycloconverter topology

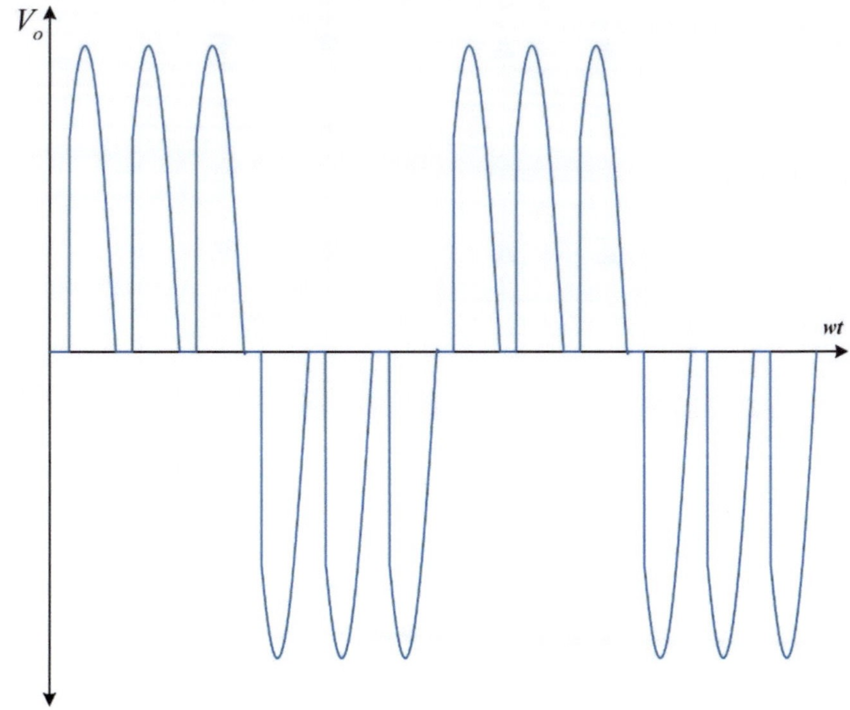

Fig. 55 Single-phase to single-phase cycle converter output waveform for resistive load

often leads to harmonic distortion, which can degrade the power quality and reduce the system's overall performance. Pulse Width Modulation (PWM) control presents a more refined alternative to phase control.

By varying the conduction times of the thyristors, PWM precisely regulates the width of the output pulses, resulting in improved voltage control and reduced harmonic distortion. This significantly enhances power quality, but it comes with increased complexity in terms of control algorithms and a requirement for fast-switching devices, making it more suitable for applications where quality and precision are prioritized. For three-phase systems, vector control offers an advanced and highly effective solution. This method involves converting three-phase AC quantities into a two-dimensional vector format, which allows for enhanced dynamic performance and better suppression of harmonic effects.

The benefits of vector control are substantial, particularly in high-performance industrial applications, but they come at the cost of greater computational demands and the need for sophisticated control algorithms. Direct Torque Control (DTC), another advanced method, shifts focus from voltage and frequency regulation to the direct control of motor torque and magnetic flux. By bypassing intermediate variables and targeting these core parameters directly, DTC achieves rapid and highly accurate motor response. This makes it particularly well-suited for demanding applications such as electric vehicles and robotics, where fast and precise control is critical. Similar to vector control, DTC requires powerful digital processing and advanced algorithm design.

4 Control Strategies for Power Converters

4.1 Classical Control Methods

Classical control techniques are the foundation of power electronics control due to their simplicity, ease of implementation, and well-established design principles.

4.1.1 Pulse Width Modulation (PWM) Techniques

The modulation techniques utilized to obtain the switching pulses in multilevel inverters are classified in Fig. 56. It is divided into two categories: basic switching and high switching frequency. While fundamental switching frequency methods are used in high voltage applications, high switching frequency methods are used in medium and low voltages. Selective harmonic elimination pulse width modulation (SHE-PWM), sinusoidal pulse width modulation (SPWM), and space vector pulse width modulation (SVPWM) are the widely preferred modulation methods for obtaining switching pulses in multilevel inverters (Colak et al. 2011).

Sinusoidal Pulse Width Modulation (SPWM)

In power converter systems, sinusoidal pulse width modulation is a common switching technique. In this method, the reference sinusoidal signal (V_{ref}) and the carrier triangular signal (V_{tri}) are compared to create switching signals. Switching

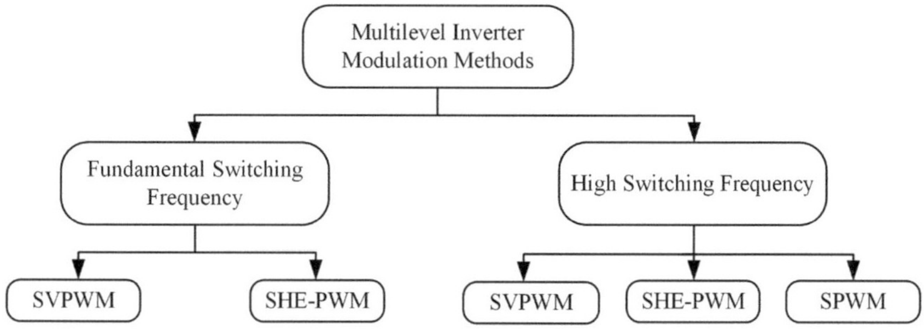

Fig. 56 Modulation methods commonly used in multilevel inverters

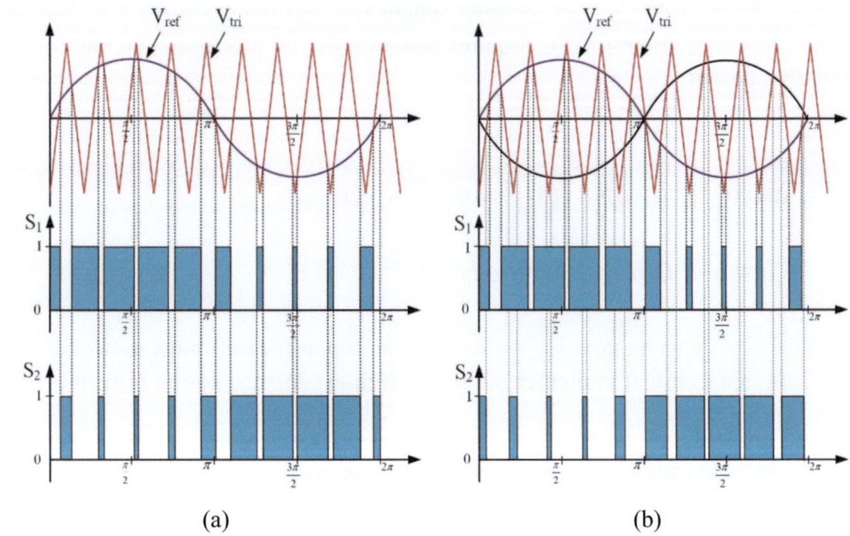

Fig. 57 SPWM methods **a** bipolar SPWM **b** unipolar SPWM

signals are obtained with SPWM using two different methods, namely unipolar and bipolar, as seen in Fig. 57. In the bipolar SPWM method, the S_1 switching pulse is obtained by comparing a single sinusoidal signal with the carrier triangle. The S_2 switching pulse is acquired by taking the inverse of the S_1 switching pulse. In the unipolar SPWM method, two sinusoidal signals are compared with the carrier triangle. While the S_1 switching pulse is acquired by comparing the first sinusoidal signal with the carrier triangle, the S_2 switching pulse is acquired with the second sinusoidal signal (Kabalcı et al. 2021).

The value of the modulation index (m_i), defined as the ratio of the amplitude value of the reference sinusoidal signal to the amplitude value of the carrier triangular signal, is obtained with Eq. (37). The modulation frequency index (m_f) is calculated by the ratio of the carrier triangular signal frequency (f_{tri}) to the reference sinusoidal signal frequency (f_{ref}) with Eq. (38) (Çolak and Kabalci 2008).

$$m_i = \frac{V_{ref}}{V_{\text{tri}}} \tag{37}$$

$$m_f = \frac{f_{\text{tri}}}{f_{ref}} \tag{38}$$

There are also vertical and horizontal multi-carrier methods of SDGM method used to obtain switching signals to develop inverter performance. Phase dissipation SPWM (PD-SPWM), phase opposition dissipation SPWM (POD-SPWM), and alternative phase opposition dissipation SPWM (APOD-SPWM) methods are vertical multi-carrier, while the phase shifted SPWM (PS-SPWM) method is a horizontal

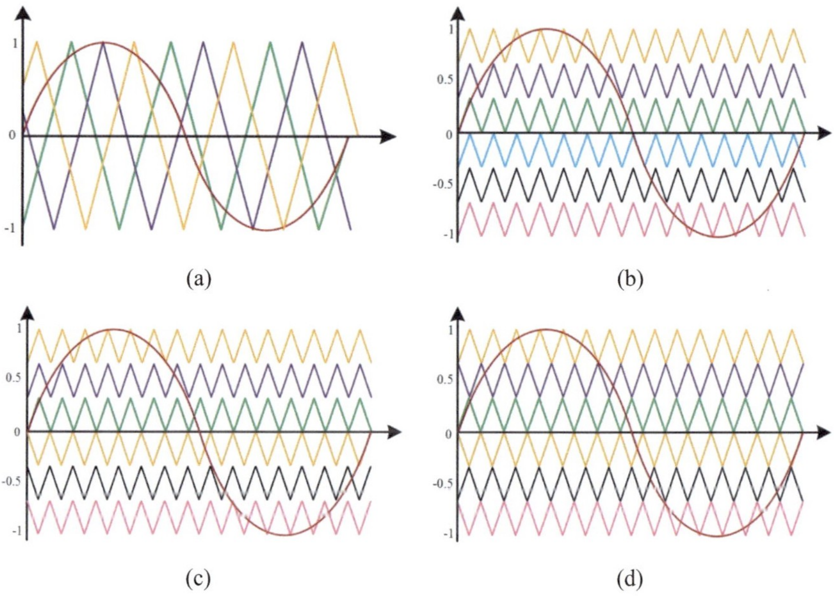

Fig. 58 Multi-carrier SPWM methods. **a** PS-SPWM, **b** PD-SPWM, **c** POD-SPWM, and **d** APOD-SPWM

multi-carrier method. The PS-SPWM method is depended on the use of phase-shifted carrier signals. Figure 58a shows the sinusoidal and carrier signals of the PS-SPWM method. The carrier signals differ in phase, but they share the same frequency and amplitude. For the m-level inverter, the phase difference between the numbers of carrier signals in the PS-SPWM method should be 180/m. In the PD-SPWM method, only the levels of the carrier signals are different and are shown in Fig. 58b. In Fig. 58c, the POD-SPWM method is seen and there is an 180° phase difference between the carrier signals in the positive level and those in the negative level. In the APOD-SPWM method, there is the 180° phase difference between each carrier triangle and the switching pulses are obtained with Fig. 58d (Boyar 2024; Kabalcı et al. 2021; Pilli and Singh 2019).

Selective Harmonic Elimination Pulse Width Modulation (SHE-PWM)

At low switching frequencies, the selective harmonic elimination pulse wide modulation method outperforms other classic carrier-based PWM methods. The output voltage waveform of the inverter is analyzed by Fourier series expansion with this technique. Switching angles are computed to remove certain low-order harmonics. With the SHE-DGM method, harmonic distortion at the inverter output, switching losses, and filter dimensions are reduced. For example, the output voltage of an 11-level multilevel inverter is expressed by Eq. (39) where *h* represents the harmonic degree and five switching signal angles (θ_1–θ_5) must be calculated. In case of suppressing the 5th, 7th, 11th, and 13th order harmonics, the required switching angles are obtained by Eqs. (40)–(44). These equations are solved with the Newton–Raphson iteration method and θ_1–θ_5 are the switching angles that will minimize the THD value of the selected harmonics. If the modulation index is 0.8, $\theta_1 = 6.57°$, $\theta_2 = 18.94°$, $\theta_3 = 27.18°$, $\theta_4 = 45.14°$, $\theta_5 = 62.24°$ (Kabalcı et al. 2021; Rai and Chakravorty 2019; Sadoughi et al. 2021; Tolbert and Shi 2018).

$$V(\omega t) = \sum_{h=1,3,5,\ldots}^{\infty} \frac{4V_{DC}}{h\pi}(\cos(h\theta_1) + \cos(h\theta_2) + \ldots \cos(h\theta_5) \cdot \sin(h\omega t) \quad (39)$$

$$\cos(\theta_1) + \cos(\theta_2) + \cos(\theta_3) + \cos(\theta_4) + \cos(\theta_5) = 5 \cdot m_a \quad (40)$$

$$\cos(5\theta_1) + \cos(5\theta_2) \ldots + \cos(5\theta_5) = 0 \quad (41)$$

$$\cos(7\theta_1) + \cos(7\theta_2) \ldots + \cos(7\theta_5) = 0 \quad (42)$$

$$\cos(11\theta_1) + \cos(11\theta_2) \ldots + \cos(11\theta_5) = 0 \quad (43)$$

$$\cos(13\theta_1) + \cos(13\theta_2) \ldots + \cos(13\theta_5) = 0 \quad (44)$$

Space Vector Pulse Width Modulation (SVPWM)

Space vector pulse width modulation method is a widely used digital modulation method in obtaining inverter switching signals. It is a vector-based switching method and the switching vectors are placed in the hexagonal structure in the α–β plane. Compared to SPWM, the SVPWM method gives a 15.5% higher output voltage amplitude. For the m-level inverter, the total number of switching is calculated as m^3, and the number of independent switching is calculated as $(3m^2 - 3m + 1)$ (Boyar 2024; Guo et al. 2018; Jayakumar et al. 2021). The transformation of the three-phase a, b, c voltage vectors seen in Fig. 59 from the abc plane to the α–β plane is achieved by the Clark transformation given in Eq. (45). The reference vector (V_{ref}) is calculated in the α–β plane by Eq. (46). The θ expressed in Eq. (47) is the angle that V_{ref} has in the $\alpha\beta$ plane. The calculation of V_{ref} with phase voltages V_a, V_b, and V_c is given in Eq. (48) (Boyar 2024; Rashid 2014).

$$\begin{bmatrix} V_\alpha \\ V_\beta \end{bmatrix} = \frac{2}{3}\begin{bmatrix} 1 & -\frac{1}{2} & -\frac{1}{2} \\ 0 & \frac{\sqrt{3}}{2} & -\frac{\sqrt{3}}{\sqrt{3}} \end{bmatrix}\begin{bmatrix} V_a \\ V_b \\ V_c \end{bmatrix} \quad (45)$$

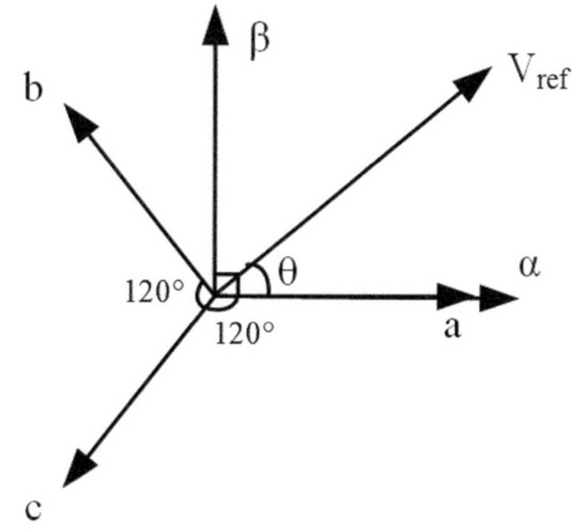

Fig. 59 *abc-αβ* transformation

$$|V_{\text{ref}}| = \sqrt{V_\alpha^2 + V_\beta^2} \tag{46}$$

$$\theta = \tan^{-1}\frac{V_\beta}{V_\alpha} \tag{47}$$

$$V_{\text{ref}} = \frac{2}{3}(V_a + V_b \cdot e^{j2\pi/3} + V_c \cdot e^{j4\pi/3}) \tag{48}$$

The two-level inverter has 8 switching states in the SVPWM method and the representation of these switching states in the space vector diagram is given in Fig. 60. There are 6 active vectors and 2 zero vectors for the two-level inverter (Jayakumar et al. 2021; Yaramasu and Wu 2017).

The sectors where V_{ref} is located according to the angle θ in the α–β plane are given in Table 6.

Sectors A, B, C, D, E, and F are formed according to the θ. When V_{ref} is located in sector A, the vectors that will generate it during the sampling time (T_s) are V_1, V_2 and V_0 or V_7. If the switching times are expressed as T_1, T_2, and T_0 these times are calculated with Eq. (49) and (50). V_{ref} in sector A can be expressed as two components by Eq. (51). Using this equation, the switching times T_1, T_2, and T_0 are calculated with Eqs. (52)–(54) where a represents $|V_{\text{ref}}|/\frac{2}{3}V_{\text{DC}}$ (Jayakumar et al. 2021; Kim 2017).

$$V_{\text{ref}} \cdot T_s = T_1 \cdot V_1 + T_2 \cdot V_2 + T_0 \cdot V_0 \tag{49}$$

$$T_s = T_1 + T_2 + T_0 \tag{50}$$

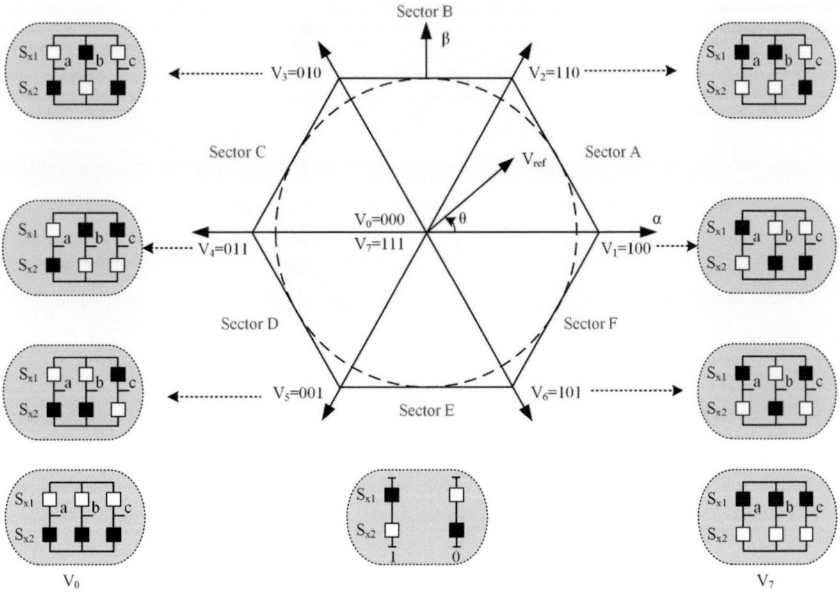

Fig. 60 Space vector diagram of two-level inverter

Table 6 Sectors according to θ

Angle (θ)	Sector
$0° < \theta < 60°$	Sector A
$60° < \theta < 120°$	Sector B
$120° < \theta < 180°$	Sector C
$180° < \theta < 240°$	Sector D
$240° < \theta < 300°$	Sector E
$300° < \theta < 360°$	Sector F

$$\begin{aligned} T_s|V_{ref}|\cos\theta &= T_1\left(\frac{2}{3}V_{DC}\right) + T_2\left(\frac{2}{3}V_{DC}\right)\cos 60° \\ T_s|V_{ref}|\sin\theta &= T_2\left(\frac{2}{3}V_{DC}\right)\sin 60° \end{aligned} \tag{51}$$

$$T_1 = aT_s\frac{\sin(60° - \theta)}{\sin 60°} \tag{52}$$

$$T_2 = aT_s\frac{\sin\theta}{\sin 60°} \tag{53}$$

$$T_0 = T_s - (T_1 + T_2) \tag{54}$$

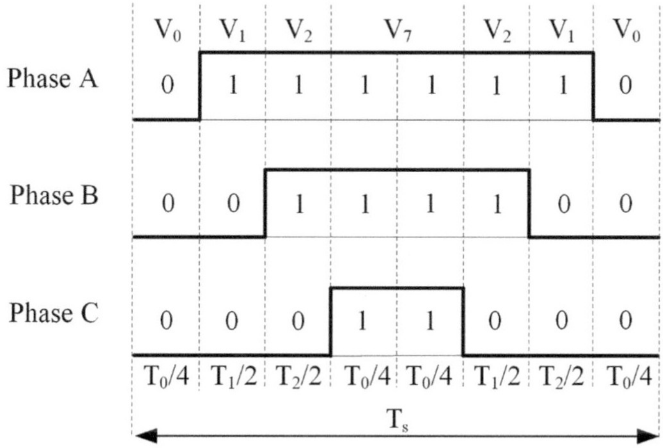

Fig. 61 Seven-state switching model in Sector A

Switching states and times in the switching period are shown in Fig. 61 for Sector A. The switching order is V_0–V_1–V_2–V_7–V_2–V_1–V_0. In the half of the switching period, which is the $T_s/2$ period, the V_0 vector is applied for the period $T_0/4$, the V_1 vector for the $T_1/2$, the V_2 vector for the $T_2/2$, and the V_7 vector for the $T_0/4$. In the other half of the switching period, the opposite situation occurs. When compared to the SPWM method, the SVPWM method achieves a 55% lower THD rate in terms of harmonics (Islam et al. 2014).

The vectors that provide the necessary switching positions for phases *a*, *b*, and *c* for a three-level inverter are in the hexagon structure in Fig. 62. There are 27 switching states: 3 zero vectors, 12 small vectors, 6 medium vectors, and 6 large vectors. In addition, there are 6 sectors and each sector consists of 4 sub-sectors (Boyar et al. 2023; Jayakumar et al. 2021).

4.2 *Model Predictive Control (MPC)*

Predictive control is a method that predicts the future behavior of a system to determine the most appropriate control within a sampling period (Valencia et al. 2021). The Model Predictive Control (MPC) method emerged as a control method in the process industry in the 1970s. Despite some initial research in the 1980s, no progress was made in research due to the lack of sufficient computational capability. The MPC method is widely utilized in the petroleum, chemical, automotive and aviation industries. Microprocessors have become widely utilized in the field of power electronics since the 2000s as their computing capabilities have increased (Kabalci et al. 2023; Karamanakos and Geyer 2020). The MPC method is preferred in industrial applications because it does not require modulation compared to linear control

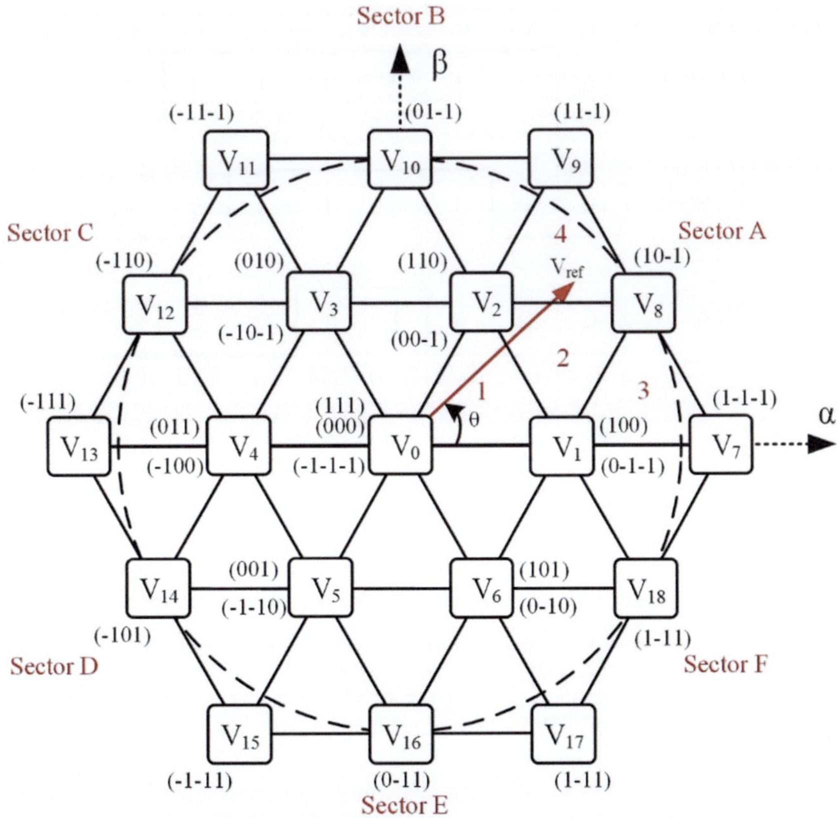

Fig. 62 Space vector diagram of three level inverter

methods, is easy to apply, and has a fast response (Boyar et al. 2022). Because of its improved performance, ease of use, and capacity to regulate nonlinear systems, the MPC has recently attracted attention in the fields of power electronics and AC motor drives (Valencia et al. 2021). Continuous control set MPC (CCS-MPC) and finite control set MPC (FCS-MPC) are the two types of MPC, the most important predictive control technology for power electronics applications. The FCS-MPC approach can create the switching state without a modulator stage, whereas the CCS-MPC method needs a modulator to create the switching states. In the FCS-MPC method, a finite number of switching states of the power converters are used to solve the optimization problem. Therefore, a discrete-time model of the system must be utilized to predict the future behavior of the system. Then, the switching pulses that optimize the predefined cost function are determined. Although the FCS-MPC can generate switching pulses without a modulator, the variable switching frequency is a disadvantage of this method (Bayhan and Abu-Rub 2024).

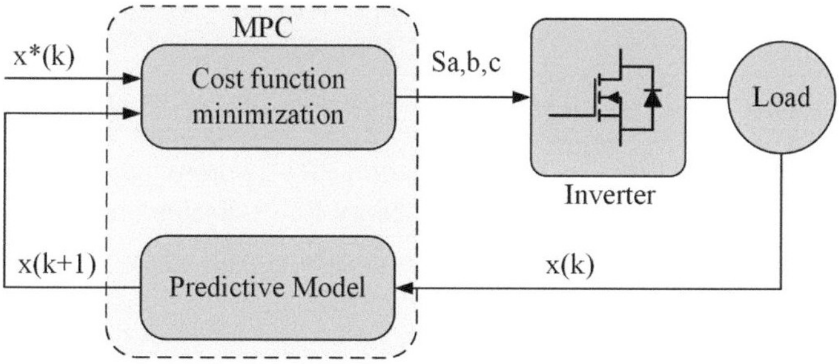

Fig. 63 General structure of MPC

The general structure of the MPC method created to control the $x(k)$ parameter of the load fed by the inverter is seen in Fig. 63. The general structure of the MPC method includes the predictive model and the minimization of the cost function. The necessary steps to perform this method can be summarized as follows (Boyar 2024; Bayhan and Abu-Rub 2024).

- The inverter model should be built and possible switching states should be determined.
- The load model must be built for the predictive model.
- The cost function that includes the parameter to be controlled must be determined.

In the MPC method, estimates are made using the state matrices of a system given in Eqs. (55) and (56) or by using the control variable equation. x, y, and u are the state, input, and output vectors, and A, B, C, and D represent the system, control, output, and feed matrix, respectively.

$$x(k+1) = Ax(k) + Bu(k) \tag{55}$$

$$y(k) = Cx(k) + Du(k) \tag{56}$$

The cost function (g) that determines the control signal is generally expressed by Eq. (57). The optimum control signal is acquired by evaluating the estimation results with the cost function. The control signal that minimizes the cost function is obtained by Eq. (58) in the MPC method (Boyar and Kabalci 2021).

$$g = f(x(k), u(k), \ldots, u(k+N)) \tag{57}$$

$$u(k) = [10\ldots\ldots0] \arg\min_{u} g \tag{58}$$

Discretization methods are utilized to acquire the estimation functions. The discretization process is easily obtained by the Euler method expressed in Eq. (59) according to the sampling period (Boyar and Kabalci 2021).

$$\frac{dx}{dt} \approx \frac{x(k+1) - x(k)}{T_s} \tag{59}$$

The voltage of the *R*, *L* load fed by the ANPC inverter can be expressed with Eq. (60) in (Boyar and Kabalci 2021). Equation (61) represents the derivative of the output current and the next value is obtained by Euler discretization method using Eq. (62).

$$v = Ri + L\frac{di}{dt} \tag{60}$$

$$\frac{di}{dt} = \frac{1}{L}(v - Ri) \tag{61}$$

$$i(k+1) = \left(1 - \frac{RT_s}{L}\right)i(k) + \frac{Ts}{L}v(k) \tag{62}$$

Equation (63) is used to determine the voltage vectors for each of the 27 switching states needed for the switching components in the ANPC inverter topology to be utilized to convert DC to AC voltage. S_a, S_b, and S_c represent the switching states of each phase, and the locations of each voltage vector are given in Fig. 16 (Boyar and Kabalci 2021).

$$V_{(n)} = \frac{2}{3}V_{\mathrm{DC}}\left[S_{a(n)} + \left(-\frac{1}{2} + j\frac{\sqrt{3}}{2}\right)S_{b(n)} + \left(-\frac{1}{2} - j\frac{\sqrt{3}}{2}\right)S_{c(n)}\right]$$
$$n = \{1, 2, \ldots, 27\} \tag{63}$$

The value of the capacitor voltages at the ANPC inverter input should be half of the input voltage. In this case, the difference of the input capacitor voltages (ΔV_c) is expressed by Eq. (64) where *C* is input capacitor value and in the balanced case this difference should be zero (Boyar and Kabalci 2021).

$$\Delta V_c = \Delta V_{c1} - \Delta V_{c2} = -\frac{T_s}{C}(S_a^2 i_a + S_b^2 i_b + S_c^2 i_c) \tag{64}$$

The cost function created to control the balance of input capacitor voltages and load current is found in Eq. (65) where i_α and i_β denote the $\alpha\beta$ components of the current. The difference between the reference current (i_α^*, i_β^*) and the predicted current and also the difference in capacitor voltages are evaluated in the cost function. The switching signal that minimizes the cost function is determined with the MPC method (Boyar and Kabalci 2021).

$$g = \left|i_\alpha^* - i_\alpha(k+1)\right| + \left|i_\beta^* - i_\beta(k+1)\right| + \lambda|\Delta v_c(k+1)| \tag{65}$$

The flowchart for the current control of the inverter with R, L load using the MPC method is shown in Fig. 64. First, the inverter's output current and the input capacitors' voltage values are measured. Next, the voltage differential between the input capacitors is computed. After the next value of the load current is obtained, the value of the cost function is calculated. The switching signal applied to the inverter is the one that minimizes the cost function (Boyar and Kabalci 2021).

4.3 Grid Synchronization and Power Quality Improvement

With the integration of renewable energy sources and power electronic systems into the grid, power quality and synchronization problems have emerged in electrical power systems. System stability and reliability are important parameters during synchronous operation with the grid. Voltage distortions, harmonics and sudden load changes can negatively affect power quality and cause system performance to decrease. Therefore, Phase Locked Loop (PLL) which provide phase and frequency synchronization in systems operating synchronously with the grid, and virtual inertia control methods, which allow the system to react to sudden frequency changes, play an important role in ensuring synchronization and improving power quality. In this section, PLL and virtual inertia methods will be discussed.

4.3.1 Phase Locked Loop (PLL) Controllers

PLL controller is generally utilized to monitor the output waveform of a system and to follow the grid reference for grid integration of inverters. There are many types of PLL controllers and the basic structure is displayed in Fig. 65 where V_{abc} stands for the three-phase grid voltage, and f_o is the frequency obtained from the loop filter output. Phase detection, loop filtering, and voltage-controlled oscillation are its three main functions. Phase detector (PD), loop filter, and voltage controlled oscillator (VCO) components are all part of the PLL controller. The reference and measured voltage values are compared and applied to the PD which generates an error signal ($\varepsilon_{\mathrm{pd}}$) for the loop filter that attenuates the high-frequency AC components. The loop filter contributes to the increase of the dynamic response of the PLL and plays an important role in its reliability and stability (Kabalcı et al. 2021; Chatterjee et al. 2018).

In the VCO unit, it generates the necessary phase angle to minimize the error between the reference and measured signal and to reduce the PLL error in the next cycle. PLL structures are widely preferred in three-phase grid-connection applications. PLL noise and harmonic elimination in grid-connected systems provide high dynamic response. Various PLL controller structures will be discussed in the following sections (Kabalcı et al. 2021; Chatterjee et al. 2018).

Fig. 64 Flowchart of MPC

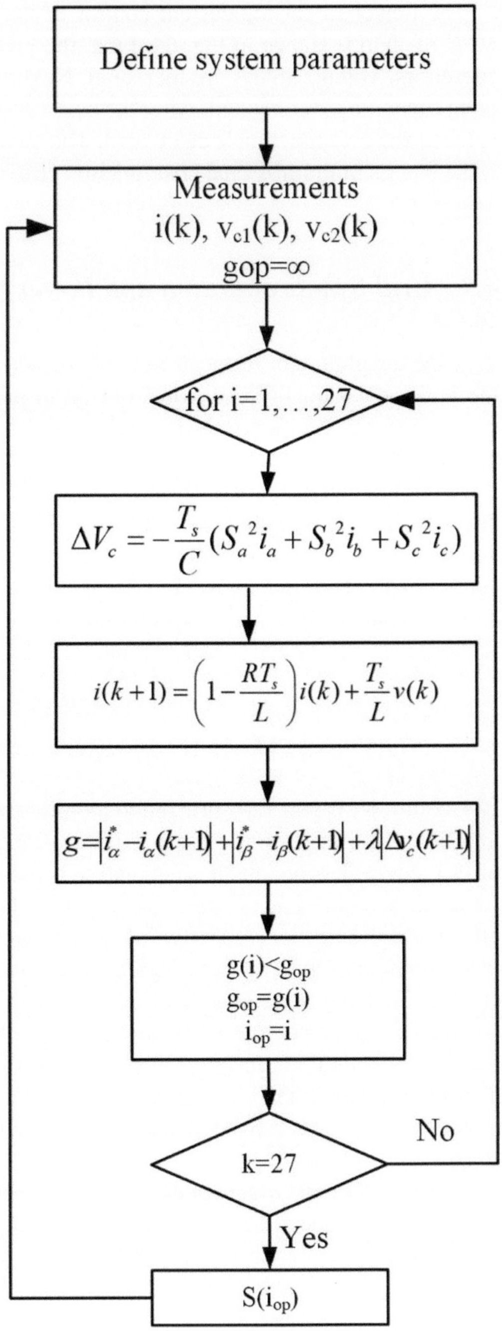

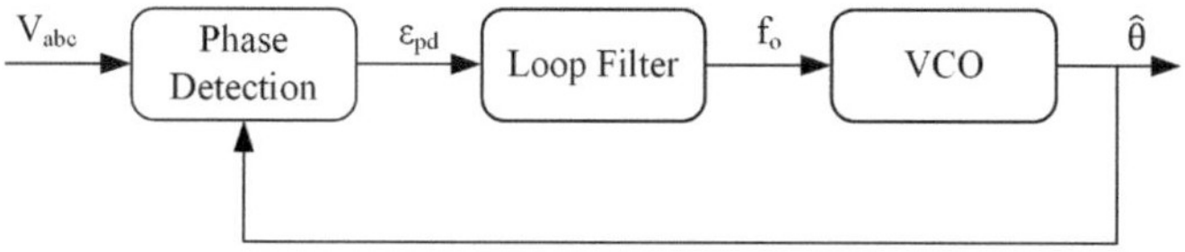

Fig. 65 PLL control structure

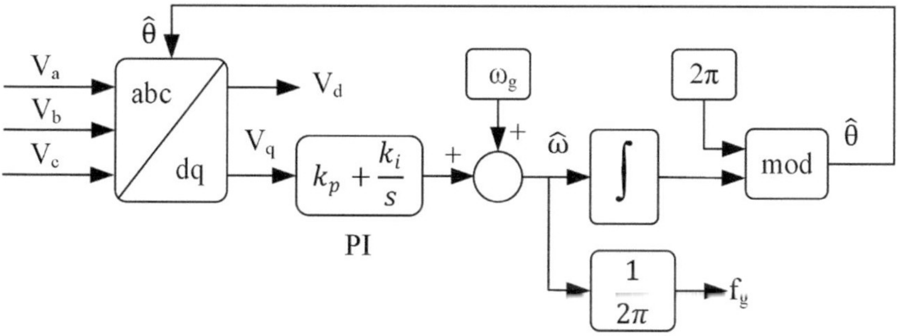

Fig. 66 SRF-PLL structure

Synchronous Reference Frame PLL (SRF-PLL)

Among PLL control methods, synchronous reference frame PLL (SRF-PLL) is commonly preferred methods in controlling three-phase systems. The SRF PLL structure is seen in the Fig. 66 where V_{abc} is three-phase voltage. After the three-phase voltage rotational reference frame is obtained, it is converted to the stationary frame. Similar to the SVPWM method, the rotational frame to stationary frame transformation is performed.

Then, the Park transformation given in Eq. (66) is performed to obtain the *dq* axis components (V_d, V_q). In addition, under ideal conditions, such as no harmonics and balanced grid voltages, the V_q value is zero while the V_d value is equal to the maximum value of the grid voltage. V_d and V_q voltages are utilized to control the system more easily (Boyar and Kabalci 2018; Sevilmiş and Karaca 2015).

$$\begin{bmatrix} V_d \\ V_q \end{bmatrix} = \frac{2}{3}\begin{bmatrix} \cos(\hat{\theta}) & \cos(\hat{\theta} - 120) & \cos(\hat{\theta} + 120) \\ -\sin(\hat{\theta}) & -\sin(\hat{\theta} - 120) & \sin(\hat{\theta} + 120) \end{bmatrix}\begin{bmatrix} V_a \\ V_b \\ V_c \end{bmatrix} \tag{66}$$

Under ideal grid conditions, where phase voltages are balanced and there are no harmonic components, the estimated grid phase angle ($\hat{\theta}$) is equal to the actual phase angle (θ) of the grid. As can be seen from Eqs. (67) and (68), when V_q is zero, V_d is equal to the amplitude value of the grid voltage (V_m) in an ideal system (Kabalcı et al. 2021; Sevilmiş and Karaca 2019; Golestan et al. 2017).

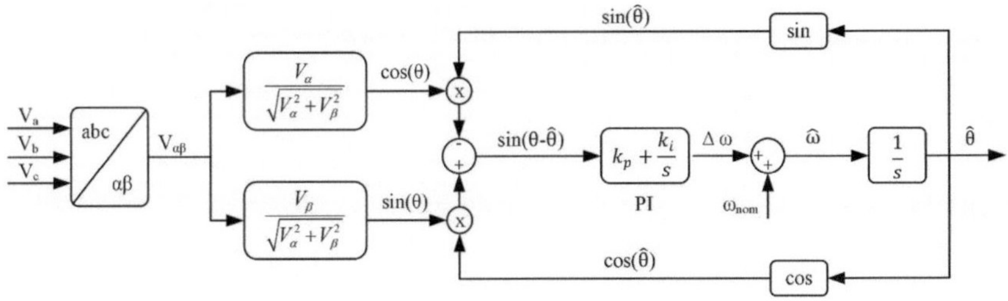

Fig. 67 *αβ*-PLL structure

$$V_d = V_m \sin(\theta - \hat{\theta}) \tag{67}$$

$$V_q = V_m \sin(\theta - \hat{\theta}) \tag{68}$$

The V_q voltage containing angle information is applied to the PI controller input that determines the system dynamics. The frequency change from the PI controller output is added to the fundamental grid angular frequency value (ω_g) to obtain the angular frequency ($\widehat{\omega}$). Finally, the integral of the obtained angular frequency is taken to determine the grid phase angle $\widehat{\theta}$. The estimated grid phase angle is continuously fed back to the *abc-dq* transformation block. Also, the grid frequency (f_g) can be defined with the data obtained in this method (Sevilmiş and Karaca 2015).

Stationary Reference Frame PLL (*αβ*-PLL)

The general structure of the *αβ*-PLL is shown in Fig. 67. In this method, first, the three-phase voltage natural *abc* reference frame is transformed to the stationary *αβ*-reference frame by the Clarke transformation using the matrix transform method. The transformation can be done by setting the phase angle of the PLL as $\widehat{\theta} = 0$ with Eq. (30). The phase angle $\widehat{\theta}$ is estimated by the trigonometric equations depending on the actual grid phase angle of θ, as seen in Eq. (69) (Hadjidemetriou et al. 2013; Ali et al. 2018).

$$\theta_{\text{error}} = \theta - \hat{\theta} \approx \sin(\theta - \hat{\theta}) \Leftrightarrow \theta_{\text{error}} \approx \sin(\theta)\cos(\hat{\theta}) - \sin(\hat{\theta})\cos(\theta) \tag{69}$$

Dual Second Order Generalized Integrator PLL (DSOGI-PLL)

The DSOGI-PLL consists of four main units, namely Clark transform, Positive–Negative Signal Calculator (PNSC), Second Order Generalized Integrator-Quadrature Signal Generator (SOGI-QSG) and Synchronous Reference Frame-Phase Locked Loop (SRF-PLL) for positive and negative sequence component of

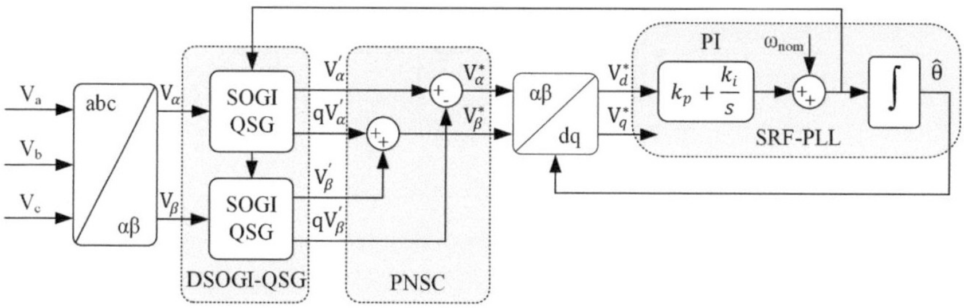

Fig. 68 DSOGI-PLL block diagram

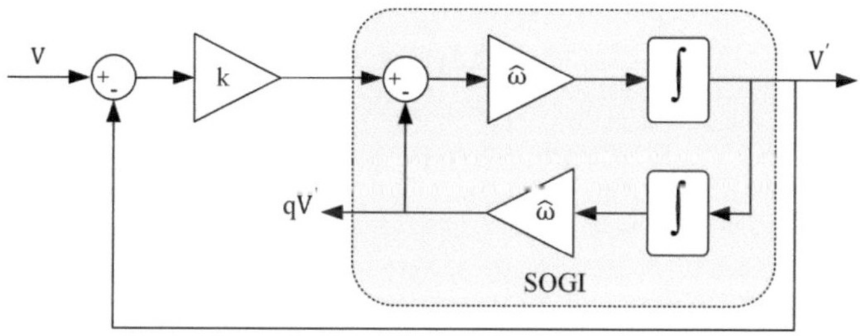

Fig. 69 SOGI-QSG block diagram

the grid voltage. Figure 68 displays the DSOGI-PLL block diagram (Sevilmiş and Karaca 2022).

During the pre-filtering phase, two second-order generalized integrators (SOGIs) are used to filter out V_α and V_β signals. Additionally, the SOGIs function as a quadrature signal generator (QSG), producing as qV'_α and qV'_β, which are quadrature filtered versions of the V_α and V_β signals. The SOGI-QSG structure is shown in Fig. 69, where k and $\widehat{\omega}$ stand for the damping factor and predicted frequency. The SOGI-QSG transfer functions are expressed with Eq. (70) (Sevilmiş and Karaca 2022; Zhang et al. 2021).

$$D(s) = \frac{V'(s)}{V(s)} = \frac{k\hat{\omega}s}{s^2 + k\hat{\omega}s + \hat{\omega}^2} \tag{70}$$

$$Q(s) = \frac{qV'(s)}{V(s)} = \frac{\hat{\omega}}{s}D(s) = \frac{k\hat{\omega}^2}{s^2 + k\hat{\omega}s + \hat{\omega}^2} \tag{71}$$

The V'_α, V'_β, qV'_α, and qV'_β signals acquired from SOGI-QSG blocks are passed through the PNSC to obtain the fundamental frequency positive sequence (FFPS) components V^*_α and V^*_β, as given in Eqs. (72) and (73). The grid frequency $\widehat{\omega}$ obtained by applying V^*_α and V^*_β to the SRF-PLL is fed back to the SOGI-QSG blocks (Sevilmiş and Karaca 2022).

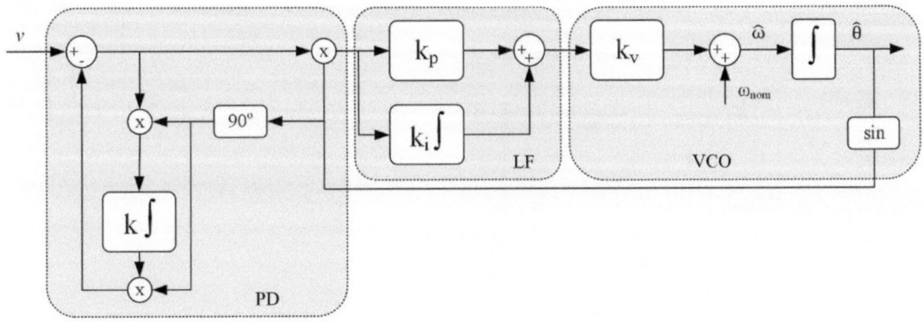

Fig. 70 Single-phase EPLL structure

$$V_{\alpha}^{*} = 0.5(V_{\alpha}' - qV_{\beta}') \tag{72}$$

$$V_{\beta}^{*} = 0.5(V_{\alpha}' + qV_{\beta}') \tag{73}$$

Enhanced PLL (EPLL)

The Enhanced Phase-Locked Loop's (EPLL) phase detection method, which offers more information like amplitude and phase angle, is the main way that it differs from traditional PLLs. The EPLL exhibits stronger resilience to noise and harmonics in the utility voltage than conventional PLLs. Furthermore, it can function dependably in situations when the grid is imbalanced. The three-phase EPLL proposed in Karimi-Ghartemani and Iravani (2004) consists of four separate EPLL blocks shown in Fig. 70. It consists of PD, LF, and VCO found in the traditional PLL structure. Its basic structure contains three independent internal parameters k, $k_p k_v$, and $k_i k_v$. The amplitude convergence rate is primarily controlled by k, whereas the phase and frequency convergence rates are controlled by $k_p k_v$ and $k_i k_v$, respectively. Based on the error signal, the EPLL dynamically adjusts its transfer function, acting as an adaptive band-pass filter (Ali et al. 2018; Karimi-Ghartemani and Iravani 2004).

In the three phase EPLL structure, each phase voltage is processed by an independent EPLL module. Each module produces two sinusoidal signals equal in amplitude and frequency but shifted 90° in phase; the second of these signals is 90° ahead of the first. The output signals from the three EPLL modules are fed into the unit of computation block to generate the positive phase sequence V_{abc}'. This positive phase sequence is then fed into a fourth EPLL to obtain the phase angle and amplitude of the fundamental component. The general schematic of the three phase EPLL structure is displayed in Fig. 71. Compared to other PLL techniques, the three phase EPLL has been found to perform correctly under unstable grid scenarios, but has slow dynamics among the benchmarks (Ali et al. 2018).

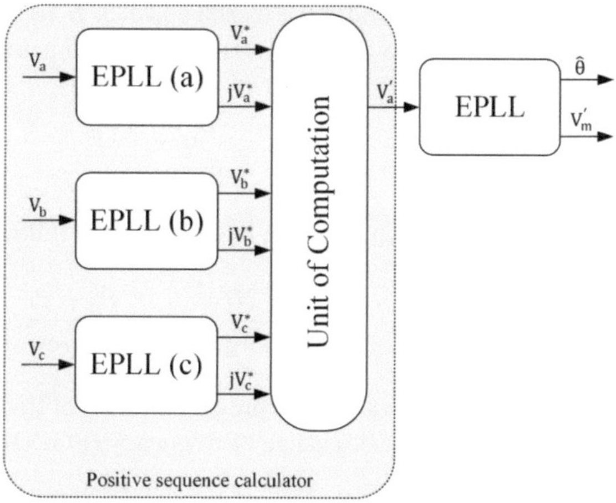

Fig. 71 Three phase EPLL structure

4.3.2 Virtual Inertia Control Methods

Concerns are increasing since fossil fuels such as coal, oil, and natural gas, which have limited reserves, cannot be used in the long term. Environmental problems arise with the widespread use of these resources in electricity generation. The use of renewable energy sources (RESs) such as solar and wind in electricity generation is increasing. The integration of these increasing systems into power systems has become inevitable to prevent increasing environmental problems (Kerdphol et al. 2021). While there are rotary generators in the traditional power grid, the use of inverters has become widespread in the modern power grid with the increasing RES. The connections of PV and wind energy systems are realized through inverters. Although this situation seems advantageous to benefit from RES resources, inverter-based systems have a disadvantage in terms of frequency stability since they do not ensure mechanical inertia response. According to the research results, it was determined that RESs have a significant lack of frequency stability thanks to the lack of inertial response. The independent system operator, Electric Reliability Council of Texas (ERCOT), reported that there was a decrease in inertial response and that additional inertial response was required (Tamrakar et al. 2017).

The frequency change in the power system with frequency distortion is approximately calculated by the swing equation given in Eqs. (74) and (75) where P_l is the power demand, P_g represents the generated power, ω_g is the system frequency, and J is the total system inertia (Tamrakar et al. 2017).

$$P_g - P_l = \frac{d\left(\frac{1}{2}J\omega_g^2\right)}{dt} \tag{74}$$

$$P_g - P_l = J\omega_g\frac{d\omega_g}{dt} \tag{75}$$

Equation (76) provides the inertia constant H of the power system in relation to the apparent power S_g of the system's connected generators.

$$H = \frac{J\omega_g^2}{2S_g} \tag{76}$$

Using Eq. (76), Eq. (75) can be expressed as follows:

$$\frac{2H}{\omega_g}\frac{d\omega_g}{dt} = \frac{P_g - P_l}{S_g} \tag{77}$$

Equation (77) can also be expressed in terms of frequency (Hz) by Eq. (78) where df/dt represents rate of change of frequency (ROCOF).

$$\frac{2H}{f}\frac{df}{dt} = \frac{P_g - P_l}{S_g} \tag{78}$$

As a result of the reduced inertia, ROCOF increases and can cause large changes in the system frequency. Therefore, with the RES integrated circuits, the need for additional inertia in the system arises. Virtual inertia application using power electronics was first proposed by Beck and Hesse. Since then, a wide variety of alternative topologies and methodologies have been suggested in the literature (Tamrakar et al. 2017; Beck and Hesse 2007). The virtual inertia topologies can be separated into three primary categories such as synchronous generator model based topology, swing equation-based topology, and frequency power response-based topology (Kerdphol et al. 2021).

Synchronous Generator Model Based Topology (Synchronverter)

Synchronverters, which operate equivalently to a combination of Synchronous Generator (SG) and small capacitor bank, control inverter based Distributed Generator (DG) units as SGs, represent the same dynamics. It can be utilized as grid-forming units without requiring significant changes to the existing system structure, and it is particularly well suited for simulating the inertia impact of DG that is not connected to the grid. The frequency drop-based control algorithm used in this structure reduces the noise in the system and effectively controls the output power of the inverter, since it does not require the derivative of the frequency. In addition, the virtual moment of inertia and the damping coefficient must be adjusted in order for the system to provide the desired dynamic response. The general and detailed structure of synchronverters is shown in Figs. 72 and 73 (Tamrakar et al. 2017; Shadoul et al. 2022).

Equations (79)–(81) basically describe the dynamics of SG. M_f defines the mutual inductance between the stator coil and the field coil size, i_f presents field excitation current, P and P^* represents to the reference and measured active power, Q and Q^*

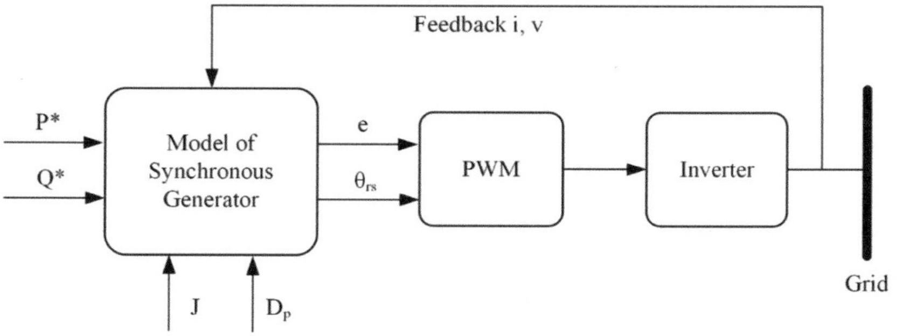

Fig. 72 The general structure of synchronverters

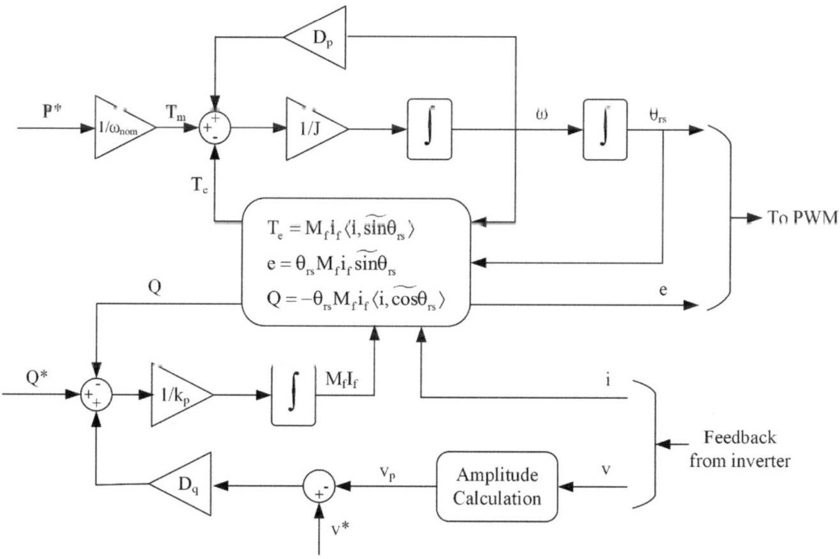

Fig. 73 The detailed structure of synchronverters

refers to the reference and measured reactive power, θ_{rs} means the angle between the rotor axis and stator, e presents the no load voltage generated, T_e stands for the electromagnetic torque of the synchronverter and the standard inner product of two vectors in three dimensions is represented by <.,.>. Equation (82) expresses the three phase stator current (i), $\widetilde{\sin}\ \theta_{rs}$ and $\widetilde{\cos}\ \theta_{rs}$ vectors (Shadoul et al. 2022; Jafari et al. 2024).

$$T_e = M_f i_f \langle i, \widetilde{sin}\theta_{rs} \rangle \tag{79}$$

$$e = \theta_{rs} M_f i_f \widetilde{sin}\theta_{rs} \tag{80}$$

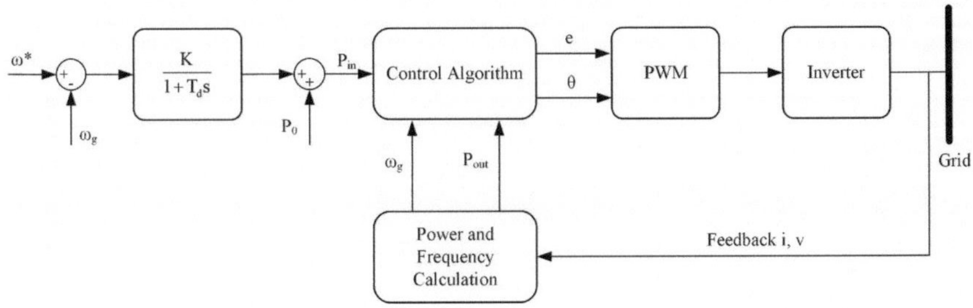

Fig. 74 The block diagram of Ise Lab's topology

$$Q = -\theta_{\mathrm{rs}} M_f i_f \langle i, \widetilde{\cos}\theta_{rs} \rangle \tag{81}$$

$$i = \begin{bmatrix} i_a \\ i_b \\ i_c \end{bmatrix};\ \widetilde{\sin}\theta_{rs} = \begin{bmatrix} \sin\theta_{rs} \\ \sin\left(\theta_{rs} - \frac{2\pi}{3}\right) \\ \sin\left(\theta_{rs} - \frac{4\pi}{3}\right) \end{bmatrix};\ \widetilde{\cos}\theta_{rs} = \begin{bmatrix} \cos\theta_{rs} \\ cos\left(\theta_{rs} - \frac{2\pi}{3}\right) \\ cos\left(\theta_{rs} - \frac{4\pi}{3}\right) \end{bmatrix} \tag{82}$$

In order to solve the differential equation in the controller, the inverter grid voltage and inverter output current signals are used as feedback. The damping factor (D_p) and the moment of inertia J can be adjusted. The voltage and frequency loop play a role in the production of mechanical torque (T_m) and $M_f I_f$ control parameters. T_m is produced in the frequency loop using the reference active power in accordance with the grid's nominal angular frequency, ω_{nom}. Thanks to this loop, it generates the virtual angular frequency of the integrated synchronverter to calculate the phase command θ_{rs} required for PWM. Likewise, within the voltage loop, the voltage drop constant between the grid voltage and the reference voltage is multiplied by D_q and added to the difference between the reference power Q^* and the calculated Q. Then, $1/k_p$ is multiplied to produce $M_f I_f$. The e and θ_{rs} values required to produce PWM are obtained from the controller output (Tamrakar et al. 2017; Jafari et al. 2024).

Swing Equation-Based Topology

This method similar to the synchronverters, which solves the power and frequency swing equation in each control loop to simulate the inertia without the need for a detailed model of the SG, was improved by Ise Lab for virtual inertia application. The general block structure of this method is shown in Fig. 74 (Tamrakar et al. 2017; Shadoul et al. 2022).

The voltage of the gird and the current value of the inverter are obtained from the common connection point and applied to the power and frequency calculation unit. In this unit, the grid frequency (ω_g) and the active power output (P_{out}) of the inverter are calculated. These two parameters obtained together with P_{in} constitute the inputs of the main control algorithm block. The solution of the swing equation in each cycle of the control algorithm is performed by Eq. (83) and determines the required θ for the PWM block. This strategy is applied similarly to the synchronverters method without the need for frequency derivatives. Frequency derivatives cause noise to be added to the system and create a disadvantage in managing the system. Equations (83) and (84) give typical swing equation of SG (Tamrakar et al. 2017; Shadoul et al. 2022; Jafari et al. 2024).

$$P_{\text{in}} - P_{\text{out}} = J\omega_m\left(\frac{d\omega_m}{dt}\right) + D_p\Delta\omega_m \tag{83}$$

$$\Delta\omega = \omega_m - \omega_g \tag{84}$$

where ω_m represents the virtual angular frequency, P_{in} and P_{out} refer to the input and output powers, ω_g represents grid frequency. Using a regulator model, the input power P_{in} is determined by the amount of frequency deviation from the reference frequency (ω^*) in Fig. 29. P_0 refer to continuous power reference of DG unit. With time constant (T_d) and gain K, the regulator is modeled as a first-order lag element. ROCOF increases with the delay of the regulator model, causing higher frequency nadirs (Tamrakar et al. 2017; Shadoul et al. 2022; Jafari et al. 2024).

Frequency Power Response-Based Topology

The basis of this method is based on the imitation of the inertial response properties of the SG in the DG system, particularly the capacity to react to variations in frequency, by Virtual Synchronous Generators (VSG). This allows the distribution of DG units by simulating the release or absorption of kinetic energy similar to the SG. Compared to conventional droop controllers that only ensure frequency regulation, the VSG method can realize frequency control. According to conventional droop controllers that only provide frequency regulation, the VSG method can realize frequency control. This control approach is based on the frequency derivative and operates similarly to inertial power generation or absorption by the SG in the event of a power imbalance. The VSG is therefore a dispatchable generator that responds to changes in frequency by modifying its output. The VSG is one of the simplest methods of implementing virtual inertia since it does not contain all the detailed equations needed in a SG. It is known that using more than one DG unit as a current source can cause instability. Equation (85) is utilized to control the VSG converter output power (Tamrakar et al. 2017; Shadoul et al. 2022; Jafari et al. 2024).

$$P_{\text{VSG}} = k_D\Delta\omega + k_I\frac{d\Delta\omega}{dt} \tag{85}$$

where k_D and k_I stand for the damping and inertial constant, $\Delta\omega$ represents the change in angular frequency, $d\Delta\omega$ refer to rate-of change in angular frequency. By providing a rapid dynamic frequency response through the frequency derivative, the

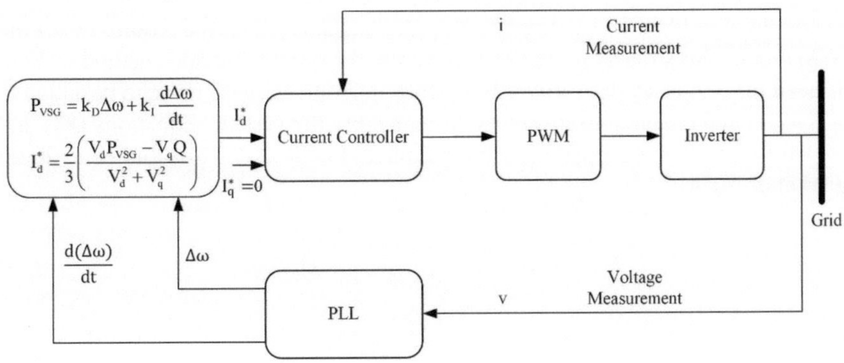

Fig. 75 The block diagram of VSG topology

damping constant aids in restoring the frequency to the steady-state value and halting the ROCOF. This is especially critical in an isolated grid, where a high initial ROCOF can trigger unneeded protective relays. The general structure of the VSG topology is displayed in Fig. 75 (Tamrakar et al. 2017; Shadoul et al. 2022; Jafari et al. 2024).

Using a PLL, the change in system frequency and ROCOF is measured. The active power reference of the inverter is calculated with Eq. (85) and the current references for current control are determined accordingly. In this topology, a quadrature based current control can be used, as well as other current controls. In d_q control, the d axis reference of the current is calculated with Eq. (86) (Tamrakar et al. 2017; Shadoul et al. 2022; Jafari et al. 2024).

$$I_d^* = \frac{2}{3}\left(\frac{V_d P_{\text{VSG}} - V_q Q}{V_d^2 + V_q^2}\right) \tag{86}$$

where V_d and V_q stand for the d-axis and q-axis components of the grid voltage. Reactive power Q and the q-axis reference current component (I_q^*) are set to zero since only active power is under control. Considering feedback from grid current, the current controller generates gate signals to operate the inverter. In actuality, the inverter works as a current-controlled voltage source inverter (Tamrakar et al. 2017; Shadoul et al. 2022; Jafari et al. 2024).

5 Conclusion

In this chapter, a comprehensive overview of advanced power converter technologies has been presented, encompassing the fundamental principles, topologies, and application areas of DC–DC, AC–DC, DC–AC, and AC–AC converters. Starting with the classification of DC–DC converters, both non-isolated and isolated structures were explored, highlighting their critical role in efficient energy conversion for modern electronic systems such as electric vehicles, renewable energy systems, and portable electronics. Then, the advantages and limitations of AC–DC rectifiers of both uncontrolled and controlled topologies are discussed. Power factor correction techniques were also examined to demonstrate how these systems align with power quality standards. Furthermore, the chapter addressed the significance of DC–AC inverters and AC–AC converters, including multilevel inverters, matrix and cycloconverter topologies, which provide flexible and high-performance solutions for industrial and smart grid applications. Through detailed analyses of converter structures, switching techniques, and emerging technologies such as soft-switching and resonant methods, this chapter aims to equip readers with a solid foundation for understanding the ongoing innovations and future challenges in the field of power electronics.

References

Abolghasemi M, Soltani I, Shivaie M, Vahedi H (2025) Recent advances of step-up multi-stage DC-DC converters: a review on classifications, structures and grid applications. Energy Rep 13:3050–3081. https://doi.org/10.1016/j.egyr.2025.02.025

Ali Z, Christofides N, Hadjidemetriou L, Kyriakides E, Yang Y, Blaabjerg F (2018) Three-phase phase-locked loop synchronization algorithms for grid-connected renewable energy systems: a review. Renew Sustain Energy Rev 90:434–452. https://doi.org/10.1016/j.rser.2018.03.086

Ali M, Iqbal A, Khan MR, Khalid M (2024) Chapter 13—AC-AC converters. In: Rashid MH (ed) Power electronics handbook, 5th edn. Butterworth-Heinemann, pp 437–480. https://doi.org/10.1016/B978-0-323-99216-9.00024-X

Asl ES, Babaei E, Sabahi M, Hasan Babayi Nozadian M, Cecati C (2018) New half-bridge and full-bridge topologies for a switched-boost inverter with continuous input current. IEEE Trans Ind Electron 65(4):3188–3197. https://doi.org/10.1109/TIE.2017.2752118

Azer P, Emadi A (2020) Generalized state space average model for multi-phase interleaved buck, boost and buck-boost DC-DC converters: transient, steady-state and switching dynamics. IEEE Access 8:77735–77745. https://doi.org/10.1109/ACCESS.2020.2987277

Bahravar S, Mahery HM, Babaei E, Sabahi M (2012) Mathematical modeling and transient analysis of dc-dc buck-boost converter in CCM. In: 2012 IEEE 5th India international conference on power electronics (IICPE), pp 1–6. https://doi.org/10.1109/IICPE.2012.6450404

Banaei MR, Ardi H, Farakhor A (2014) Analysis and implementation of a new single-switch buck–boost DC/DC converter. IET Power Electron 7(7):1906–1914. https://doi.org/10.1049/iet-pel.2013.0762

Bayhan S, Abu-Rub H (2024) Chapter 38—Predictive control of power electronic converters. In: Rashid MH (ed) Power electronics handbook, 5th edn. Butterworth-Heinemann, pp 1283–1297. https://doi.org/10.1016/B978-0-323-99216-9.00034-2

Beck H-P, Hesse R (2007) Virtual synchronous machine. In: 2007 9th International conference on electrical power quality and utilisation, pp 1–6. https://doi.org/10.1109/EPQU.2007.4424220

Bilhan AK, Kabalci E (2024) Chapter 9—DC-DC converters. In: Rashid MH (ed) Power electronics handbook, 5th edn. Butterworth-Heinemann, pp 269–291. https://doi.org/10.1016/B978-0-323-99216-9.00018-4

Biswas M, Majhi S, Nemade H (2019) Two-phase high efficiency interleaved buck converter with improved step-down conversion ratio and low voltage stress. IET Power Electron 12(15):3942–3952. https://doi.org/10.1049/iet-pel.2019.0547

Bose BK (2007) Neural network applications in power electronics and motor drives—an introduction and perspective. IEEE Trans Ind Electron 54(1):14–33. https://doi.org/10.1109/TIE.2006.888683

Bose BK (2021a) Chapter 4—Voltage-source converters. In: Bose BK (ed) Power electronics and motor drives, 2nd edn. Academic Press, pp 197–345. https://doi.org/10.1016/B978-0-12-821360-5.00004-X

Bose BK (2021b) Chapter 3—Phase-controlled converters and cycloconverters. In: Bose BK (ed) Power electronics and motor drives, 2nd edn. Academic Press, pp 111–195. https://doi.org/10.1016/B978-0-12-821360-5.00003-8

Boyar A (2024) Improvement of IoT-based industrial motor drive including model predictive control and multilevel inverter. Nigde Omer Halisdemir University, Graduate School of Natural and Applied Sciences

Boyar A, Kabalci E (2018) Comparison of a two-phase interleaved boost converter and flyback converter. In: 2018 IEEE 18th international power electronics and motion control conference (PEMC), pp 352–356. https://doi.org/10.1109/EPEPEMC.2018.8521891

Boyar A, Kabalci E (2021) Design of model predictive controller for current control of ANPC inverter. In: 2021 3rd Global power, energy and communication conference (GPECOM), pp 50–55. https://doi.org/10.1109/GPECOM52585.2021.9587777

Boyar S, Kabalci E, Kabalci Y (2022) MRAS based model predictive torque control of induction motor drive for electric vehicles. In: 2022 4th Global power, energy and communication conference (GPECOM), pp 51–56. https://doi.org/10.1109/GPECOM55404.2022.9815689

Boyar A, Kabalcı E, Kabalcı Y (2023) Model predictive torque control-based induction motor drive with remote control and monitoring interface for electric vehicles. Electr Power Compon Syst 1–12. https://doi.org/10.1080/15325008.2023.2211581

Bruckner T, Bemet S (2001) Loss balancing in three-level voltage source inverters applying active NPC switches. In: 2001 IEEE 32nd annual power electronics specialists conference (IEEE Cat. No.01CH37230), vol 2, pp 1135–1140. https://doi.org/10.1109/PESC.2001.954272

Chatterjee S, Kumar P, Chatterjee S (2018) A techno-commercial review on grid connected photovoltaic system. Renew Sustain Energy Rev 81:2371–2397. https://doi.org/10.1016/j.rser.2017.06.045

Cheng H, Jung S, Kim Y-B (2025) A novel reinforcement learning controller for the DC-DC boost converter. Energy 321:135479. https://doi.org/10.1016/j.energy.2025.135479

Çolak İ, Kabalci E (2008) Evirici Topolojileri ve Gelişimleri Üzerine Bir İnceleme. A review on inverter topologies and developments, pp 1–5

Colak I, Kabalci E, Bayindir R (2011) Review of multilevel voltage source inverter topologies and control schemes. Energy Convers Manage 52(2):1114–1128. https://doi.org/10.1016/j.enconman.2010.09.006

Deshmukh (Gore) S et al (2022) Review on classification of resonant converters for electric vehicle application. Energy Rep 8:1091–1113. https://doi.org/10.1016/j.egyr.2021.12.013

El-Hosainy A, Hamed HA, Azazi HZ, El-Kholy EE (2017) A review of multilevel inverter topologies, control techniques, and applications. In: 2017 Nineteenth international middle east power systems conference (MEPCON), pp 1265–1275. https://doi.org/10.1109/MEPCON.2017.8301344

Emon AI, Yuan Z, Mirza AB, Deshpande A, Hassan MU, Luo F (2021) 1200 V/650 V/160 A SiC+Si IGBT 3L hybrid T-type NPC power module with enhanced EMI shielding. IEEE Trans Power Electron 36(12):13660–13673. https://doi.org/10.1109/TPEL.2021.3089578

Erickson RW, Maksimović D (2001) Fundamentals of power electronics, 2nd edn. Softcover reprint of the hardcover 2nd edition 2001. Springer Science+Business Media, LLC, New York, NY

Feng Z, Zhang X, Wang J, Yu S (2020) A high-efficiency three-level ANPC inverter based on hybrid SiC and Si devices. Energies 13(5), Art. no. 5. https://doi.org/10.3390/en13051159

Golestan S, Guerrero JM, Vasquez JC (2017) Three-phase PLLs: a review of recent advances. IEEE Trans Power Electron 32(3):1894–1907. https://doi.org/10.1109/TPEL.2016.2565642

Guo X, He M, Yang Y (2018) Over modulation strategy of power converters: a review. IEEE Access 6:69528–69544. https://doi.org/10.1109/ACCESS.2018.2872776

Gupta KK, Ranjan A, Bhatnagar P, Sahu LK, Jain S (2016) Multilevel inverter topologies with reduced device count: a review. IEEE Trans Power Electron 31(1):135–151. https://doi.org/10.1109/TPEL.2015.2405012

Gürçam K, Nuri Almalı M (2025) Single-stage modified Vienna rectifier SEPIC AC-DC LED driver. Ain Shams Eng J 16(1):103200. https://doi.org/10.1016/j.asej.2024.103200

Hadjidemetriou L, Kyriakides E, Blaabjerg F (2013) A new hybrid PLL for interconnecting renewable energy systems to the grid. IEEE Trans Ind Appl 49(6):2709–2719. https://doi.org/10.1109/TIA.2013.2265252

Hossain MZ, Rahim NA, a/l Selvaraj J (2018) Recent progress and development on power DC-DC converter topology, control, design and applications: a review. Renew Sustain Energy Rev 81:205–230. https://doi.org/10.1016/j.rser.2017.07.017

Hu C et al (2017) An improved virtual space vector modulation scheme for three-level active neutral-point-clamped inverter. IEEE Trans Power Electron 32(10):7419–7434. https://doi.org/10.1109/TPEL.2016.2621776

Hui (Ron) SY, Chung HSH (2007) 16—Resonant and soft-switching converters. In: Rashid MH (ed) Power electronics handbook, 2nd edn. Engineering. Academic Press, Burlington, pp 405–449. https://doi.org/10.1016/B978-012088479-7/50034-1

Islam MdR, Guo Y, Zhu J (2014) Power converter topologies for grid-integrated medium-voltage applications. In: Islam MdR, Guo Y, Zhu J (eds) Power converters for medium voltage networks. Green energy and technology. Springer, Berlin, Heidelberg, pp 51–107. https://doi.org/10.1007/978-3-662-44529-7_3

Ivanovic Z, Knezic M (2022) Modeling push–pull converter for efficiency improvement. Electronics 11(17), Art. no. 17. https://doi.org/10.3390/electronics11172713

Jafari M, Gharehpetian GB, Anvari-Moghaddam A (2024) On the role of virtual inertia units in modern power systems: a review of control strategies, applications and recent developments. Int J Electr Power Energy Syst 159:110067. https://doi.org/10.1016/j.ijepes.2024.110067

Jayakumar V, Chokkalingam B, Munda JL (2021) A comprehensive review on space vector modulation techniques for neutral point clamped multi-level inverters. IEEE Access 9:112104–112144. https://doi.org/10.1109/ACCESS.2021.3100346

Jovanovic MM, Irving BT (2016) On-the-fly topology-morphing control—efficiency optimization method for *LLC* resonant converters operating in wide input- and/or output-voltage range. IEEE Trans Power Electron 31(3):2596–2608. https://doi.org/10.1109/TPEL.2015.2440099

Kabalci E (2017) Maximum power point tracking (MPPT) algorithms for photovoltaic systems. In: Bizon N, Mahdavi Tabatabaei N, Blaabjerg F, Kurt E (eds) Energy harvesting and energy efficiency, vol 37. Springer International Publishing, Cham, pp 205–234. https://doi.org/10.1007/978-3-319-49875-1_8

Kabalci E (2019) From smart grid to internet of energy. Academic Press, an imprint of Elsevier, London, United Kingdom; San Diego, CA, United States

Kabalci E (2021a) Multilevel inverters: control methods and advanced power electronic applications. Academic Press, London. https://search.ebscohost.com/login.aspx?direct=true&scope=site&db=nlebk&db=nlabk&AN=2712992. Accessed 2 Aug 2022

Kabalcı E (2021b) Chapter 1—Introduction to multilevel voltage source inverters. In: Kabalcı E (ed) Multilevel inverters. Academic Press, pp 1–27. https://doi.org/10.1016/B978-0-12-821668-2.00015-5

Kabalci E, Boyar A (2018) Design and analysis of a single phase flyback micro inverter. In: 2018 6th International conference on control engineering information technology (CEIT), pp 1–6. https://doi.org/10.1109/CEIT.2018.8751843

Kabalci E, Boyar A (2019) Design and analysis of two-phase interleaved boost converter and H5 inverter based microinverter. In: 2019 1st Global power, energy and communication conference (GPECOM), pp 122–127. https://doi.org/10.1109/GPECOM.2019.8778497

Kabalcı E, Boyar A (2021) Chapter 7—Multilevel inverter applications for electric vehicle drives. In: Kabalcı E (ed) Multilevel inverters. Academic Press, pp 185–208. https://doi.org/10.1016/B978-0-323-90217-5.00006-X

Kabalci E, Kabalci Y (2018) Smart grids and their communication systems. Springer Berlin Heidelberg, New York, NY

Kabalcı E, Boyar A, Kabalcı Y (2021) Chapter 1—Pulse width modulation and control methods for multilevel inverters. In: Kabalcı E (ed) Multilevel inverters. Academic Press, pp 1–33. https://doi.org/10.1016/B978-0-323-90217-5.00009-5

Kabalci E, Kabalci Y, Boyar A (2023) Comparison of model predictive torque control based NPC and ANPC inverters for IM drive. In: 2023 5th Global power, energy and communication conference (GPECOM), pp 27–32. https://doi.org/10.1109/GPECOM58364.2023.10175830

Karamanakos P, Geyer T (2020) Guidelines for the design of finite control set model predictive controllers. IEEE Trans Power Electron 35(7):7434–7450. https://doi.org/10.1109/TPEL.2019.2954357

Karimi-Ghartemani M, Iravani MR (2004) A method for synchronization of power electronic converters in polluted and variable-frequency environments. IEEE Trans Power Syst 19(3):1263–1270. https://doi.org/10.1109/TPWRS.2004.831280

Kerdphol T, Rahman FS, Watanabe M, Mitani Y (2021) An overview of virtual inertia and its control. In: Kerdphol T, Rahman FS, Watanabe M, Mitani Y (eds) Virtual inertia synthesis and control. Springer International Publishing, Cham, pp 1–11. https://doi.org/10.1007/978-3-030-57961-6_1

Kheraluwala MN, Gascoigne RW, Divan DM, Baumann ED (1992) Performance characterization of a high-power dual active bridge DC-to-DC converter. IEEE Trans Ind Appl 28(6):1294–1301. https://doi.org/10.1109/28.175280

Kim S-H (2017) Chapter 7—Pulse width modulation inverters. In: Kim S-H (ed) Electric motor control. Elsevier, pp 265–340. https://doi.org/10.1016/B978-0-12-812138-2.00007-6

Lakshmi M, Hemamalini S (2018) Nonisolated high gain DC–DC converter for DC microgrids. IEEE Trans Industr Electron 65(2):1205–1212. https://doi.org/10.1109/TIE.2017.2733463

Lavieville J-P, Carrere P, Meynard T (1997) Electronic circuit for converting electrical energy, and a power supply installation making use thereof. US5668711A. https://patents.google.com/patent/US5668711/en. Accessed 16 Apr 2020

Li L, Xu G, Sha D, Liu Y, Sun Y, Su M (2023) Review of dual-active-bridge converters with topological modifications. IEEE Trans Power Electron 38(7):9046–9076. https://doi.org/10.1109/TPEL.2023.3258418

Liang T-J, Cheng M-H, Huang W-Y, Tseng W-J (2019) Interleaved half-bridge flyback converter with zero-current switching. IEEE Trans Power Electron 34(4):3370–3383. https://doi.org/10.1109/TPEL.2018.2852332

Liu C et al (2013) High-efficiency hybrid full-bridge–half-bridge converter with shared ZVS lagging leg and dual outputs in series. IEEE Trans Power Electron 28(2):849–861. https://doi.org/10.1109/TPEL.2012.2205019

Liu H, Zhao T, Wu X (2022) Performance evaluation of Si/SiC hybrid switch-based three-level active neutral-point-clamped inverter. IEEE Open J Ind Appl 3:90–103. https://doi.org/10.1109/OJIA.2022.3179225

Mchaouar Y, Abouobaida H, Abouelmahjoub Y, Abouloifa A, Lachkar I, Guerrero JM (2025) Sensorless non-linear control with MPPT for a hybrid wind-photovoltaic grid-connected system, based on a single DC-DC converter and an inverter with LCL filter. J Franklin Inst 107679. https://doi.org/10.1016/j.jfranklin.2025.107679

Meynard TA, Foch H (1992) Multi-level conversion: high voltage choppers and voltage-source inverters. In: PESC'92 record. 23rd Annual IEEE power electronics specialists conference, vol 1, pp 397–403. https://doi.org/10.1109/PESC.1992.254717

Mohan N, Undeland TM, Robbins WP (2002) Power electronics: converters, applications, and design, 3rd edn. Wiley. https://www.wiley.com/en-us/Power+Electronics%3A+Converters%2C+Applications%2C+and+Design%2C+3rd+Edition-p-9780471226932. Accessed 14 Apr 2025

Mudiyanselage GA, Keshmiri N, Emadi A (2023) A review of DC-DC resonant converter topologies and control techniques for electric vehicle applications. IEEE Open J Power Electron 4:945–964. https://doi.org/10.1109/OJPEL.2023.3331180

Nabae A, Takahashi I, Akagi H (1981) A new neutral-point-clamped PWM inverter. IEEE Trans Ind Appl IA-17(5):518–523. https://doi.org/10.1109/TIA.1981.4503992

Nguyen AD, Chen S-X, Chen Y, Chen C-W, Han B, Lai J-S (2021) Four-phase interleaved TCM DC-DC buck converter with matrix inductor in battery charging application. IET Power Electron 14(1):132–139. https://doi.org/10.1049/pel2.12017

Park J-E, Han J-K, Choi S-H, Moon G-W (2023) Two-switch forward converter with an integrated buck converter for high bus voltage in satellites. IEEE Trans Power Electron 38(2):2041–2051. https://doi.org/10.1109/TPEL.2022.3197610

Patin N (2015) 3—Resonant converters and soft switching. In: Patin N (ed) Power electronics applied to industrial systems and transports, vol 3. Elsevier, pp 41–58. https://doi.org/10.1016/B978-1-78548-002-7.50003-X

Petit P, Aillerie M, Sawicki J-P, Charles J-P (2012) Push-pull converter for high efficiency photovoltaic conversion. Energy Procedia 18:1583–1592. https://doi.org/10.1016/j.egypro.2012.05.174

Pilli NK, Singh SK (2019) Comparative analysis of CB-PWM techniques in modified multilevel DC link inverter for PV applications. IET Power Electron 12(14):3802–3809. https://doi.org/10.1049/iet-pel.2019.0273

Rai N, Chakravorty S (2019) Generalized formulations and solving techniques for selective harmonic elimination PWM strategy: a review. J Inst Eng India Ser B 100(6):649–664. https://doi.org/10.1007/s40031-019-00411-1

Raju S, Mohan N (2024) Switching optimized modulation for direct three-level matrix converter. IEEE Trans Power Electron 39(3):3313–3328. https://doi.org/10.1109/TPEL.2023.3339480

Ramos-Paja CA, Bastidas-Rodriguez JD, Saavedra-Montes AJ (2022) Sliding-mode control of a photovoltaic system based on a flyback converter for microinverter applications. Appl Sci 12(3), Art. no. 3. https://doi.org/10.3390/app12031399

Rana RA, Patel SA, Muthusamy A, Woo Lee C, Kim H-J (2019) Review of multilevel voltage source inverter topologies and analysis of harmonics distortions in FC-MLI. Electronics 8(11), Art. no. 11. https://doi.org/10.3390/electronics8111329

Rashid MH (2014) Power electronics: devices, circuits, and applications, 4th edn. Pearson, Upper Saddle River, NJ

Rashid MH (ed) (2024a) Power electronics handbook, 5th edn. Butterworth-Heinemann, Amsterdam

Rashid MH (2024b) Power electronics handbook, 5th edn. Elsevier Shop. https://shop.elsevier.com/books/power-electronics-handbook/rashid/978-0-323-99216-9. Accessed 17 Mar 2024

Rashid MH, (Ron) Hui SY, Chung HS-H, Madichetty S, Kumar NBS, Krishna BM (2024) Chapter 11—Resonant and soft-switching converters. In: Rashid MH (ed) Power electronics handbook, 5th edn. Butterworth-Heinemann, pp 345–405. https://doi.org/10.1016/B978-0-323-99216-9.00022-6

Ravindrababu M, Nagaraju M, Sukumar GD, Syed MS (2025) ANFIS based SVPWM controlled indirect matrix converter fed single sided linear induction motor drive. e-Prime Adv Electr Eng Electron Energy 11:100932. https://doi.org/10.1016/j.prime.2025.100932

Sadoughi M, Zakerian A, Pourdadashnia A, Farhadi-Kangarlu M (2021) Selective harmonic elimination PWM for cascaded H-bridge multilevel inverter with wide output voltage range using

PSO algorithm. In: 2021 IEEE Texas power and energy conference (TPEC), pp 1–6. https://doi.org/10.1109/TPEC51183.2021.9384945

Salimi M, Soltani J, Markadeh GA, Abjadi NR (2013) Adaptive nonlinear control of the DC-DC buck converters operating in CCM and DCM. Int Trans Electr Energy Syst 23(8):1536–1547. https://doi.org/10.1002/etep.1679

Saravanan S, Pandiyan P, Chinnadurai T, Tiwari R, Prabaharan N (2022) Overview of bidirectional DC–DC converters topologies for electric vehicle and renewable energy system. In: Priyadarshi N, Bhoi AK, Bansal RC, Kalam A (eds) DC–DC converters for future renewable energy systems. Springer, Singapore, pp 433–458. https://doi.org/10.1007/978-981-16-4388-0_24

Schweizer M, Kolar JW (2013) Design and implementation of a highly efficient three-level T-type converter for low-voltage applications. IEEE Trans Power Electron 28(2):899–907. https://doi.org/10.1109/TPEL.2012.2203151

Sevilmiş F, Karaca H (2015) Simulation of three-phase grid interactive inverter for wind energy systems. In: 2015 IEEE 15th international conference on environment and electrical engineering (EEEIC), pp 1169–1174. https://doi.org/10.1109/EEEIC.2015.7165333

Sevi̇lmiş F, Karaca H (2019) Performance analysis of SRF-PLL and DDSRF-PLL algorithms for grid interactive inverters. Int Adv Res Eng J 3(2), Art. no. 2. https://doi.org/10.35860/iarej.412250

Sevilmiş F, Karaca H (2022) Performance enhancement of DSOGI-PLL with a simple approach in grid-connected applications. Energy Rep 8:9–18. https://doi.org/10.1016/j.egyr.2021.11.186

Shadoul M, Ahshan R, AlAbri RS, Al-Badi A, Albadi M, Jamil M (2022) A comprehensive review on a virtual-synchronous generator: topologies, control orders and techniques, energy storages, and applications. Energies 15(22), Art. no. 22. https://doi.org/10.3390/en15228406

Shao S et al (2022) Modeling and advanced control of dual-active-bridge DC–DC converters: a review. IEEE Trans Power Electron 37(2):1524–1547. https://doi.org/10.1109/TPEL.2021.3108157

Tahmid MR, Malek MdA, Islam MdR, Khan MAG (2025) Advanced thyristor-based cycloconverter for efficient three-phase conversion with low THD. e-Prime Adv Electr Eng Electron Energy 11:100918. https://doi.org/10.1016/j.prime.2025.100918

Tamrakar U, Shrestha D, Maharjan M, Bhattarai BP, Hansen TM, Tonkoski R (2017) Virtual inertia: current trends and future directions. Appl Sci 7(7), Art. no. 7. https://doi.org/10.3390/app7070654

Tan FD (2002) The forward converter: from the classic to the contemporary. In: APEC. Seventeenth annual IEEE applied power electronics conference and exposition (Cat. No.02CH37335), vol 2, pp 857–863. https://doi.org/10.1109/APEC.2002.989344

Tarter RE (1993) Solid-state power conversion handbook. Wiley. Wiley-Interscience Publication. https://www.wiley.com/en-us/Solid-State+Power+Conversion+Handbook-p-9780471572435. Accessed 15 Apr 2025

Tolbert LM, Shi X (2018) 13—Multilevel power converters. In: Rashid MH (ed) Power electronics handbook, 4th edn. Butterworth-Heinemann, pp 385–416. https://doi.org/10.1016/B978-0-12-811407-0.00013-1

Valencia DF, Tarvirdilu-Asl R, Garcia C, Rodriguez J, Emadi A (2021) A review of predictive control techniques for switched reluctance machine drives. Part II: torque control, assessment and challenges. IEEE Trans Energy Convers 36(2):1323–1335. https://doi.org/10.1109/TEC.2020.3047981

Vázquez N, Vaquero J (2024) Chapter 10—Inverters. In: Rashid MH (ed) Power electronics handbook, 5th edn. Butterworth-Heinemann, pp 293–343. https://doi.org/10.1016/B978-0-323-99216-9.00001-9

Vijaya Sambhavi Y, Ramachandran V (2023) A technical review of modern traction inverter systems used in electric vehicle application. Energy Rep 10:3882–3907. https://doi.org/10.1016/j.egyr.2023.10.056

Wang Z, Wu Y, Mahmud MH, Zhao Z, Zhao Y, Mantooth HA (2020) Design and validation of A 250-kW all-silicon carbide high-density three-level T-type inverter. IEEE J Emerg Sel Top Power Electron 8(1):578–588. https://doi.org/10.1109/JESTPE.2019.2951625

Wang G, Wen H, Liu W, Li F (2025) Physical structure, characteristics, and applications of monolithic bidirectional switches: a comprehensive review. IEEE Trans Power Electron 40(7):9187–9199. https://doi.org/10.1109/TPEL.2025.3543635

Wheeler PW, Rodriguez J, Clare JC, Empringham L, Weinstein A (2002) Matrix converters: a technology review. IEEE Trans Industr Electron 49(2):276–288. https://doi.org/10.1109/41.993260

Xu S-Z, Wang C-J, Wang Y (2019) An improved fault-tolerant control strategy for high-power ANPC three-level inverter under short-circuit fault of power devices. IEEE Access 7:55443–55457. https://doi.org/10.1109/ACCESS.2019.2913594

Yadeo D, Chaturvedi P, Suryawanshi H, Atkar D, Saketi SK (2021) Transistor clamped dual active bridge DC-DC converter to reduce voltage and current stress in low voltage distribution network, vol 31, no 1. https://onlinelibrary.wiley.com/doi/10.1002/2050-7038.12665. Accessed 13 Apr 2025

Yano M, Matsui M, Chin T-H, Kuroki K, Shimizu T (2000) Recent trends in power conversions. In: 2000 26th Annual conference of the IEEE industrial electronics society. IECON 2000. 2000 IEEE international conference on industrial electronics, control and instrumentation. 21st century technologies, vol 2, pp 1340–1346. https://doi.org/10.1109/IECON.2000.972316

Yaramasu V, Wu B (2017) Modeling of power converters for model predictive control. In: Model predictive control of wind energy conversion systems. IEEE, pp 149–176. https://doi.org/10.1002/9781119082989.ch5

Zhang Y, Li F, Zhang L, Wen S, Ma M, Zhang X (2021) Modeling and frequency characteristic analysis of DSOGI-PLL in *dq* reference frame. Energy Rep 7:545–551. https://doi.org/10.1016/j.egyr.2021.08.007

Zhang G, Wang H, Zhang C, Yang Y (2024) Chapter 4—Matrix converter: model and control. In: Blaabjerg F (ed) Control of power electronic converters and systems, vol 4. Academic Press, pp 101–125. https://doi.org/10.1016/B978-0-323-85622-5.00013-4

Zhu B, Yang Y, Wang K, Liu J, Vilathgamuwa DM (2024) High transformer utilization ratio and high voltage conversion gain flyback converter for photovoltaic application. IEEE Trans Ind Appl 60(2):2840–2851. https://doi.org/10.1109/TIA.2023.3310488

Linear and Nonlinear Control for Power Converters

Mohammed Kabir Billal Boumegouas, Mohamed Trabelsi, and Hani Vahedi

Abstract This chapter provides an in-depth assessment of common control strategies for power converters. First, the classical linear technique, proportional-integral (PI) controller, is discussed. Although widely adopted for its simplicity and ease of implementation, this method often falls short when dealing with inherent nonlinearities and parameter uncertainties in power electronic systems. To address these challenges, advanced nonlinear control frameworks are explored, with a particular focus on fuzzy logic control (FLC), sliding mode control (SMC), synergetic control (SC), and model predictive control (MPC). These nonlinear methods are recognized for their enhanced robustness to disturbances, adaptability to system variations, and superior dynamic performance. SMC is noted for its robustness and fast response but suffers from chattering effects, while SC offers smoother control signals and hardware-friendly implementation without compromising stability. The chapter presents a comparative study supported by case examples to illustrate the operational benefits and practical trade-offs between linear and nonlinear control approaches. The analysis underscores the relevance of nonlinear strategies in advancing converter performance under diverse operating conditions.

Keyword Power converters · Proportional-integral (PI) · Fuzzy logic control (FLC) · Sliding mode control (SMC) · Synergetic control (SC)

M. K. B. Boumegouas
Laboratory of Semiconductors and Functional Materials, Amar Telidji University of Laghouat, Laghouat, Algeria
e-mail: mk.boumegouas@lagh-univ.dz

M. Trabelsi
Electronics and Communications Engineering Department, Kuwait College of Science and Technology, Doha, Kuwait
e-mail: m.trabelsi@kcst.edu.kw

H. Vahedi (✉)
Delft University of Technology, Delft, Netherlands
e-mail: Hani.Vahedi@ieee.org

E. Kabalci (ed.), *Modern Control Methods for Power Electronics*, Energy Systems in Electrical Engineering, https://doi.org/10.1007/978-981-92-1974-2_2

Nomenclature

Symbols	Description
K_P	Proportional gain of PI regulator
K_i	Integral gain of PI regulator
K_d	Derivative gain of PI regulator
$e(t)$	Error between the reference and the actual value
S	Sliding surface in sliding mode control
$V(x)$	Lyapunov function
ψ	Synergetic control's manifold
L	Buck-boost inductance
C	Buck-boost capacitor
R	Buck-boost load
V_0	Buck-boost output voltage
D	Buck-boost diode
E	Buck-boost source voltage
I_L	Inductor current
V_c	Capacitor voltage
u	Control input

1 Introduction

Power electronics is a multidisciplinary field that merges concepts from electrical engineering, electronics, and control systems to study the conversion, regulation, and efficient management of electrical power, with the core aim of developing compact, cost-effective, and energy-efficient methods for transforming and controlling electrical energy (Rashid 2013). The basic building blocks of a static converter consist of nonlinear elements (primarily semiconductor switches operating in switching mode) and linear reactive components (such as capacitors, inductors, and transformers), which provide intermediate energy storage and filter voltage and current waveforms while often contributing significantly to the converter's size, weight, and cost (Carrol 2004; Rufer 2004).

There are several topologies of power converters, each classified according to their energy conversion function, whether transforming AC to AC, AC to DC, DC to AC, or DC to DC. These topologies are fundamental in adapting voltage levels, frequencies, and waveforms to suit specific application requirements in modern electrical and electronic systems. For instance, AC/AC converters such as matrix converters modify both voltage and frequency, while AC/DC rectifiers convert alternating current into direct current. DC/DC converters adjust the voltage level of a DC supply, and DC/AC inverters produce AC output from a DC source (Chung et al. 2024).

The control strategies are essential to ensure stable, efficient, and accurate operation. These control methods can generally be categorized into linear and nonlinear approaches. The choice between them depends on the converter's dynamic behavior, operating conditions, and performance requirements.

Linear control strategies, such as proportional-integral (PI) control, have been widely implemented due to their simplicity, reliability, and ease of design. PI controllers are particularly favored in power converters for their ability to regulate output voltage or current with minimal steady-state error. By adjusting the proportional and integral gains, PI controllers effectively address disturbances and maintain stability in systems with relatively linear operating conditions (Boumegouas et al. 2023). However, despite their widespread use, PI controllers exhibit inherent limitations, such as susceptibility to overshoots, slow dynamic response, and the inability to handle significant parameter variations or nonlinear dynamics. These demerits often manifest as performance bottlenecks in applications requiring high precision or rapid adaptability.

Fuzzy logic control, considered as nonlinear control approach, offers an intuitive design framework based on linguistic rules rather than mathematical models. This control method is particularly advantageous for systems with incomplete or imprecise models, as it relies on expert knowledge to define input–output relationships. In power converter applications, fuzzy controllers demonstrate acceptable performance in regulating output parameters and accommodating minor nonlinearities (Tang and Ahmad 2024; Liu et al. 2022). However, similar to PI controllers, fuzzy logic control may struggle with overshoots and ripples, especially under dynamic or highly nonlinear conditions. The rule-based structure, while flexible, can also become complex and less efficient for systems requiring rapid response or operating in highly variable environments.

To overcome these limitations, the sliding mode control (SMC) and synergetic control are emerging as prominent solutions. These methods are particularly effective in managing the inherent nonlinearities of power converters, ensuring robust performance against parameter variations and external disturbances.

SMC operates by defining a sliding surface that governs the system's trajectory, compelling it to converge to the desired operating point despite system uncertainties. This approach offers numerous advantages, including reduced sensitivity to model inaccuracies, rapid dynamic response, and improved system stability (Boumegouas et al. 2023). In power converter applications, SMC has demonstrated superior performance compared to linear control methods, effectively mitigating overshoots and ripples, ensuring smoother output voltage or current profiles. Additionally, its robustness extends to handling abrupt load changes, making it suitable for high-performance and mission-critical applications (Komurcugil et al. 2020; Wu et al. 2021). However, SMC is not without challenges, as the phenomenon of chattering, characterized by high-frequency oscillations around the sliding surface due to the switching nature of the control law, remains a notable drawback. This issue can be mitigated through advanced techniques such as boundary layer modifications or higher-order sliding modes, though these approaches may increase the complexity of the controller design.

On the other hand, synergetic control presents an alternative nonlinear strategy that addresses some of the challenges associated with SMC. Unlike SMC, which relies on discontinuous control laws, synergetic control employs a continuously differentiable control function derived from invariant manifolds and system dynamics. This results in a smoother control action, reducing chattering effects while maintaining robustness to disturbances and parameter variations. Synergetic control is particularly advantageous in applications requiring high precision and minimal switching-induced stress on power electronic components. By constructing a suitable macro-variable and designing a corresponding control law, synergetic control ensures global system stability and convergence without the need for switching-based control strategies. Additionally, its systematic design methodology facilitates implementation in complex nonlinear systems while offering improved adaptability to varying operating conditions (Huang et al. 2025; Boumegouas and Kouzi 2022).

By integrating both SMC and synergetic control, power converter applications can leverage the strengths of each approach, balancing robustness, dynamic performance, and reduced chattering. The selection of the appropriate control method depends on the specific application requirements, including stability, response speed, and implementation complexity.

Thus, the control of power converters necessitates a careful balance between simplicity, performance, and robustness. While linear control methods like PI provide straightforward solutions with acceptable performance under certain conditions, their limitations in handling nonlinearities and dynamic changes highlight the need for advanced strategies. Nonlinear control methods, exemplified by FLC, SMC, SC, and MPC, address these challenges by offering enhanced performance, robustness, and adaptability.

The ongoing development and integration of these control techniques are essential for advancing power converter technology and meeting the increasing demands of modern electronic systems. Future research should focus on hybrid approaches that combine the strengths of linear and nonlinear controls, leveraging the simplicity of linear methods and the robustness of nonlinear strategies to achieve optimal performance in diverse applications. By integrating SMC and synergetic control, power converter applications can benefit from improved stability, dynamic response, and reduced chattering, paving the way for more efficient and reliable electronic systems.

2 Linear Control Method for Power Converters

As the demand for efficient and reliable power electronic systems continues to grow, stimulated by the spread of renewable energy sources, electric vehicles, and advanced industrial automation, control has become a pivotal element (Maroti et al. 2022). The control system governs the behavior of a power converter by regulating its output voltage/current in response to varying load conditions and input disturbances, ultimately ensuring stability, dynamic performance, and high efficiency. Selecting the appropriate control approach depends on several factors, including system complexity, dynamic requirements, implementation cost, and sensitivity to parameter variations (Al-Dabbagh et al. 2024).

In practice, control strategies are broadly categorized into two classes: linear and nonlinear methods (Boumegouas and Kouzi 2022). Linear control approaches have been the cornerstone of industrial control systems. They have been a preferred option in many applications due to their ease of design and computational simplicity. These characteristics make linear controllers particularly attractive in systems where operating conditions remain within predictable bounds, and model linearity can be reasonably assumed. However, linear controllers often face performance degradation when confronted to highly nonlinear dynamics, large parameter uncertainties, or rapid transients (common features in many real-world power converter applications) (Boumegouas et al. 2023; Boumegouas and Kouzi 2022).

Nonetheless, linear control methods continue to play a significant role, particularly in systems where robustness to nonlinearity is not critical, or where low-complexity implementation is essential. In this context, proportional-integral-derivative (PID) controllers have emerged as widely adopted solutions (Boumegouas and Kouzi 2022). These methods have found success in a range of converter topologies due to their intuitive tuning mechanisms and relatively modest computational demands.

The following provides a detailed exploration of these linear control techniques, focusing on their operating principles, practical advantages, and inherent limitations within the domain of power electronics. This discussion lays the foundation for the subsequent evaluation of nonlinear control strategies, highlighting the motivations behind the transition from conventional linear approaches to more advanced and adaptive control paradigms.

2.1 Deadbeat Control

The deadbeat control method is a digital control strategy derived from circuit state equations. Known for its exceptional dynamic response, this technique achieves rapid regulation by adjusting the switching duty cycle to reach the reference value within just one sampling period (Mizushima et al. 2015). These fast-response characteristics make deadbeat control particularly effective for generating high–low (HL) pulses. As a result, it has been widely implemented in DC-DC converters and inverters (Mizushima et al. 2015).

To implement deadbeat control in a power converter, the following steps must be executed (Kawamura et al. 1988):

- **Modelling the Converter**: Develop a mathematical model of the converter topology (e.g., buck, boost, or buck-boost) and derive its state-space representation to capture the dynamics of inductor currents and capacitor voltages.
- **Discretization**: Convert the continuous-time state-space model into discrete time using methods such as zero-order hold (ZOH) or Euler approximation, ensuring compatibility with digital control implementation.

- **Reference Tracking**: Define the reference trajectory such that the output voltage converges to the desired value within one sampling period, a key feature of deadbeat control.
- **Duty Cycle Calculation**: Solve the discrete-time equations to compute the required duty cycle, enabling precise and rapid regulation.

Figure 1 presents a simple flowchart of the deadbeat control for the power converter.

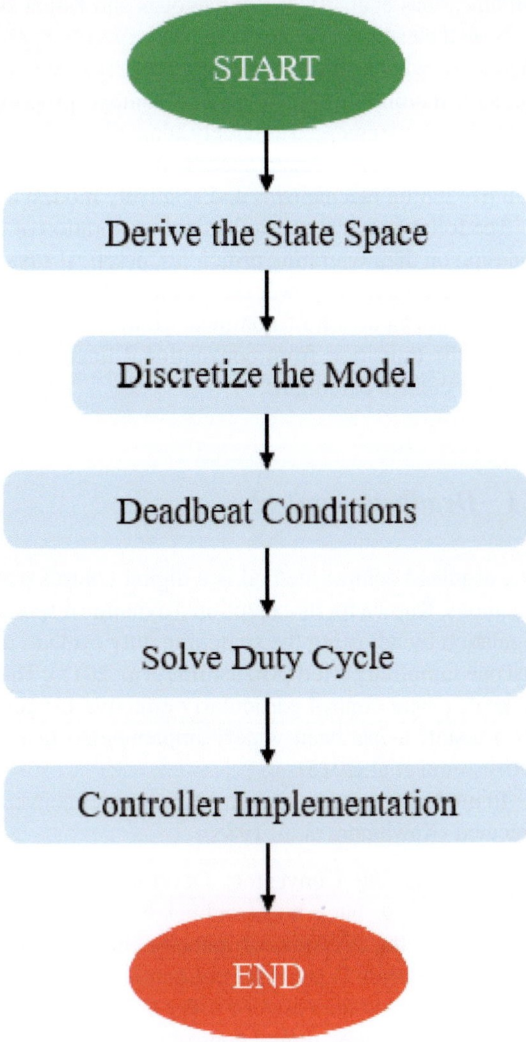

Fig. 1 Deadbeat control flowchart for power converter

2.2 PID Control

This is the most commonly used linear control strategy, relying on proportional, integral, and derivative actions to regulate system behavior (Butterworth-Heinemann 2017). The structure of this control is expressed as:

$$u(t) = K_p e(t) + K_i \int e(t)dt + K_d \frac{de(t)}{dt}. \quad (1)$$

The proportional-integral (PI) controller is widely recognized as a standard and reliable solution in industrial control systems. Figure 4 illustrates the block diagram of a typical PI controller (Seo and Choi 2019), which operates within a feedback loop framework. Its primary function is to minimize the error arising from discrepancies between the measured process variable and the desired setpoint. Based on this error, the controller executes corrective actions to drive the system toward the target value (Sasikala and Kumar 2019).

The controller output is determined by the weighted combination of these two terms, which is then used to modulate the control element and regulate the process. While the proportional term generates an output directly proportional to the instantaneous error, it can lead to a steady-state offset, where the system fails to reach the exact setpoint. Additionally, increasing the proportional gain to improve response speed can risk instability in the control loop. To address these limitations, the integral term is incorporated.

The integral action contributes a correction based on the duration and magnitude of the error, effectively eliminating the steady-state offset and ensuring long-term accuracy. However, it also reduces the system's forward gain and can slow down the transient response if not properly tuned. Thus, the PI controller offers a balanced approach by combining immediate error correction with long-term stability, making it a dependable choice for many industrial automation applications (Sasikala and Kumar 2019).

2.2.1 Proportional (P) Controller

A proportional (P) controller, as illustrated in Fig. 2, generates a control signal directly proportional to the instantaneous error, which is defined as the difference between the reference (desired) value and the system output (feedback). This controller essentially amplifies the error signal through a gain factor known as K_p. The merit of this control is its simplicity, ability to respond quickly to deviations, and error increase will provide a powerful corrective action. Achieving faster response and improving transient performance. Yet, the overshoot significantly increases. Figure 2 illustrates the structure of this technique for the power converter.

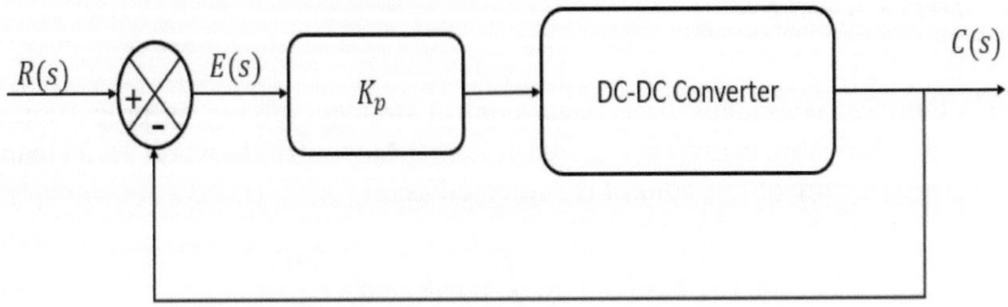

Fig. 2 P regulator structure for DC-DC converter

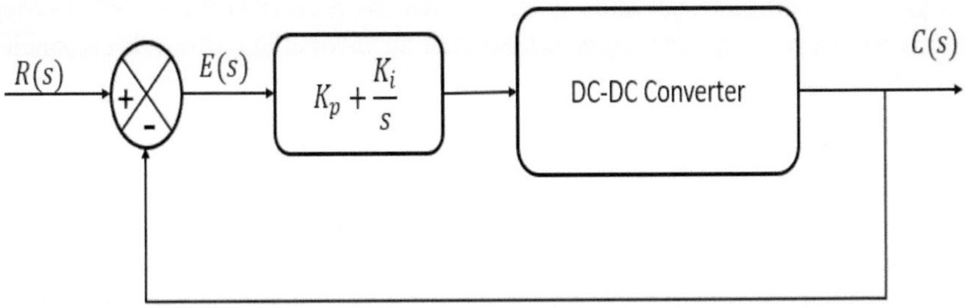

Fig. 3 PI regulator structure for DC-DC converter

2.2.2 Integral (I) Controller

Unlike the P regulator, the integral controller provides its output by computing the area under the curve of the error signal from the start of operation up to the current time. Means the integral removes the steady state error completely by minimizing the error toward zero. This makes it particularly effective in correcting persistent offsets that the P controller alone cannot handle. Nevertheless, integral control offers drawbacks such as overshoot and slower response. Additionally, because the integral action does not inherently include predictive behavior or damping, it is incapable of eliminating oscillations or stabilizing rapidly changing error conditions. Thus, in practical applications, the I controller is typically not used on its own but is combined with P or D components for more balanced performance. Figure 3 exhibits the PI regulator design structure for a power converter.

2.2.3 Derivative (D) Controller

The derivative (D) controller, depicted in Fig. 4, focuses on the rate at which the error signal changes rather than the magnitude of the error itself. Its main function is to anticipate future errors by observing how fast the current error is increasing or

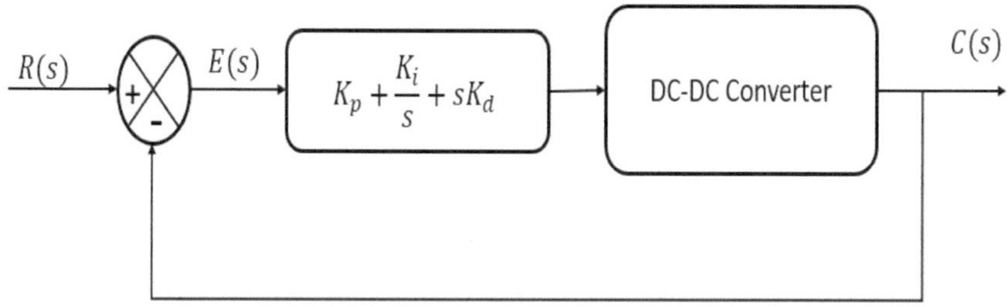

Fig. 4 PID regulator structure for DC-DC converter

decreasing. This predictive nature allows the controller to take pre-emptive corrective actions before the error becomes significant. In essence, it acts as a form of "foresight," smoothing the response and helping to dampen potential oscillations. By acting on the slope of the error curve, the derivative controller enhances the system's stability and reduces both the overshoot and the settling time. Furthermore, it is responsible for increasing the stability of the control system due to its merits above and is never used independently because it cannot minimize the error, respond to constant varying errors, and is sensitive to noise in the measurements due to the absence of P and I components. Hence, it is most effective when combined with a P or PI controller, forming PD or PID structures. Figure 4 shows the PID regulator design structure for a power converter.

Based on the comparative table, the P controller offers a fast rise time and is relatively simple and reliable for basic systems, but it suffers from increased overshoot and cannot fully eliminate the steady-state error, making it less suitable for precision–critical applications. The I controller also reduces the rise time and eliminates the steady-state error by integrating past errors, enhancing accuracy and reliability; however, it increases both overshoot and settling time, which can make the system sluggish and less robust against fast-changing dynamics.

The D controller, on the other hand, improves robustness by damping oscillations and reducing overshoot and settling time without significantly affecting the rise time or steady-state error, but it is highly sensitive to noise and is rarely used alone. When combined in a PID controller, these three actions balance speed, accuracy, and damping—offering the best overall performance in terms of response time, reliability, and robustness, though at the cost of more complex tuning and implementation.

Comparing PI and PID controllers, both eliminate steady-state error due to the integral action, making them more accurate than P alone. However, the PI controller tends to have higher overshoot and longer settling times because it lacks the derivative term to anticipate and dampen rapid error changes. In contrast, the PID controller includes this predictive component, significantly enhancing robustness and stability by reducing overshoot and speeding up the settling process.

This makes the PID more suitable for dynamic or sensitive systems that require fast, precise, and stable responses, while the PI controller remains preferable for

Table 1 Advantage and disadvantage of of P, I, and D regulator

	Overshoot	Settling Time	Rise time	Steady state error
P	↑	~	↓	↓
I	↑	↑	↓	↓
D	↓	↓	~	~

slower, less complex systems where tuning simplicity and elimination of steady-state error are the main goals. Table 1 summarizes the advantages and disadvantages of each action of the PID regulator.

As explained above, each action of the PID regulator plays a significant role in the control design. However, regulator gains depend on the engineer's expertise or the use of tuning methods such as Ziegler–Nichols, Cohen Coon,…,etc.

2.2.4 Zielger–Nichols

This method is particularly suited for processes that exhibit a step response resembling a first-order system with a transportation delay, as illustrated in Fig. 5.

The step response is characterized by two critical parameters: the delay time (L), which represents the interval between the step input application and the point where the process output begins its significant rise, and the time constant (T), corresponding to the time required for the output to reach approximately 63.2% of its steady-state value after the initial delay. To determine these parameters, the steepest slope of the step response curve, identified as the point of inflection (where the second derivative is zero), is used to construct a tangent line. This tangent is extended to intersect both the time axis (baseline) and the steady-state value line, allowing L to be derived from the tangent's intersection with the time axis and T from the time difference between

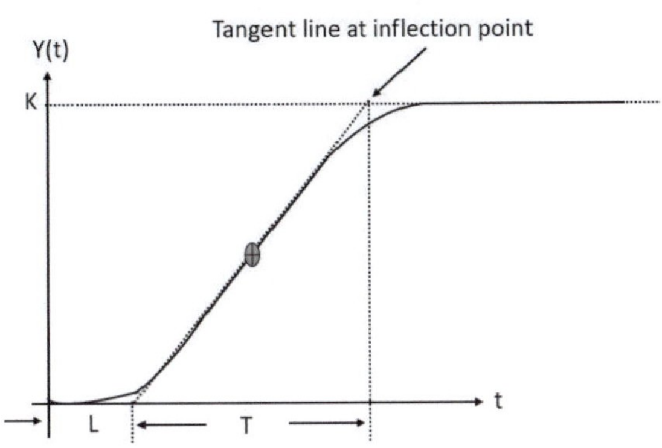

Fig. 5 Ziegler–Nichols method curve response

Table 2 Zielger–Nichols tuning method

	K_c	T_i	T_d
P	$\frac{T}{L}$		
PI	$\frac{0.9T}{L}$	$\frac{L}{0.3}$	
PID	$\frac{1.2T}{L}$	$2L$	$0.5L$

the tangent's steady-state intersection and the delay time (Nippatla and Mandava 2025).

While the graphical tangent method provides a straightforward way to estimate these parameters from inflection point analysis, more advanced techniques like nonlinear regression or system identification algorithms may yield higher accuracy for complex dynamics. Once L and T are determined, classical Ziegler–Nichols tuning rules (Table 2) can be directly applied to calculate initial PID controller parameters, providing a systematic approach for achieving baseline control performance before further refinement (Abbas and Mustafa 2024).

2.2.5 Cohen Coon

Various techniques exist for determining the appropriate values of the proportional, integral, and derivative gains in a PID controller. One widely used approach is the Cohen-Coon tuning method. This method models the system's response to a step input as a first-order system with dead time (Suksawat and Kaewpradit 2021), as illustrated in Fig. 6.

From this response curve, three key parameters are extracted (Suksawat and Kaewpradit 2021):

- $\mathbf{K_c}$: the steady-state gain, calculated as the ratio of the steady-state output to the magnitude of the step input.
- $\boldsymbol{\tau_m}$: the effective time constant representing the speed of the first-order response.
- $\boldsymbol{\tau_d}$: the process dead time, indicating the delay before the system begins to respond.

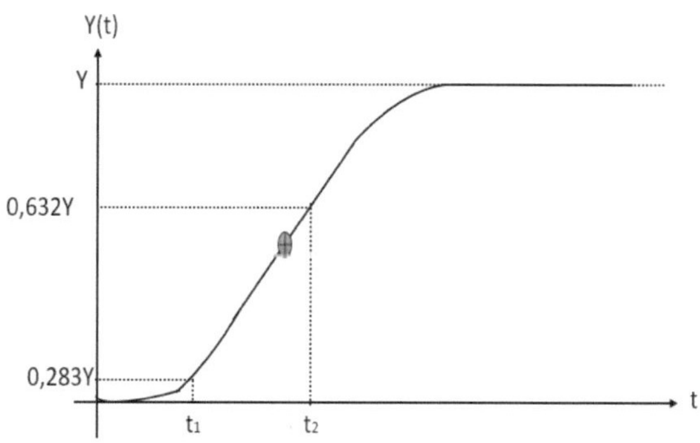

Fig. 6 Cohen-Coon method curve response

Table 3 Cohen-Coon tuning rules

	K_P	T_i	T_d
P	$\frac{\tau_m}{Kt_d}(1+\frac{t_d}{3\tau_m})$		
PI	$\frac{\tau_m}{Kt_d}(0.9+\frac{t_d}{12\tau_m})$	$t_d(\frac{30+3\frac{t_d}{\tau_m}}{9+20\frac{t_d}{\tau_m}})$	
PID	$\frac{\tau_m}{Kt_d}(1+\frac{t_d}{3\tau_m})$	$t_d(\frac{32+6\frac{t_d}{\tau_m}}{13+8\frac{t_d}{\tau_m}})$	$t_d(\frac{4}{11+2\frac{t_d}{\tau_m}})$

These parameters serve as the foundation for computing the PID controller gains using the Cohen-Coon tuning formulas expressed as:

$$\tau_m = \frac{3}{2}(t_2 - t_1),$$

$$\tau_d = t_2 - \tau_m.$$

Cohen-Coon tuning rules, presented in Table 3, offer an empirical approach for PID controller design that builds upon the Ziegler–Nichols method. These rules are specifically applicable to processes where the dead time (L) is less than twice the dominant time constant (T), a condition expressed mathematically as L < 2T. The methodology was developed through extensive experimental studies to achieve consistent closed-loop performance characterized by a quarter decay ratio (25% amplitude reduction between successive oscillations), representing an optimal balance between response speed and stability (Firmansyah and Irmawanto 2021).

In control systems, PID regulator remains most utilized for their simplicity, adaptability, and effectiveness. However, tuning the gains of this technique has a critical impact on control system. Therefore, several methods have been introduced as Ziegler–Nichols, which offers a straightforward approach for initial tuning, providing aggressive control for fast-responding processes with minimal dead time. However, it tends to produce higher overshoot and may lack robustness in systems with significant delays. Secondly, the Cohen-Coon method refines this approach by optimizing for smoother control, better disturbance rejection, and more conservative gains, making it better suited for processes with moderate time delays (Isdaryani et al. 2020).

Table 4 summarizes the PID regulator's and deadbeat controller's advantages and limitations in power converter applications.

3 Nonlinear Control Techniques

This section presents several nonlinear control strategies for power converters while introducing the advantages and limitations of each technique.

Table 4 Advantages and limitations of the PID regulator for power converters

	Advantages	Limitations
PID regulator	• Simplicity and ease of implementation • Good performance in linear systems • Steady-state error elimination	• Poor performance in nonlinear systems • Limited robustness • Overshoot and oscillations
Deadbeat control	• Fast transient response • No steady-state error • Suitable for digital Implementation	• Highly sensitive to model accurate • Computational complexity • Limited robustness

3.1 Fuzzy Logic Control (FLC)

Unlike PID, FLC does not require an explicit mathematical model of the system. Instead, it uses rule-based decision-making to handle uncertainties and nonlinearities. While more adaptable than PID, it lacks the rigorous stability guarantees of model-based approaches.

Fuzzy logic control is widely utilized in various DC-DC converter topologies, including buck, boost, and buck-boost converters, to regulate output voltage or current with high precision. Fuzzy logic control incorporated in these systems offers significant performance enhancements, particularly in improving transient response characteristics. This includes minimizing overshoots, reducing settling time, and ensuring stable operation under varying load conditions. Moreover, fuzzy controllers demonstrate superior robustness against parameter uncertainties and external disturbances, making them highly effective for dynamic and nonlinear power conversion environments (Sahoo and Kishore 2020; Lakshmi et al. 2024).

To implement fuzzy logic control in power converter systems, several key steps are involved, beginning with the design of the fuzzy controller and its seamless integration with the target hardware. An overview of the complete implementation process of this technique is given as follows (Sahoo and Kishore 2020; Lakshmi et al. 2024; Bendaoud, et al. 2017):

(a)-Fuzzification: Converts numerical inputs (e.g., error and error derivative) into fuzzy variables using membership functions that are given as linguistic terms negative big (NB), negative medium (M), zero (Z), positive medium (PM), positive big (PB), and based on the inputs equation that expressed as:

$$\begin{aligned} e(n) &= V_{ref} - V_{out} \\ \Delta e(n) &= e(n) - e(n-1). \end{aligned} \tag{2}$$

(b)-Rule Base: A set of predefined IF–THEN rules derived from expert knowledge (e.g., "If error is high and increasing, decrease output voltage rapidly"). And determines the control action based on the rule base and the current fuzzy variables. The rule-based table is given as (Table 5).

Table 5 Fuzzy logic control table of inference rules

e_n Δe_n	NB	NS	ZE	PS	PB
NB	NB	NB	NB	NS	ZE
NS	NB	NS	NS	ZE	PS
ZE	NB	NS	ZE	PS	PB
PS	NS	ZE	PS	PS	PB
PB	ZE	PS	PB	PB	PB

The fuzzy logic system utilizes five linguistic variables (fuzzy sets), defined as follows: negative big (NB), negative small (NS), zero (EZ), positive small (PS), and positive big (PB). Based on these fuzzy sets, a total of 25 fuzzy inference rules were formulated to capture the relationship between input and output variables. The inference mechanism is implemented using the sum-product algorithm, which effectively aggregates and evaluates the rule base. For defuzzification, the center of gravity (centroid) method is adopted to convert the fuzzy output into a crisp control signal.

(c)-Defuzzification: Converts the fuzzy control action back into a precise numerical output for the power converter. Duty cycle of output is expressed as:

$$d_n = d_{n-1} + \eta \times \Delta d. \tag{3}$$

The advantages and limitations of fuzzy logic control (FLC) in power converter applications are summarized in Table 6.

The implementation of fuzzy logic control (FLC) in power converters, it begins by computing the control error, defined as the difference between the reference voltage and the actual output voltage of the converter. This error is then processed through standard fuzzy logic steps: fuzzification, rule-based (predefined membership functions), and the application of a rule table to determine the system's behavior. After deriving the fuzzy output, defuzzification is applied to convert it into a crisp value, typically the duty cycle, which is then used to control the power converter, ensuring that it tracks the desired reference voltage.

The overall FLC structure is illustrated in the accompanying figure, specifically for DC-DC converter applications, as discussed in references (Lee 1990; Liu and Sen 2005) (Fig. 7).

Table 6 Advantages and limitations of the FLC regulator for power converters

Advantages	Limitations
• No need for a precise system model • Nonlinearity and robustness • Adaptive behavior • Linguistic interpretation	• Complex design process • Higher computational cost • No formal stability guarantees

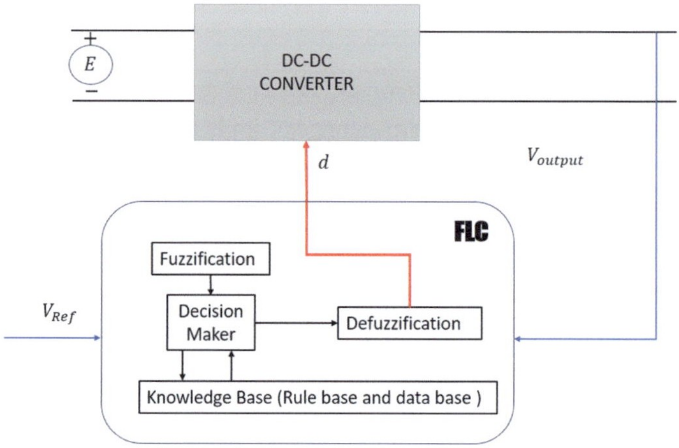

Fig. 7 Fuzzy logic controller (FLC) block diagram for DC-DC converters

3.1.1 Comparison of PID and FLC in Power Converters

Table 7 resumes the comparison of both techniques above for power converter applications.

Both PID and FLC offer valuable solutions for power converter control, each with distinct advantages and drawbacks. While PID control is easy to implement and performs well in stable systems, it struggles with nonlinearities and dynamic variations. Fuzzy logic control, on the other hand, excels in handling nonlinearity and uncertainties but comes with increased computational complexity and design challenges.

Integrating nonlinear control techniques, such as sliding mode control (SMC) and synergetic control, can be explored to balance robustness, adaptability, and efficiency and achieve optimal performance in power converters.

Table 7 Comparison of PID control and FLC for power converters

Feature	PID control	Fuzzy logic control
Mathematical model	Requires a well-defined model	No explicit model needed
Adaptability	Fixed parameters, limited adaptability	Can handle parameter variations and uncertainties
Performance in nonlinear systems	Poor	Good
Computational complexity	Low	Medium to high
Tuning complexity	Requires precise tuning of Kp, Ki, Kd	Requires expert knowledge for rule definition

3.2 Sliding Mode Control

Sliding mode control (SMC) is widely recognized as one of the most powerful and reliable control strategies across a broad range of applications due to its outstanding dynamic performance, inherent robustness, rapid response, and relatively straightforward implementation. These characteristics make SMC particularly well-suited for power converters (Utkin et al. 2017; Utkin 2013). Notably, SMC maintains high control accuracy and system stability even in the presence of external disturbances and model uncertainties, a capability first introduced in the 1970s by Russian mathematician and engineer Vadim Utkin (Linares-Flores et al. 2021; Utkin 1978).

The core principle of sliding mode control lies in driving the system's state trajectory toward a predefined sliding surface and ensuring it remains constrained to this surface throughout the control process. This involves two essential conditions: convergence, where the system state is guided toward the sliding surface, and attractiveness, which ensures the state persists on the surface, thus achieving robust and optimal control performance during the sliding phase (Utkin et al. 2017; Mohd Zaihidee et al. 2019). Figure 8 exhibits the general structure of SMC.

Consider the following nonlinear system

$$\dot{X} = [A(x)]\,[X] + [B]\,[U]. \tag{4}$$

It is necessary to choose "s" sliding surfaces for a vector $[U]$ of dimension "s." Researcher J.J. Slotine proposed a general equation form to determine the sliding surface that ensures the convergence of a state variable x toward its reference value x_{ref}.

$$S = e(x) + \lambda_1 \frac{d}{dt}e(x) + \lambda_2 \frac{d^2}{dt^2}e(x) + \lambda_3 \frac{d^3}{dt^3}e(x) + \cdots\cdots + \lambda_m \frac{d^m}{dt^m}e(x). \tag{5}$$

where:

- x is the variable to be regulated.
- $e(x)$ is the deviation of the variable to be regulated $e = x_{ref} - x$.
- $\lambda_i (i = 1....m)$ are positive constants.

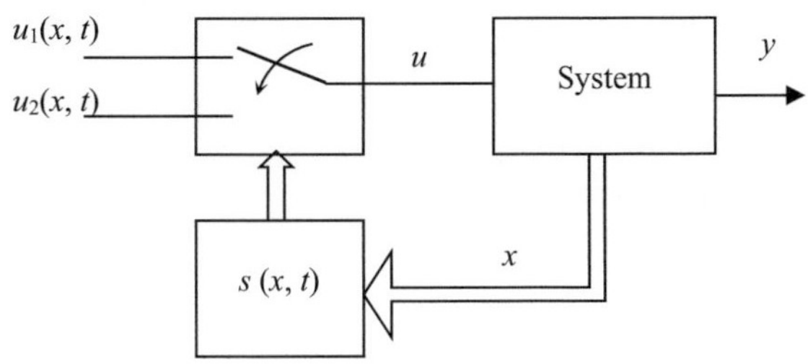

Fig. 8 General structure of SMC

3.3 Convergence Conditions

The convergence conditions allow the system's variables to be controlled so that they converge toward the sliding surface in steady state. Thus, it involves formulating a positive scalar function $V(x)$, called a Lyapunov function for the system's state variables, and selecting a control law u that will decrease this function.

$$\left(\dot{V}(x) < 0\right). \tag{6}$$

Based on the definition of Lyapunov:

$$V(x) = \frac{1}{2}S^2(x). \tag{7}$$

And it derives as:

$$V(x) = S(x).\dot{S}(x). \tag{8}$$

Thus, to obtain a decrease in the Lyapunov equation, it must guarantee the following condition

$$\dot{V}(x) = S(x).\dot{S}(x) < 0. \tag{9}$$

Once the sliding surface is selected, the next step is to determine the control law required to drive the controlled variable toward the surface and subsequently to its equilibrium point (the origin of the phase plane), while ensuring the existence condition of the sliding mode (Lakshmi et al. 2024).

One of the fundamental assumptions in the design of variable structure systems is that the control must switch between the extrema of the control law, $U_{\max}$, $U_{\min}$.

$$u(x,t) = \begin{cases} u_i^+(x,t) & si \quad S(x) > 0 \\ u_i^-(x,t) & si \quad S(x) < 0 \end{cases}. \tag{10}$$

SMC is widely used in power converters for its robustness and fast dynamic response. However, it faces several limitations, the most critical being chattering, high-frequency oscillations caused by discontinuous control switching. Chattering increases switching losses, generates electromagnetic interference (EMI), and degrades efficiency. Other drawbacks include sensitivity to unmatched disturbances, design complexity for high-order systems, steady-state errors under parameter variations, and challenges in low-switching-frequency applications. To address these issues, researchers have developed various chattering reduction methods, each with trade-offs in implementation complexity, steady-state accuracy, and switching behavior.

Several approaches were introduced to mitigate chattering, a high-frequency oscillation issue in sliding mode control (SMC) of power converters (Wu et al.

2021). Hysteresis function and boundary layer methods (Al-Ogaili et al. 2018; Gorji et al. 2019) are simple to implement but suffer from drawbacks like time-varying switching frequencies (hysteresis) or steady-state errors if design parameters (hysteresis band h or boundary layer thickness φ) are poorly chosen. The state observer technique avoids chattering by generating an ideal sliding mode in a secondary observer loop, ensuring the main converter follows a smoothed reference (Wu et al. 2021; Al-Ogaili et al. 2018).

Meanwhile, the super-twisting algorithm (STA) is effective for systems with relative degree one, where the control input directly influences the first derivative of the sliding surface, reducing chattering through continuous control law (Al-Ogaili et al. 2018). These methods offer simplicity and performance, with observer-based and STA approaches offering better precision at the cost of computational complexity.

Power converter-specific methods (Villalón et al. 2023) which provide flexibility but require intricate implementation and lack comprehensive theoretical stability proofs. Fixed switching frequency techniques (Komurcugil et al. 2020) simplify hardware design but may introduce steady-state deviations. Overall, the optimal method depends on the application: hysteresis and boundary layer suit low-cost systems where slight errors are tolerable, while observer-based and STA methods are preferable for high-precision, EMI-sensitive applications. Future research could focus on hybrid approaches (e.g., AI-tuned SMC) to balance robustness and chattering suppression.

Hence, designing a novel technique that keeps the advantage of SMC and reduces chattering much better is required. Thus, a synergetic approach is introduced in the following section.

3.4 Synergetic Control

Synergetic control has emerged as a prominent nonlinear control strategy in recent decades, gaining widespread recognition not only within the robust control community but also in industrial applications. Its practical implementation in power electronics and battery charging systems demonstrates its effectiveness and adaptability in real-world scenarios (Huang et al. 2025; Boumegouas et al. 2024).

At its core, synergetic control theory (SACT) operates on the principle of directional self-organization, leveraging manifold theory to reduce system order through carefully designed manifolds. A key objective of this approach is to ensure that closed-loop systems exhibit desired transient performance while maintaining asymptotic stability near attractors. By integrating a synergetic control algorithm with the system's mathematical model, it becomes possible to develop robust nonlinear controllers tailored to specific dynamic requirements (Kolesnikov 2002; Guermit et al. 2019).

3.4.1 Concept of Synergetic Control

A nonlinear system nth order is described as:

$$\dot{x} = f(x, u, t), \tag{11}$$

where x denotes the state vector, u represents the control input vector, and $f(x, u, t)$ is a nonlinear function describing the system dynamics. The synthesis of a synergetic controller follows a fully analytical procedure, which involves the following key steps (Boumegouas and Kouzi 2022; Kolesnikov 2002; Guermit et al. 2019):

The first step in synergetic control design is the formulation of macro-variables. These are constructed as algebraic functions of the system's state variables, capturing the essential design requirements. The macro-variables guide the system from an arbitrary initial state toward a desired equilibrium, defined by the manifold.

$$\psi = \psi(x, t). \tag{12}$$

With ψ representing a user-defined macro-variable expressed as $\psi(x, t)$. The synergetic controller drives the system to operate on the manifold $\psi = 0$, with the designer able to tailor the macro-variable's characteristics to meet specific control requirements.

In the second design step, the dynamic evolution of the macro-variable on the manifold ($\psi = 0$) is governed by a functional equation. This equation, which defines the system's convergence behavior, typically takes the following general form (Kolesnikov 1994; Boumegouas et al. 2022):

$$T\dot{\psi} + \psi = 0. \tag{13}$$

With T > 0.

T: designates the designer chosen speed convergence to the desired manifold.

Differentiating the macro-variable (1.10) along (1.9) leads to (1.12):

$$\dot{\psi} = \frac{d\psi}{dx}\dot{x}. \tag{14}$$

By combining (1.9), (1.11), and (1.12), the following equation is obtained:

$$T\frac{d\psi}{dx}f(x, u, t). \tag{15}$$

The control law is synthesized by combining Eq. (1.13) with the system's dynamic model, yielding the resulting expression (1.14):

$$u = u(x, \psi(x, t), T, t). \tag{16}$$

Equation (1.12) reveals that the control action depends on three key elements: (1) the system's state variables, (2) the defined macro-variable, and (3) the control parameter T. This relationship provides the designer with two adjustable parameters to shape the controller's performance: the macro-variable configuration and the time constant T.

The described methodology offers several practical advantages:

- The systematic design procedure lends itself to straightforward implementation for automated control law synthesis.
- Proper selection of macro-variables can ensure:
  - Global stability of the control system.
  - Robustness to parameter variations.
  - Effective noise suppression capabilities.

3.5 Model Predictive

Model predictive control (MPC) has become a cornerstone in power converter applications due to its ability to handle multivariable systems with constraints and nonlinearities, offering fast dynamic responses and precise control. In power electronics, MPC is particularly advantageous for managing converters with multiple switches and states, such as current, voltage, power, and torque, while easily incorporating system constraints and nonlinearities. This makes MPC suitable for high-power converters and industrial drives, where it enhances performance by optimizing control actions over a finite horizon (Domański 2020; Quevedo et al. 2019).

Generalized predictive control (GPC), a variant of MPC, utilizes a transfer function model, specifically, the Controlled Auto-Regressive Integrated Moving Average (CARIMA) model, to predict future outputs and compute control actions by minimizing a quadratic cost function over a specified prediction horizon. GPC's structure allows for straightforward tuning of control and prediction horizons, providing flexibility in controller design. While both MPC and GPC share the receding horizon principle, GPC distinguishes itself by its analytical solution without constraints and its suitability for systems where an explicit state-space model is challenging to obtain. In contrast, standard MPC often requires solving a constrained optimization problem at each control step, which can be computationally intensive. Therefore, the choice between MPC and GPC depends on the specific application requirements, including model availability, computational resources, and the nature of system constraints (Domański 2020; Quevedo et al. 2019; Yu and Long 2024). Figure 9 illustrates the predictive control structure design for power converter applications.

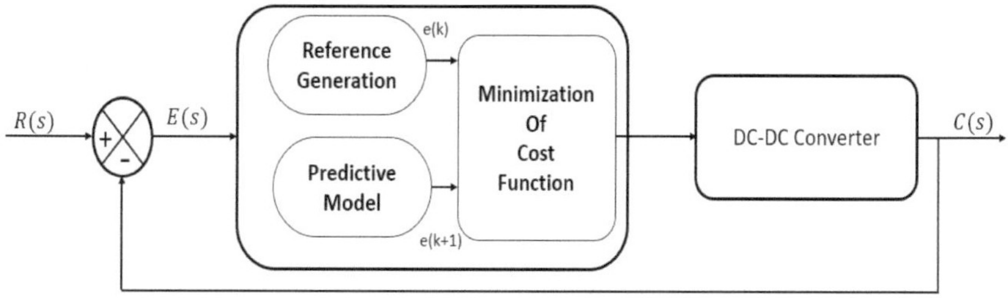

Fig. 9 Predictive control structure for power converter applications

3.6 Comparison of the Nonlinear Controls for Power Converters

These advanced control strategies demonstrate the significant potential of intelligent control systems in improving the performance, robustness, and adaptability of power converters. A key insight from contemporary research is that the efficiency and stability of power electronic systems are strongly influenced by the suitability and effectiveness of the employed control technique. Consequently, the controller design must carefully consider the full range of operating scenarios, particularly in grid-connected configurations, to ensure robust and reliable system performance.

Each control method offers distinct advantages and suffers from some limitations, making their applicability highly dependent on the requirements of the target application. MPC excels in providing optimal control when system dynamics are well known, while FLC enables intuitive management of nonlinearities and uncertain behaviors through rule-based logic. Synergetic control offers a balanced compromise, delivering smooth and globally stable control grounded in rigorous mathematical formulation. SMC, on the other hand, is widely recognized for its inherent robustness, low sensitivity to model inaccuracies, and strong disturbance rejection.

While both SMC and SC are known for their effectiveness in handling nonlinear dynamics, SC is increasingly regarded as a powerful alternative to SMC due to its continuous control law and reduced chattering. To further explore their relative merits and trade-offs, the next section presents a detailed comparison between SMC and SC in the context of power converter applications.

3.6.1 Robustness and Adaptability

Both SMC and SC are well-known for their robustness. SMC stands out for its inherent robustness to model uncertainties and external disturbances due to its discontinuous nature. However, this same property leads to high-frequency switching, known as chattering, which can damage power electronic components or cause inefficiency.

SC also offers strong robustness but relies more heavily on accurate modeling to design appropriate macro-variables and evolution laws. It tends to produce smoother control signals and thus avoids the chattering problem, making it a more hardware-friendly option, especially in converter applications where switching stress must be minimized.

3.6.2 Implementation and Real-World Suitability

In terms of implementation, SMC is moderately complex and requires the design of a suitable sliding surface, alongside stability analysis, often using Lyapunov methods. Its simplicity in theory can be complicated in practice by the need to reduce chattering.

SC has a higher initial design complexity because it involves constructing macro-variables and determining evolution laws that guarantee convergence. Nevertheless, the resulting control is typically continuous and easier on actuators and switches in real-time embedded systems.

3.6.3 Application to Power Converters

Tables 8 and 9 provide a comparative summary of the SMC and SC, highlighting their respective strengths, limitations, and suitability for power converter applications.

Table 8 Comparative features of nonlinear control for power converters

Feature	Sliding mode control (SMC)	Mode predictive control (MPC)	Synergetic control (SC)
Common use cases	Buck, boost, buck-boost, inverters, multi-level converters	Power converters, grid-connected systems, energy management, and motor control	Similar applications: DC-DC converters, inverter control, multilevel topologies
Performance	Fast response, excellent voltage/current regulation	Excellent dynamic performance, optimal control over the prediction horizon	Comparable performance, smoother trajectories, sometimes faster convergence
Implementation difficulty	Moderate; requires surface definition and stability proof via Lyapunov	High; requires accurate models, online computation, and real-time optimization	Slightly more complex due to macro-variable design and evolution law derivation
Switching harmonics	High-frequency switching (chattering) is an issue	Typically smooth switching; can be optimized for reduced harmonic distortion	Less chattering due to its continuous nature, preferable for power electronics

Table 9 Nonlinear controllers' criterion comparative

Criterion	Sliding mode control	Predictive control	Synergetic control
Design complexity	Moderate	Higher	Higher
Robustness	Very high	Moderate	High
Sensitivity to parameters	Low	High (depends on model precision)	Moderate
Chattering	Yes (main issue)	None (smooth control actions)	Minimal
Control smoothness	Discontinuous	Continuous/smooth	Continuous/smooth
Response time	Fast	Very fast (optimized for each sampling interval)	Fast/faster (in some cases)
Preferred use when	High robustness is needed with acceptable switching noise	Fast dynamic response and precise tracking are critical	High precision and smoothness are needed

4 Applications Examples

This section presents a practical example illustrating the application of both linear and nonlinear control techniques to a buck-boost converter. It begins with the mathematical modeling of the converter and subsequently examines the implementation of two control strategies: proportional-integral (PI) control and sliding mode control (SMC) (Duong et al. 2017; Guldemir 2011; Ahmed et al. 2003).

4.1 Mathematical Model of Buck-Boost Converter

A buck-boost converter is a versatile DC-DC power conversion device capable of regulating the output voltage to levels either higher or lower than the input voltage. Notably, the output voltage polarity is inverted relative to the input. The fundamental topology of the converter is illustrated in Fig. 10 (Guldemir 2011). The circuit is composed of a DC input voltage source (E), a controllable switch (S), a diode (D), a filter inductor (L), a filter capacitor (C), and a load resistor (R). Each of these components plays a critical role in shaping the energy transfer and ensuring stable voltage regulation across a wide range of operating conditions (Fig. 11).

The state-space mathematical model of the buck-boost DC-DC converter is derived using the fundamental principles governing its operation. For this purpose, the circuit is redrawn in Fig. 12 (Guldemir 2011), to facilitate the formulation of the state equations, incorporating the control input u. This control input u represents the switching function of the converter and is a discrete-valued signal that takes values from the set {0, 1}, corresponding to the ON and OFF states of the switch.

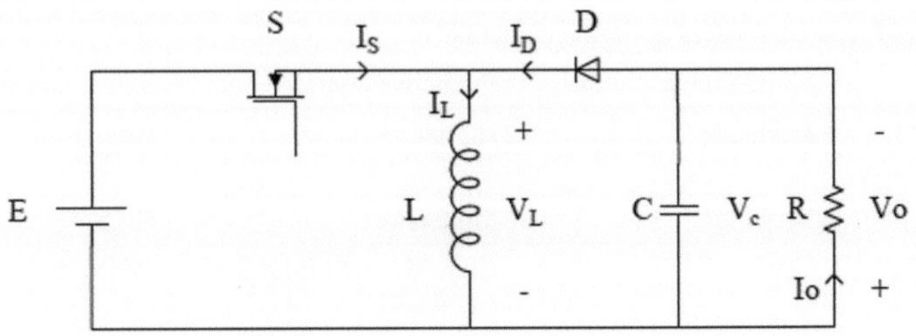

Fig. 10 Buck-boost synoptic diagram

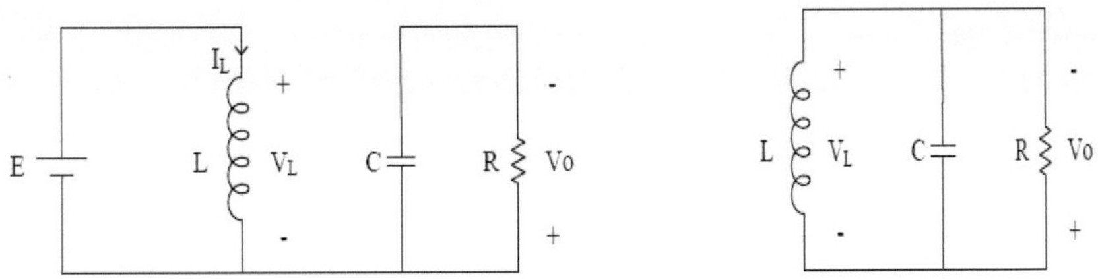

Fig. 11 Buck-boost diagram in ON/OFF switching

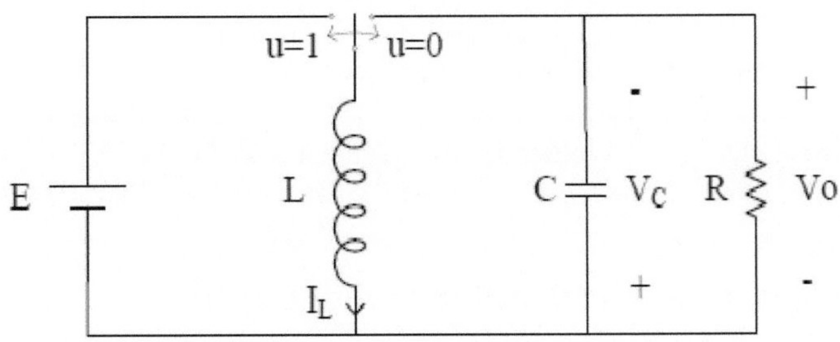

Fig. 12 Buck-boost converter diagram incorporating control signal u

The dynamic behavior of the converter is analyzed under the assumption that it operates in Continuous Conduction Mode (CCM), where the current through the inductor never falls to zero throughout the switching cycle.

The mathematical model is based on the control input (u). When (u = 1) switching is ON, the model in state space is expressed as (Guldemir 2011; Ahmed et al. 2003):

$$\begin{aligned} \frac{di_L}{dt} &= \frac{E}{L} \\ \frac{dv_c}{dt} &= \frac{V_0}{RC}. \end{aligned} \tag{17}$$

Thus, when (u = 0) switching is OFF, it is written as (Guldemir 2011; Ahmed et al. 2003):

$$\begin{aligned}\frac{di_L}{dt} &= \frac{V_c}{L} \\ \frac{dv_c}{dt} &= -\frac{i_L}{C} - \frac{V_c}{RC}.\end{aligned} \tag{18}$$

And by combining these above states, the buck-boost model in state-space is given as follows (Guldemir 2011; Ahmed et al. 2003):

$$\begin{aligned}\frac{di_L}{dt} &= (1-u)\frac{V_c}{L} + \frac{E}{L}u \\ \frac{dv_c}{dt} &= -(1-u)\frac{i_L}{C} - \frac{V_c}{RC}.\end{aligned} \tag{19}$$

4.1.1 Simulation Results

This section presents the simulation results of the buck-boost converter under a constant duty cycle, which is altered at $t = 0.5$s to observe the dynamic response. Meanwhile, Table 10 summarizes the key parameters used in the buck-boost converter model.

Figure 13 illustrates the output voltage of a buck-boost converter operating without control, with a duty cycle shift from 0.25 to 0.4 at t = 0.5 s. Initially, the output stabilizes around (-6.7 V), and after the duty cycle change, it transitions to approximately (−13.3 V). The response shows noticeable oscillations after the transition and lacks damping due to the absence of feedback. Overall, the converter follows the expected voltage levels based on duty cycle changes but suffers from poor dynamic behavior and sensitivity to disturbances, highlighting the limitations of open-loop operation for regulated applications.

Table 10 Buck-boost converter parameters

Parameter	Value
E	20 V
L	1 mH
C	110 μF
R	15 Ω
Vref	15–25 V
F	10 kHz

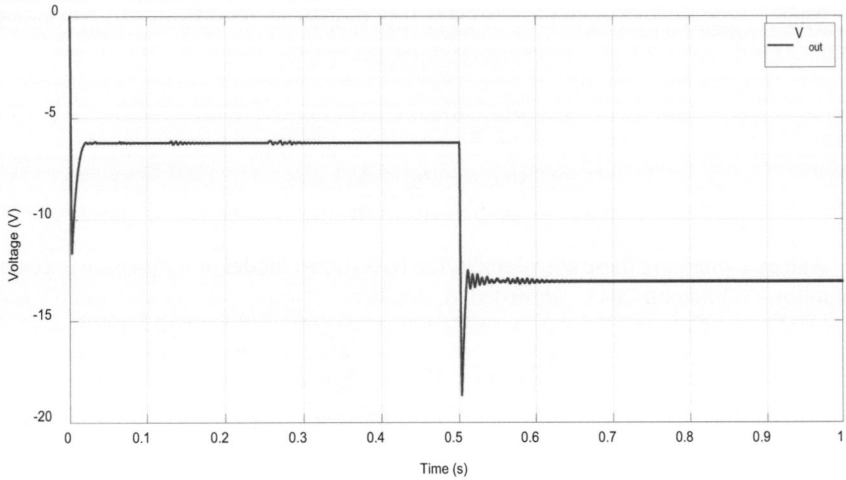

Fig. 13 Buck-boost performance with duty cycle variations

4.2 PI Regulator

The PI regulator operates by adjusting the duty cycle to maintain the output voltage at a desired reference level. It functions by calculating the error between the reference and the actual output voltage and then applying proportional and integral actions to minimize this error over time. Implementation involves developing a suitable system model and tuning the controller gains to achieve the desired dynamic response. In this section, the synthesis and implementation of the PI regulator for the buck-boost converter are presented in detail.

By using the Laplace transformation, the second term in Eq. (1.19) is written as:

$$\dot{\tilde{v_c}} = \frac{i_L}{C}\tilde{u} - \frac{(1-\tilde{u})}{C}i_L - \frac{1}{RC}\tilde{v_c}. \tag{20}$$

The objective is to obtain a first-order transfer function given as:

$$H_b = \frac{v_c}{u}. \tag{21}$$

Then,

$$H_b = \frac{v_c}{u} = \frac{K_b}{T_b s + 1}, \tag{22}$$

where

$$K_b = \frac{i_L}{C}/(\frac{1}{RC}), \tag{23}$$

$$T_b = \frac{1}{RC}. \tag{24}$$

Meanwhile, design the control (PI regulator) for the buck-boost converter. The law equation of the PI regulator and by using the Laplace transformation, we have:

$$u(s) = K_p e(s) + \frac{1}{s} K_i \int e(s). \tag{25}$$

Thus,

$$\frac{u(s)}{e(s)} = G_s = K_p + \frac{1}{s} K_i. \tag{26}$$

From the above equation, we can write:

$$G_s = K_P\left(1 + \frac{1}{T_i s}\right). \tag{27}$$

The closed-loop transfer function is given as:

$$TF_B = \frac{G_s H_b}{1 + G_s H_b}. \tag{28}$$

$$TF_B = \frac{K_P\left(1 + \frac{1}{T_i s}\right)\left(\frac{K_b}{T_b s+1}\right)}{1 + K_P\left(1 + \frac{1}{T_i s}\right)\left(\frac{K_b}{T_b s+1}\right)}. \tag{29}$$

After simplifying the above equation, we obtain:

$$TF_B = \frac{T_i s + 1}{\frac{T_i T_b}{K_p K_b} s^2 + T_i\left(1 + \frac{1}{K_p K_b}\right) s + 1}. \tag{30}$$

This equation is a second-order transfer function; the ordinary function that is based on bandwidth and damping ratio is expressed as:

$$TF_B = \frac{\omega_n^2}{s^2 + 2\xi\omega_n s + \omega_n^2}. \tag{31}$$

We obtained the gains of the PI regulator are written as:

$$K_p = \frac{2\xi\omega_n T_b - 1}{K_b}, \tag{32}$$

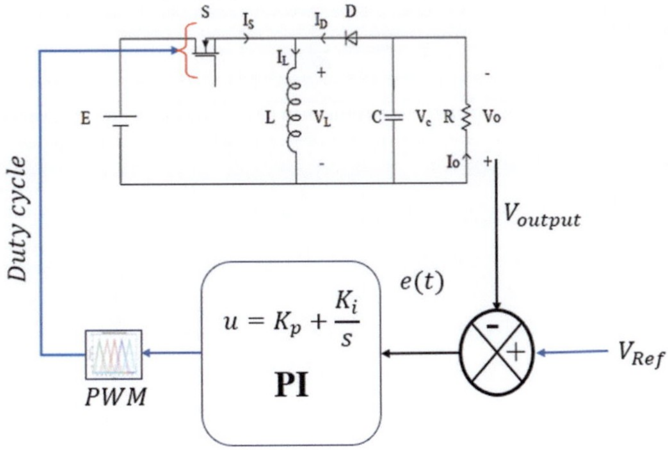

Fig. 14 PI regulator structure for DC-DC converter

$$T_i = \frac{2\xi\omega_n T_b - 1}{\omega_n^2 T_b}. \tag{33}$$

Figure 14 illustrates the structural diagram of the PI regulator as applied to the buck-boost converter, highlighting its key components and feedback control loop.

4.2.1 Simulation Results

The reference voltage was set to 15 V, while the input voltage is maintained at 20 V. To assess the robustness of the control algorithm, additional tests are conducted by varying key circuit parameters, specifically, increasing and decreasing the output capacitor and inductor values by 50%, respectively. Furthermore, scenarios involving changes in the reference voltage are examined to evaluate the controller's ability to maintain regulation and stability under dynamic operating conditions.

4.2.2 Discussion

Based on the above results, the proposed controller successfully tracks the reference voltage under nominal conditions, as shown in Fig. 15. The system exhibits a moderate overshoot and a satisfactory settling time, indicating reliable performance of the PI regulator. In scenarios involving variable reference voltages, the controller maintains consistent tracking behavior with acceptable dynamic response. As illustrated in Fig. 18, the PI regulator achieves stable operation with controlled overshoot and settling times across both buck and boost modes, demonstrating its effectiveness in handling reference changes.

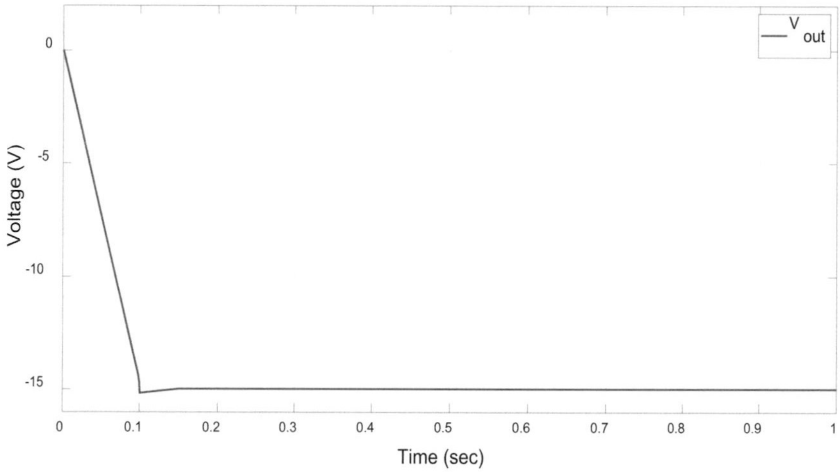

Fig. 15 PI regulator performance for buck-boost converter

These outcomes highlight the reliable performance of the buck-boost converter under linear control using the PI regulator. To further evaluate the effectiveness and durability of this control strategy, a robustness test was conducted by varying key passive components in the converter model.

As illustrated in Figs. 16 and 17, a $\pm$ 50% variation in the inductor and the output capacitor values has a noticeable effect on the system's dynamic behavior. An increase in capacitance results in a longer settling time but contributes to reduced overshoot, enhancing stability at the expense of response speed. Conversely, a reduction in capacitance shortens the settling time but leads to a more pronounced overshoot. In contrast, changes in inductance primarily influence the overshoot, with minimal impact on settling time. Despite these variations, the converter maintains acceptable performance, demonstrating a reasonable level of robustness; however, the sensitivity to parameter changes highlights certain limitations of the PI control strategy under non-ideal conditions.

From the above analysis, it can be concluded that the PI regulator ensures satisfactory performance by maintaining voltage tracking with moderate overshoot and settling time. However, its sensitivity to parameter variations, evident from the robustness test, represents a notable drawback of this linear control approach. Consequently, to address these limitations and enhance performance under varying conditions, a nonlinear control technique will be introduced in the following section for improved regulation of the buck-boost converter.

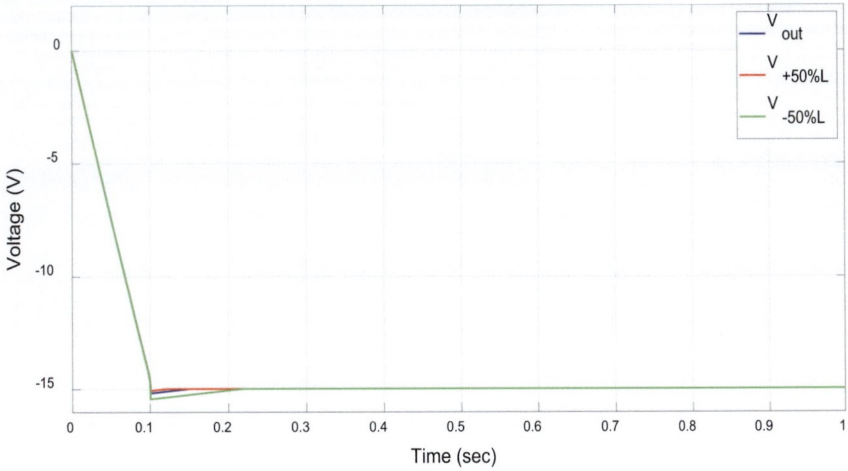

Fig. 16 PI robustness performance with inductance variations

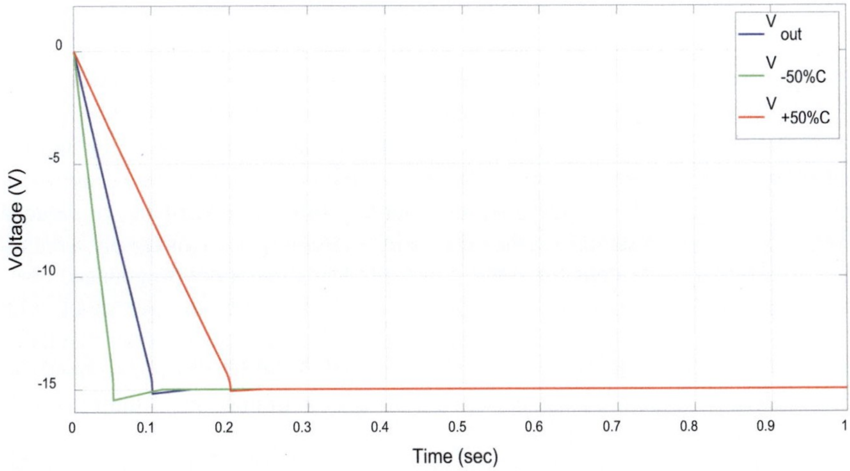

Fig. 17 PI robustness performance with capacitor variations

4.3 *Sliding Mode Control*

The sliding mode control uses its law to generate the desired duty cycle for the buck-boost converter to buck or boost the output voltage to the reference voltage, which selected as the sliding surface. And by fast tracking and high robustness by remain the system remains in this designed sliding surface.

The sliding mode synthesis is carried out, beginning with the design of the sliding surface, which describes the desired performance. Controlling the output voltage

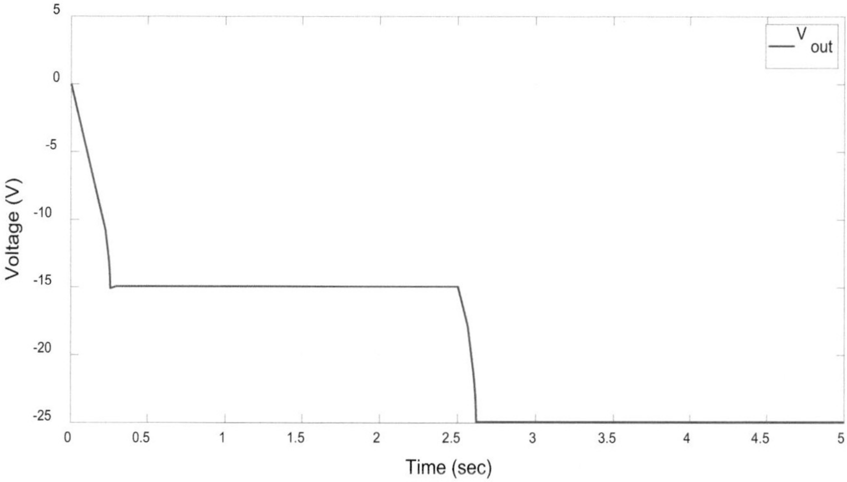

Fig. 18 PI robustness performance with voltage variations

is achieved by generating the required duty cycle for the buck-boost to follow the reference voltage. Therefore, the sliding surface is expressed as:

$$s = k_1\left(x_1 - x_{1-ref}\right) + k_2\left(x_2 - x_{2-ref}\right). \quad (34)$$

The law of the SMC is written as:

$$u = u_{eq} + u_n. \quad (35)$$

The first term is based on the system model. Meanwhile, the second term is responsible for the robustness of this technique by using the continuous function, which is given as:

$$u_n = k_3 sign(s). \quad (36)$$

The equivalent term is obtained by:

$$s\dot{s} = 0 \quad (37)$$

By replacing Eq. 1.19 into Eq. 1.36, we have:

$$u_{eq} = \frac{-k_1 RCx_2 + k_2 L(Rx_1 - x_2)}{k_1 RC(E - x_2) + k_2 LRx_1}. \quad (38)$$

Therefore, the control law of the SMC for the buck-boost converter is expressed as:

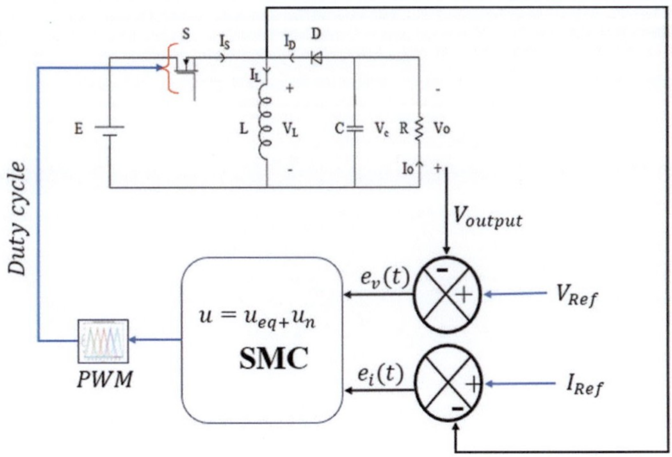

Fig. 19 SMC structure for DC-DC converter

$$u_{control} = \frac{-k_1 RCx_2 + k_2 L(Rx_1 - x_2)}{k_1 RC(E - x_2) + k_2 LRx_1} + k_3 sign(s). \tag{39}$$

Figure 19 presents the structural block diagram of the SMC scheme, outlining the key components involved in the control loop.

4.3.1 Simulation Results

The simulation is conducted under the same operating conditions. Furthermore, the reference voltage is varied to examine the controller's ability to maintain accurate voltage regulation and dynamic stability under changing operating conditions.

4.3.2 Discussion

The results obtained indicate that the buck-boost converter, under sliding mode control (SMC), accurately follows the reference voltage without exhibiting any overshoot, as shown in Fig. 20. The response is characterized by improved settling time, demonstrating fast convergence to the desired output. During the variable reference voltage test, the controller maintains precise tracking performance across both buck and boost modes, ensuring stable output voltage transitions with high dynamic response and no overshoot, as observed in Fig. 21.

This behavior confirms the effectiveness of SMC in ensuring high-performance regulation for the buck-boost converter. To further assess the controller's resilience, a robustness test was performed by introducing variations in the system parameters, including output capacitance, inductance, and input speed.

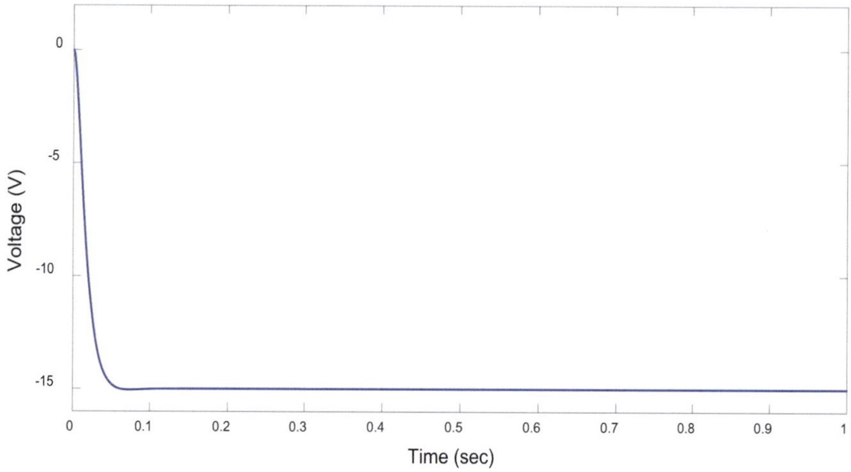

Fig. 20 SMC performance for buck-boost converter

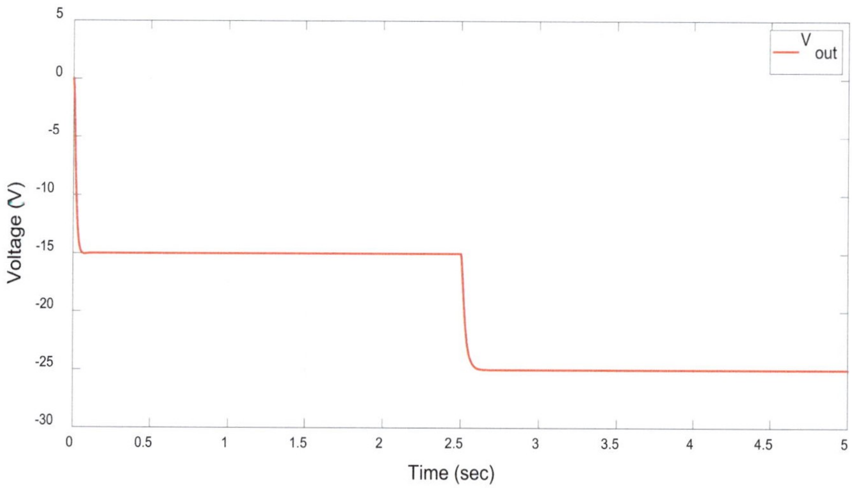

Fig. 21 SMC robustness performance with voltage variations

As illustrated in Figs. 22 and 23, a 50% variation in the output capacitor shows minimal impact on the settling time. However, when the capacitance is increased, a slight overshoot appears, still within acceptable limits. Conversely, a reduction in capacitance does not produce noticeable deterioration in performance. In the case of inductance variation, the response remains largely unaffected, with almost no change in both settling time and overshoot, indicating a strong tolerance to inductive parameter shifts.

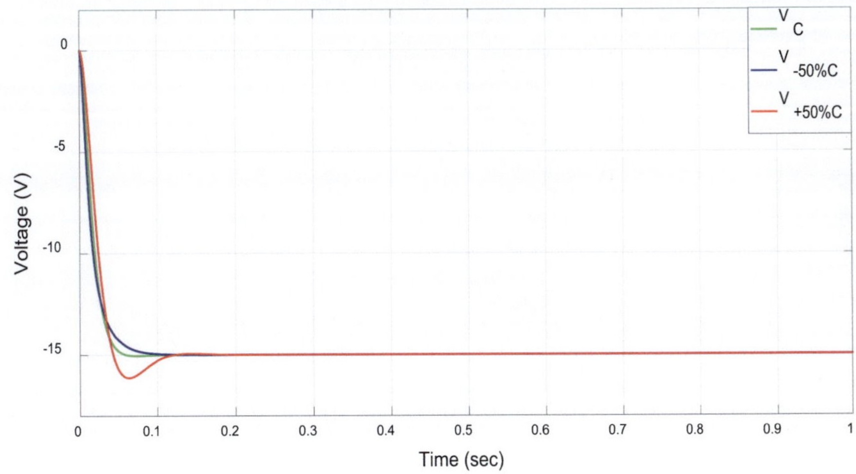

Fig. 22 SMC robustness performance with capacitor variations

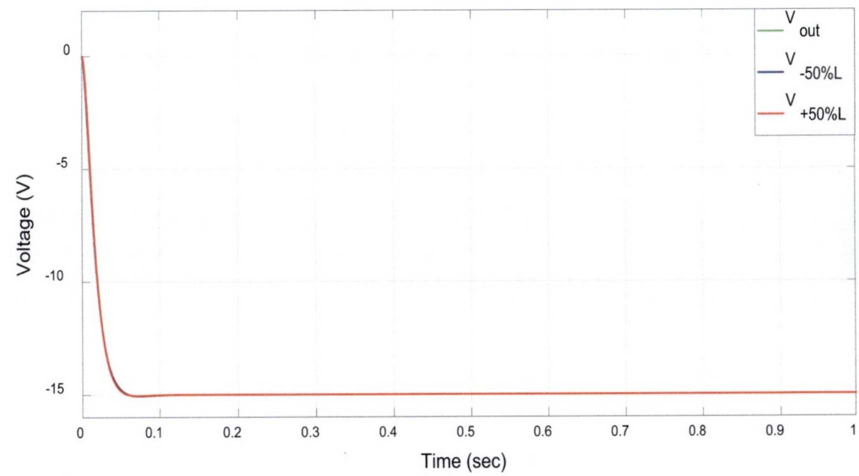

Fig. 23 SMC robustness performance with inductance variations

Additionally, in tests involving variations in input speed, the controller maintains consistent tracking and dynamic behavior. The voltage output remains stable, and the system continues to operate with high precision and fast response, confirming the robustness of the control strategy under different operating conditions.

Overall, the results validate the suitability of sliding mode control for the buck-boost converter application, showcasing its ability to deliver fast, stable, and robust performance even in the presence of system uncertainties and parameter variations.

4.3.3 Comparison Between PI and SMC

The PI controller applied to the buck-boost converter demonstrates acceptable performance in tracking the reference voltage under both steady-state and variable reference conditions. It provides moderate settling time and maintains overshoot within acceptable limits during both buck and boost modes. However, its performance degrades under parameter variations, as revealed in the robustness tests. Specifically, changes in the output capacitance significantly affect the dynamic response: Increasing capacitance results in a longer settling time and reduced overshoot, while decreasing it causes faster settling but with higher overshoot. The inductance variation has a minimal effect on settling time but introduces small changes in overshoot. These results highlight a key limitation of the PI controller, its sensitivity to model uncertainties and parameter drift, which makes it less reliable in dynamically changing environments without retuning.

In contrast, the sliding mode controller (SMC) exhibits superior dynamic performance and robustness. It ensures accurate voltage tracking with no overshoot and a faster settling time, even under variable reference conditions. The robustness tests further confirm its resilience: Variations in the capacitor have minimal impact on settling time, and any introduced overshoot remains within acceptable bounds when the capacitance is increased. Changes in inductance and input speed produce negligible effects on both settling time and overshoot, demonstrating the controller's strong immunity to system uncertainties. This behavior reflects the inherent robustness and high-performance characteristics of SMC, making it a more suitable choice for applications requiring fast, stable, and reliable voltage regulation in the presence of disturbances or parameter changes.

5 Conclusion

In this chapter, an overview is carried out on various control techniques applied to power converters, namely PID, fuzzy logic control (FLC), sliding mode control (SMC), synergetic control (SC), and model predictive control (MPC). These techniques demonstrate remarkable potential in significantly improving the performance, robustness, and adaptability of power converters. Therefore, it is imperative to design control systems that comprehensively address a wide range of operating scenarios, especially in grid-connected applications, to ensure optimal reliability, resilience, and system integrity. Furthermore, each technique exhibits distinct advantages and limitations in terms of complexity, robustness, adaptability, and implementation feasibility.

PID control is widely used in industrial power converter applications due to its simplicity, ease of implementation, and reliable performance under nominal conditions. However, its effectiveness diminishes in the presence of nonlinearities, parameter variations, and external disturbances. Fuzzy logic control (FLC) offers a model-free alternative that emulates human reasoning, providing improved adaptability and robustness in uncertain or nonlinear environments. Despite these strengths, FLC may fall short in achieving high dynamic precision and rapid response.

Advanced techniques like model predictive control (MPC), synergetic control (SC), and sliding mode control (SMC) address many limitations of PID and FLC. Synergetic control overcomes this issue by providing a smoother, globally stable response through invariant manifold design. Though more complex to implement, SC combines high performance and hardware compatibility, making it a promising approach for modern converter systems. While model predictive control (MPC) offers optimal performance under known operating conditions, it also excels in handling multi-variable systems with constraints. Its predictive nature enables proactive adjustment, enhancing the converter's dynamic response and efficiency. Meanwhile, sliding mode control (SMC) surpasses many control methods with its exceptional robustness against system uncertainties and external disturbances. It ensures fast, precise regulation and maintains high performance even under varying or harsh operating conditions.

In summary, while PID remains popular for its simplicity, ease of implementation, and intuitive design, its performance can be limited under nonlinear, time-varying, or uncertain conditions. In contrast, advanced nonlinear control techniques, such as fuzzy logic control (FLC), sliding mode control (SMC), synergetic control (SC), and model predictive control (MPC), offer significantly enhanced robustness, adaptability, and dynamic precision. SMC stands out for its resilience to uncertainties and disturbances, SC ensures globally stable and smooth responses, and MPC excels in handling constrained multivariable systems through predictive optimization. Therefore, the selection of an appropriate control strategy should be guided by the specific demands of the power converter application, including dynamic performance, implementation complexity, and required robustness under real-world operating conditions.

References

Abbas IA, Mustafa MK (2024) A review of adaptive tuning of PID-controller: optimization techniques and applications. Int J Nonlinear Anal Appl 15(2):29–37

Ahmed M, Kuisma M, Pyrhonen O, Silventoinen P (2003) Sliding mode control for buck-boost converter using control desk dSPACE. In: Proceedings of the international conference on power electronics and drive systems (PEDS), vol 2, pp 1491–1494

Al-Dabbagh ZA, Shneen SW, Hanfesh AO (2024) Fuzzy logic-based PI controller with PWM for buck-boost converter. J Fuzzy Syst Control 2(3):147–159

Al-Ogaili AS et al (2018) A review of different control methods in power electronics. J Eng Appl Sci 13:2242–2253

Bendaoud K et al (2017) Fuzzy logic controller (FLC): application to control DC-DC buck converter. In: Proceedings of international conference on engineering management and information science (ICEMIS), pp 1–5

Boumegouas MKB, Kouzi K (2022) A new synergetic scheme control of electric vehicle propelled by six-phase permanent magnet synchronous motor. Adv Electr Electron Eng 20(1):1

Boumegouas MKB, Kouzi K, Birame MH (2023) Design of a robust nonlinear controller for electric vehicles driven by a six-phase permanent-magnet synchronous motor. Elektrotehniski Vestnik 90(1/2):23–31

Boumegouas MKB, Kouzi K, Birame MH (2022) Investigate the performance of synergetic control for battery electric vehicle. In: The international conference on advanced renewable energy systems. Springer Nature Singapore, Singapore, pp. 431–440

Boumegouas MKB, Ilten E, Kouzi K, Demirtas M, M'hamed B (2024) Application of a novel synergetic observer for PMSM in electrical vehicle. Electr Eng 106(5):5507–5521

Carrol E (2004) In: Brandt D (ed) Proceedings of the CERN-CAS accelerator school on power converters for particle accelerators. Warrington, UK. CERN-2006–010 (CERN, Geneva, 2006), pp 43–56

Chung HSH, Wang H, Blaabjerg F, Pecht M (eds) (2024) Reliability of power electronic converter systems. The Institution of Engineering and Technology

Domański PD (2020) Performance assessment of predictive control—a survey. Algorithms 13(4):97

Duong MQ, Sava GN, Scripcariu M, Mussetta M (2017) Design and simulation of PI-type control for the Buck Boost converter. In: Proceedings of the international conference on energy and environment (CIEM), pp 79–82

Firmansyah R, Irmawanto R (2021) Comparison study of PI controller tuning method to regulate the DC motor speed. In: 2021 3rd International conference on research and academic community services (ICRACOS), pp 93–97

Gorji SA, Sahebi HG, Ektesabi M, Rad AB (2019) Topologies and control schemes of bidirectional DC–DC power converters: an overview. IEEE Access 7:117997–118019

Guermit H, Kouzi K, Bessedik SA (2019) Novel design of an optimized synergetic control for dual stator induction motor. COMPEL 38(6):1828–1845

Guldemir H (2011) Modeling and sliding mode control of dc-dc buck-boost converter. In: Proceedings of the international advanced technology symposium (IATS), vol 4, pp 475–480

Huang D, Zhao Q, Li R et al (2025) Fixed-time synergetic control of multi-interior permanent magnet synchronous motor traction system with dynamic adhesion. IET Control Theory Appl 19(1):e70030

Isdaryani F, Feriyonika F, Ferdiansyah R (2020) Comparison of Ziegler-Nichols and Cohen Coon tuning method for magnetic levitation control system. In: Journal of physics: conference series, vol 1450, no 1. IOP Publishing, pp 012033

Kawamura A, Haneyoshi T, Hoft RG (1988) Deadbeat controlled PWM Inverter with parameter estimation using only voltage sensor. IEEE Trans Power Electron 3(2):118–125

Kolesnikov AA (1994) Synergetic control theory. Moscow, Energoatomizdat, pp 5–34

Kolesnikov A et al (2002) Synergetic synthesis of DC-DC boost converter controllers: Theory and experimental analysis. In: Proceedings of IEEE applied power electronics conference and exposition (APEC), vol 1, pp 409–415

Komurcugil H, Biricik S, Bayhan S, Zhang Z (2020) Sliding mode control: overview of its applications in power converters. IEEE Ind Electron Mag 15(1):40–49

Lakshmi GS, Sreeshobha E, Swarupa ML, Pandraka VK (2024) Bi-directional DC-DC converter and fuzzy logic controller based switched reluctance motor for EV applications. Przegląd Elektrotechniczny 100

Lee CC (1990) Fuzzy logic in control systems: fuzzy logic controller. I. IEEE Trans Syst Man Cybern 20(2):404–418

Linares-Flores J, Juárez-Abad JA, Hernandez-Mendez A, Castro-Heredia O, Guerrero-Castellanos JF, Heredia-Barba R, Curiel-Olivares G (2021) Sliding mode control based on linear extended state observer for DC-to-DC buck–boost power converter system with mismatched disturbances. IEEE Trans Ind Appl 58(1):940–950

Liu YF, Sen PC (2005) Digital control of switching power converters. In: Proceedings of 2005 IEEE conference on control applications, pp 635–640

Liu X, Qiu L, Fang Y et al (2022) A fuzzy approximation for FCS-MPC in power converters. IEEE Trans Power Electron 37(8):9153–9163

Maroti PK, Padmanaban S, Bhaskar MS, Ramachandaramurthy VK, Blaabjerg F (2022) The state-of-the-art of power electronics converters configurations in electric vehicle technologies. Power Electron Devices Compon 1:100001

Mizushima S, Kawamura A, Yuzurihara I, Takayanagi A, Ohma R (2015) Development of a DC-DC converter for high/low pulse voltage output using deadbeat control. In: 2015 IEEE 2nd international future energy electronics conference (IFEEC), pp 1–6

Mohd Zaihidee F, Mekhilef S, Mubin M (2019) Robust speed control of PMSM using sliding mode control (SMC)—a review. Energies 12(9):1669

Nippatla VR, Mandava S (2025) Performance analysis of permanent magnet synchronous motor based on transfer function model using PID Controller tuned by Ziegler-Nichols method. Results Eng 105460

Quevedo DE, Aguilera RP, Geyer T (2019) Model predictive control for power electronics applications. Handbook of model predictive control, pp 551–580

Rashid MH (ed) (2017) Power electronics handbook. Butterworth-Heinemann

Rashid MH (2013) Power electronics: devices, circuits, and applications, 4th edn. Pearson Education, Harlow

Rufer, A (2004) In: Brandt D (ed) Proceedings of the CERN-CAS accelerator school on power converters for particle accelerators, Warrington. CERN-2006–010, CERN, Geneva, 2006, pp 57–102

Sahoo SK, Kishore NK (2020) Battery state-of-charge-based control and frequency regulation in the MMG system using fuzzy logic. IET Gener Transm Distrib 14:2698–2709

Sasikala K, Kumar RK (2019) Design and analysis of boost DC-DC converter with PI controller

Seo S-W, Choi HH (2019) Digital implementation of fractional order PID-type controller for boost DC–DC converter. IEEE Access 7:142652–142662

Suksawat T, Kaewpradit P (2021) Comparison of Ziegler-Nichols and Cohen-Coon tuning methods: implementation to water level control-based MATLAB and Arduino. Eng J Chiang Mai Univ 28(1):153–168

Tang HH, Ahmad NS (2024) Fuzzy logic approach for controlling uncertain and nonlinear systems: a comprehensive review of applications and advances. Syst Sci Control Eng 12(1):2394429

Utkin VI (1978) Sliding modes and their applications in variable structure systems. Mir, Moscow

Utkin VI (2013) Sliding modes in control and optimization. Springer

Utkin V, Guldner J, Shi J (2017) Sliding mode control in electro-mechanical systems. CRC Press

Villalón A, Muñoz C, Muñoz J, Rivera M (2023) Fixed-switching-frequency modulated model predictive control for islanded AC microgrid applications. Mathematics 11(3):672

Wu L, Liu J, Vazquez S, Mazumder SK (2021) Sliding mode control in power converters and drives: a review. IEEE/CAA J Autom Sinica 9(3):392–406

Wu L, Liu J, Vazquez S, Mazumder SK (2021) Sliding mode control in power converters and drives: a review. IEEE/CAA J Autom Sin 9(3):392–406

Yu Z, Long J (2024) Review on advanced model predictive control technologies for high-power converters and industrial drives. Electronics 13(24):4969

Digital Control of Power Electronic Systems

Tiago Davi Curi Busarello

Abstract This chapter presents fundamental and practical concepts for understanding and implementing digital controllers in power electronic systems. The discussion focuses on two representative case studies: a digital phase-lead controller applied to a Buck converter and a digital proportional-resonant (PR) controller for a grid-connected inverter with an LCL filter. Key topics, including a review of digital signal processing, discretization methods, and the transition from analog-to-digital control, are carefully described. Special attention is given to the impact of sampling delays, modeled using Padé approximations. The effectiveness of the proposed approaches is demonstrated through Controller Hardware-in-the-Loop (C-HIL) testing using a Typhoon HIL402 real-time simulator and a TMS320F28335 microcontroller.

1 Introduction

When we begin discussing digital control for power electronic systems, it quickly becomes apparent that there is a vast number of controller types. Simultaneously, there are hundreds of power electronic converter typologies. Consequently, it would be impossible to cover all controller types and their applications to various converters within a single chapter. Therefore, we must limit our discussion to a select few controllers. However, the goal is to present control strategies and design methodologies that are easily extendable to other types of controllers and converter typologies. To achieve this, we have chosen to focus on two examples: a phase lead and lag controller applied to a Buck converter, and a proportional-resonant controller applied to a grid-connected inverter with an LCL filter.

Although power electronic systems are a broad and complex subject, most applications can essentially be described by a common set of components: (i) power semiconductors, such as diodes and transistors; (ii) passive components, including inductors, capacitors, and resistors; (iii) power sources, such as DC or AC supplies; (iv) loads, including electric motors and inductive heaters; (v) sensors, such as volt-

T. D. C. Busarello (✉)
Federal University of Santa Catarina, Florianópolis, Brazil
e-mail: tiago.busarello@ufsc.br

E. Kabalci (ed.), *Modern Control Methods for Power Electronics*, Energy Systems in Electrical Engineering, https://doi.org/10.1007/978-981-92-1974-2_3

age, current, and speed sensors; (vi) a conditioning circuit, responsible for adjusting the offset and amplitude of the sensed variables; and (vii) a digital control unit, typically implemented using a microcontroller, FPGA, or processor.

In power electronics, controllers are broadly classified into linear and nonlinear types. Linear controllers operate based on linear system assumptions and are widely used due to their simplicity and ease of implementation. Common examples include proportional–integral, proportional–integral–derivative, PR controllers [1, 2], as well as phase-lead and phase-lead-lag controllers, which are effective in regulating voltage, current, and power with predictable dynamic behavior.

Nonlinear controllers, on the other hand, are designed to handle system nonlinearities and complex dynamics that linear methods cannot manage effectively. Examples include model predictive control, fuzzy logic control, sliding mode control, and controllers based on artificial neural network techniques. These methods offer higher performance in systems with uncertainties or variable operating conditions, but often require more computational resources and complex design procedures.

Figure 1 presents a flowchart of digital controller design approaches from s-domain modeling to digital implementation, valid for linear controllers. The process of designing a digital linear controller typically begins with system modeling in the s-domain, where the system is represented in continuous time using the Laplace transform. From this starting point, there are several design paths available.

One common approach is to first design a controller using conventional continuous-time methods directly in the s-domain. Once the continuous-time controller is obtained, it is converted into a digital controller using a discretization method such as the backward Euler or the Bilinear (Tustin) transform. This results in a digital controller that can be implemented in a digital system.

Another approach involves transforming the continuous-time system model into the z-domain using the Z-transform. This discretizes the system itself, allowing the designer to work entirely within the z-domain. Once the system is in the z-domain, control design techniques such as the pole placement method can be applied directly. After designing the controller in this domain, the final result is a digital controller ready for implementation.

A third design path starts with applying the bilinear transform to the s-domain model to obtain a system model in a fictitious domain known as the w-domain. The w-domain provides a mapping that preserves the frequency response characteristics of the system. In this domain, conventional control design methods are used to synthesize the controller. After the design is complete, the inverse bilinear transform is applied to convert the controller back into the z-domain. The output of this path is also a digital controller.

Additionally, once the system is in the z-domain, a state-space-based design can be carried out using the state-feedback method. This technique allows the designer to place closed-loop poles and shape system dynamics using full-state feedback. The result is another form of digital controller.

In all cases, the result of the process is a digital controller, but the path taken can differ significantly depending on whether the designer begins with a focus on

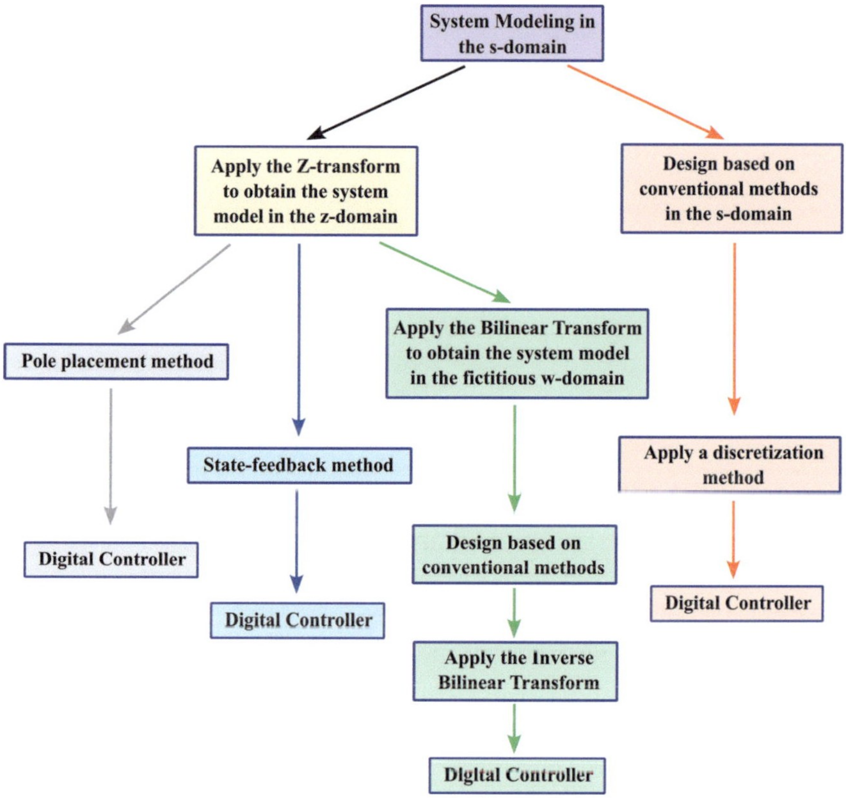

Fig. 1 Flowchart of digital controller design approaches from *s*-domain modeling to digital implementation

continuous-time or discrete-time methods, and on the specific transformation or design methodology used.

Several books on digital control for power electronics are available in the scientific literature. Table 1 lists some of these books, along with other relevant works closely related to digital control.

2 Review of Digital Signal Processing

This section provides a brief review of fundamental concepts in digital signal processing. If you are already familiar with this subject, you may skip ahead to the next section. The aim here is to offer a concise, non-mathematical overview of the following topics: (i) the basics of Fourier Series, (ii) time-domain versus frequency-domain analysis, and (iii) evaluating a signal process.

Table 1 Books about digital control and their features

Title	Some features
Digital Control in Power Electronics [3]	This book presents a great approach to digital control, emphasizing the practical challenges of implementing controllers on real-time digital platforms such as DSPs and microcontrollers. It introduces the multi-sampling approach to enhance system performance, particularly in high-frequency switching converters, improving tracking accuracy and dynamic response without increasing computational load
Digital Control of Dynamic Systems [4]	Although this book is not specifically focused on power electronics, it is highly recommended for anyone interested in digital control. It provides a comprehensive understanding of digital control theory, effectively demonstrating theoretical foundations with practical insights, making it applicable to a wide range of systems. Key topics such as system modeling, design techniques, and stability analysis are clearly and thoroughly presented
Fundamentals of Power Electronics [5]	Although this book does not cover digital control, it is a must-read for its rigorous treatment of power conversion principles, with a strong emphasis on the analysis, modeling, and design of switching converters. The material is presented in a clear, systematic, and accessible manner, making it an essential resource for anyone working in power electronics
Digital Control of High-Frequency Switched-Mode Power Converters [6]	This book provides an in-depth and application-oriented approach of digital control techniques linked to the challenges of high-frequency switching converters. It covers key topics such as sampling effects, quantization, delays, and implementation on digital platforms, all presented in a structured and accessible way
Model Predictive Control of Wind Energy Conversion Systems [7]	This is one of the best books on model predictive control. It offers a comprehensive approach to advanced control strategies for renewable energy systems. The book focuses on the application of model predictive control to wind energy conversion, covering key topics such as system modeling, constraint handling, optimization, and real-time control

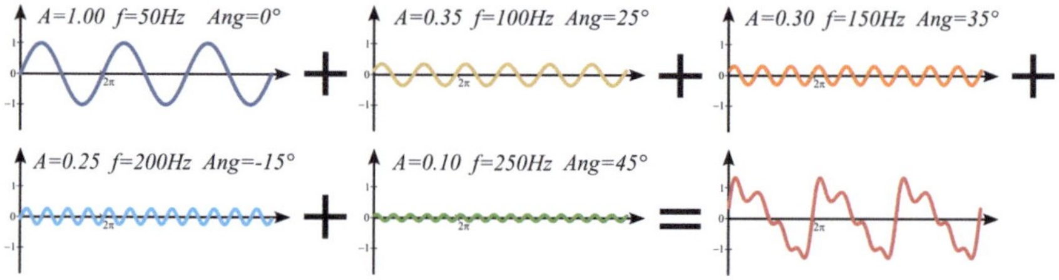

Fig. 2 Five individual sinusoidal signals, as well as the resulting signal formed by their sum

2.1 *Basics of Fourier Series*

Let's explore the basics of the Fourier series through a simple example. In this, we sum five sinusoidal signals, each with its own amplitude, frequency, and phase angle. Figure 2 illustrates the five individual sinusoidal signals, as well as the resulting signal formed by their sum.

The sinusoidal signals we are combining are as follows:

- A sine wave with an amplitude of 1.00, a frequency of 50 Hz, and a phase angle of 0°.
- A sine wave with an amplitude of 0.35, a frequency of 100 Hz, and a phase angle of 25°.
- A sine wave with an amplitude of 0.30, a frequency of 150 Hz, and a phase angle of 35°.
- A sine wave with an amplitude of 0.25, a frequency of 200 Hz, and a phase angle of −15°.
- A sine wave with an amplitude of 0.10, a frequency of 250 Hz, and a phase angle of 45°.

Notice that the resulting waveform has an unusual shape. As stated earlier, we know it is composed of five individual sine waves. But imagine we didn't know this. If we only looked at the resulting signal, it would appear as a single, complex waveform. Hidden within it, however, there are the five sinusoidal signals we summed earlier.

This idea leads us to the core principle of the Fourier series: any periodic signal can be represented as the sum of sinusoidal functions. In theory, this may involve an infinite number of sine waves. However, as demonstrated in our exercise, some signals can be accurately represented using only a few sinusoidal signals.

The Fourier series is a powerful tool that allows us, through its mathematical framework, to extract the amplitude, frequency, and phase of each sinusoidal component from any periodic signal. The process of performing this decomposition is beyond the scope of this chapter but can be found in standard textbooks on signals and systems [8].

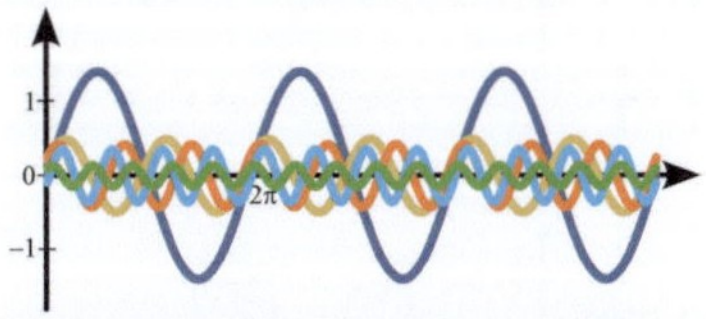

(a) Time-domain plot of the five sine signals on a single chart.

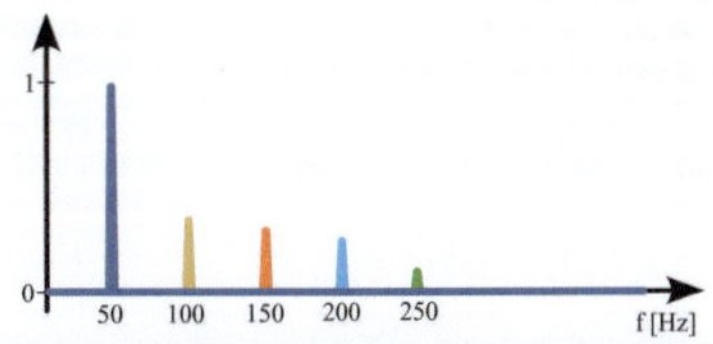

(b) Frequency-domain plot of the five sine signals on a single chart.

Fig. 3 Representing a signal in the time domain and in the frequency domain

2.2 *Time Domain Versus Frequency Domain*

Let's continue with the example discussed in the previous section. Suppose we want to plot all five sine waves on a single chart, as shown in Fig. 3a. This chart is in the time domain, but as you can see, it appears visually cluttered and not very informative. It becomes difficult to distinguish the amplitude and frequency of each individual sine wave. It's important to emphasize that our example includes only five sine waves, so imagine the complexity if we had to display dozens of signals. The figure would become an indistinct blur, making analysis nearly impossible.

To address this, we can switch to the frequency domain, as shown in Fig. 3b. In this representation, we plot the Fourier spectrum, where each peak (or "trace") corresponds to a sinusoidal component. From this chart, we can easily identify both the amplitude and the frequency of each sine wave. Although a phase spectrum could also be plotted to provide information about the phase angle of each sine, it has been omitted here for simplicity.

2.3 *Evaluating a Signal Process*

Representing and analyzing signals is often more effective in the frequency domain, as briefly demonstrated in the previous section. To further illustrate the advantages of frequency-domain analysis, let us evaluate a signal processing operation from both the time and frequency perspectives. Suppose we do not know the nature of the process applied to the input signal, and we only have access to the input and output signals.

Figures 4 and 5 compare the time-domain and frequency-domain representations of a signal processing operation, highlighting how each domain offers distinct insights into how a signal is transformed by a process such as filtering or control. In both cases, the same input signal is applied, and the resulting output signals are identical. However, Fig. 4 presents the signals in the time domain, while Fig. 5 shows their frequency-domain representation.

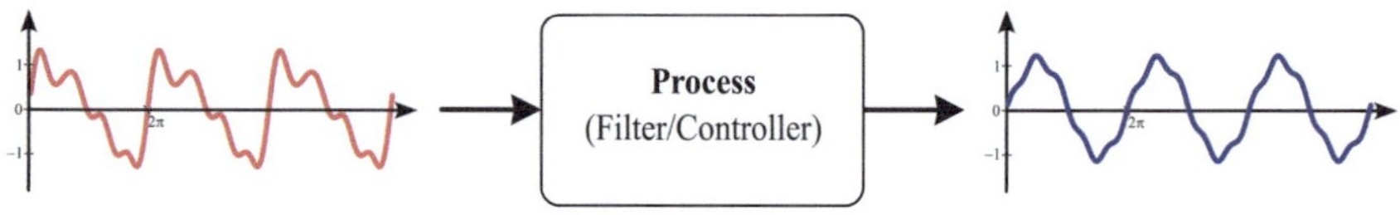

Fig. 4 Example of evaluating a signal processing in the time domain

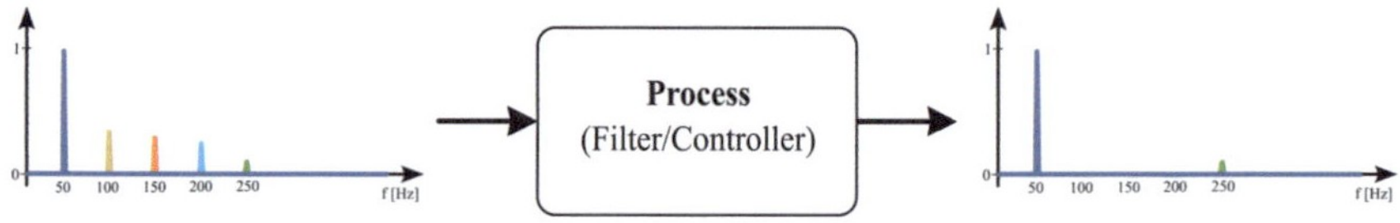

Fig. 5 Example of evaluating a signal processing in the frequency domain

From the time-domain view in Fig. 4, it is difficult, if not impossible, to determine what transformation occurred. Analyzing this scenario in the time domain provides limited insight into the underlying process.

In contrast, the frequency-domain analysis in Fig. 5, which shows the Fourier spectra of both the input and output signals, makes it immediately clear that 100 Hz, 150 Hz, and 200 Hz components have been removed. This allows us to confidently conclude that the system acts as a band-rejection filter. Reaching the same conclusion through time-domain analysis alone would be challenging, unless one is an expert in signal processing and already familiar with the specific signals involved.

3 Discretization Methods

In designing digital control systems for power electronics applications, discretization methods play a crucial role in ensuring that continuous-time controllers can be accurately implemented on digital platforms, such as microcontrollers or digital signal processors. As shown in Fig. 1, all paths require a process of discretization. Discretization methods convert continuous-time models, often described by differential equations, into discrete-time models that are suitable for implementation in digital control systems. One common discretization method is the Zero-Order Hold (ZOH), which approximates the system's input as being held constant between sampling intervals. This method is straightforward but may not capture higher-order dynamics well, especially when the sampling rate is low. Another basic method, the First-Order Hold (FOH), assumes that the input changes linearly between sampling instants, providing a more accurate approximation for systems with faster dynamics. While these methods are useful for basic applications, they may not provide the desired precision in more complex power electronics systems where higher accuracy is needed.

For more accurate discretization, methods like the forward Euler and backward Euler techniques are employed. Forward Euler approximates the derivative at the cur-

Table 2 Equivalences of different discretization methods

Discretization method	Equivalence
Forward Euler	$s = \frac{z-1}{T_s}$
Backward Euler	$s = \frac{z-1}{zT_s}$
Trapezoid (Tustin)	$s = \frac{2}{T_s}\frac{z-1}{z+1}$
Tustin with pre-warping	$s = \frac{\omega_0}{\tan(\frac{\omega_0 T_s}{2})}\frac{z-1}{z+1}$
Zero-order hold	$X(z) = (1 - z^{-1})Z\left\{\frac{X(s)}{s}\right\}$
First-order hold	$X(z) = \frac{(z-1)^2}{zT_s}Z\left\{\frac{X(s)}{s^2}\right\}$
Zero-pole matching	$z = e^{sT_s}$
Impulse invariant	$X(z) = Z\{X(s)\}$

rent sampling point, leading to a simple but less stable approximation, particularly for systems with fast dynamics [9]. On the other hand, the backward Euler method approximates the derivative at the next sampling point, providing better stability but introducing some delay, which may be undesirable in some applications. The Trapezoidal method (Tustin) improves upon these methods by averaging the forward and backward Euler approximations, offering a balanced solution with better accuracy and stability for power electronics control systems.

In cases where very high precision is needed, more advanced methods like Tustin with pre-warping, zero-pole matching, and impulse invariant methods are often used. The Tustin with pre-warping method compensates for the frequency distortion that may occur when discretizing a system, which is particularly useful for control systems involving high-frequency switching in power electronics. The zero-pole matching method is highly effective for systems where the transfer function of the continuous-time system is known, and the aim is to preserve the frequency characteristics of the original system. The impulse invariant method, on the other hand, ensures that the impulse response of the continuous-time system is matched in the discrete domain, making it suitable for applications requiring accurate transient behavior.

Each of these discretization methods offers unique advantages and trade-offs, and the choice of method depends on the specific power electronics application, the desired performance characteristics, and the complexity of the system being controlled. By selecting the appropriate discretization technique, we can design digital controllers that replicate the performance of analog controllers while taking full advantage of the benefits of digital systems, such as flexibility, precision, and ease of implementation.

Table 2 presents the discretization method and demonstrates its equivalence in obtaining a discrete-time domain transfer function $X(z)$ from a continuous-frequency domain transfer function $X(s)$. Z means Z-transform.

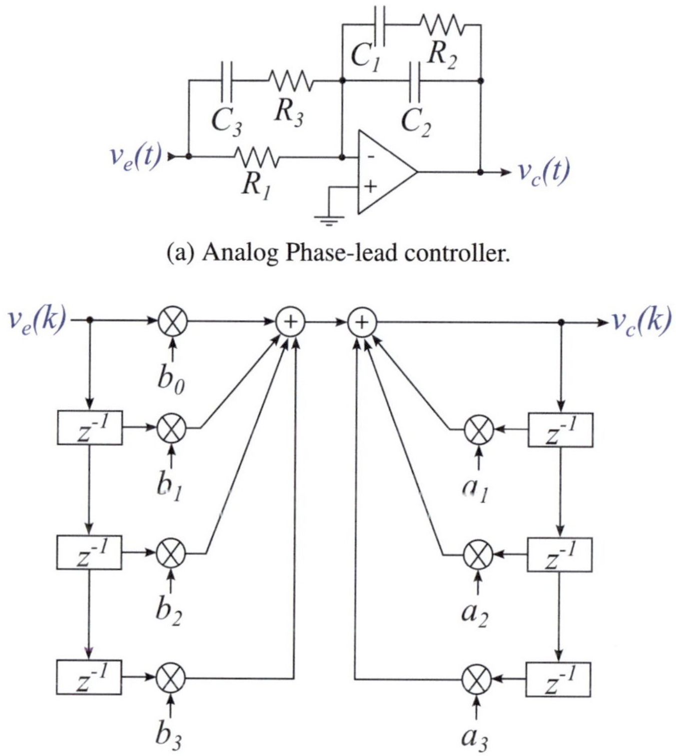

(a) Analog Phase-lead controller.

(b) Digital Phase-lead controller.

Fig. 6 Comparison of how an analog and a digital phase-lead controller is realized

4 Analog Versus Digital Controller

This section briefly compares the implementation approaches of analog and digital controllers. Figure 6 presents a comparison between analog and digital implementations of a phase-lead controller. In Fig. 6a, the analog implementation uses operational amplifiers, resistors, and capacitors to realize the desired phase compensation. In Fig. 6b, the digital implementation is realized using a discrete-time filter structure, involving delays, multipliers, and adders to compute the control signal based on past and current input and output values [10].

In an analog controller, we can move to the s-domain by replacing the capacitor with its impedance, allowing us to derive the controller's transfer function. As a result, the transfer function of the analog controller is given by (1)

$$C_v(s) = \frac{V_c(s)}{V_e(t)} = \frac{s^2 R_2 C_1 C_3 (R_1 + R_3) + s(R_2 C_1 + R_1 C_3 + R_3 C_3) + 1}{s^3 R_1 R_2 R_3 C_1 C_2 C_3 + s^2 (R_1 R_3 C_3 (C_1 + C_2) + R_1 R_2 C_1 C_2) + s R_1 (C_1 + C_2)} \quad (1)$$

Similarly, for a digital controller, we can move to the z-domain to obtain its transfer function, which is given by (2)

$$C_v(z) = \frac{V_c(z)}{V_e(z)} = \frac{b_0 + b_1 z^{-1} + b_2 z^{-2} + b_3 z^{-3}}{1 + a_1 z^{-1} + a_2 z^{-2} + a_3 z^{-3}} \tag{2}$$

To implement the digital controller on a microcontroller, we must express the transfer function $C_v(z)$ in the form of a difference equation. This equation defines how the controller output ($v_c[n]$) is computed based on past outputs and current and past input error signal ($v_e[n]$). The corresponding difference equation is given by (3)

$$\begin{aligned} v_c[n] = b_0 v_e[n] + b_1 v_e[n-1] + b_2 v_e[n-2] + b_3 v_e[n-3] \\ -a_1 v_c[n-1] - a_2 v_c[n-2] - a_3 v_c[n-3] \end{aligned} \tag{3}$$

This equation should be computed at each sampling instant (n), using stored values of previous inputs and outputs. By updating these values iteratively, the controller can be efficiently implemented on any microcontroller.

Later in this chapter, the digital controller will be designed by first synthesizing the analog controller in the continuous-time domain, followed by its discretization to derive the corresponding discrete-time transfer function.

5 Controller Performance Testing Procedures

After designing a controller, it is essential to test its performance to verify that it meets the desired specifications. The following procedures are commonly recommended:

- Set a reference value (setpoint) and verify that the controlled variable accurately tracks it.
- Monitor the control error. Since the controller is digital, accessing the error signal typically requires reading an internal variable or outputting the value through a digital-to-analog converter.
- Change the setpoint and observe whether the output variable correctly follows the new reference. It is important that the controller is designed to operate over the expected range of reference values.
- Apply a load change and verify that the controlled variable remains regulated at its setpoint despite the variation in load conditions.
- Introduce disturbances into the system, particularly in the input voltage source. The controlled variable should remain stable and follow the setpoint if the controller is properly designed to reject such disturbances.
- Evaluate controller response time, such as rise time, settling time, and delay between command and response.

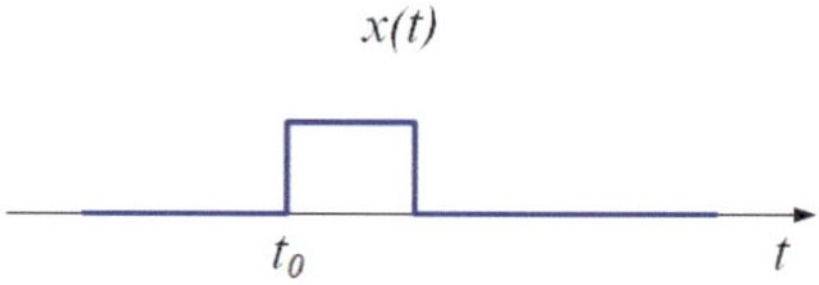

Fig. 7 Time-domain signal consisting of a single pulse occurring at the instant t_0

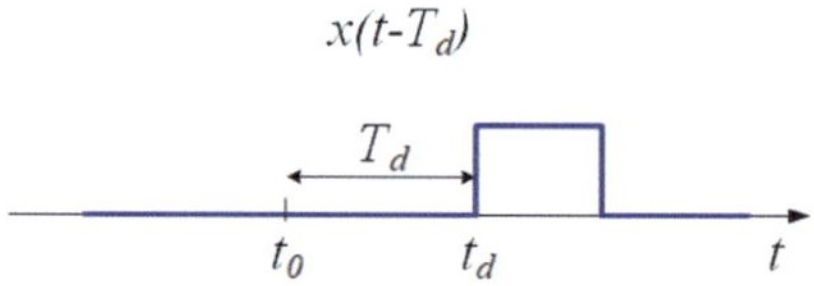

Fig. 8 The same signal of Fig. 7, but now with the pulse occurring at t_d

- Test under worst-case operating conditions, such as minimum and maximum input voltage and extreme load conditions.
- Validate behavior during startup and shutdown, especially in power converters, where inrush currents may exist.

6 Understanding the Delay Caused by Digital Control Strategies

This section presents a comprehensive approach to understanding the delays introduced by digital control strategies.

6.1 Representing a Delayed Signal in Time Domain

Figure 7 presents a time-domain signal consisting of a single pulse occurring at the instant t_0. Note that the variable on the x-axis is time (t), therefore the signal is represented by the function $x(t)$.

If, for any reason, the signal mentioned above is delayed, causing the pulse to occur at time t_d, the result is illustrated in Fig. 8. Note that the duration of the delay is represented by T_d, which is a constant value.

Considering the signals illustrated in Figs. 7 and 8, we can conclude

- the shape of the signal did not change;
- the signal is now described by the function $x(t - T_d)$;
- $t_d > t_0 > 0$).

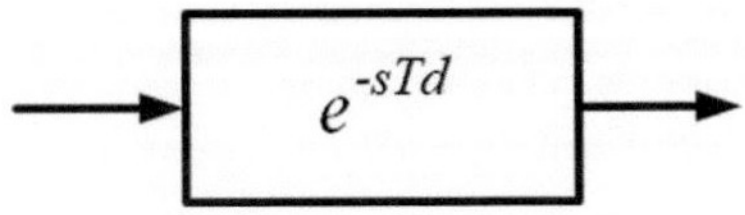

Fig. 9 Representation in s-domain of the time delay caused by the digital implementation

6.2 Representing a Delayed Signal in Frequency Domain

Since most linear controllers for power electronic applications are designed in the frequency domain, specifically in the s-domain, the signals discussed in the previous section must be transformed accordingly.

To transfer a signal from time domain to the frequency domain, the Laplace transform is typically used. The formal definition and properties of the Laplace transform are beyond the scope of this chapter. Nevertheless, a very popular property is the *Time Shifting*.

Equations (4) and (5) show how the signals $x(t)$ and $x(t - T_d)$ are represented in the frequency domain through the Laplace transform.

$$x(t) \Leftrightarrow X(s) \tag{4}$$

$$x(t - T_d) \Leftrightarrow X(s)e^{-sT_d}. \tag{5}$$

With these equations, we can conclude

- The delay in a signal of duration T_d appeared as e^{-sT_d} in the frequency domain.
- The term e^{-sT_d} represents a phase shift in the system, as will be observed later.

6.3 Using Padé's approximation to Replace e^{-sT_d}

As mentioned earlier, linear controllers are typically designed in the frequency domain (the s-domain). These controllers are later implemented on digital devices, such as microcontrollers or FPGA boards. Therefore, during the design process, it is very important to account for the delay introduced by the digital implementation.

The representation in s-domain of the time delay caused by the digital implementation could be done by inserting a representative transfer function as shown in the block of Fig. 9.

The problem of using the block shown in Fig. 9 is the fact that its transfer function is irrational, bringing complexities to the design process of the controllers. So, one option is to use Padé's approximation to replace the original transfer function.

The detailed derivation of Padé approximation equations is also beyond the scope of this chapter, but it is important to note that this approximation is derived from power series concept. Moreover, Padé's approximation can represent several functions like

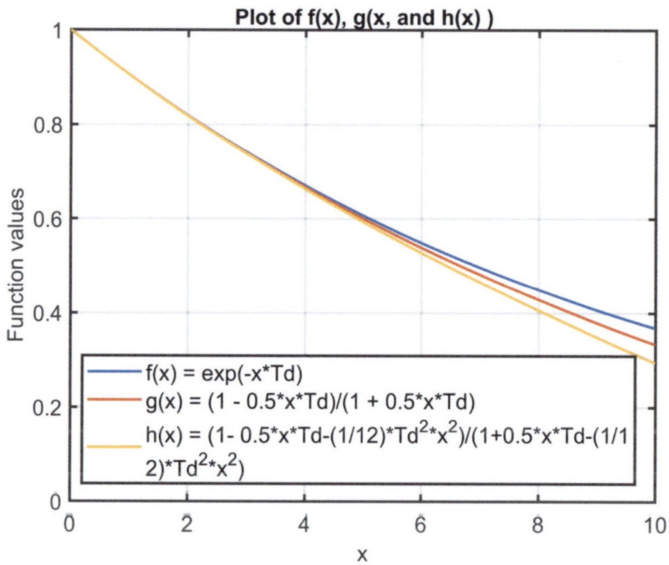

Fig. 10 The exponential function and their first- and second-order Padé's approximations

the sine, Bessel, and exponential function as a rational function of polynomials. The most common approximation used for the exponential function is the first-order approximation. Using high-order Padé's approximation is quite rare in power electronic applications, but one example can be found in [11].

Let's first examine Padé's approximation in continuous-time domain. Equations (6) and (7) present the first- and second-order Padé's approximation for the exponential function:

$$e^{-x\sigma} \approx \frac{1-\frac{\sigma x}{2}}{1+\frac{\sigma x}{2}} \tag{6}$$

$$e^{-x\sigma} \approx \frac{1-\frac{\sigma x}{2}+\frac{\sigma^2 x^2}{12}}{1+\frac{\sigma x}{2}-\frac{\sigma^2 x^2}{12}} \tag{7}$$

where σ is a constant.

Figure 10 shows the exponential function along with its first- and second-order Padé approximations, with $\sigma = 0.1$. The x-axis represents time. From the figure, it is evident that the three functions coincide over a certain range of x. The smaller σ is, the larger the region of coincidence becomes. This is advantageous because, in digital control, we aim for a high sampling frequency, which results in a short sampling period and minimal time delay. Consequently, representing the delay using Padé approximations provides an accurate solution.

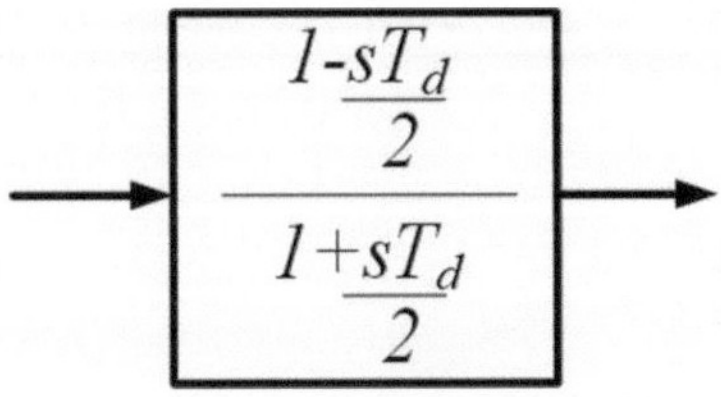

Fig. 11 Representation of the time delay by the first-order Padé's approximation

Moving back to the *s*-domain, now we can replace the block shown in Fig. 9 to that shown in Fig. 11. As a result, the block shows the representation of the time delay by the first-order Padé's approximation.

To verify the effect that the inclusion of the above transfer function can cause in a control strategy, let's plot Bode diagrams of two transfer functions. The following transfer functions are taken as example. The first is a unitary transfer function and the second is the product of the unitary transfer function and Padé's approximation, as given by (8) and (9), respectively.

$$TF(s) = 1 \tag{8}$$

$$TF_{delayed}(s) = TF(s).\frac{1 - \frac{sT_d}{2}}{1 + \frac{sT_d}{2}} \tag{9}$$

Figure 12 presents the frequency response of a unitary transfer function and its delayed version.

Looking at the frequency response that has just been plotted, we can conclude

- In the magnitude diagram, both responses are coincident. This was expected, since a time delay in a signal does not change its shape, as previously concluded.
- In the phase diagram, the delayed transfer function coincides with the unitary transfer function at low frequencies but begins to diverge significantly at higher frequencies.
- To plot these Bode diagrams, it was adopted 10 kHz as sampling frequency and $T_d = 1.5T_s$, where T_s is the sampling period.
- If the sampling frequency increases, the time delay (T_d) decreases, resulting in a larger region of coincidence in the phase frequency response. This also leads to a larger region of overlap in the graphs shown in Fig. 10.

6.4 What Value to Use in the Constant T_d?

In the previous section, it was adopted $T_d = 1.5T_s$ [3, 4]. The 1.5 is actually $(1 + 0.5)T_s$, consisting of

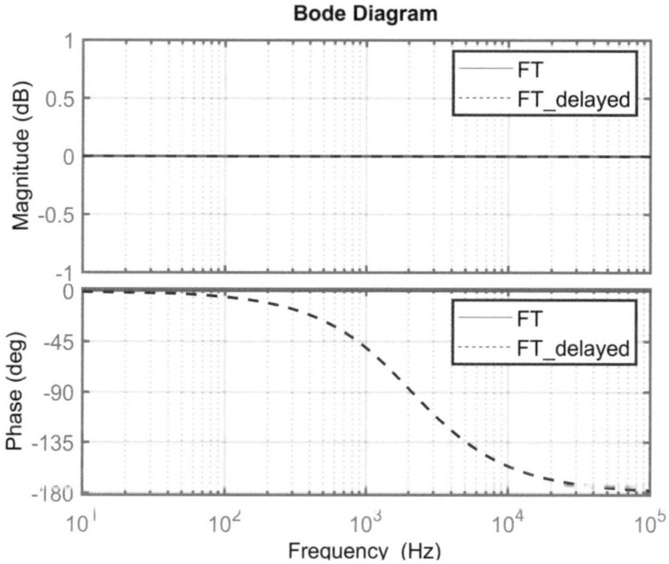

Fig. 12 Frequency response of a unitary transfer function and its delayed version

- $0.5T_s$ is attributed to the sampling and holding process, as illustrated in Fig. 13. The figure presents the original signal (a sinusoidal signal in this case) and its sampled and held version. Additionally, the figure includes the *First Harmonic* representation. Within the microcontroller, only the sampled and held signal exists. Notice also that the first harmonic is delayed by half related to the sampling period. That's the reason the $0.5T_s$ is adopted. This is valid for any signal and any sampling frequency.
- $1T_s$ is attributed to the time required to compute the entire control algorithm. In summary, the control algorithm is a sequential process composed of the following steps: (*i*) sampling and holding the variables, (*ii*) computing the control algorithm, (*iii*) updating the PWM, and (*iv*) applying the result to the output. Therefore, one sampling period ($1T_s$) is the delay caused by the control algorithm.

7 Digital Voltage Controller with Lead-Phase Compensation for a Buck Converter

This section presents the methodology for designing a digital voltage controller with lead-phase compensation. A Buck DC–DC converter is selected as the application for this controller. Although a Buck converter can be controlled using simpler control strategies, the objective here is to outline a general design procedure that can be readily applied to other types of converters as well.

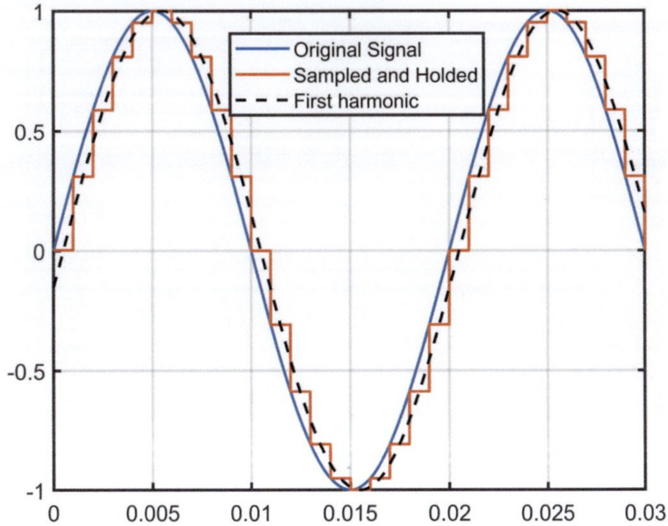

Fig. 13 The original signal and its sampled and held version, as well as the first harmonic

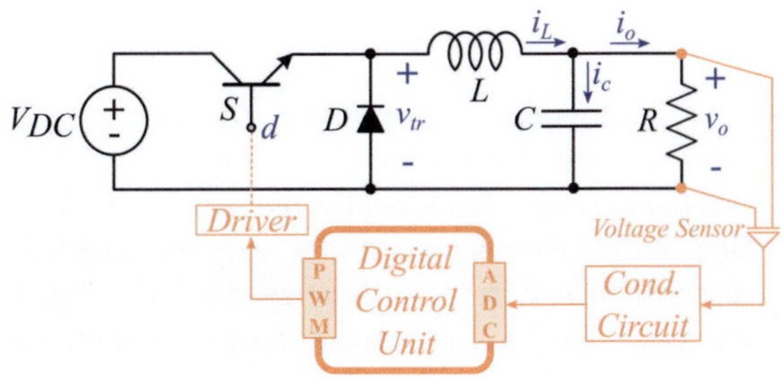

Fig. 14 Simplified closed-loop diagram of the Buck converter

Figure 14 shows a simplified closed-loop diagram of the Buck converter. The output voltage is sensed via a voltage sensor and sent to the Digital Control Unit through a signal conditioning circuit. This conditioning circuit is assumed to be ideal, meaning it introduces no phase shift or amplitude distortion. The Digital Control Unit executes the control algorithm and generates the corresponding PWM signal for the Buck converter. This signal is passed through a driver circuit before reaching the converter switching transistor.

7.1 System Modeling

As shown in Fig. 1, the first step in designing a digital linear controller is to model the system in the s-domain. This requires representing all components of the control strategy within this domain. Based on the configuration presented in Fig. 14, we

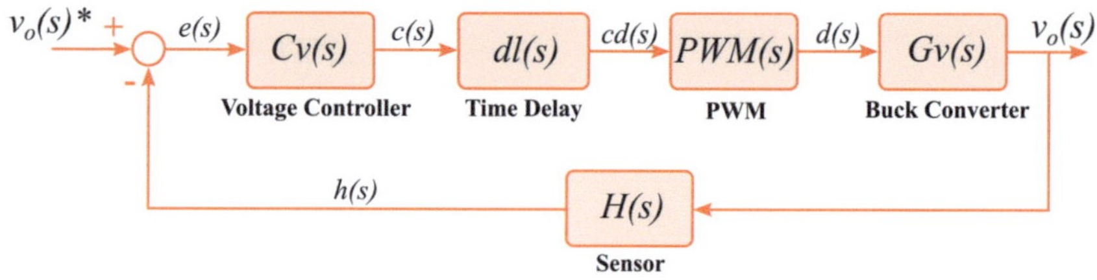

Fig. 15 Control system block diagram

can construct the control system block diagram as depicted in Fig. 15. The entire diagram is expressed in the s-domain because the digital controller will be designed using conventional continuous-time techniques, as outlined on the right side of Fig. 1, and subsequently discretized using the Z-transform. The block diagram includes the controller, the time delay, the PWM stage, the Buck converter, and the voltage sensor.

Let us now derive all the transfer functions shown in Fig. 15. Starting with the sensor, its transfer function can be modeled as a simple gain, as given in (10).

$$H(s) = H_v \tag{10}$$

The transfer function of the time delay is that shown in Sect. 4 and repeated here for convenience.

$$dl(s) = \frac{1 - \frac{sT_d}{2}}{1 + \frac{sT_d}{2}} \tag{11}$$

Since the transfer function of the time delay accounts for the delay introduced by both the PWM and the sampling-and-hold process, the PWM block in Fig. 15 can be modeled as a simple gain, as shown in (12).

$$PWM(s) = \frac{1}{c_{pk}} \tag{12}$$

where c_{pk} is the peak amplitude of the carrier signal, which is typically equal to 1 in digital control strategies.

Moving on to derive the transfer function of the Buck converter, we can redraw the circuit as shown in Fig. 16. In this representation, the DC source, the transistor, and the diode from the original circuit are replaced by a voltage-controlled voltage source. This substitution is valid because the new source represents the average value of the voltage across the diode, as expressed in Eqs. (13) to (14). The variable d is the duty cycle.

$$\overline{v_{tr}}(t) = \frac{1}{T}\int_0^{dT} V_{DC}dt \tag{13}$$

$$\overline{v_{tr}}(t) = V_{DC}d \tag{14}$$

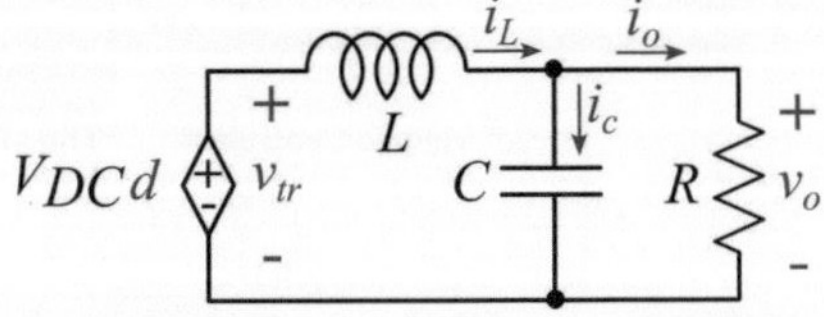

Fig. 16 Buck converter circuit with a voltage-controlled voltage source replacing the DC source, the transistor, and the diode

Since the duty cycle d varies as a result of the closed-loop control strategy, the voltage-controlled voltage source is now defined by (15).

$$\overline{v_{tr}}(t) = V_{DC}d(t) \tag{15}$$

Applying the Kirchhoff current law in the circuit of Fig. 16, it results in (16).

$$i_L(t) = i_c(t) + i_o(t) \tag{16}$$

The load current as well as the inductor and capacitor currents are defined by (17).

$$i_o(t) = \frac{v_o(t)}{R} \qquad i_c(t) = C\frac{dv_o(t)}{dt} \qquad i_L(t) = \frac{dv_L(t)}{dt} \tag{17}$$

According to Fig. 15, the transfer function of the Buck converter relates the output voltage to the duty cycle, and it is given by (18).

$$Gv(s) = \frac{v_o(s)}{d(s)} \tag{18}$$

We now derive the differential equation describing the dynamic behavior of the Buck converter, aiming to obtain an expression in which only the output voltage and the duty cycle appear as time-varying variables. This procedure is demonstrated in Eqs. (19) to (21).

$$V_{DC}d(t) - v_o(t) = L\frac{d[i_c(t) + i_o(t)]}{dt} \tag{19}$$

$$V_{DC}d(t) - v_o(t) = L\frac{d\left[C\frac{dv_o(t)}{dt} + \frac{v_o(t)}{R}\right]}{dt} \tag{20}$$

$$V_{DC}d(t) = LC\frac{d^2v_o(t)}{dt^2} + \frac{L}{R}\frac{dv_o(t)}{dt} + v_o(t) \tag{21}$$

Equation (21) should be linearized using small-signal analysis around the operating point. This is done by introducing the definitions in (22) and (23). In both cases, the time-varying variable is expressed as the sum of its steady-state (average) value and a small time-varying perturbation.

$$d(t) = D + \widetilde{d}(t) \tag{22}$$

$$v_o(t) = V_o + \widetilde{v_o}(t) \tag{23}$$

Replacing them in (21), it results in (24).

$$V_{DC}(D + \widetilde{d}(t)) = LC\frac{d^2\left[V_o + \widetilde{v_o}(t)\right]}{dt^2} + \frac{L}{R}\frac{d\left[V_o + \widetilde{v_o}(t)\right]}{dt} + V_o + \widetilde{v_o}(t) \tag{24}$$

Since the derivative of a constant is zero, this leads to Eq. (25).

$$V_{DC}(D + \widetilde{d}(t)) = LC\frac{d^2\widetilde{v_o}(t)}{dt^2} + \frac{L}{R}\frac{d\widetilde{v_o}(t)}{dt} + V_o + \widetilde{v_o}(t) \tag{25}$$

By separating the terms corresponding to average values from those representing small-signal variations, we can express the results as in (26) and (27), respectively.

$$V_{DC}D = V_o \tag{26}$$

$$V_{DC}\widetilde{d}(t) = LC\frac{d^2\widetilde{v_o}(t)}{dt^2} + \frac{L}{R}\frac{d\widetilde{v_o}(t)}{dt} + \widetilde{v_o}(t) \tag{27}$$

Notice that (26) resulted in the steady-state relation for the Buck converter. By applying the Laplace transform to (27) and rearranging the equation, we obtain the transfer function of the Buck converter, as shown in (30).

$$V_{DC}\widetilde{D}(s) = LCs^2\widetilde{V_o}(s) + \frac{L}{R}s\widetilde{V_o}(s) + \widetilde{V_o}(s) \tag{28}$$

$$\frac{\widetilde{V_o}(s)}{\widetilde{D}(s)} = \frac{V_{DC}}{LCs^2 + \frac{L}{R}s + 1} \tag{29}$$

$$G_v(s) = \frac{V_{dc}}{s^2\,LC + s\frac{L}{R} + 1} \tag{30}$$

7.2 Controller Design

Since all the transfer functions of the control strategy are known, the digital controller can be designed using conventional s-domain methods. The design approach adopted here is based on frequency response analysis and follows the K-type methodology [12].

The process begins by specifying the desired cut-off frequency and phase margin. The cut-off frequency determines the bandwidth of the closed-loop system and is directly related to how quickly the controller can respond to changes or disturbances. A higher cut-off frequency results in a faster response, but it is constrained by the system's switching frequency.

As a rule of thumb, the cut-off frequency should be set to one-tenth of the switching frequency to ensure the controller can accurately detect and respond to disturbances. Accordingly, the desired cut-off frequency is given by (31).

$$f_c = \frac{f_s}{10} \tag{31}$$

where f_s is the switching frequency.

The phase margin of a closed-loop system indicates how oscillatory the system's response will be when subjected to disturbances or variations. A low phase margin, typically between 5° and 15°, places the system close to the stability limit, often resulting in highly oscillatory behavior during disturbance rejection. Conversely, a high phase margin, such as 100°, produces a more stable and less oscillatory response, but may lead to slower convergence to the steady state. A commonly recommended range for the phase margin is a value between 30° and 70°, which offers a balanced trade-off between response behavior and stability. Therefore, the desired phase margin is given by (32).

$$PM_d = 60^\circ \tag{32}$$

With the desired cut-off frequency and phase margin defined, the next step is to write the uncontrolled open-loop transfer function ($OLTF_u(s)$), which is given by (33). Notice that this transfer function is defined by the product of the time delay, the PWM, the buck converter, and the sensor. Hereinafter, the transfer function of the time delay is assumed to be unitary for simplicity.

$$OLTF_u(s) = dl(s)PWM(s)Gv(s)H(s) \tag{33}$$

The next step is to determine the phase and real gain of the uncontrolled open-loop transfer function at the desired cut-off frequency. This is achieved using (34) to (36).

$$\theta_{fc} = \angle OLTFu(f_c)\,[\text{deg}] \tag{34}$$

$$G_{dB_{fc}} = |OLTF_u(f_c)|\,[dB] \tag{35}$$

$$G_{fc} = 10^{\frac{|G_{dBfc}|}{20}} \tag{36}$$

The required phase lead introduced by the controller, as well as the value of the K-factor, is given by Eqs. (37) and (38), respectively.

$$\alpha = PM_d - \theta_{fc} - 90^\circ \tag{37}$$

$$K = \left[\tan\left(\frac{\alpha}{4} + 45^\circ\right)\right]^2 \tag{38}$$

Now, we can compute the parameters of the lead-phase controller corresponding to its analog version, using Eqs. (39) to (44):

$$C_2 = 100x10^{-9} \text{ (this is a defined value)} \tag{39}$$

$$R_1 = \frac{1}{2\pi f_c G_{fc} C_2} \tag{40}$$

$$C_1 = C_2(K - 1) \tag{41}$$

$$R_2 = \frac{\sqrt{K}}{2\pi f_c C_1} \tag{42}$$

$$R_3 = \frac{R_1}{K - 1} \tag{43}$$

$$C_3 = \frac{1}{2\pi f_c R_3 \sqrt{K}} \tag{44}$$

By computing the parameters above, we conclude the design of the analog version of the controller. Before proceeding with discretization, we verify whether the controller meets the desired cut-off frequency and phase margin in the overall system. Then, we define the controlled open-loop transfer function, given by (45).

$$OLTF_c(s) = OLTF_u(s) C_v(s) \tag{45}$$

where $C_v(s)$ is the transfer function of the controller, as presented in (1) and reproduced here for convenience.

$$C_v(s) = \frac{V_c(s)}{V_e(t)} = \frac{s^2 R_2 C_1 C_3 (R_1 + R_3) + s(R_2 C_1 + R_1 C_3 + R_3 C_3) + 1}{s^3 R_1 R_2 R_3 C_1 C_2 C_3 + s^2 (R_1 R_3 C_3 (C_1 + C_2) + R_1 R_2 C_1 C_2) + s R_1 (C_1 + C_2)} \tag{46}$$

Finally, the closed-loop transfer function can be expressed as shown in (47).

$$CLTF(s) = \frac{OLTF_c(s)}{1 + OLTF_c(s)} \tag{47}$$

7.3 Discretization of the Designed Controller

The discretization is performed by the backward Euler method, given by (48).

$$C_v(z) = C_v(s)|_{s \to \frac{z-1}{zT_s}} \tag{48}$$

Let's defining the following variable to use (48) in (46).

$$N_1 = R_2C_1C_3(R_1 + R_3) \tag{49}$$

$$N_2 = R_2C_1 + R_1C_3 + R_3C_3 \tag{50}$$

$$N_3 = R_1R_2R_3C_1C_2C_3 \tag{51}$$

$$N_4 = R_1R_3C_1(C_1 + C_2) + R_1R_2C_1C_2 \tag{52}$$

$$N_5 = R_1(C_1 + C_2) \tag{53}$$

$$P_1 = N_3 + N_4T_s + N_5T_s^2 \tag{54}$$

Then, the coefficients of the digital controller are given by

$$b_0 = \frac{T_sN_1 + N_2T_s^2 + T_s^3}{P_1} \tag{55}$$

$$b_1 = -\frac{2T_sN_1 + N_2T_s^2}{P_1} \tag{56}$$

$$b_2 = \frac{N_1T_s}{P_1} \tag{57}$$

$$b_3 = 0 \tag{58}$$

$$a_0 = 1 \tag{59}$$

$$a_1 = -\frac{3N_3 + 2N_4T_s + N_5T_s^2}{P_1} \tag{60}$$

$$a_2 = \frac{3N_3 + N_4T_s}{P_1} \tag{61}$$

$$a_3 = -\frac{N_3}{P_1} \tag{62}$$

7.4 Effects of Designing in s-Domain and Discretizing

This section illustrates what happens when a digital controller is designed simply by discretizing an analog controller. Figure 17 presents a generic frequency response for both the analog and the corresponding digital controller. In this case, the cut-off frequency is assumed to be 2 kHz. As shown in the magnitude response, the analog and digital controllers provide nearly identical gain characteristics. However, a mismatch appears in the phase response. While the controller was designed to provide a certain phase lead, the actual phase lead introduced by the digital controller is a little lower. This mismatch indicates that discretization affects the phase behavior of the controller. In this example, discretization was performed using the Backward Euler method. It is worth noting that other discretization methods could yield different frequency responses.

7.5 Controller Hardware-in-the-Loop Results

The system shown in Fig. 14 was tested using a Controller Hardware-in-the-Loop (C-HIL) setup. Complete details of this setup are provided in Sect. 9. In this configuration, the Buck converter is emulated in the real-time simulator Typhoon HIL402, while the digital control strategy is executed on the physical microcontroller TMS320F28335. The system parameters are listed in Table 3.

The desired cut-off frequency and phase margin are set to 2 kHz and 60°, respectively. Using the previously described methodology for designing the digital phase-lead controller, the resulting filter coefficients are presented in Table 4. The high values obtained for the b coefficients were expected due to the low value of the sensor gain.

Once the phase-lead controller has been designed, its performance can be assessed through Bode diagrams to verify the effectiveness of the design. Figure 18 shows the Bode plots used for this evaluation. Figure 18a presents the Bode diagrams for the uncontrolled open-loop transfer function ($OLTF_u(s)$) and the controller transfer function ($C_v(s)$). The uncontrolled open-loop transfer function has a cut-off fre-

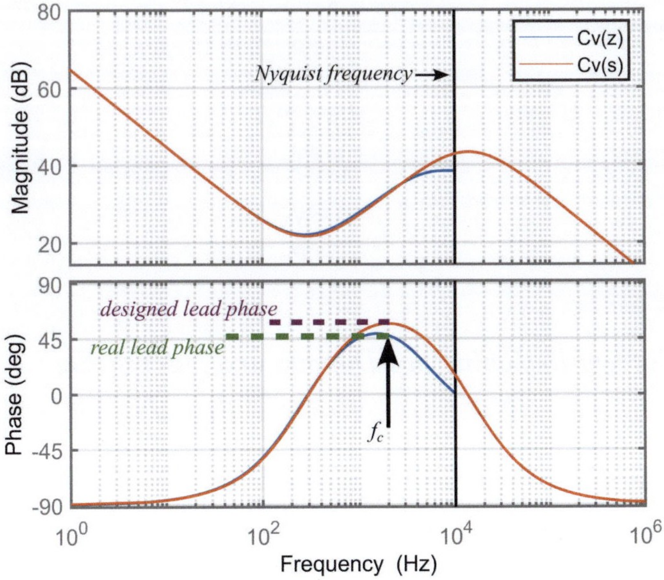

Fig. 17 Frequency response of the designed phase-lead controller, valid for analog and its discretized version

Table 3 System parameters of the Buck converter

Parameter	Description	Value
V_{dc}	Input voltage	100 V
V_o	Output voltage	50 V
f_s	Switching and sampling frequencies	20 kHz
H_v	Voltage sensor gain	$\frac{1}{100}$
t_s	Sampling period	$1/f_s$
D	Duty cycle at operating point	0.5
ΔV_o	Maximum voltage ripple	2.5 V
R	Load resistance	2.5 Ω
L_{min}	Minimum inductance for continuous conduction	$\frac{V_{dc}(1-D)D}{2 f_s I_o}$
L	Adopted inductance	0.4 mH
C_{min}	Minimum capacitance to ensure ΔV_o	$\frac{V_o(1-D)}{8 f_s^2 L\left(\frac{\Delta V_o}{100}\right)}$
C	Adopted capacitance	680 μF

Table 4 Coefficients of the digital phase-lead voltage controller

Parameter	Value
b_0	602.42949115
b_1	−110.49849329
b_2	506.69485808
b_3	0.0000000000
a_0	1.0000000000
a_1	−1.2817306391
a_2	0.32146289882
a_3	−0.0397322596

quency around 300 Hz and does not exhibit a decreasing behavior at low frequencies, indicating that the steady-state error wouldn't be negligible if a controller was not implemented. In the phase response, it can be observed that the controller provides phase lead around the desired cut-off frequency.

Figure 18b shows the Bode diagram of the controlled open-loop transfer function ($OLTF_c(s)$). The achieved cut-off frequency and phase margin, which match the specified desired values, are indicated. This confirms that the controller was properly designed. Finally, Fig. 18c presents the closed-loop transfer function ($CLTF(s)$), showing that both the magnitude and phase responses remain flat up to the desired cut-off frequency. This indicates that the controller does not introduce magnitude or phase errors in the controlled voltage.

Figure 19 presents the C-HIL results for the Buck converter using the designed digital phase-lead voltage controller. In Fig. 19a, the system's initialization behavior is shown. Initially, the output voltage (v_o), output current (i_o), and inductor current (i_L) are all zero, while the input voltage (V_{DC}) is held constant at 100 V. When the controller is activated, a significant overshoot in the output voltage occurs, followed by a high initial inductor current. These peaks arise because the output capacitor is fully discharged at startup. Once steady state is reached, the controller successfully regulates the output voltage to its reference setpoint. It is important to note that the controller was designed with steady-state operation condition, meaning that the transient response was not modeled. As a result, overshoot and oscillations are expected during startup. To improve this behavior, one approach would be to model the transient response and design the controller to meet specific overshoot and settling time requirements. However, this is a complex task and is rarely done in the design of voltage controllers for DC–DC converters, as a more straightforward solution exists, which is to use a ramp in the voltage reference during initialization.

Figure 19b shows the system response when the controller is initialized with a ramped reference signal. In this case, the output voltage increases linearly without overshoot, demonstrating the effectiveness of this approach in managing startup transients.

Figure 19c illustrates the system's behavior during a step change in the output voltage reference from 50 V to 75 V. By checking the response, the output voltage

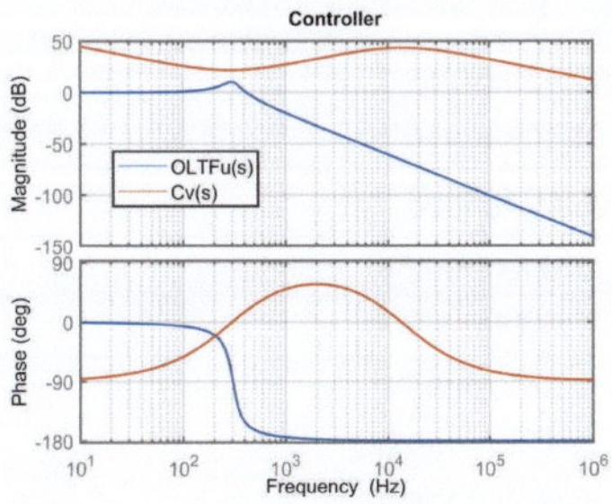

(a) Bode diagrams of the uncontrolled open-loop transfer function and the controller transfer function.

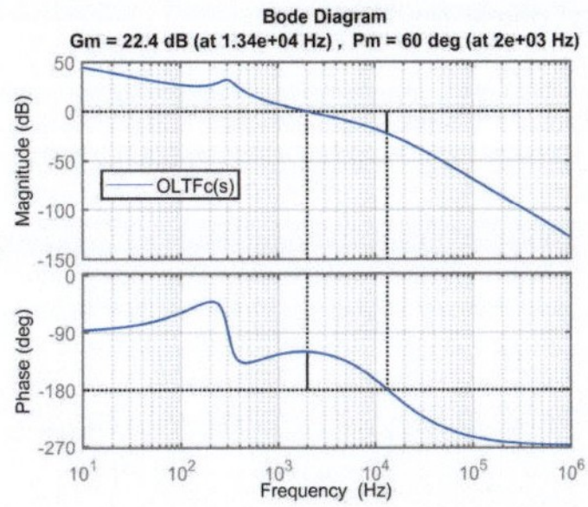

(b) Bode diagram for the controlled open-loop transfer function.

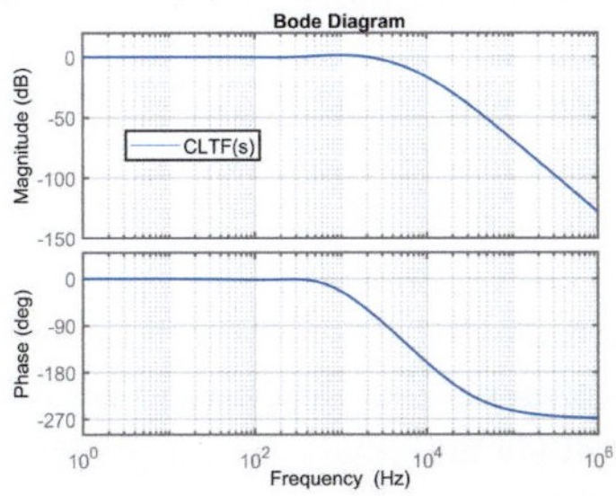

(c) Bode diagram of the closed-loop transfer function.

Fig. 18 Bode plots used to evaluate the designed phase-lead controller

exhibits a slight overshoot before quickly settling at the new setpoint. The inductor current experiences a sharp peak due to the additional current required to charge the output capacitor to the new steady-state voltage level.

Figure 19d shows the system's response to a load step, where an additional 2.5Ω resistor is connected in parallel with the existing load. As expected, both the output current and inductor current increase due to the added load. After a brief transient response, the output voltage is successfully regulated back to its reference value, demonstrating the controller's ability to maintain voltage regulation under varying load conditions.

Figure 19e and f presents the system's response to disturbances in the input voltage. In Fig. 19e, a low-frequency oscillation is introduced in the input voltage, while in Fig. 19f, the input voltage undergoes a step change from 100V to 125V. In both scenarios, the output voltage remains well regulated at its reference setpoint, demonstrating the controller's robustness and its ability to reject disturbances originating from the input voltage.

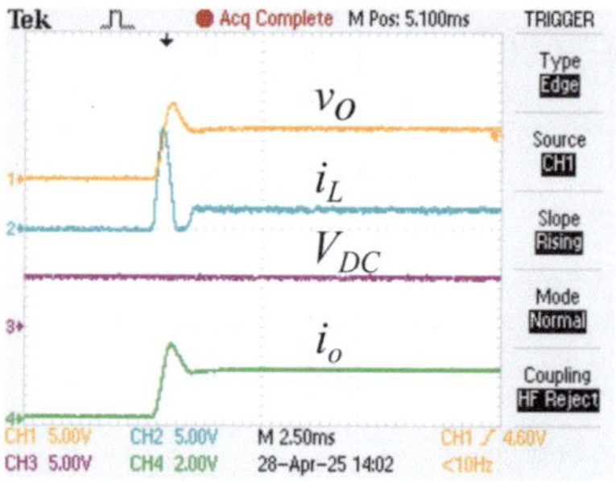

(a) The system's initialization behavior.

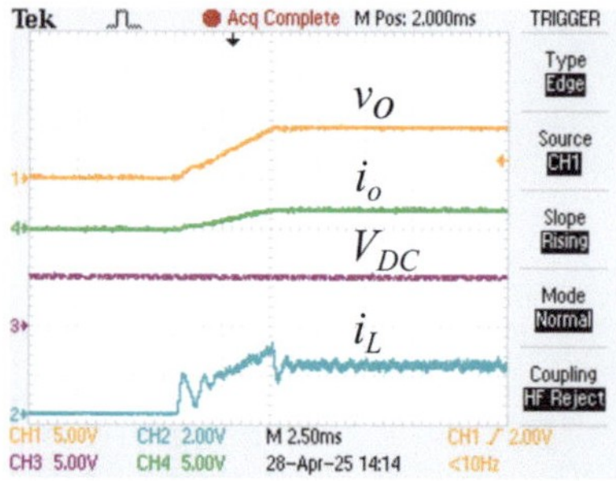

(b) System response when the controller is initialized with a ramped reference signal.

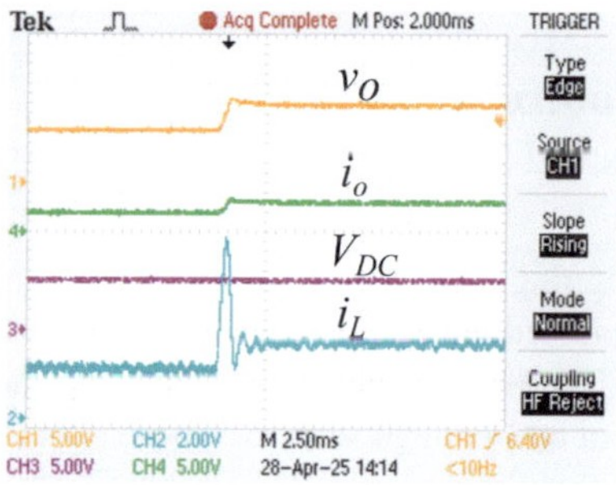

(c) The system's behavior during a step change in the output voltage reference from 50V to 75V.

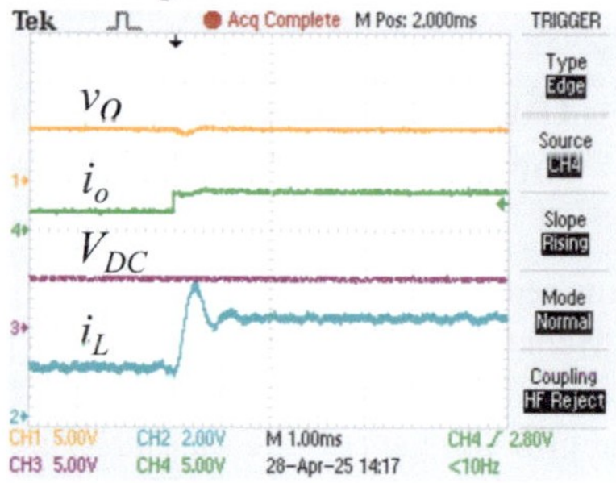

(d) The system's response to a load step, where an additional 2.5Ω resistor is connected in parallel with the existing load.

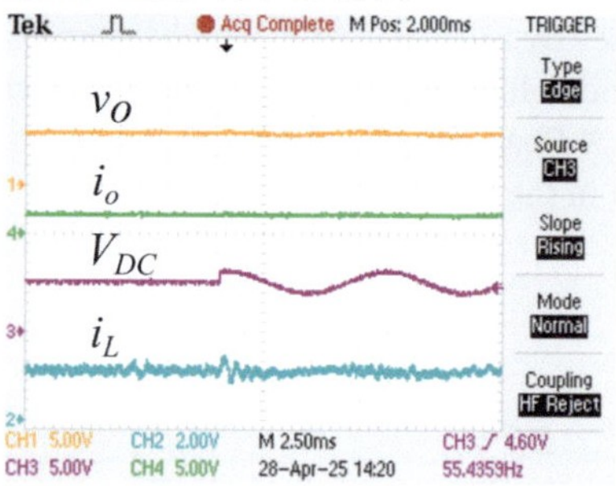

(e) The system's response to a step change from 100V to 125V in the input voltage.

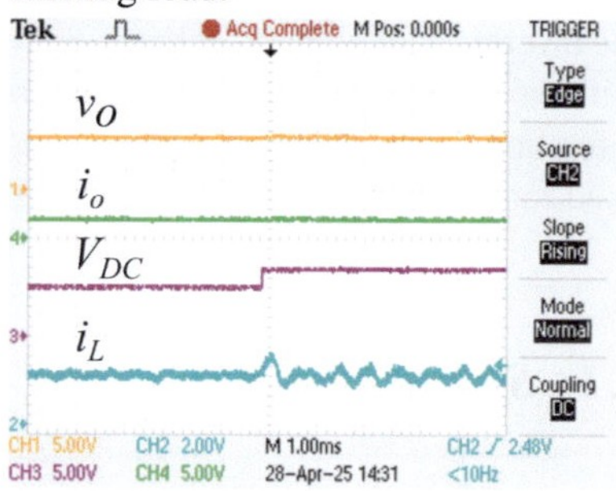

f) The system's response to a low-frequency oscillation disturbance in the input voltage.

Fig. 19 C-HIL results for the Buck converter using the designed digital phase-lead voltage controller

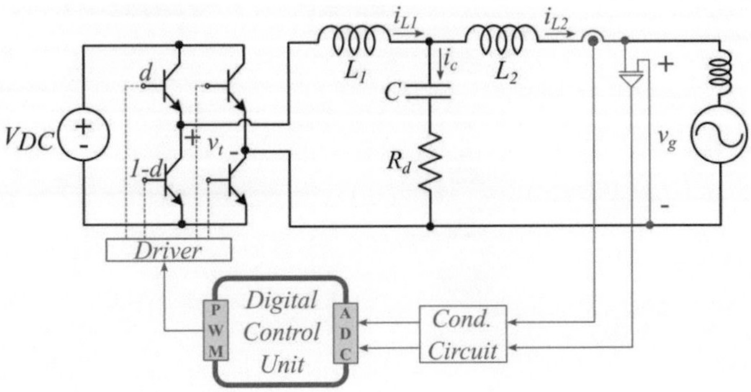

Fig. 20 A simplified closed-loop diagram of the LCL-filtered grid-connected inverter

8 Digital Proportional-Resonant Controller for an LCL-Filtered Grid-Connected Inverter

This section presents the methodology for designing a digital proportional-resonant current controller for an LCL-filtered grid-connected inverter. Figure 20 illustrates a simplified closed-loop diagram of the system. The inverter output current (i_{L2}) and the grid voltage (v_g) are measured using current and voltage sensors. These signals are then passed through a conditioning circuit before being sent to the Digital Control Unit.

As in the case of the Buck converter, the conditioning circuit is assumed to be ideal, introducing neither phase shift nor amplitude distortion. The Digital Control Unit implements the control algorithm and generates the corresponding PWM signal for the inverter. This signal is subsequently processed by a driver circuit, which interfaces with the inverter's switching transistors. The grid voltage is used to generate a current reference signal that is synchronized with it. A Phase-Locked Loop (PLL) [13] is used for this purpose, although it is not shown in the figure.

8.1 System Modeling

Figure 21 presents the control system block diagram for the grid-connected inverter. The PR current controller is characterized by the proportional gain (k_p), the integral gain (k_i), and the resonant filter ($R(s)$).

The transfer function of the PR current controller is given by (63).

$$C_{PR}(s) = k_p + k_i R(s) \tag{63}$$

The transfer function of the sensor is given by (64), where h_i is the current sensor gain.

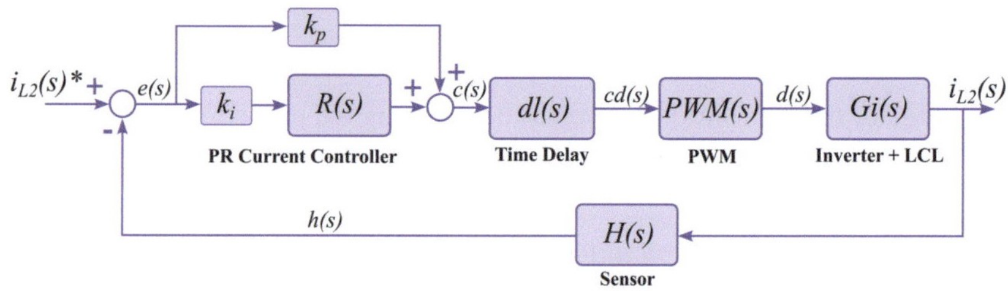

Fig. 21 Control system block diagram for the grid-connected inverter

$$H(s) = h_i \tag{64}$$

To derive the transfer function of the inverter combined with the LCL filter, a procedure similar to that used for the Buck converter can be followed. This results in a transfer function that relates the inverter output current to the transistor duty cycle. For convenience, the full derivation is omitted here, and only the resulting transfer function is provided, which is given by (65).

$$G_i(s) = \frac{i_{L2}(s)}{d(s)} = K_{inv}\frac{sC_fR_d + 1}{s^3L_gL_cC_f + s^2C_fR_d(L_c + L_g) + s(L_c + L_g)} \tag{65}$$

where K_{inv} is the inverter gain, given by the value of the DC source (V_{DC}).

The previous equation is valid for an LCL filter with passive damping. When active damping is employed via capacitor current feedback, the transfer function becomes what is shown in (66).

$$G_{i_ad}(s) = \frac{i_{inv}(s)}{d(s)} = K_{inv}\frac{1}{s^3L_gL_cC_f + s^2C_fL_gK_{inv} + s(L_c + L_g)k_a} \tag{66}$$

where k_a is the active damping gain.

Finally, we can define the controlled open-loop transfer function as

$$OLTFc_{PR}(s) = C_{PR}(s)dl(s)PWM(s)G_i(s) \tag{67}$$

With such an equation, we can design the digital PR current controller, as demonstrated in the next section.

8.2 Controller Design

The design of the PR controller involves determining the proportional and integral gains, as well as designing the digital resonant filter. The values of the proportional and integral gains are given by (68) and (69), respectively [14–17].

$$k_p = \frac{(2\xi + 1)(\sqrt{(2\xi + 1)}\omega_r L_{eq} - R_{eq}}{V_{DC} h_i} \tag{68}$$

$$k_i = \frac{w_r^2 L_{eq}[(2\xi + 1)^2 - 1]}{2 V_{DC} h_i} \tag{69}$$

Regarding these equations, we have

- L_{eq} represents the total inductance of the LCL filter, defined as $L_{eq} = L_1 + L_2$.
- R_{eq} denotes the total resistance of the inductors in the LCL filter, given by $R_{eq} = R_1 + R_2$.
- ξ is the damping ratio, which should lie within the range of 0 to 1. A value between 0.5 and 1 is generally considered appropriate.
- The calculation of the proportional and resonant gains neglects the effect of the filter capacitance. Consequently, the LCL filter is approximated as a single inductance filter with inductance L_{eq}. This simplification is acceptable depending on the specific values of the LCL components, as discussed in [1]. It is worth noting that at low frequencies, the dynamic behavior of the LCL and L filters is equivalent.
- ω_r refers to the angular resonant frequency, which is selected to match the grid voltage frequency.

The transfer function of the resonant filter in s-domain is given by (70).

$$R(s) = \frac{B_r s}{s^2 + 2\zeta B_r s + w_r^2} \tag{70}$$

where B_r represents the resonant bandwidth and ζ is the damping factor. The value of B_r should be on the order of a few Hz, while ζ can range between 0.1 and 1. Choosing $\zeta = 0.5$ ensures that the resonant filter exhibits 0 dB gain at the resonant frequency. Therefore, this value is adopted in the present design. The digital resonant filter is then given by

$$R(z) = Z\{R(s)\} = \frac{b_0 + b_1 z^{-1} + b_2 z^{-2}}{a_0 + a_1 z^{-1} + a_2 z^{-2}} \tag{71}$$

where $Z\{\cdot\}$ means the Z-transform and it is computed by its original definition [18].

By defining ω as

$$\omega = \sqrt{\omega_r^2 - 0.25 B_r^2} \tag{72}$$

one can write (70) as

$$R(s) = B_r \frac{s + 0.5B_r}{(s + 0.5B_r)^2 + w^2} - \frac{0.5B_r^2}{w} \frac{w}{(s + 0.5B_r)^2 + w^2} \tag{73}$$

The purpose of expressing $H(s)$ as shown in (73) is to facilitate the application of the following pair of equations derived from the Z-transform table.

$$\frac{\omega}{(s + a)^2 + w^2} \Leftrightarrow \frac{e^{-aT_s} z^{-1} sin(\omega T_s)}{1 - 2e^{-aT_s} z^{-1} cos(\omega T_s) + e^{-2aT_s} z^{-2}} \tag{74}$$

$$\frac{s + a}{(s + a)^2 + w^2} \Leftrightarrow \frac{e^{-aT_s} z^{-1} cos(\omega T_s)}{1 - 2e^{-aT_s} z^{-1} cos(\omega T) + e^{-2aT_s} z^{-2}} \tag{75}$$

Using (74) and (75) in (71), the coefficients of the digital resonant filter are given by the following equations:

$$b_0 = B_r T_S \tag{76}$$

$$b_1 = (-B_r e^{-0.5B_r T} cos(T \sqrt{\omega_r^2 - 0.25B_r^2} \quad \psi)) T_S \tag{77}$$

$$b_2 = 0 \tag{78}$$

$$a_0 = 1 \tag{79}$$

$$a_1 = -2e^{-0.5B_r T_S} cos(T_S \sqrt{\omega_r^2 \quad 0.25B_r^2}) \tag{80}$$

$$a_2 = e^{-B_r T_S} \tag{81}$$

where

$$\psi = \frac{0.5B_r^2}{\sqrt{\omega_r^2 - 0.25B_r^2}} e^{-0.5B_r T} sin(T_S \sqrt{\omega_r^2 - 0.25B_r^2}) \tag{82}$$

The T_S in (76) and (77) was included to compensate the process of sampling.

8.3 Results of PR

The system shown in Fig. 20 was tested using a Controller Hardware-in-the-Loop (C-HIL) setup. The power structure is emulated in the real-time simulator Typhoon HIL402, while the digital control strategy is executed on the physical microcontroller TMS320F28335. The system parameters are listed in Table 5 while Table 6 presents the chosen parameter to design the resonant filter.

Table 5 System parameters of the grid-connected inverter

Parameter	Description	Value
U_{rms}	Line-to-line RMS voltage	127 V
P_n	Rated power	1000 W
f_0	Grid frequency	60 Hz
ω_0	Angular grid frequency	$2\pi 60$
f_{sw}	Switching and sampling frequencies	$24k$ Hz
T_s	Sampling period	$\frac{1}{24000}$
h_i	Current sensor gain	$\frac{1}{10}$
V_{DC}	DC-link voltage	225 V
L_1	Converter-side inductance (LCL)	1 mH
L_2	Grid-side inductance (LCL)	300 uH
C_f	Filter capacitance (LCL)	5 uF
R_1	Converter-side resistance (LCL)	0.1Ω
R_2	Grid-side resistance (LCL)	0.1Ω

Table 6 Resonant controller parameters

Parameter	Description	Value
f_r	Desired resonant frequency	60 Hz
ω_r	Desired resonant angular frequency	$2\pi 60$
ζ	Resonant damping factor	0.5
B_s	Desired resonant frequency bandwidth	1.5 Hz
B_r	Desired resonant angular frequency bandwidth	$2\pi 1.5$

Using the parameters from Tables 5 and 6, and following the procedure to design the digital PR current controller, the resulting digital PR controller parameters are listed in Table 7.

The procedure used to tune the digital PR current controller for 60 Hz can also be applied to design a PR controller for a harmonic frequency of interest. To demonstrate this, let us design a PR controller to inject the 5th harmonic into the grid. The equations remain the same, except that the fundamental frequency f_0 is replaced with $5 f_0$. Figure 22 presents the PR controller to inject the fundamental and the 5th harmonic into the grid.

Table 8 presents the coefficients of the designed PR controller tuned at 5th harmonic.

Table 7 Coefficients of the designed digital resonant filter

Parameter	Value
k_p	0.101474487082
k_i	31.62458120614
b_0	0.000392699082
b_1	−0.000392650642
b_2	0.000000000000
a_0	1.000000000000
a_1	−1.999360691418
a_2	0.999607378014

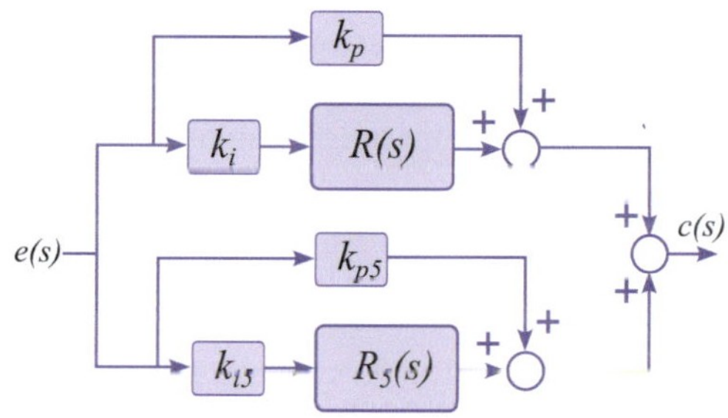

Fig. 22 PR current controller to inject the fundamental frequency and 5th harmonic

Table 8 Coefficients of the digital resonant filter tuned at 5th harmonic

Parameter	Value
k_{p5}	0.542927990968
k_{i5}	790.6145301536
b_0	0.000392699082
b_1	−0.000391488680
b_2	0.000000000000
a_0	1.000000000000
a_1	−1.993443255883
a_2	0.999607378014

After designing the PR controllers, we can analyze their performance using Bode diagrams to verify whether they were properly designed. Figure 23 presents the Bode plots used to evaluate the controllers.

In Fig. 23a, the analog and digital resonant filters are plotted for comparison. It can be observed that the magnitude responses of both controllers are nearly identical around the resonant frequency. Additionally, the gain at 60 Hz is 0 dB, as designed and expected. Although the phase responses differ significantly, they match closely near 60 Hz which is the frequency of interest. This confirms that designing the resonant filter in the s-domain and then discretizing it is a valid and effective methodology for implementing a digital PR controller.

Figure 23b shows the open-loop transfer functions with the PR controller, both considering and neglecting the delay transfer function. As expected, the magnitude responses are coincident, and the phase difference at the cut-off frequency is minimal. This confirms that the delay effect can be reasonably neglected during the design process.

Finally, Fig. 23c illustrates the Bode diagram of a PR controller designed to operate at two resonant frequencies: the fundamental and the 5th harmonic. As shown, the frequency response exhibits two distinct peaks at 0 dB, indicating that the controller can effectively operate at both frequencies.

Figure 24 presents the C-HIL results for the grid-connected inverter with the designed digital PR controller. Figure 24a illustrates the system's response during the startup process. At the beginning, both the current reference signal ($i{*}_{L2}$) and the inverter's output current (i_{L2}) are zero. Then, the reference signal abruptly transitions to a sinusoidal waveform, representing an active power injection with unitary power factor. The inverter's output current quickly aligns with the new reference without exhibiting significant oscillations. Once steady-state conditions are reached, the tracking error becomes minimal. The grid voltage (v_g) as well as the DC voltage (v_{DC}) remains unaffected by the inverter's operation. Figure 24b illustrates the same result, but over a longer time scale.

Figure 24c depicts the system's performance during a step increase in the current reference signal, corresponding to an addition in injected active power. Figure 24d illustrates the system's response when the current reference signal is inverted, indicating a reversal in power flow where the inverter begins to drain active power from the grid. In both scenarios, the inverter's output current successfully tracks the reference signal with minimal steady-state error following the transitions. In the second case, the transition is more abrupt due to the reversal in power direction.

Figure 24e shows the system's response when an oscillation appears in the DC-link voltage. The disturbance has an amplitude of 22.5 V and a frequency of 120 Hz. The inverter's output current remains unaffected by this oscillation, indicating that the designed PR controller is capable of maintaining proper operation even in the presence of such disturbances.

Figure 24f shows the system's response when the inverter is injecting a current containing both the fundamental and 5th harmonic components. As expected, the reference signal includes both frequency components. The inverter output current follows the reference with negligible error, demonstrating the accuracy of the design methodology presented in this chapter. The result also includes the initialization process, which did not exhibit any unpredictable behavior.

9 Controller Hardware-in-the-Loop Setup

Hardware-in-the-Loop technology involves a configuration that combines a Real-Time Simulator (RTS) with an external physical component. This external component can be either a control unit or a hardware device, giving rise to two specific

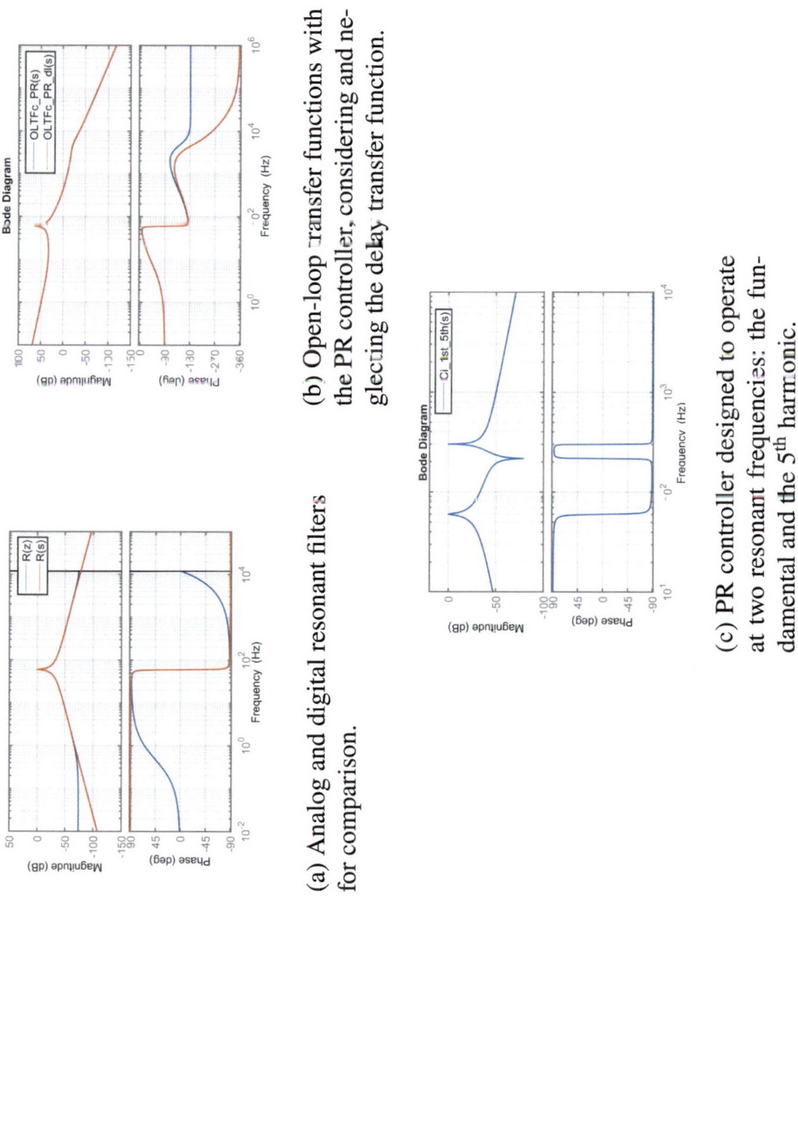

(a) Analog and digital resonant filters for comparison.

(b) Open-loop transfer functions with the PR controller, considering and neglecting the delay transfer function.

(c) PR controller designed to operate at two resonant frequencies: the fundamental and the 5th harmonic.

Fig. 23 Bode plots used to evaluate the PR controllers

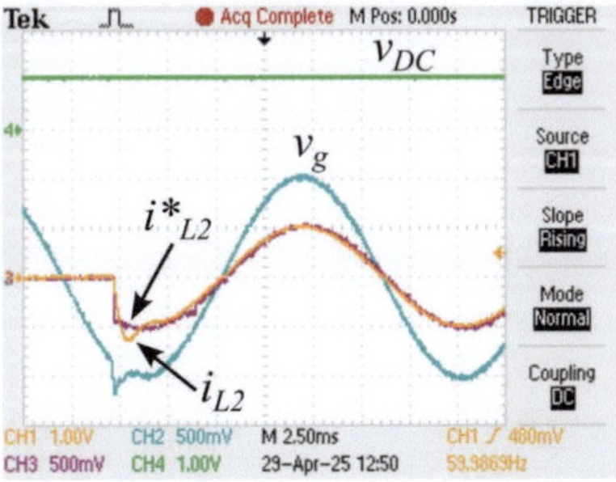

(a) System's response during the startup process.

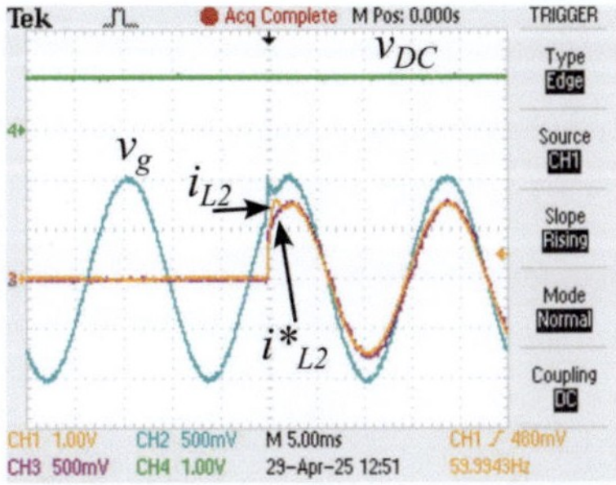

(b) System's response during the startup process over a longer time scale.

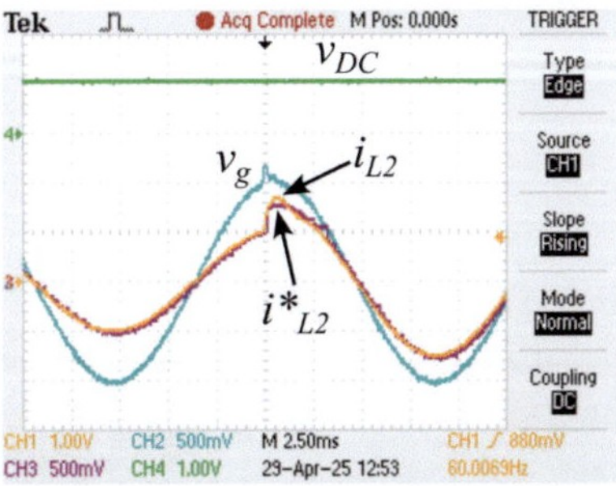

(c) System's performance during a step increase in the current reference signal.

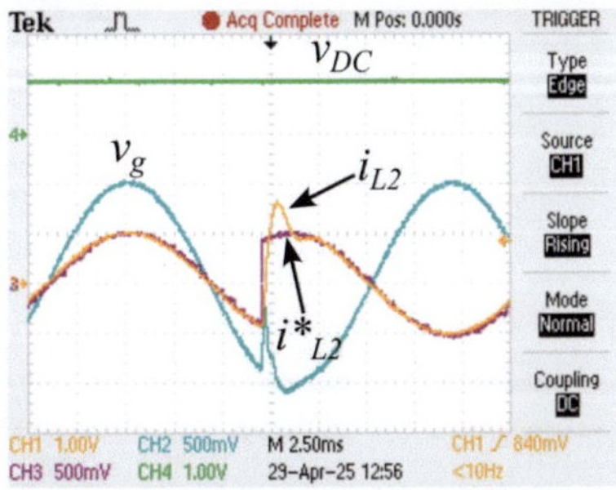

(d) System's response when the current reference signal is inverted, indicating a reversal in power flow.

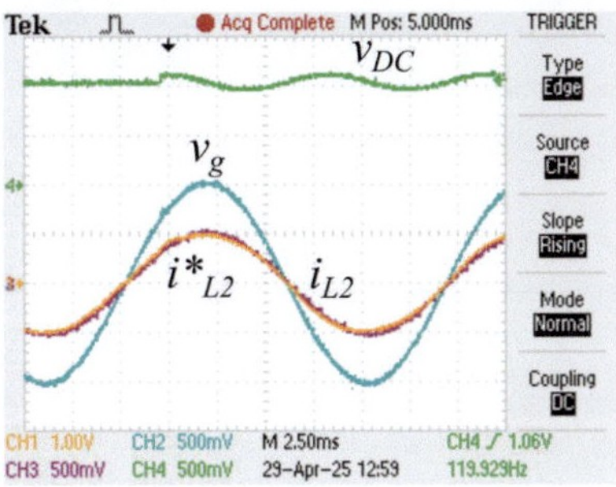

(e) System's response when an oscillation appears in the DC-link voltage.

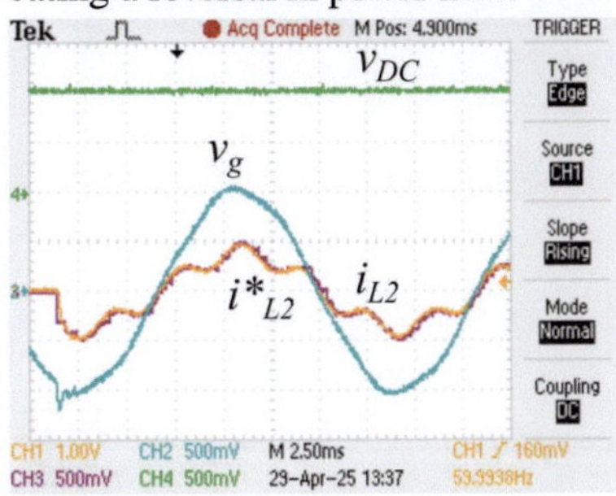

f) System's response when the inverter is injecting a current containing both the fundamental and 5th harmonic components.

Fig. 24 C-HIL results for the grid-connected inverter with the designed digital PR controller

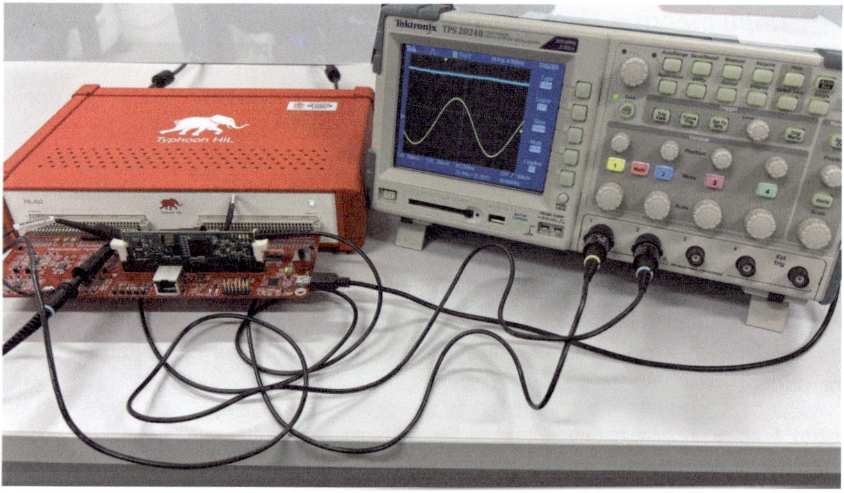

Fig. 25 A picture of the controller hardware in the loop setup

categories: Controller Hardware-in-the-Loop (C-HIL) and Power Hardware-in-the-Loop (P-HIL). The external hardware is typically referred to as the Hardware under Test (HUT) or Device Under Test (DUT) [19, 20].

In a C-HIL configuration, the RTS is paired with an external control unit, which serves as the DUT (commonly a microcontroller, digital signal controller, or digital signal processor). The connection between the RTS and the control unit may be established either directly or via an interface board, depending on the voltage compatibility of their I/O terminals. In a C-HIL setup, the RTS performs real-time simulation of a system, producing waveforms that reflect the dynamic response and behavior of the simulated model. These waveforms are fed into the control unit, which executes its control algorithm and responds by sending command signals back to the RTS.

Validating research through a Controller Hardware-in-the-Loop (C-HIL) setup holds merit, as the control algorithm operates on an actual physical device. When the power plant is accurately represented within the real-time simulator, which features minimal latency, the microcontroller cannot easily discern whether it is interfacing with a physical system or a simulated one. For studies centered on digital control strategies, optimization techniques, or decision-making algorithms, the C-HIL approach offers an attractive option for experimental validation.

Figure 25 shows the C-HIL setup used in this chapter to validate the controllers designed for both the Buck converter and the LCL-filtered grid-connected inverter.

9.1 Implementing the Controller in C Code

The code shown in Listing 1 represents the core implementation of a proportional-resonant controller using C-code for a microcontroller. The values used in the code are those designed for the grid-connected inverter. All other essential components, such as the PLL, PWM, and ADC, must be properly configured beforehand. These subsystems are hardware-dependent and require careful initialization, including setting the appropriate microcontroller registers. These configurations are specific to each microcontroller and should be carried out in accordance with the device's datasheet and reference manuals. This modular approach ensures that the PR controller can be reused flexibly, provided the hardware peripherals are correctly initialized and operational.

Listing 1 Core implementation of a Proportional-Resonant controller using C-code for a microcontroller.

```
    {
        //Resonant Filter order
        const int order = 2;

        // Control gains
        const DefaultType Kp = 0.101474487082;
        const DefaultType Ki  = 31.62458120614;

        // Static filter buffers
        static DefaultType veBuffer[3] = {0.0, 0.0, 0.0};
        static DefaultType vcBuffer[3] = {0.0, 0.0, 0.0};

        // Filter coefficients
        const DefaultType A[3] =
            {1.000000000000,-1.999360691418,0.999607378014};
        const DefaultType B[3] = {0.392699082,-0.392650642,0.0};

        // Apply the input error
        DefaultType vc = B[0] * ve;

        for (int i = order; i > 0; --i) {
            veBuffer[i] = veBuffer[i - 1];
            vcBuffer[i] = vcBuffer[i - 1];
            vc += B[i] * veBuffer[i] - A[i] * vcBuffer[i];
        }

        // Update buffer
        vcBuffer[0] = vc;
        veBuffer[0] = ve;
    }

    // Compute control output
    Output = Kp * ve + vc * Ki;
```

References

1. Teodorescu R, Liserre M, Rodriguez P (2011) Grid converters for photovoltaic and wind power systems. Wiley, IEEE Press. ISBN: 1119957206, 9781119957201, 416 p
2. Yepes AG, Freijedo FD, Lopez Ó, Doval-Gandoy J (2011) High-performance digital resonant controllers implemented with two integrators. IEEE Trans Power Electron 26(2):563–576
3. Buso S, Mattavelli P (2015) Digital control in power electronics, 2nd edn. Wiley
4. Franklin GF, Powell JD, Workman ML (1997) Digital control of dynamic systems, 3rd edn. Addison-Wesley
5. Robert W, Erickson D, Maksimović (2020) Fundamentals of power electronics, 3rd edn. Springer
6. Corradini L, Maksimovic D, Mattavelli P, Zane R (2015) Digital control of high-frequency switched-mode power converters, 1st edn. IEEE Press
7. Yaramasu V, Wu B (2016) Model predictive control of wind energy conversion systems, 1st edn. Wiley
8. Oppenheim AV, Willsky AS, Nawab SH (1997) Signals and systems, revised. Pearson Educación, Prentice-Hall Signal Processing Series. ISBN: 970170116X, 9789701701164, 957 p
9. Yepes AG, Freijedo FD, Doval-Gandoy J, López Ó, Malvar J, Fernandez-Comesaña P (2010) Effects of discretization methods on the performance of resonant controllers. IEEE Trans Power Electron 25(7):1692–1712
10. Lyons RG (2011) Understanding digital signal processing, 3rd edn. Pearson Education International. ISBN: 0132119374, 9780132119375, 925 p
11. Liao Y, Wang X (2020) Impedance-based stability analysis for interconnected converter systems with open-loop RHP poles. IEEE Trans Power Electron 35(4):4388–4397. https://doi.org/10.1109/TPEL.2019.2939636
12. Dean Venable H (1983) The K factor: a new mathematical tool for stability analysis and synthesis. In: Proceedings of powercon 10, Rancho Palos Verdes, California, March 1983
13. Golestan S, Guerrero JM, Vasquez JC (2017) Three-phase PLLs: a review of recent advances. IEEE Trans Power Electron 32(3):1894–1907
14. Bacha S, Munteanu I, Bratcu AI (2014) Power electronic converters modeling and control: with case studies. Advanced textbooks in control and signal processing. Springer. ISBN: 978-1447154778
15. Busarello TDC, Pomilio JA, Simoes MG (2018) Design procedure for a digital proportional-resonant current controller in a grid connected inverter. In: 2018 IEEE 4th southern power electronics conference (SPEC), pp 1–8
16. Busarello TDC, Guerreiro JF, Simões MG, Pomilio JA (2021) Hardware-in-the-loop experimental setup of a LCL-filtered grid-connected inverter with digital proportional-resonant current controller. In: 2021 IEEE 22nd workshop on control and modelling of power electronics (COMPEL), pp 1–8
17. Busarello TDC, Zeb K, Simões MG (2020) Highly accurate digital current controllers for single-phase LCL-filtered grid-connected inverters. Electricity 1(1):12–36
18. Ogata K (1995) Discrete-time control systems, 2nd edn. Pearson. ISBN: 978-0130342812
19. Osório CRD, Miletic M, Zelic J, Majstorovic D, Gagrica O (2021) Advancements on real-time simulation for high switching frequency power electronics applications (invited paper). In: 2021 21st international symposium on power electronics (Ee), pp 1–6
20. Busarello TDC, Simões MG, Laaksonen H (2023) Hardware-in-the-loop technology and applications in power electronics. In: Power electronic converters and systems. Volume 2: Applications. The Institution of Engineering and Technology, Chap 15

Lyapunov-Based Control Methods for Power Converters

Khaled Rayane, Mohamed Trabelsi, and Hani Vahedi

Abstract This chapter presents a comprehensive investigation into the application of Lyapunov-based control methods for power electronic converters, with a focus on a DC–DC boost converter and a 5-level Packed U-Cell (PUC5) inverter. Lyapunov-based control offers a rigorous framework for designing nonlinear controllers that ensure global asymptotic stability, robustness against parameter variations, and fast dynamic response. Through a systematic design of Lyapunov Control Functions (LCFs), the proposed controllers regulate output voltage and current trajectories with high precision under diverse operating conditions, including load disturbances and input voltage fluctuations. The chapter details both the direct and indirect Lyapunov control approaches, highlighting their mathematical formulation, stability criteria, and practical implementation. A Lyapunov-based controller is first developed for a boost converter, demonstrating robust voltage regulation and excellent transient performance through simulation results. The proposed approach is then applied to a PUC5 inverter, where a multi-objective control scheme is developed to simultaneously regulate the capacitor voltage and the output current. The control design is validated via MATLAB/Simulink simulations and comparative studies with conventional PI controllers, showing superior performance in terms of transient behavior, steady-state accuracy, and Total Harmonic Distortion (THD) even under parameter mismatches and varying source conditions. These findings underscore the suitability of Lyapunov-based methods for next-generation power electronic systems in renewable energy, electric mobility, and smart grid applications.

K. Rayane
LACoSERE Lab., University of Laghouat, Laghouat, Algeria

M. Trabelsi
Department of Electronics and Communications Engineering, Kuwait College of Science and Technology, Doha, Kuwait
e-mail: m.trabelsi@kcst.edu.kw

H. Vahedi (✉)
Department of Electrical Engineering, Mathematics, and Computer Science, Delft University of Technology, Delft, The Netherlands
e-mail: Hani.Vahedi@ieee.org

E. Kabalci (ed.), *Modern Control Methods for Power Electronics*, Energy Systems in Electrical Engineering, https://doi.org/10.1007/978-981-92-1974-2_4

Keywords Lyapunov control · Boost converter · PUC inverter · Nonlinear control · Voltage regulation · Power electronics

1 Introduction

Lyapunov-based control methodologies have emerged as a powerful framework for designing robust controllers in nonlinear systems, particularly in power electronic converters that operate under complex, time-varying, and uncertain conditions (Kaźmierkowski et al. 2002; Kassakian et al. 2023). Unlike conventional linear controllers—such as proportional-integral-derivative (PID) or linear quadratic regulators (LQR)—Lyapunov-based techniques provide formal mathematical guarantees of stability and can handle strong nonlinearities, abrupt load changes, parameter variations, and switching behaviors typical of modern power conversion systems (Komurcugil et al. 2023; Cecati et al. 2021).

Over the past decade, increasing research has focused on Lyapunov-based control due to its capacity to systematically manage large-signal disturbances and ensure global or asymptotic stability across a wide operating range. For instance, in Li et al. (2023), adaptive Lyapunov controllers have been proposed for DC–DC converters under input and load uncertainties, while (Kabalan et al. 2016) demonstrates their use in maintaining voltage regulation in multilevel inverters under grid disturbances. Additionally, backstepping sliding mode control (Oubbati et al. 2020) and passivity-based extensions (Min et al. 2022) have all been developed from the core Lyapunov framework to enhance performance and robustness under real-world operating scenarios.

Recent studies have emphasized the limitations of classical control methods when applied to power converters with nonlinear dynamics. Linear approaches often require linearization around specific operating points, which compromises performance when the system deviates from nominal conditions. In contrast, Lyapunov-based designs directly account for the system's nonlinear behavior and can deliver better tracking, reduced harmonic distortion, and enhanced fault tolerance. As shown in Kabalan et al. (2016), Lyapunov-based methods have proven particularly effective in renewable energy systems, electric vehicles, and smart grids, where dynamic conditions and fault resilience are essential.

In practical applications, Lyapunov-based controllers have been successfully implemented in various converter topologies. For example, Ma et al. (2022) presents a globally stable controller for a boost converter using a quadratic Lyapunov function, while Shao et al. (2017) applies indirect Lyapunov design to ensure smooth voltage transitions in cascaded inverter systems. Furthermore, hybrid strategies combining Lyapunov control with adaptive estimation, neural networks, or disturbance observers have recently been proposed to improve robustness and adaptability (Abate et al. 2020; Queiroz et al. 2001).

This chapter presents a comprehensive overview of Lyapunov-based control strategies tailored for power electronic converters. It begins with foundational concepts, including Lyapunov function selection, direct and indirect methods, and conditions for system stability. The theoretical discussion is supported by detailed case studies involving a DC–DC boost converter and a five-level Packed U-Cell (PUC5) inverter. For each system, the controller design is developed from system modeling to stability analysis and implementation.

The chapter also includes extensive simulation results and comparative performance evaluations against classical control methods. Key implementation considerations—such as computational cost, sensitivity to parameter variations, and real-time applicability—are critically discussed. By integrating theoretical depth with applied insights, this chapter aims to equip readers with the tools and understanding necessary to design and implement Lyapunov-based controllers that enhance the reliability, efficiency, and stability of modern power conversion systems in renewable energy and beyond.

2 Lyapunov-Based Control Design for Power Converters

Lyapunov-based control methods guide power converters to a steady state by controlling the system's stored energy variation over time. The controller uses a designed scalar function $V(x)$ to quantify the deviation of the converter's current state from its desired equilibrium. By designing $V(x)$ to decrease along every trajectory, the controller enforces a smooth transient response and ensures global stability even under large disturbances (Lu et al. 2013).

In this framework, the Lyapunov function is constructed from the energy stored in reactive elements. For a converter with n_L inductors and n_c capacitors, one possible choice is

$$V = \frac{1}{2}\sum_{j=1}^{n_L} L_j\left(i_{L_j} - i^*_{L_j}\right)^2 + \frac{1}{2}\sum_{k=1}^{n_C} C_j\left(V_{C_k} - V^*_{C_k}\right)^2, \tag{1}$$

where $i^*_{L_j}$ and $V^*_{C_k}$ are the inductor currents and capacitor voltages at the desired operating point. The control law is then derived by requiring $\dot{V}(x) < 0$ for all non-equilibrium states, which produces an algebraic expression for the duty cycle or switching signal. Saturation limits can be handled naturally: Even when the control input saturates, $V(x)$ remains strictly decreasing, driving the system back toward its equilibrium.

Practical implementations demonstrate that Lyapunov Function Control delivers more uniform transient behavior across both current and voltage dynamics, without requiring excessive current injection. In modular or cascaded converter architectures, this approach extends straightforwardly: each stage employs its own Lyapunov function and tuning parameter, directly influencing convergence speed. Robustness to load disturbances can be enhanced by embedding simple adaptive laws or integrators into the Lyapunov framework, yielding zero steady-state error and accurate reference tracking with minimal computational overhead.

Although selecting an appropriate Lyapunov function and reference values relies on engineering insight rather than a fully automated procedure, the resulting controllers are easy to implement and offer strong, analytically guaranteed stability for a wide range of power conversion applications.

Lyapunov's stability theory provides a mathematical foundation for analyzing the stability of nonlinear dynamical systems, including power converters. The core idea is to construct a Lyapunov function $V(x)$ a scalar, positive-definite energy-like function that decreases over time $\left(\dot{V}(x) \leq 0\right)$, ensuring asymptotic stability. For a power converter described by state-space dynamics:

$$\dot{x} = f(x, u), \tag{2}$$

where x is the state vector (e.g., inductor currents, capacitor voltages) and u is the control input (e.g., duty cycle), the control objective is to derive u such that $V(x)$ decreases along system trajectories. Figure 1 shows the steps involved in the Lyapunov function-based control approach (Wu et al. 2021).

The Lyapunov-based control framework includes two primary approaches: the direct method and the indirect method, both illustrated in the conceptual flowchart presented in Fig. 1. These methods differ in how the Lyapunov function is applied to ensure system stability, particularly in nonlinear systems such as power converters.

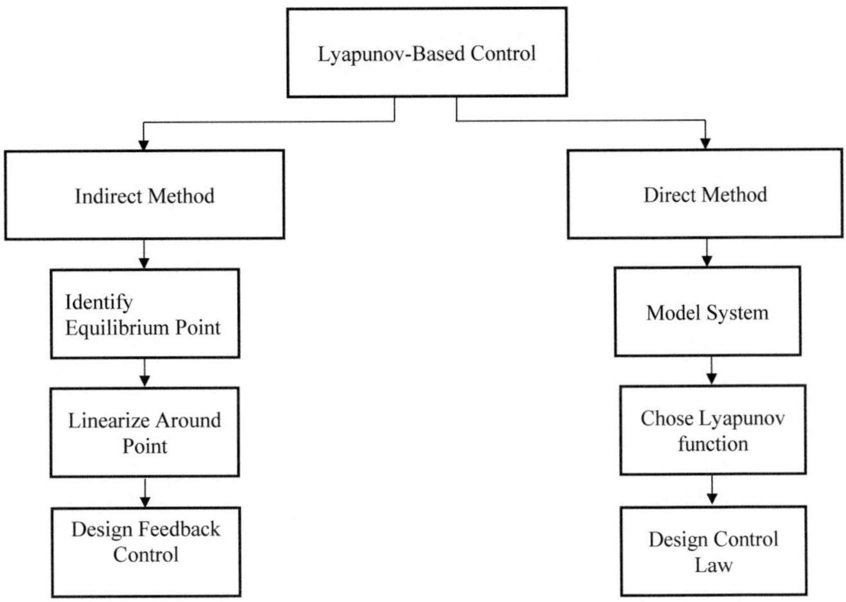

Fig. 1 Steps of direct and indirect of Lyapunov-based control

In the direct method, a suitable positive-definite Lyapunov function—commonly expressed as $V(x) = \frac{1}{2}x^T P(x)$, with $P > 0$ is explicitly constructed based on the nonlinear system dynamics. The control law is then designed to ensure that the derivative of the Lyapunov function along system trajectories is negative definite, i.e., $\dot{V}(x) < 0$. This condition guarantees asymptotic stability and is particularly well-suited for systems subject to large-signal variations or strong nonlinearities. Unlike conventional methods, the direct approach does not rely on system linearization, allowing it to offer global or semi-global stability guarantees. However, the primary challenge lies in selecting or formulating an appropriate Lyapunov function, especially for systems with complex or high-order dynamics.

In contrast, the indirect method involves linearizing the nonlinear system around an equilibrium point and applying Lyapunov stability theory to the resulting linearized model. The stability condition is typically assessed using the continuous-time Lyapunov equation:

$$A^T P + PA = -Q, Q > 0,$$

where A is the Jacobian of the system evaluated at the equilibrium point. This approach simplifies the controller design process and is easier to implement, but it only guarantees local stability around the nominal operating point. It may perform poorly under large perturbations or in systems with significant nonlinearities.

Figure 2 outlines the structured design process of both methods, showing the flow from system modeling to Lyapunov function selection, control law synthesis, and implementation. This schematic serves as a reference for the control design procedures discussed throughout the chapter. Specifically, the process of computing the nominal control input at equilibrium and designing the perturbed control input is detailed in Sect. 4 for the DC–DC boost converter and in Sect. 5 for the PUC5 inverter. These case studies demonstrate how each Lyapunov-based approach is applied in practice and highlight the trade-offs between design complexity, stability guarantees, and performance.

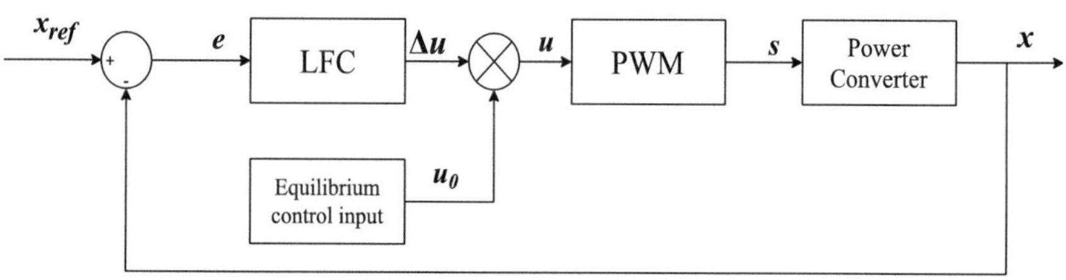

Fig. 2 Comprehensive block diagram of the Lyapunov function-based control strategy for a power converter

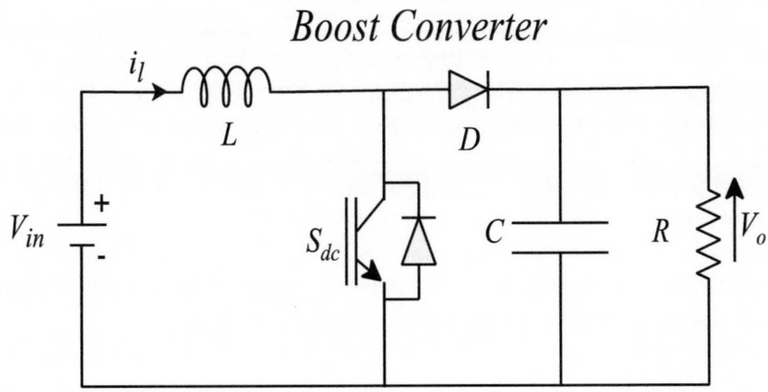

Fig. 3 Circuit diagram of a DC–DC boost converter

3 Lyapunov-Based Control for a DC–DC Boost Converter

3.1 System Modeling

A simplified averaged model of the boost converter under study (Fig. 3) is used for control design. Assume continuous conduction mode (CCM) (Forouzesh et al. 2017).

The classical differential equations describing the dynamics of the boost converter are expressed as follows:

$$\begin{cases} L\frac{di_l}{dt} = -(1-u)V_o + V_{\text{in}} \\ C\frac{dV_o}{dt} = (1-u)i_l - \frac{V_0}{R} \end{cases}. \tag{3}$$

Here, u denotes the control input, with the primary objective of regulating the load voltage (V_o) amidst variations in load (R), reference load voltage (V_o^*), and input voltage (V_{in}). Additionally, i_L represents the inductor current. The state variables and control input are defined as follows:

$$x = \begin{bmatrix} x_1 \\ x_2 \end{bmatrix} = \begin{bmatrix} i_l^* - i_l \\ V_o^* - V_o \end{bmatrix} u = u_o + \Delta u. \tag{4}$$

3.2 Control Design

First, let us define the Lyapunov function as follows:

$$V(x) = \frac{1}{2}Lx_1^2 + \frac{1}{2}Cx_2^2. \tag{5}$$

The chosen Lyapunov function directly mirrors the converter's total stored "energy." Its first term captures the energy in the inductor L, while the second term accounts for the energy in the capacitor C. By construction, it is zero at the origin ($V(0) = 0$), strictly positive for any nonzero state ($V(x) > 0$ if $x \neq 0$), and unbounded as the state norm grows $\left(V(x) \to \infty \text{ as } \sqrt{x_1^2 + x_2^2} \to \infty \right)$.

The controller inputs are defined through the fulfillment of the stability condition of the Lyapunov function, the derivative of the Lyapunov function is given as:

$$\dot{V}(x) = Lx_1\dot{x}_1 + Cx_2\dot{x}_2. \tag{6}$$

Substituting the original dynamics:

$$\dot{x}_1 = -\frac{1}{L}(-(1-u)V_o + V_{in}),\ \dot{x}_2 = -\frac{1}{C}\left((1-u)i_l - \frac{V_o}{R}\right). \tag{7}$$

Given that the actual variables align with their reference values in steady-state condition where $V_o = V_o^*$, $i_l = i_l^*$, and $u = u_0$, the following equations can be derived

$$\begin{array}{c} L\frac{di_l^*}{dt} = -(1-u_0)V_o^* + V_{in} \\ C\frac{dV_o^*}{dt} = (1-u_0)i_l - \frac{V_0^*}{R} \end{array}. \tag{8}$$

Assuming steady-state conditions $\left(\frac{di_l^*}{dt} = 0,\ \frac{dV_o^*}{dt} = 0\right)$, the error dynamics in the steady state is given as:

$$\begin{array}{c} L\dot{x}_1 = x_2u_0 + x_2\Delta u - x_2 + V_o^*\Delta u \\ C\dot{x}_2 = x_1 - x_1\Delta u - \frac{x_1}{R} - i_l^*\Delta u \end{array}. \tag{9}$$

Substituting Eqs. (8) and (7) into Eq. (6), we obtain:

$$\frac{dV(x)}{dt} = \frac{x_1^2}{R} - \left(i_l^*x_1 - V_o^*x_2\right)\Delta u. \tag{10}$$

To satisfy the Lyapunov stability condition $\dot{V}(x) < 0$, we calculate perturbation Δu control input and the equivalent control input u_0.

At equilibrium ($x_1 = 0,\ x_2 = 0$), the nominal control u_0 should satisfy:

$$V_{\text{in}} + V_o^*u_0 - V_o^* = 0 \Rightarrow u_0 = 1 - \frac{V_{in}}{V_o^*}. \tag{11}$$

This ensures a steady-state operation.

To ensure $\dot{V}(x) < 0$, we choose Δu to cancel cross terms and introduce damping:

$$\Delta u = -k_1 x_1 - k_2 x_2, \quad (12)$$

where $k_1, k_2 > 0$ are the control gain

Thus, the control input u is given by:

$$u = u_0 + \Delta u = 1 - \frac{V_{in}}{V_o^*} + -k_1 x_1 - k_2 x_2. \quad (13)$$

3.3 Simulation Results

The performance of the closed-loop system is evaluated through MATLAB/Simulink simulations, with the system and control parameters listed in Table 1.

Figure 4 presents the responses of the output voltage and inductor current during a reference output voltage change, where V_o^* is adjusted from 50 to 40 V. The results demonstrate the controller's ability to effectively track the new reference, with the output voltage smoothly transitioning to 40 V while maintaining stability. The inductor current also adjusts accordingly to support the voltage regulation, reflecting the controller's robust performance in handling reference changes.

Figure 5 depicts the responses of the output voltage and inductor current following an abrupt change in load resistance R from 100 to 50 Ω, with an input voltage V_{in} of 25 V. Although the output voltage experiences an undershoot due to the sudden load change, the controller effectively drives the output voltage to track its reference, stabilizing it at 50 V. Concurrently, both the inductor current and its reference adapt to the sharp decrease in R, exhibiting a gradual adjustment in their values. Notably, the dynamic response of the output voltage and inductor current can be further enhanced by optimizing the controller gains.

Figure 6 illustrates the dynamic responses of the output voltage and inductor current in the presence of a step change in the DC input voltage, where V_{in} drops from 30 to 20 V. This scenario emulates a sudden disturbance commonly encountered in renewable energy sources such as photovoltaic systems, where input variations may result from shading, cloud transients, or source fluctuations.

Table 1 Parameters of the system with Boost DC Converter

Parameters	Values
DC voltage V_{in}	25 V
Output voltage reference V_o^*	50 V
Boost inductor L	1 mH
Boost capacitor C	100 μF
Resistive load R	100 Ω
Switching frequency f_s	5 kHz

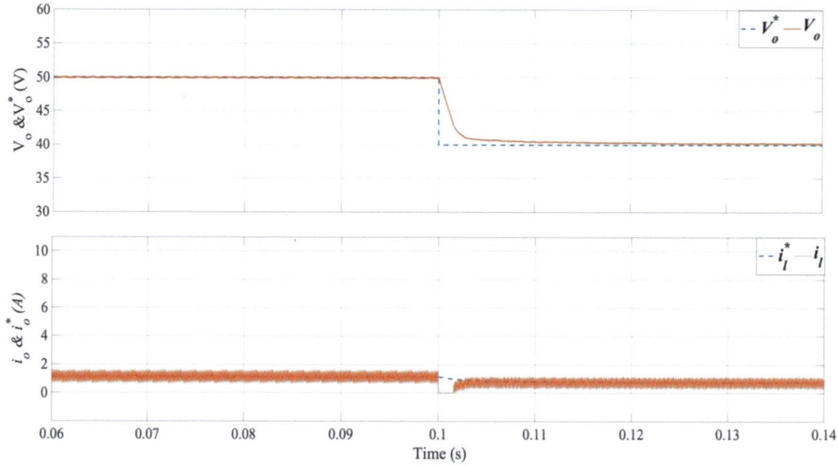

Fig. 4 Tracking performance of the output voltage and inductor current following a sudden change in the output voltage from 50 to 40 V

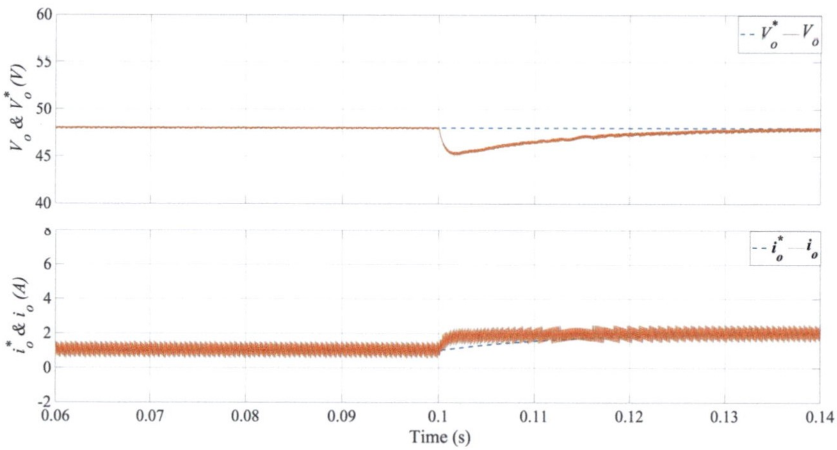

Fig. 5 Responses of the output voltage and inductor current following a sudden change in R from 100 to 50 Ω

As shown in the figure, the output voltage exhibits a minor undershoot immediately following the input drop. However, it rapidly returns to its reference value with minimal delay, indicating fast transient recovery and effective voltage regulation. The inductor current simultaneously adapts to the new operating condition, increasing its average value to compensate for the lower input and maintain power balance at the output.

These results highlight the controller's robustness and adaptability under changing input conditions. The system remains stable throughout the disturbance and continues

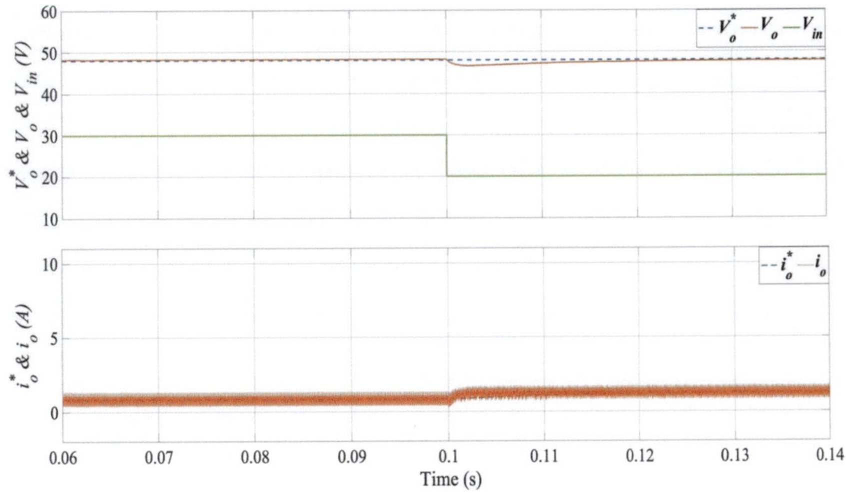

Fig. 6 Responses of the output voltage and inductor current following a sudden change in V_{in} from 30 to 20 V

to deliver a regulated output without oscillations or instability. The smooth and coordinated adjustment of both voltage and current further demonstrates the effectiveness of the Lyapunov-based control strategy in handling real-world scenarios where input variability is inevitable.

3.4 *Comparative Study of LFC and PI Controllers*

A comparative performance study was conducted between the nonlinear Lyapunov Function Control (LFC) approach and the conventional linear Proportional-Integral (PI) controller to evaluate their effectiveness in regulating the output voltage of the power converter. The PI controller parameters were obtained using MATLAB's built-in auto-tuning feature to ensure optimal performance under standard conditions.

Figure 7 presents the transient and steady-state response of the system using both control strategies. As illustrated, the Lyapunov-based controller significantly outperforms the PI controller in terms of transient behavior. Specifically, the LFC achieves a faster rise time, reduced overshoot, and quicker convergence to the desired voltage level. In addition to superior transient response, the LFC also demonstrates improved steady-state performance, maintaining the output voltage with noticeably lower ripple compared to the PI-controlled system.

These results highlight the advantages of Lyapunov-based control in handling nonlinear dynamics and disturbances more effectively than traditional linear methods. The LFC approach provides stronger stability guarantees and improved

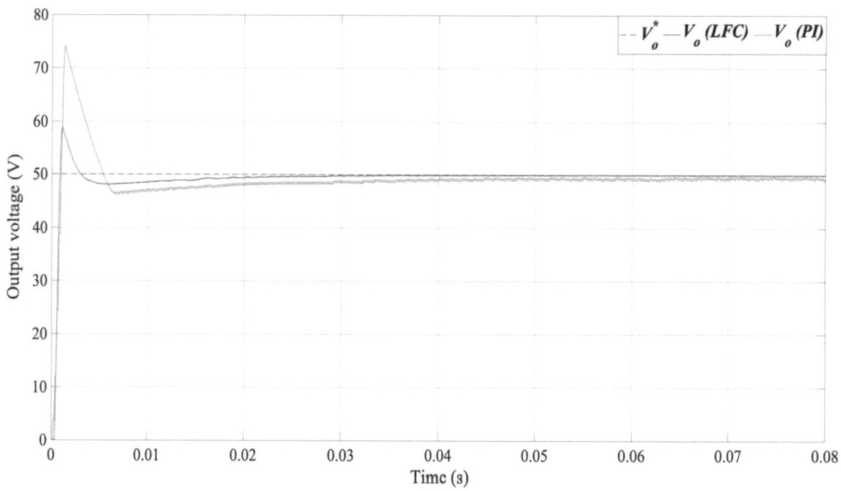

Fig. 7 Transient and steady-state performance of output voltage of the boost converter under LFC and PI controllers

dynamic response, making it particularly well-suited for applications where precise voltage regulation and fast settling time are critical.

4 Lyapunov-Based Control for Multilevel PUC5 Inverter

The general structure of an *n*-cell Packed U-Cell (PUC) inverter is illustrated in Fig. 8. This topology is built on a series connection of switching cells—each consisting of a pair of complementary switches (s_n, $\bar{s}_n$) and a capacitor (C_n). Additionally, a packed cell, made up of a DC source (E) and two complementary switches (s_p, $\bar{s}_p$), enables switching between positive and negative output voltage levels. The total number of output voltage levels is determined by the number of series-connected inverter cells (Vahedi et al. 2015; Rayane et al. 2023).

While the PUC inverter reduces the need for multiple isolated DC sources, capacitors, and switches, it faces challenges due to the varying voltage ratings required for its switches. However, the design ensures that only the two switches operating at the fundamental frequency handle the higher voltages, while the remaining high-frequency switches are subjected to lower voltage stress, matching standard semiconductor device ratings.

Due to its complex structure—especially as the number of voltage levels increases—the operation modes of a single-phase PUC inverter must be systematically analyzed. The system reaches equilibrium when each cell's voltage equals *E/n*, requiring the floating capacitor voltages to be regulated at:

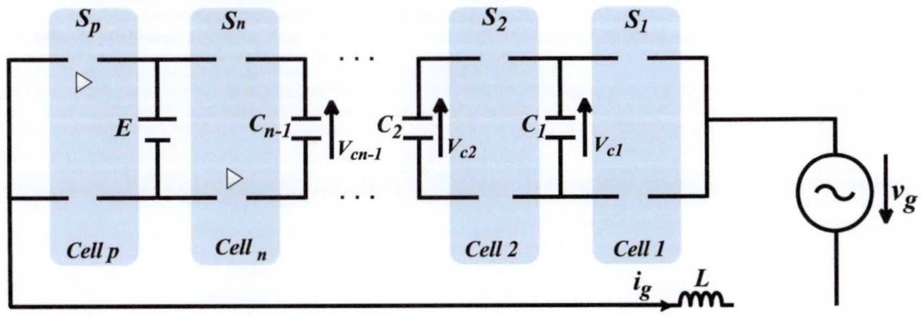

Fig. 8 General structure of PUC converter

$$V_{ck} = k \times \frac{E}{n}, \quad \text{where } k = 1 \ldots n - 1. \tag{14}$$

In theory, a multicellular PUC inverter can achieve an unlimited number of voltage levels, as the level count scales with the number of connected cells. Typically, the output voltage features $2n + 1$ levels, resulting in a lower *dv/dt* compared to traditional multilevel inverter (MLI) topologies.

In this chapter, we focus on a configuration consisting of two two-cell Packed U-Cell (PUC) inverters connected to a common inductive load. This setup, known as the PUC5 inverter, is capable of generating five distinct output voltage levels, making it a suitable candidate for medium-voltage and high-performance power conversion applications. The five-level output is achieved through strategic switching of the PUC topology, which offers advantages such as reduced voltage stress on components, improved output waveform quality, and enhanced power conversion efficiency.

The following section presents a comprehensive mathematical model of the PUC5 inverter operating in stand-alone mode. This model forms the foundation for the design and implementation of a Lyapunov-based control strategy. The proposed controller is specifically developed to meet three key objectives: (1) to accurately synthesize five-level stepped output voltages, (2) to regulate and stabilize the floating capacitor voltage at its desired reference value, and (3) to shape the load current into a sinusoidal waveform with minimal THD. By leveraging the nonlinear control capabilities of Lyapunov theory, the controller ensures robust performance, even under dynamic load conditions and parameter uncertainties, thus enhancing the overall stability and efficiency of the PUC5 inverter system (Vahedi et al. 2017; Rayane et al. 2020).

4.1 System Modeling

The configuration of single-phase stand-alone PUC5 inverter is presented in Fig. 9.

The dynamic behavior of the two-cell Packed U-Cell (PUC) inverter is described by the following mathematical model:

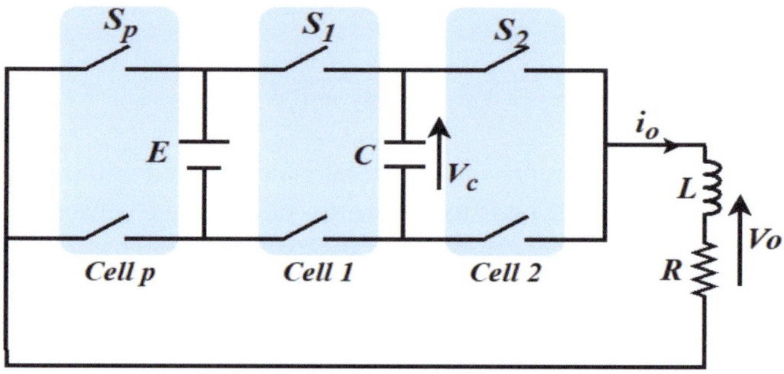

Fig. 9 Five-level packed U-Cell converter

$$\begin{aligned} \frac{dVc}{dt} &= \left(\frac{i_o}{C}\right) \times (u_2 - u_1) \\ \frac{di_o}{dt} &= \left((u_1 - u_2) \times \frac{V_c}{L}\right) + \left(\frac{E}{L}\right) \times \left(u_p - u_1\right) - \left(\frac{R}{L}\right) i_o \text{'} \end{aligned} \tag{15}$$

where V_c and V_o represent the capacitor and output voltages, respectively. i_o is the output current, and E is the DC source voltage. u_1 and u_2 denote the average control inputs. u_p is the control input that generates the negative output voltage, toggling between 1 and 0 according to the reference current i_o^*, as defined below:

$$\begin{aligned} u_p = \{1 \text{ if } i_o^* \geq 0 \\ 0 \text{ if } i_o^* < 0\}. \end{aligned} \tag{16}$$

The system can be represented using a state-space model as follows:

$$\frac{dx}{dt} = Ax + B(x)u, \tag{17}$$

where $x = [V_c i_o]^{\mathrm{T}}$ is the continuous state vector, and $u = [u_1 u_2]^{\mathrm{T}}$ is the control input vector.

4.2 Control Design

The objective of this work is to develop a robust controller capable of regulating the capacitor voltage (V_c) and the load current (i_o) to their desired reference values. To address this multi-objective control challenge in the PUC5 converter, a Lyapunov-based control strategy is proposed, ensuring both high-quality voltage conversion and strong robustness against disturbances. We define the Lyapunov function as follows:

$$V(x) = \frac{1}{2} C x_1^2 + \frac{1}{2} L x_2^2, \tag{18}$$

where:

$$x_1 = V_c^* - V_c, x_2 = i_o^* - i_o \tag{19}$$

This Lyapunov function represents the total energy stored within the converter system. The first term corresponds to the energy stored in the capacitor (C), while the second term reflects the energy stored in the load inductor (L).

The proposed control strategy can be developed according to the stability condition of Lyapunov function described by:

$$\dot{V}(x) < 0. \tag{20}$$

We first compute the time derivative of $V(x)$:

$$\frac{dV(x)}{dt} = Cx_1\frac{dx_1}{dt} + Lx_2\frac{dx_2}{dt}. \tag{21}$$

Substituting the error dynamics will lead to:

$$\begin{aligned} \tfrac{dx_1}{dt} &= -\tfrac{i_o}{C}(u_2 - u_1) \\ \tfrac{dx_2}{dt} &= \tfrac{di_{oref}}{dt} - \left(\tfrac{V_c}{L}(u_1 - u_2) + \tfrac{E}{L}\big(u_p - u_1\big) - \tfrac{R}{L}i_o\right) \end{aligned}. \tag{22}$$

Substitute the given system dynamics in Eqs. (19) and (22) to Eq. (21), we have:

$$\frac{dV(x)}{dt} = -x_1 i_o((u_2 - u_1)) + x_2\left(L\frac{di_{oref}}{dt} - (u_1 - u_2)V_c + E\big(u_p - u_1\big) - Ri_o\right). \tag{23}$$

The stability of the converter can be achieved if $\frac{dV(x)}{dt} < 0$ holds.

To enforce $\frac{dV}{dt} = -K_v x_1^2 - K_i x_2^2$ (where $K_v, K_i > 0$), we design u_1 and u_2.

Step 1: Voltage Tracking Term

Choose $u_2 - u_1$ to cancel e_V coupling:

$$u_2 - u_1 = K_V x_1 + \phi(x_2), \tag{24}$$

where $\phi(x_2)$ is a stabilizing term for e_i.

Step 2: Current Tracking Term

Substitute $u_2 - u_1$ into $\frac{dV}{dt}$ and solve for u_1:

$$u_1 = u_p + \frac{Ri_o + L\frac{di_{oref}}{dt} - V_c(K_V x_1 + \phi(x_2)) + K_i x_2}{E}. \tag{25}$$

Let $(x_2) = K_i x_2$, ensuring damping for x_2.
The final control inputs are than give as:

$$
\begin{aligned}
u_1 = u_p + \frac{Ri_o + L\frac{di_{oref}}{dt} - V_c(K_V x_1 + K_i x_2) + K_i x_2}{E} \\
u_2 = u_1 + K_v x_1 + K_i x_2
\end{aligned}
. \tag{26}
$$

4.3 Simulation Results

The schematic of the Lyapunov control on the PUC5 inverter is shown in Fig. 10. The parameters of the inverter are listed in Table 2.

Figure 11 illustrates the transient dynamics and steady-state voltage of the capacitor, the output voltage, and the load current. The capacitor voltage and the load current converge to their respective references successively with high performance and low steady-state error. Additionally, it should be noted that the capacitor voltage ripples are approximately ±2 V, while the Total Harmonic Distortion (THD) of the injected grid current is around 2.03%, as indicated by the FFT harmonic spectrum shown in Fig. 12. Figure 13 shows the performance of the current tracking capability when there a sudden change in the load resistor.

A test was performed by incrementally varying the DC source voltage E from 200 to 240 V over a controlled period (see Fig. 14). The resulting graph highlights that the capacitor voltage rapidly and reliably aligns with the updated reference, maintaining stability throughout the transition. These outcomes underscore the controller's exceptional dynamic performance, showcasing its ability to handle voltage fluctuations effectively while ensuring consistent system response under varying conditions.

Figure 15 illustrates the performance comparison between the Lyapunov controller and the PI controller for a system where a switch from the PI to the Lyapunov controller occurs at time t_1. The results highlight the Lyapunov controller's superior performance, demonstrating reduced error and enhanced stability. Notably,

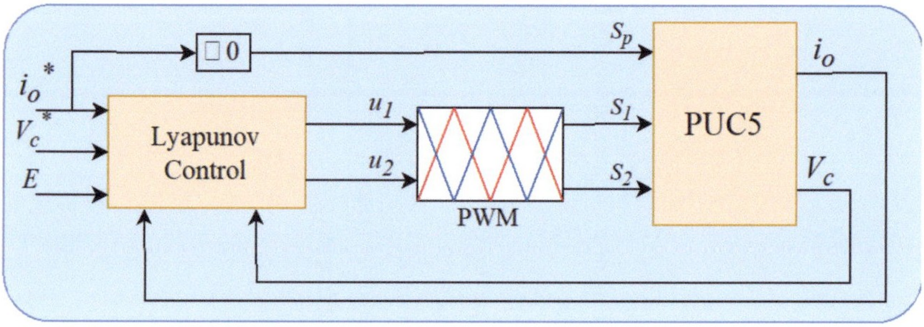

Fig. 10 Lyapunov control of PUC5 inverter

Table 2 Parameters of the system with PUC5 converter

Parameters	Values
DC voltage E	200 V
PUC5 capacitor C	100 μF
Load inductor L	5 mH
Load resistor R	40 Ω
Fundamental frequency f_n	50 Hz
Switching frequency f_s	2 kHz

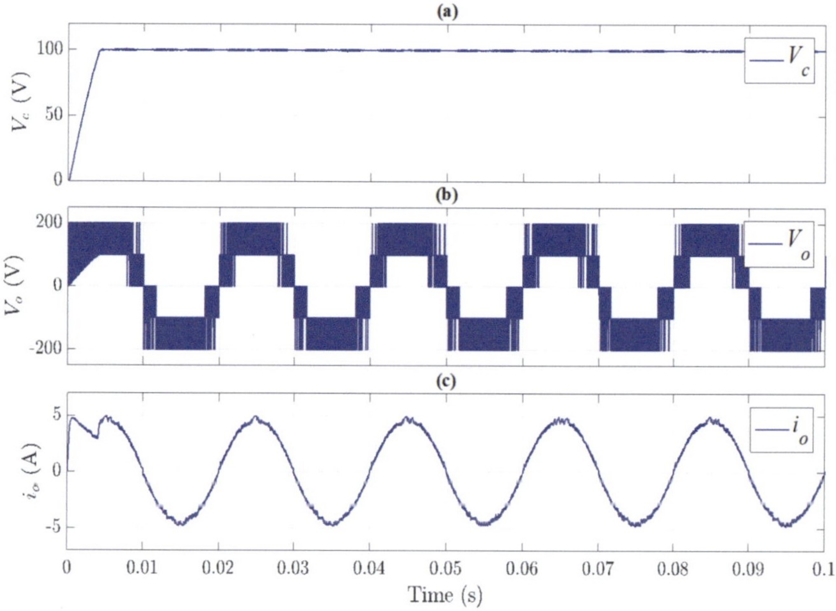

Fig. 11 Transient and steady state performance

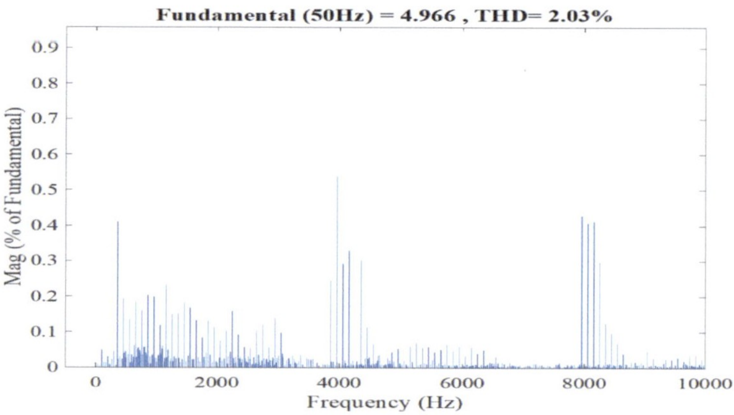

Fig. 12 FFT analysis of load current

the Lyapunov controller achieves a lower Total Harmonic Distortion (THD) of 1.55%, compared to the PI controller's 3.22%, indicating improved waveform quality and efficiency.

Figure 16 presents a detailed performance comparison between the Lyapunov Function Controller (LFC) and the conventional PI controller under conditions of

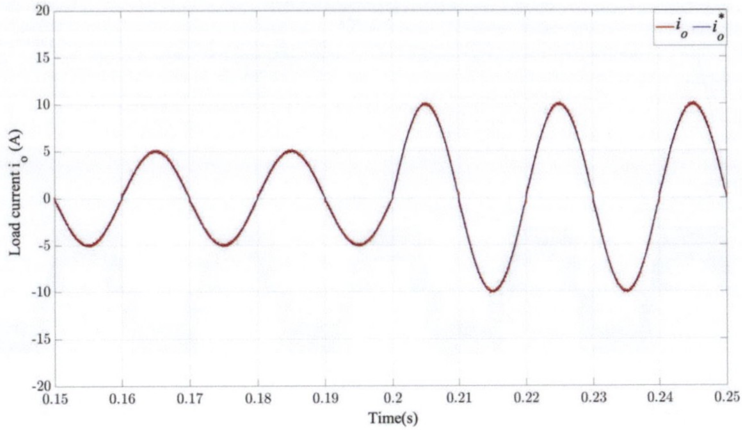

Fig. 13 Responses of the Load current following a sudden change in R from 40 to 20 Ω

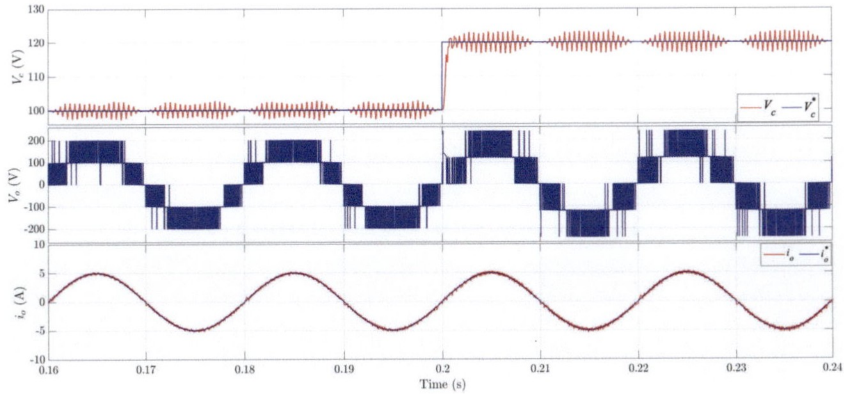

Fig. 14 Simulation under DC voltage source change

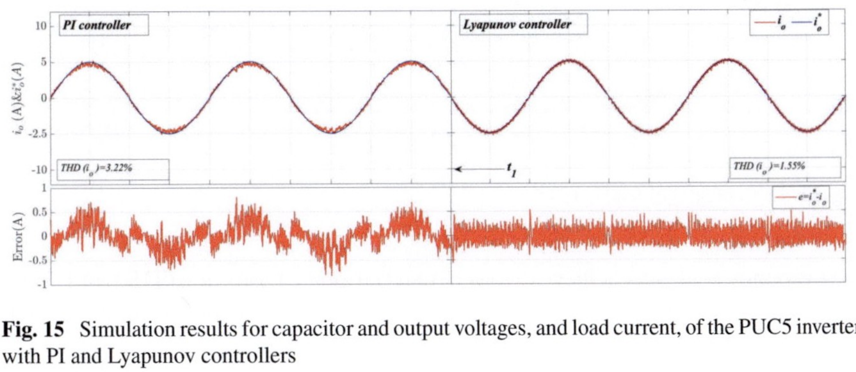

Fig. 15 Simulation results for capacitor and output voltages, and load current, of the PUC5 inverter with PI and Lyapunov controllers

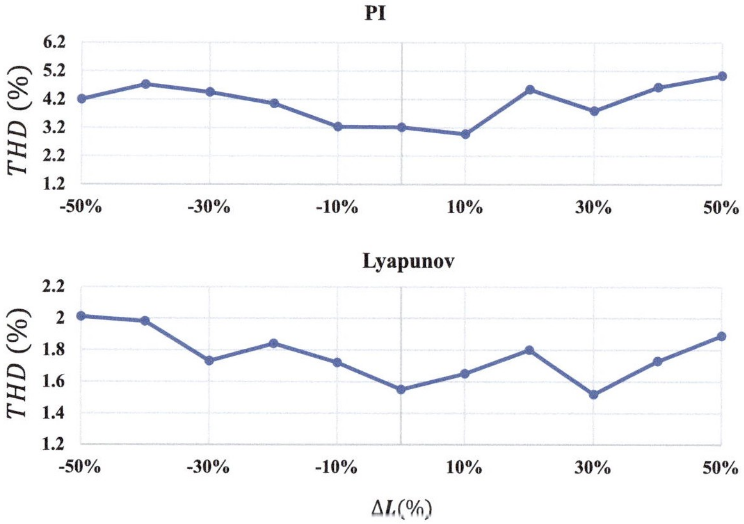

Fig. 16 Performance comparison of Lyapunov and PI controllers under load inductor mismatch

parameter mismatch, specifically targeting the load inductor. The inductance value was varied from −50 to +50% of its nominal value to simulate realistic deviations caused by manufacturing tolerances, thermal effects, or aging.

The results clearly demonstrate the superior robustness of the Lyapunov controller in handling such uncertainties. Across the entire range of inductance variations, the LFC consistently maintains system stability and accurate output voltage tracking. In contrast, the PI controller shows noticeable degradation in performance, particularly under large deviations, where increased overshoot, prolonged settling time, or even instability may occur.

Furthermore, the Lyapunov controller exhibits significantly lower Total Harmonic Distortion (THD) across all tested cases. This indicates its ability to preserve output waveform quality even under adverse conditions, contributing to higher overall power quality and system efficiency. The LFC's performance advantage stems from its nonlinear design framework, which explicitly accounts for system dynamics and ensures stability regardless of parameter variations.

These findings underscore the importance of adopting Lyapunov-based control in applications requiring high reliability and consistent performance under uncertain or variable operating condition.

5 Conclusion

This chapter has highlighted the transformative potential of Lyapunov-based control methods in the field of power electronics, particularly for applications involving nonlinear, time-varying, and uncertain operating conditions. By leveraging Lyapunov's stability theory, these control strategies offer a mathematically rigorous approach to ensuring global or regional asymptotic stability in power converters, including DC–DC and multilevel inverter topologies.

Through theoretical analysis and practical implementation, the chapter illustrated how Lyapunov-based controllers outperform classical linear PI controllers in terms of dynamic response, robustness, and harmonic suppression, especially under challenging scenarios such as load inductance mismatches and input voltage disturbances. Simulation studies for the DC–DC boost converter and the 5-level PUC inverter confirmed the advantages of these nonlinear controllers in achieving accurate voltage regulation, low Total Harmonic Distortion (THD), and strong disturbance rejection without requiring extensive tuning.

In addition to presenting foundational design methodologies, the chapter addressed practical aspects of controller implementation, including considerations related to computational complexity and parameter selection. While Lyapunov-based methods offer significant benefits, challenges remain, particularly in scaling these controllers for large, interconnected systems or implementing them in real-time with limited hardware resources. This underlines the growing importance of adaptive control, observer-based estimation, and machine learning-assisted tuning techniques as future enhancements.

As power electronics systems continue to play a pivotal role in renewable energy integration, electric mobility, and smart grid infrastructures, Lyapunov-based control is well-positioned to meet the demands of high performance, reliability, and resilience. Future research directions include improving computational efficiency, extending applicability to multi-converter systems, and integrating Lyapunov-based designs with emerging technologies such as digital twins and edge computing platforms.

References

Abate A, Ahmed D, Giacobbe M, Peruffo A (2020) Formal synthesis of Lyapunov neural networks. IEEE Control Syst Lett 5(3):773–778

Cecati F, Zhu R, Liserre M, Wang X (2021) Nonlinear modular state-space modeling of power-electronics-based power systems. IEEE Trans Power Electron 37(5):6102–6115

De Queiroz MS, Dawson DM, Nagarkatti SP, Zhang F, Bentsman J (2001) Lyapunov-based control of mechanical systems. Appl Mech Rev 54(5):B81–B81

Forouzesh M, Siwakoti YP, Gorji SA, Blaabjerg F, Lehman B (2017) Step-up DC–DC converters: a comprehensive review of voltage-boosting techniques, topologies, and applications. IEEE Trans Power Electron 32(12):9143–9178

Kabalan M, Singh P, Niebur D (2016) Large signal Lyapunov-based stability studies in microgrids: a review. IEEE Trans Smart Grid 8(5):2287–2295

Kassakian JG, Perreault DJ, Verghese GC, Schlecht MF (2023) Principles of power electronics. Cambridge University Press

Kaźmierkowski MP, Krishnan R, Blaabjerg F (eds) (2002) Control in power electronics: selected problems. Academic Press

Komurcugil H, Bayhan S, Guzman R, Malinowski M, Abu-Rub H (2023) Advanced control of power converters: techniques and matlab/simulink implementation. Wiley

Li B, Wen S, Yan Z, Wen G, Huang T (2023) A survey on the control Lyapunov function and control barrier function for nonlinear-affine control systems. IEEE/CAA J Autom Sin 10(3):584–602

Lu Q, Sun Y, Mei S (2013) Nonlinear control systems and power system dynamics, vol 10. Springer Science & Business Media

Ma W, Zhang B, Qiu D, Sun H (2022) Switching control strategy for DC–DC converters based on polynomial Lyapunov function and sum-of-squares approach. IEEE Trans Industr Electron 70(4):3663–3673

Min H, Shengyuan X, Zhang B, Ma Q, Yuan D (2022) Fixed-time Lyapunov criteria and state-feedback controller design for stochastic nonlinear systems. IEEE/CAA J Autom Sin 9(6):1005–1014

Oubbati BK, Boutoubat M, Rabhi A, Belkheiri M (2020) Experiential integral backstepping sliding mode controller to achieve the maximum power point of a PV system. Control Eng Pract 102:104570

Rayane K, Bougrine M, Krama A, Benalia A, Trabelsi M, Abu-Rub H (2020) Average model-based feedforward and feedback control for PUC5 inverter. IEEE Access 8:172962–172971

Rayane K, Bougrine M, Abu-Rub H, Benalia A, Trabelsi M (2023) Average model-based sliding mode control for quality improvement of a grid-connected PUC inverter. IEEE J Emerg Sel Topics Power Electron 11(4):3765–3774

Shao K, Gu X, Han F, Yang G, Gao W, Zhang Y (2017) An indirect Lyapunov approach of stability of fractional order linear system with and without time delay. In: 2017 29th Chinese control and decision conference (CCDC). IEEE, pp 1010–1014

Vahedi H, Labbé PA, Al-Haddad K (2015) Sensor-less five-level packed U-cell (PUC5) inverter operating in stand-alone and grid-connected modes. IEEE Trans Industr Inf 12(1):361–370

Vahedi H, Shojaei AA, Dessaint L-A, Al-Haddad K (2017) Reduced DC-link voltage active power filter using modified PUC5 converter. IEEE Trans Power Electron 33(2):943–947

Wu L, Liu J, Vazquez S, Mazumder SK (2021) Sliding mode control in power converters and drives: a review. IEEE/CAA J Autom Sin 9(3):392–406

Extended Kalman Filter

Doğan Çelik and Mehmet Emin Meral

Abstract This chapter presents a comprehensive study of the Extended Kalman Filter (EKF) and its application in nonlinear state estimation for power electronics and Electric Vehicle (EV) systems. Building upon the limitations of the standard Kalman Filter (KF) in linear systems, the EKF is introduced as an advanced solution capable of handling nonlinearity through Jacobian-based linearization. The chapter systematically covers the derivation of the EKF, nonlinear state-space modeling, and the recursive prediction-correction process. Key implementation challenges such as model uncertainty, Jacobian computation overhead, and stability concerns are discussed in detail. Practical case studies, particularly in three-phase Shunt Active Power Filters (SAPFs) used in EVs, highlight the EKF's robustness under harmonic distortion, faulted grid voltages, and highly inductive and unbalanced loads. Advanced applications include frequency tracking, harmonic component separation, and integration with orthogonal vector-based structures. Simulation results demonstrate the EKF's high accuracy, adaptability, and effectiveness in improving power quality and control reliability. This chapter establishes the EKF as a powerful tool for dynamic estimation and real-time control in complex power electronic environments.

Keywords Kalman filter · Extended Kalman filter · Harmonic estimation · Orthogonal Vector (OV) estimation · Power electronics

Nomenclature

Symbols	Description
$\mathbf{x}_n$	State vector at time step n
$\mathbf{A}_n$	State transition matrix n

D. Çelik (✉) · M. E. Meral
Department of Electrical and Electronics Engineering, Van Yuzuncu Yil University, Van, Turkey
e-mail: dogancelik@yyu.edu.tr

M. E. Meral
e-mail: emeral@yyu.edu.tr

E. Kabalci (ed.), *Modern Control Methods for Power Electronics*, Energy Systems in Electrical Engineering, https://doi.org/10.1007/978-981-92-1974-2_5

$\mathbf{w}_n$	Process noise
y_n	Measurement vector at time step n
$\mathbf{H}_n$	Measurement matrix
$\mathbf{v}_n$	Measurement noise
$\mathbf{K}(n)$	Kalman gain at time step n
$\mathbf{I}$	Identity matrix
θ	Phase offset
$\mathcal{M}_0$	Amplitude of the waveform
ω_0	Angular frequency
T_s	Sampling period
ϕ_0	Fundamental phase angle of the sinusoidal signal
$\mathbf{r}$	Measurement noise variance
$\mathbf{L}$	Observation matrix
k_s and k_q	Steady-state Kalman gains
q_1 and q_2	Process noise covariances
$\mathbf{F}(n)$	State transition matrix
$\mathbf{P}(n)$	Predicted error covariance matrix
$\mathbf{f}(\cdot)$ and $\mathbf{h}(\cdot)$	Nonlinear vector-valued functions
$\lambda \in \{\alpha, \beta\}$	Orthogonal Clarke components
$\mathbf{u}_{n-1}$	Control input vector
$\mathbf{B}_{n-1}$	Control input matrix
$\mathbf{F}_{n-1}$	State transition matrix

1 Chapter Overview

This chapter provides a comprehensive examination of the Extended Kalman Filter (EKF), a powerful nonlinear state estimation method widely employed in power systems, motor control, and power electronic converter applications (Zhao et al. 2025; Song et al. 2024; Davari et al. 2024). As an extension of the standard Kalman Filter (KF), which is limited to linear systems with Gaussian noise, the EKF adapts the filtering process for nonlinear environments by linearizing the system and measurement models using Jacobian matrices. The EKF's recursive prediction and correction structure allows it to operate effectively in real-time, even under uncertain, noisy, and dynamic conditions (Moreno and Pigazo 2009). In the context of power systems, the EKF is frequently used for Dynamic State Estimation (DSE) with Phasor Measurement Unit (PMU) data, enabling accurate and timely tracking of system states such as voltage and current phasors (Pesend and Taherkhani 2021; Ghahremani and Kamwa 2011; Khodaparast 2022). This capability is crucial for grid stability monitoring, fault detection, and real-time control. Additionally, the EKF plays a central role in frequency and harmonic estimation, where it is applied to track and isolate signal components under distorted waveforms and fluctuating operating conditions. The EKF proves to be a highly reliable method for the precise estimation of harmonic content and decaying DC components, even within environments characterized by nonlinearity and substantial signal noise.

The EKF also sees extensive use in power electronic converter systems, including voltage source inverters, multilevel converters, and grid-connected PV systems (Dou 2021; Tashakor et al. 2022). In these applications, the EKF enables real-time estimation of internal states such as inductor currents, capacitor voltages, and switching behavior, supporting robust control and monitoring without the need for extensive sensing infrastructure. An EKF-assisted dead-beat power control strategy has been successfully applied to PV systems, enhancing the system's dynamic response and maintaining stable grid interaction despite fluctuations in irradiance and load demand. In motor control applications, particularly for permanent magnet synchronous motors and brushless DC motors, the EKF facilitates sensorless control by estimating rotor position and speed with high accuracy. This is especially valuable in industrial applications where physical sensors are prone to failure or increase system cost. A significant benchmark showcasing the EKF's robustness in ensuring accurate motor control under varying load conditions and system parameter uncertainties (Dhaouadi et al 1991; Ahmed et al. 2023; Simhamed et al. 2022).

This chapter presents a comprehensive overview of the EKF, beginning with its derivation from the standard KF and the formulation of nonlinear state-space models. It provides a detailed explanation of the EKF's core operations, specifically the two main steps: the prediction phase, which estimates the current system state, and the correction phase, where this estimate is refined using incoming measurements. Implementation challenges are thoroughly examined, including linearization errors arising from Jacobian matrix evaluations, the associated computational burden, and stability concerns that may lead to filter divergence in highly nonlinear systems. These issues necessitate careful consideration of robustness and real-time feasibility, especially in embedded applications such as motor drives and Photovoltaic (PV) inverters. The chapter also highlights EKF's practical significance through its application in power quality monitoring and renewable energy system management, illustrating its ability to operate effectively in noisy, complex environments where timely and accurate estimation is crucial. Furthermore, recent innovations are discussed, including hybrid estimation approaches that integrate EKF with machine learning techniques like Support Vector Machines (SVMs) for enhanced fault classification and system identification. In summary, the EKF stands out as a robust and flexible estimation method, capable of delivering precise and adaptive performance in diverse nonlinear control and monitoring scenarios within power electronics and energy systems.

2 Historical Overview and Types of Kalman Filters

The foundational principles of the KF can be traced back to the early nineteenth century, when Carl Friedrich Gauss developed the method of least squares for estimating parameters in systems with measurement noise. He notably applied this approach to predict planetary orbits with improved accuracy. Building on these ideas, the modern KF was formally introduced by Rudolf E. Kálmán in his seminal 1960 paper, *"A new approach to linear filtering and prediction problems"* (Kalman 1960). This work presented a mathematically rigorous, recursive algorithm for linear filtering and prediction, combining model-based prediction with measurement-based correction in a statistically optimal framework for linear systems with Gaussian noise. However, most real-world systems inherently exhibit nonlinear behavior, thereby violating the core assumptions underpinning the classical KF. To address this limitation, several advanced variants have been developed to extend the applicability of Kalman filtering techniques to nonlinear estimation problems (Peyman et al. 2022). The KF, also known as Linear Quadratic Estimation (LQE), is a powerful recursive algorithm used for estimating the internal state of a dynamic system over time by optimally integrating noisy input measurements with a predictive mathematical model. The filter operates in two fundamental steps: In the prediction phase, it generates an a priori estimate of the current state based on the system's previous state and model dynamics, taking into account process noise and uncertainties. In the update (or correction) phase, the filter refines this prediction by incorporating new measurement data, using a statistically weighted averaging process to minimize the estimation error covariance (Khodarahmi and Maihami 2023). The core state transition model of the KF is expressed as follows:

$$\mathbf{x}_n = \mathbf{F}_{n-1}\mathbf{x}_{n-1} + \mathbf{B}_{n-1}\mathbf{u}_{n-1} + \mathbf{w}_{n-1} \tag{1}$$

where $\mathbf{x}_n$ is the state vector at time step n, $\mathbf{F}_{n-1}$ is the state transition matrix, $\mathbf{u}_{n-1}$ is the control input vector, $\mathbf{B}_{n-1}$ is the control input matrix, and $\mathbf{w}_{n-1}$ represents the process noise, typically modeled as zero-mean Gaussian noise. This recursive framework makes the KF especially suitable for real-time applications in fields such as control systems, signal processing, robotics, and navigation. The KFs are particularly well-suited for systems that undergo constant changes. One of their key advantages is that they require minimal memory, as they only store information about previous states. Due to their speed, KFs are ideal for real-time applications and embedded systems. While the filter is most effective for systems with linear dynamics to estimate the optimal state, there have been various advancements over time to enhance its performance. For instance, while the basic method may not work well with nonlinear systems, the EKF provides improved results in such cases.

As dynamic systems become more complex, the EKF may struggle with convergence or become difficult to implement. Additionally, the linearization process requires calculating partial derivatives at each iteration of the filter, which results in high computational costs. As shown in Fig. 1, to address these issues and enhance EKF performance, the Unscented KF (UKF) has been developed as a more effective alternative (Varnoosfaderani and Khorsandi 2025).

Since the standard KF may not perform well under model uncertainty, the Robust Kalman Filter (RKF) was introduced for discrete-time systems. Estimating states in high-dimensional spaces was a significant challenge, but the Cubature Kalman Filter (CKF) provided a systematic solution to nonlinear filtering problems (Davari et al.

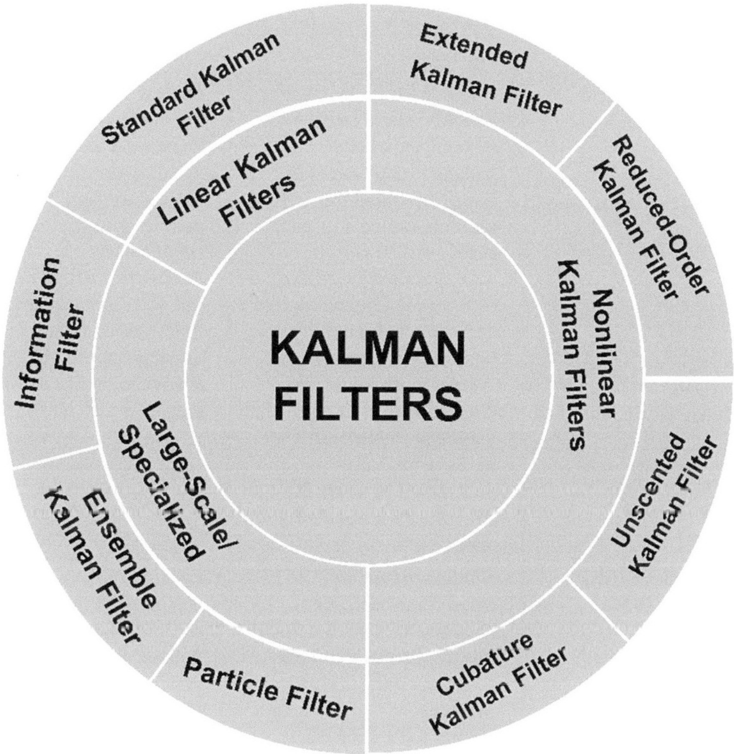

Fig. 1 Most known KF types

2024). Table 1 lists the advantages and disadvantages of the most common filter types.

3 Mathematical Modeling of Kalman Filters

The standard KF provides optimal state estimation for linear discrete-time systems and is widely applied in power electronics for enhancing control accuracy and system robustness. Its application is particularly effective in scenarios where accurate real-time estimation of internal states such as voltages, currents, and harmonics is critical for system performance. In power electronic systems, the KF is employed to refine feedback signals for control strategies, mitigate the effects of measurement noise, and estimate unmeasurable states in converters and electric drives. The state-space model typically takes the following linear form, which serves as the foundation for KF-based estimation in applications ranging from PWM rectifier control to harmonic compensation in cascaded H-bridge converters, and advanced sliding mode control

Table 1 Advantages and disadvantages of KFs

Year-Type	Advantages	Disadvantages
1960-Basic KF (Kalman 1960)	• *Optimal for linear systems*: The KF is best suited for systems with linear dynamics and Gaussian noise, providing the most accurate estimate in such cases • *Computationally efficient*: Its recursive nature makes it fast and computationally lightweight, especially for systems with lower dimensions • *Simple to implement*: The basic form and understanding of the KF are straightforward and easy to apply	• *Limited to linear systems*: The KF is designed only for linear systems. If the system dynamics or measurement models are nonlinear, its performance declines significantly • *Assumes Gaussian noise*: The filter assumes that noise follows a Gaussian distribution, which may not always be true in real-world scenarios
1966-extended KF (EKF) (Ding and Wen 2024)	• *Nonlinear systems*: The EKF extends the basic KF to handle nonlinear systems by linearizing the nonlinear functions using Jacobian matrices • *Widely used*: EKF is commonly used in various fields such as robotics, control systems, and navigation for nonlinear filtering problems	• *Linearization errors*: When dealing with highly nonlinear systems, the process of linearizing the system may lead to significant errors • *High computational cost*: The calculation of Jacobian matrices and their inverses can be computationally expensive • *Instability risks*: If the system is too nonlinear, linearization can cause instability, making the filter unreliable
1994-robust Kalman filter (RKF) (Rocha and Terra 2021)	• *Resistant to outliers*: The RKF is designed to handle outliers or unexpected large deviations in the data, making it useful when the measurement noise is not Gaussian • *Better performance in Noisy environments*: It performs better than the standard KF in situations where the noise is non-Gaussian or contains significant disturbances	• *Increased complexity*: The addition of robustness mechanisms, like outlier detection and weighting functions, adds complexity to the filter's design and implementation • *Tuning challenges*: Properly tuning the filter to achieve an optimal balance between robustness and accuracy can be difficult

(continued)

Table 1 (continued)

Year-Type	Advantages	Disadvantages
1997-unscented Kalman filter (UKF) (Biswas et al. 2017)	• *Improved nonlinear performance*: The UKF uses a set of strategically chosen "sigma points" to approximate nonlinearities, leading to better performance in systems with strong nonlinearities • *No need for Jacobians*: Unlike the EKF, the UKF does not require the calculation of Jacobian matrices, avoiding potential errors from linearization	• *Higher computational cost*: Although the UKF performs better for nonlinear systems, it requires the computation of multiple sigma points, which increases its computational cost • *Assumes* Gaussian noise: Like other KFs, the UKF still assumes Gaussian noise, limiting its effectiveness for systems with non-Gaussian noise
2009-cubature Kalman filter (CKF) (Yu et al. 2014)	• *Better performance in high dimensions*: The CKF improves upon the UKF by using cubature points instead of sigma points to better approximate nonlinear transformations • *More accurate approximation*: In certain high-dimensional problems, the CKF provides a more accurate approximation of the true mean and covariance	• *Computationally expensive*: Like the UKF, the CKF requires more computational resources, especially in higher-dimensional systems • *Limited to* Gaussian *noise*: The CKF still assumes Gaussian noise, which makes it less effective for systems with non-Gaussian noise distributions

in active filters for EV charging stations (Centi et al. 2024; Ofoli 2024; Celik et al. 2023).

The standard KF provides optimal state estimation for linear discrete-time systems, described by the following state transition model (Samantaray and Dash 2010; Kazemi et al. 2021):

$$\mathbf{x}_{n+1} = \mathbf{A}_n\mathbf{x}_n + \mathbf{w}_n \tag{2}$$

where $\mathbf{x}_n$ is the state vector at time step n, $\mathbf{A}_n$ is the state transition matrix, and $\mathbf{w}_n$ represents the process noise. The measurement model for the KF is expressed as:

$$\mathbf{y}_n = \mathbf{H}_n\mathbf{x}_n + \mathbf{v}_n \tag{3}$$

where y_n is the measurement vector at time step n, $\mathbf{H}_n$ is the measurement matrix, and $\mathbf{v}_n$ denotes the measurement noise. $\mathbf{w}_n \sim \mathcal{N}(0, \mathbf{Q}_n)$ and $\mathbf{v}_n \sim \mathcal{N}(0, \mathbf{R}_n)$ are mutually uncorrelated white Gaussian noise processes.

In practical power systems and signal estimation problems, dynamics are often nonlinear. For such systems, the state and measurement models are given by

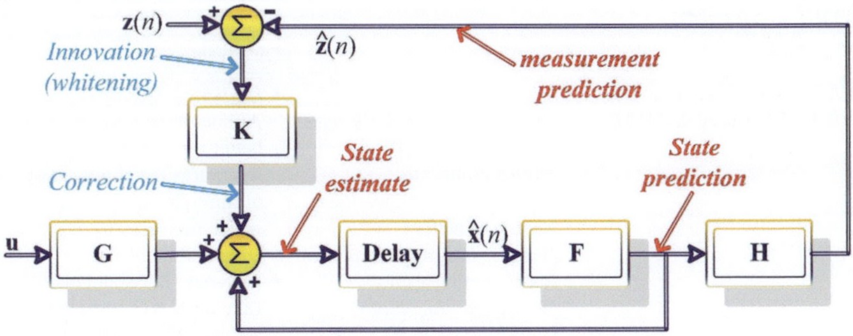

Fig. 2 Block diagram of the discrete-time KF

$$\mathbf{x}_{n+1} = \mathbf{f}(\mathbf{x}_n) + \mathbf{w}_n \tag{4a}$$

$$\mathbf{y}_n = \mathbf{h}(\mathbf{x}_n) + \mathbf{v}_n \tag{4b}$$

where $\mathbf{f}(\mathbf{x}_n)$ and $\mathbf{h}(\mathbf{x}_n)$ are nonlinear vector-valued functions. The EKF addresses this nonlinearity by linearizing $\mathbf{f}$ and $\mathbf{h}$ around the latest state estimate using first-order Taylor series expansions, thereby enabling recursive estimation in real-time.

This rotation-based transition captures the sinusoidal dynamics. The measurement model is as follows:

$$y_n = h(\mathbf{x}(n)) + v(n) = \begin{bmatrix} 0 & 1 \end{bmatrix} \begin{bmatrix} x_1(n) \\ x_2(n) \end{bmatrix} + v(n) \tag{5}$$

Figure 2 provides a block diagram representation of the discrete-time KF algorithm, clearly illustrating each stage of the process including state prediction, error covariance propagation, Kalman gain computation, measurement update, and state correction. It facilitates a comprehensive understanding of the sequential data flow and highlights the interaction between the prediction and update phases of the filter (Reid et al. 2001).

4 Extended Kalman Filter (EKF)

The initialization step of the EKF involves setting the initial estimate of the system state as follows:

$$\hat{\mathbf{x}}(0|0) = \Xi[\mathbf{x}(0)] \tag{6}$$

where $\hat{\mathbf{x}}(0|0)$ denotes the initial estimate of the state vector at time $n = 0$, and $\Xi[\mathbf{x}(0)]$ represents a prior knowledge-based function or approximation of the true initial state $\mathbf{x}(0)$. This initial estimate is critical, as it serves as the starting point for the recursive EKF algorithm, influencing the accuracy of future predictions and updates.

$$\mathbf{P}(0|0) = \Xi\left[(\mathbf{x}(0) - \hat{\mathbf{x}}(0|0))(\mathbf{x}(0) - \hat{\mathbf{x}}(0|0))^T\right] \tag{7}$$

In the prediction step (also known as the time update) of the EKF, the state estimate at time step n is predicted based on the nonlinear state transition function and the previously estimated state as follows:

$$\hat{\mathbf{x}}(n|n-1) = \mathbf{f}(\hat{\mathbf{x}}(n-1|n-1)) \tag{8}$$

where $\hat{\mathbf{x}}(n|n-1)$ is the predicted state estimate at time step n, based on information available up to time $n-1$; $\hat{\mathbf{x}}(n-1|n-1)$ is the estimated state at the previous time step. This step propagates the previous estimate forward in time using the system's nonlinear dynamics.

The Jacobian matrix of the nonlinear state transition function $\mathbf{f}$ is computed to linearize the system dynamics around the current state estimate. It is given by

$$\mathbf{F}(n-1) = \left.\frac{\partial \mathbf{f}}{\partial \mathbf{x}}\right|_{\hat{\mathbf{x}}(n-1|n-1)} \tag{9}$$

where $\mathbf{F}(n-1)$ is the Jacobian of $f(\mathbf{x})$ evaluated at the previous estimated state. This matrix captures the local linear approximation of the nonlinear state transition function and is used in the prediction of the error covariance in the EKF algorithm.

In the prediction step of the EKF, the error covariance matrix is updated to reflect the uncertainty in the predicted state estimate as given by

$$\mathbf{P}(n|n-1) = \mathbf{F}(n-1)\mathbf{P}(n|n-1)\mathbf{F}^T(n-1) + \mathbf{Q}(n-1) \tag{10}$$

where $\mathbf{P}(n|n-1)$ represents the predicted error covariance matrix at time step nnn, which quantifies the uncertainty associated with the predicted state estimate. The matrix $\mathbf{F}(n-1)$ is the Jacobian of the nonlinear state transition function, evaluated at the previous state estimate and it linearizes the system dynamics around that point. $\mathbf{Q}(n-1)$ denotes the process noise covariance matrix, which models the uncertainty and disturbances in the system's dynamics. This covariance prediction step propagates the estimation uncertainty forward in time, taking into account both the linearized dynamics and the inherent process noise, and plays a crucial role in weighting the trust between the predicted state and incoming measurements during the update phase.

In the measurement update (correction) step of the EKF, the nonlinear measurement function $h(\mathbf{x})$ is linearized using its Jacobian matrix. This Jacobian is computed as follows:

$$\mathbf{H}(n-1) = \left.\frac{\partial \mathbf{h}}{\partial \mathbf{x}}\right|_{\hat{\mathbf{x}}(n|n-1)} \tag{11}$$

This Jacobian matrix $\mathbf{H}(n)$ captures the local linear behavior of the nonlinear measurement model around the predicted state and is essential for computing the Kalman gain and updating the state estimate with measurement information. By linearizing the measurement function, the EKF can apply standard KF equations to nonlinear systems.

The Kalman gain $\mathbf{K}(n)$ in the EKF is a key quantity used to optimally blend the predicted state with the new measurement. It is computed as follows:

$$\mathbf{K}(n|n-1) = \mathbf{P}(n|n-1)\mathbf{H}^T(n)\left[\mathbf{H}(n)\mathbf{P}(n|n-1)\mathbf{H}^T(n) + \mathbf{R}(n)\right]^{-1} \tag{12}$$

In the Kalman gain equation, $\mathbf{K}(n)$ represents the Kalman gain at time step n, which determines the weight assigned to the measurement in updating the state estimate. The term $\mathbf{P}(n|n-1)$ is the predicted error covariance matrix, reflecting the uncertainty in the predicted state before incorporating the new measurement. $\mathbf{R}(n)$ is the measurement noise covariance matrix, which characterizes the uncertainty associated with the sensor measurements. Together, these terms determine how much the predicted state should be corrected based on the new measurement, balancing prediction confidence against measurement reliability. This gain determines how much the new measurement influences the updated state estimate. A higher Kalman Gain means the filter gives more weight to the measurement, while a lower gain gives more trust to the prediction. The denominator term in the equation represents the innovation (or residual) covariance, accounting for both predicted uncertainty and measurement noise.

In the measurement update (correction) step of the EKF, the state estimate is refined by incorporating the latest measurement. This state update is given by

$$\hat{\mathbf{x}}(n|n) = \hat{\mathbf{x}}(n|n-1) + \mathbf{K}(n)\left[y(n) - \mathbf{h}(\hat{\mathbf{x}}(n|n-1))\right] \tag{13}$$

where, $\mathbf{h}(\hat{\mathbf{x}}(n|n-1))$ represents the predicted measurement, which is obtained by applying the nonlinear measurement function to the predicted state estimate. By combining the predicted state with this weighted innovation term, the EKF achieves an optimal balance between the model's forecast and the new sensor information.

After updating the state estimate in the EKF, the error covariance matrix is also updated to reflect the improved confidence in the new estimate. The covariance update equation is given by

$$\mathbf{P}(n|n) = [\mathbf{I} - \mathbf{K}(n)\mathbf{H}(n)]\mathbf{P}(n|n-1) \tag{14}$$

where, **I** denotes the identity matrix. This covariance update step refines the predicted error covariance $\mathbf{P}(n|n-1)$ by incorporating the effect of the new measurement. It effectively reduces the uncertainty in the state estimate based on the degree of correction introduced by the Kalman Gain. A larger Kalman gain, reflecting higher confidence in the measurement, leads to a more significant reduction in estimation uncertainty. This process ensures that the filter dynamically adjusts its confidence in the state estimate by balancing model predictions with real-time sensor observations.

Consider the problem of tracking a sinusoidal signal that is contaminated by additive measurement noise. The observed signal can be mathematically expressed as follows:

$$y(n) = A\sin(\omega n + \theta) + v(n) \tag{15}$$

where, $y(n)$ represents the noisy measured signal at discrete-time step n, while A is the amplitude of the underlying sinusoidal signal, ω denotes the angular frequency, and θ is the phase offset. The term $v(n)$ accounts for zero-mean additive white Gaussian noise, which introduces uncertainty into the measurement. This sinusoidal tracking problem serves as a standard benchmark for evaluating the performance of the EKF, as it requires the filter to estimate hidden nonlinear parameters such as amplitude, frequency, or phase, while effectively suppressing measurement noise and adapting to time-varying signal behavior.

To formulate an EKF for amplitude and phase tracking, define the system states as follows:

$$x_1(n) = A\sin(\omega n + \theta) \tag{16a}$$

$$\boldsymbol{x_2(n) = A\cos(\omega n + \theta)} \tag{16b}$$

Using trigonometric identities, the state-space transition becomes

$$\begin{pmatrix} x_1(n+1) \\ x_2(n+1) \end{pmatrix} = \begin{pmatrix} \cos(\omega) & -\sin(\omega) \\ \sin(\omega) & \cos(\omega) \end{pmatrix} \begin{pmatrix} x_1(n) \\ x_2(n) \end{pmatrix} + \mathbf{w}_n \tag{17}$$

Given the updated state vector $\hat{\mathbf{x}}(n|n) = \left[\hat{x}_1(n)\ \hat{x}_2(n)\right]^T$, the instantaneous signal amplitude can be estimated as follows:

$$\hat{A}(n) = \sqrt{\hat{x}_1^2(n) + \hat{x}_1^2(n)} \tag{18}$$

As illustrated in Fig. 3, the filter follows a predictor–corrector structure, where the system state is first estimated using a prediction model and subsequently refined through correction based on incoming measurement data.

Figure 4 highlights the key components of the EKF framework, including the prediction and correction (update) stages. It illustrates the data flow between the

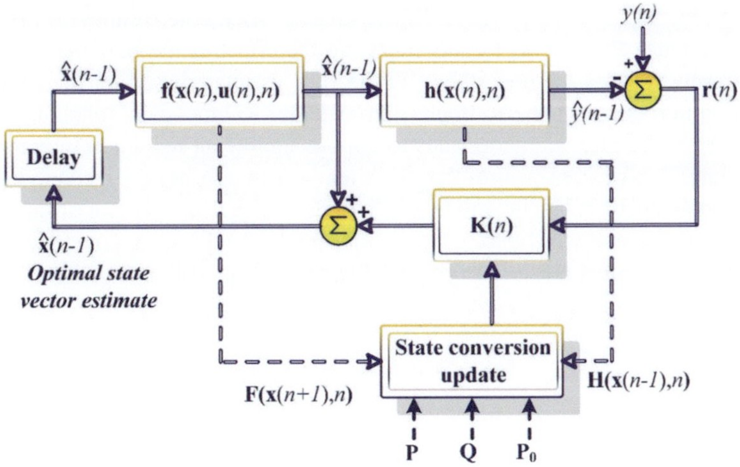

Fig. 3 Functional block diagram of the EKF for nonlinear state estimation

system model, measurement model, and EKF algorithm, enabling accurate state estimation despite the presence of both process and measurement noise. The EKF has the capability to effectively handle system nonlinearities and inherent uncertainties.

5 Advanced Kalman Filtering Approaches for Power Applications

To accurately track and extract the Fundamental Frequency Positive Sequence (FFPS) components of sinusoidal grid signals, we adopt a multiple Orthogonal Vector (OV)-based EKF structure. Unlike the LKF, which assumes linear dynamics, the EKF accounts for the nonlinear nature of the sinusoidal signals, making it more suitable for grid-connected systems with frequency variations and harmonic distortions (Xu et al. 2022; Golestan et al. 2018, 2019).

5.1 Orthogonal Vector-Based on EKF

Consider the single-phase voltage waveform:

$$v(n) = \mathcal{M}_0 \sin(\omega_0 n T_s + \phi_0) \tag{19}$$

where $v(n)$ denotes the instantaneous voltage at the discrete-time index n, where $\mathcal{M}_0$ is the amplitude of the waveform, ω_0 is the angular frequency in radians per

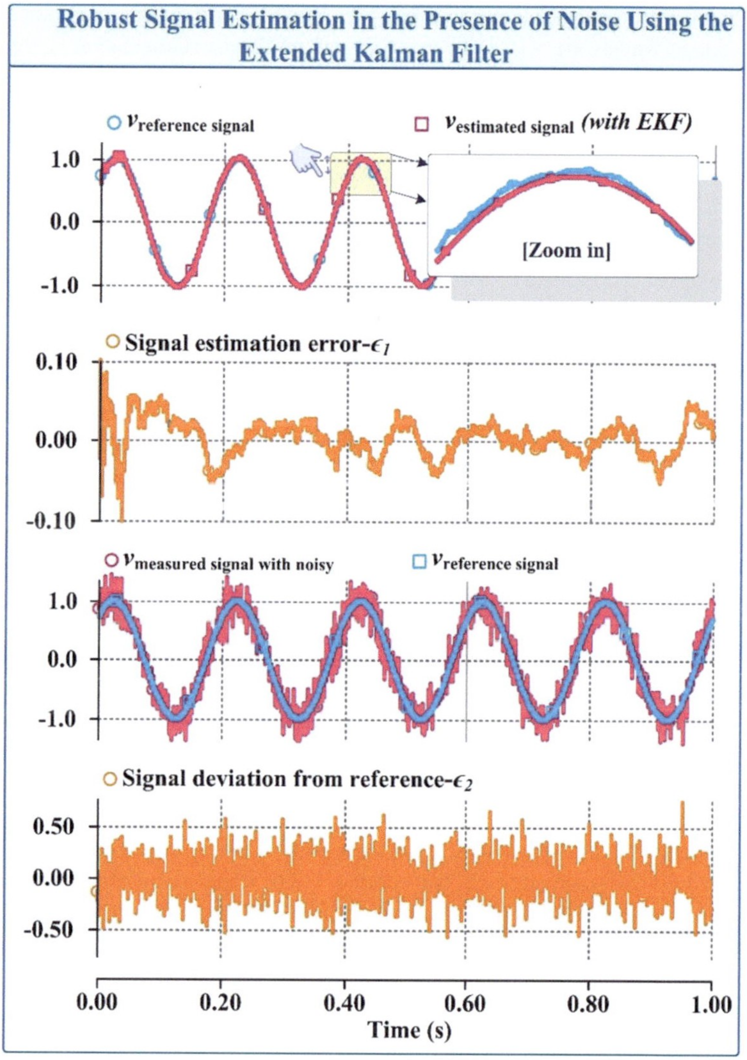

Fig. 4 Signal estimation under noisy conditions using the EKF

second, T_s represents the sampling period (i.e., the time interval between successive samples), and ϕ_0 is the fundamental phase angle of the sinusoidal signal.

To facilitate state-space modeling and filtering, the single-phase sinusoidal voltage signal can be decomposed into orthogonal components and represented as a state vector:

$$\mathbf{x}(n) = \begin{pmatrix} x_1(n) \\ x_2(n) \end{pmatrix} = \begin{pmatrix} \mathcal{M}_0 \sin(\theta(n)) \\ \mathcal{M}_0 \sin(\theta(n)) \end{pmatrix} \tag{20}$$

where $\theta(n) = \omega(n)nT_s + \phi_0$. $\mathbf{x}(n)$ captures the in-phase and quadrature components of the voltage waveform, with $\mathcal{M}_0$ as the signal amplitude (Golestan et al. 2019).

The nonlinear state evolution describing how the system propagates from one time step to the next is given by

$$\mathbf{x}(n+1) = \begin{pmatrix} \cos(\omega(n)T_s) & -\sin(\omega(n)T_s) \\ \sin(\omega(n)T_s) & \cos(\omega(n)T_s) \end{pmatrix} \mathbf{x}(n) + \mathbf{w}_n \tag{21}$$

where the term $\mathbf{w}_n$ is the process noise, typically modeled as a zero-mean white Gaussian vector that accounts for modeling inaccuracies or system perturbations.

The measurement model, which relates the observable voltage sample $v(n)$ to the internal state, is expressed as follows:

$$v(n) = \mathbf{L} \cdot \mathbf{x}(n) + \gamma_n \tag{22}$$

where $\mathbf{L} = \begin{bmatrix} 1 & 0 \end{bmatrix}$. In (22), only the in-phase component $\mathbf{x}(n)$ contributes directly to the observed signal. The term γ_n represents the measurement noise, also assumed to be zero-mean. The expression of (22) reflects the fact that voltage measurements capture only a projection of the full state vector, necessitating the use of filtering techniques to infer the complete state from noisy observations.

In this context, the instantaneous angular frequency $\omega(n)$ is dynamically estimated using a Frequency-Locked Loop (FLL), which tracks changes in frequency over time. This framework provides the foundation for applying the EKF, enabling the estimation of both the sinusoidal signal's internal state and its frequency characteristics under noisy, time-varying conditions.

To estimate the nonlinear state vector $\mathbf{x}(n)$ from the observed measurements, the EKF algorithm is employed, which operates in two main stages: prediction (time update) and update (measurement correction). These steps are tailored to accommodate the nonlinear dynamics of sinusoidal signal tracking. In the prediction step, the prior state estimate $\hat{\mathbf{x}}(n|n-1)$ and its associated error covariance matrix $\mathbf{P}(n|n-1)$ are computed using the nonlinear state transition model and its linearization via the Jacobian as follows:

$$\hat{\mathbf{x}}(n|n-1) = \mathbf{f}(\hat{\mathbf{x}}(n|n-1), \omega(n-1)) = \mathbf{J}(\omega(n-1)) \cdot \hat{\mathbf{x}}(n-1|n-1) \tag{23}$$

$$\mathbf{P}(n| \cdot n-1) = \mathbf{JP}(n-1|n-1) \cdot \mathbf{J}^T + \mathbf{Q} \tag{24}$$

where $\mathbf{Q}$ is the process noise covariance matrix. Jacobian matrix, $\mathbf{J}(\omega(n))$ of the nonlinear state transition function with respect to $\mathbf{x}(n)$, has been evaluated at the current estimate of the angular frequency as follows:

$$\mathbf{J}(\omega(n)) = \begin{bmatrix} \cos(\omega(n)T_s) & -\sin(\omega(n)T_s) \\ \sin(\omega(n)T_s) & \cos(\omega(n)T_s) \end{bmatrix} \tag{25}$$

The update step refines the predicted state estimate by incorporating the actual voltage measurement $v(n)$. This process begins with the computation of the Kalman gain, which is defined as follows:

$$\mathbf{K}(n) = \mathbf{P}(n|n-1) \cdot \mathbf{L}^T \left(\mathbf{L} \cdot \mathbf{P}(n|n-1) \cdot \mathbf{L}^T + r\right)^{-1} \tag{26}$$

where $\mathbf{K}(n)$ is the Kalman gain matrix, $\mathbf{L}$ is the observation matrix, mapping the state vector to the measured output, $\mathbf{r}$ is the measurement noise variance. Then, the updated state estimate and its covariance are calculated as follows:

$$\hat{\mathbf{x}}(n|n) = \hat{\mathbf{x}}(n|n-1) + \mathbf{K}(n) \cdot [v(n) - \mathbf{L} \cdot \hat{\mathbf{x}}(n|n-1)] \tag{27}$$

$$\mathbf{P}(n|n) = [\mathbf{I} - \mathbf{K}(n) \cdot \mathbf{L}] \cdot \mathbf{P}(n|n-1) \tag{28}$$

The estimated frequency $\hat{\omega}(n)$ is provided by the FLL, which receives the estimated in-phase and quadrature voltage components:

$$\hat{v}(n) = x_1(n),\ \hat{v}_q(n) = x_2(n) \tag{29}$$

The magnitude and phase can be reconstructed as follows:

$$\widehat{\mathcal{M}}_0(n) = \sqrt{x_1^2(n) + x_1^2(n)} \tag{30}$$

$$\hat{\phi}_0(n) = \tan^{-1}\left(\frac{x_2(n)}{x_1(n)}\right) \tag{31}$$

The continuous-time transfer functions of the EKF-FLL can be expressed as follows (Çelik 2024):

$$\mathcal{G}(s) = \frac{\hat{v}(s)}{v(s)} = \frac{k_s s - k_q \hat{\omega}}{s^2 + k_s s + \hat{\omega}\left(\hat{\omega} - k_q\right)} \tag{32a}$$

$$\mathcal{G}_q(s) = \frac{\hat{v}_q(s)}{v(s)} = \frac{\partial \hat{\omega} + k_q s}{s^2 + k_s s + \hat{\omega}\left(\hat{\omega} - k_q\right)} \tag{32b}$$

where k_s and k_q represent the steady-state Kalman gains in the continuous domain. The Kalman gains can be optimized based on the noise covariance matrices (Xu et al. 2022; Çelik 2024):

$$k_s = \sqrt{\frac{q_1}{r} - 2\hat{\omega}^2 + 2\hat{\omega}^2\sqrt{2\hat{\omega}^2 + \frac{q_2}{r}}} \tag{33a}$$

$$k_q = \hat{\omega} - \sqrt{2\hat{\omega}^2 + \frac{q_2}{r}} \tag{33b}$$

where, q_1 and q_2 are process noise covariances, and r is the measurement noise variance.

Once the fundamental in-phase and quadrature components are estimated, the grid voltages and load currents can be transformed into the αβ-reference frame:

$$X_\alpha(t) = \mathcal{M}_\alpha^+ \cos\left(\omega_0 t + \phi_0^+\right) + \mathcal{M}_\alpha^- \sin\left(\omega_0 t + \phi_0^-\right) \tag{34a}$$

$$X_\beta(t) = \mathcal{M}_\beta^+ \sin\left(\omega_0 t + \phi_0^+\right) - \mathcal{M}_\beta^- \sin\left(\omega_0 t + \phi_0^-\right) \tag{34b}$$

The symmetrical components (positive and negative sequences) are extracted as follows:

$$\begin{bmatrix} v_{\alpha 0}^+ \\ v_{\beta 0}^+ \\ v_{\alpha 0}^- \\ v_{\beta 0}^- \end{bmatrix} = \begin{bmatrix} 1/2 & 0 & 0 & 1/2 \\ 0 & 1/2 & 1/2 & 0 \\ 1/2 & 0 & 0 & -1/2 \\ 0 & 1/2 & -1/2 & 0 \end{bmatrix} \begin{bmatrix} \hat{v}_\alpha \\ \hat{v}_{\alpha q} \\ \hat{v}_\beta \\ \hat{v}_{\beta q} \end{bmatrix} \tag{35}$$

The separation of harmonic components using the EKF can be expressed in the Laplace domain as follows:

$$\mathcal{G}_{\lambda,h}(s) = \frac{\hat{v}_{\lambda,h}(s)}{v(s)} = g_{\lambda,h}(s) \prod_{i \neq h}^{\mathcal{H}} \left(\frac{1 - g_{\lambda,h}(s)}{1 - g_{\lambda,h}(s) g_{\lambda,i}(s)} \right) \tag{36}$$

where $\lambda \in \{\alpha, \beta\}$ denotes the orthogonal Clarke components. h denotes the order of the harmonic component being analyzed, while $\mathcal{H}$ represents the highest harmonic order considered in the system.

Each $g_{\lambda,h}(s)$ serves as a bandpass filter centered around the *hth* harmonic frequency and is defined differently for the α and β components as follows (Golestan et al. 2019; Çelik 2024):

$$g_{\lambda,h}(s) = \begin{cases} g_{\alpha,h}(s) = \frac{\hat{v}_{\alpha,h}(s)}{v(s)} = \frac{k_s s - k_q h\hat{\omega}}{s^2 + k_s s + h\hat{\omega}\left(h\hat{\omega} - k_q\right)} \\ g_{\beta,h}(s) = \frac{\hat{v}_{\beta,h}(s)}{v(s)} = \frac{\partial h\hat{\omega} + k_q s}{s^2 + k_s s + h\hat{\omega}\left(h\hat{\omega} - k_q\right)} \end{cases} \tag{37}$$

These filters are designed to extract specific harmonic components from the signal in the $\alpha\beta$-reference frame, enabling the EKF to effectively track and estimate individual harmonics in noisy and dynamic conditions.

The performance of the EKF-based OV estimation, including both positive and negative sequence components, has been assessed under normal and abnormal grid voltage conditions. These conditions include harmonic distortions and fault-induced

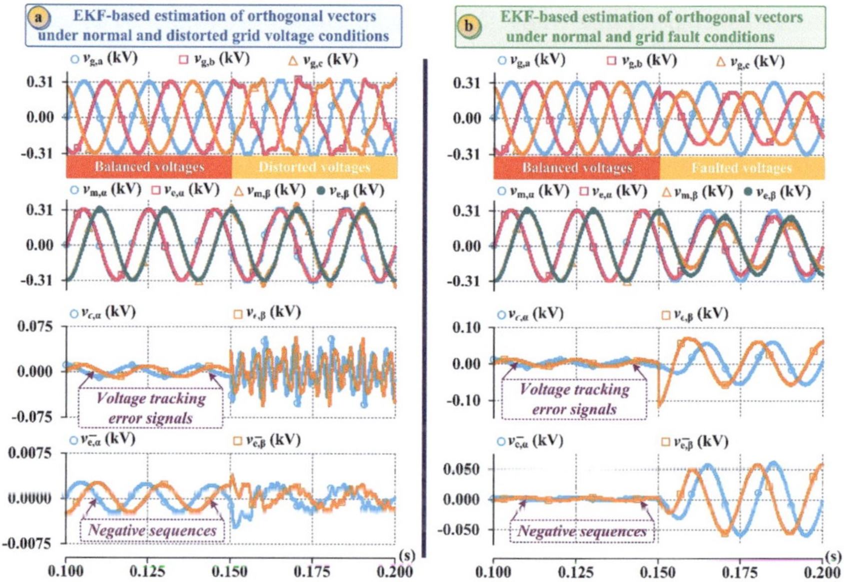

Fig. 5 Performance of EKF-based OV estimation under **a** harmonic-distorted grid voltages and **b** fault-induced voltage disturbances

voltage disturbances. As shown in Fig. 5a and b, harmonic components are injected into the grid voltage at $t = 0.15s$, simulating a transition from ideal to distorted operating conditions. Additionally, faulted voltage waveforms are introduced to evaluate the robustness of the estimation algorithm. Despite these disturbances, the EKF demonstrates accurate and stable tracking of the voltage signals, effectively estimating both sequence components under non-ideal conditions.

5.2 *Frequency and Harmonic Estimation Extended Kalman Filter for Power Electronic Systems*

The discrete-time measured signal $S(n)$ is modeled as a combination of multiple harmonic components and a decaying Direct Current (DC) offset. The mathematical expression for the signal at the *hth* sample is given by

$$S(n) = \sum_{h=1}^{\infty} A_h \sin(h\omega(n)T_s + \phi_h) + A_0 e^{-\alpha(n)T_s} \tag{38}$$

where the parameter α represents a positive decay constant that governs the rate at which the DC component of the signal exponentially diminishes over time. The term ϕ_h denotes the phase angle (in radians) associated with the *hth* harmonic component, capturing its temporal shift within the signal waveform. Meanwhile, A_0 defines the initial amplitude of the decaying DC offset. The second term in the signal model, $A_0 e^{-\alpha(n)T_s}$, specifically accounts for this decaying DC component. Such a component is commonly observed during transient events or as a result of magnetic saturation in inductive elements like transformers or machines. It gradually fades as time progresses due to system damping or natural dissipation. Accurate estimation of this decaying DC element is essential for reliable signal reconstruction, improved power quality monitoring, and precise harmonic analysis in power systems and power electronic applications.

The signal is represented using cosine and sine terms for each harmonic, with two states for the decaying DC component:

$$\mathbf{x}(n) = \begin{bmatrix} \{A_h \sin(\phi_h)\}_{h=1}^{\infty} \\ \{A_h \cos(\phi_h)\}_{h=1}^{\infty} \\ e^{-\alpha(n)T_s} \\ A_0 e^{-\alpha(n)T_s} \end{bmatrix} \tag{39}$$

The state evolution of the system is governed by the discrete-time state-space model:

$$\mathbf{x}(n+1) = \mathbf{F}(n+1)\mathbf{x}(n) + \mathbf{w}(n) \tag{40}$$

where $\mathbf{x}(n+1)$ represents the state vector at time step $n+1$, $\mathbf{F}(n)$ is the state transition matrix at time n, and $\mathbf{w}(n)$ denotes the process noise.

For each harmonic component h, the corresponding dynamic behavior is modeled using a 2×2 rotation matrix given by

$$\mathbf{F}_h = \begin{bmatrix} \cos(h\omega T_s) & -\sin(h\omega T_s) \\ \sin(h\omega T_s) & \cos(h\omega T_s) \end{bmatrix} \tag{41}$$

The measurement model used in the estimation process is defined as follows:

$$\mathbf{S}(n) = \mathbf{H}(n)\mathbf{x}(\mathrm{n}) + v(n) \tag{42}$$

where $S(n)$ denotes the measurement vector, while $\mathbf{H}(n)$ represents the observation matrix, defined as follows:

$$\mathbf{H}(n) = \left[\sin(\omega(n)T_s)\ \cos(\omega(n)T_s)\ \ \sin(3\omega(n)T_s)\ \cos(3\omega(n)T_s)\ \ \cdots 0\ 1 \right] \tag{43}$$

The prediction step estimates the next state based on the current state and system dynamics. It also projects the error covariance forward to account for process noise and is mathematically expressed as follows:

$$\hat{\mathbf{x}}(n|n-1) = \mathbf{F}(n-1)\hat{\mathbf{x}}(n-1|n-1) \tag{44}$$

$$\mathbf{P}(n|n-1) = \mathbf{F}(n-1)\mathbf{P}(n-1|n-1)\mathbf{F}(n-1)^T + \mathbf{Q} \tag{45}$$

The Kalman gain determines the balance between the predicted state and the new measurement. It is calculated to minimize the posterior error covariance, ensuring optimal state estimation. A higher gain places greater trust in the measurement, while a lower gain relies more heavily on the prediction, as expressed by

$$\mathbf{K}(n) = \mathbf{P}(n|n-1)\mathbf{H}(n)^T + \left(\mathbf{H}(n)\mathbf{P}(n|n-1)\mathbf{H}(n)^T + \mathbf{R}\right)^{-1} \tag{46}$$

The state update step refines the predicted state by incorporating the new measurement weighted by the Kalman gain. This adjustment corrects the estimate to more accurately reflect the observed data. As a result, the state estimate becomes more precise and its uncertainty is reduced, as expressed by

$$\hat{\mathbf{x}}(n|n) = \hat{\mathbf{x}}(n|n-1) + \mathbf{K}(n)(S(n) - \mathbf{H}(n)\hat{\mathbf{x}}(n|n-1)) \tag{47}$$

Error covariance update can be given by

$$\mathbf{P}(n|n) = (\mathbf{I} - \mathbf{K}(n)\mathbf{H}(n))\mathbf{P}(n|n-1) \tag{48}$$

To enhance estimation accuracy in dynamic conditions, the measurement noise covariance $\mathbf{R}(n)$ is adaptively updated at each time step as expressed by

$$\mathbf{R}(n) = \xi_n \mathbf{R}(n-1) + (1-\xi_n)\varepsilon^2(n) \tag{49}$$

where $\varepsilon(n) = \mathbf{H}(n)\hat{\mathbf{x}}(n|n)$ denotes the measurement residual, $\xi_n = \frac{1}{1+\left|\frac{\mathcal{R}(n)}{\mathcal{R}_0}\right|}$ is a forgetting factor with $\xi_n \in (0, 1)$, and $\mathcal{R}_0$ is initial covariance.

The model error covariance $c_e(n)$ is updated recursively as follows:

$$c_e(n) = l_q c_e(n-1) + \left(1 - l_q\right)\varepsilon(n)\varepsilon(n-1) \tag{50}$$

where l_q is a forgetting factor for the model error covariance. l_q controls the influence of past errors on the current estimate. Based on the value of $c_e(n)$ compared to a predefined threshold c_{th}, the process noise covariance $Q(n)$ is adjusted adaptively:

$$Q(n) = \begin{cases} Q_0, \ c_e(n) > c_{th} \\ Q_1, \ \ otherwise \end{cases} \tag{51}$$

Alternatively, $Q(n)$ can be updated continuously using a weighted average:

$$\mathbf{Q}(n) = \left(1 - \sigma_q\right)\mathbf{Q}(n-1) + \sigma_q \mathbf{P}(n-1)\mathbf{P}(n-1)^{\mathrm{T}} c_e(n) \tag{52}$$

where σ_q is an adaptation rate parameter that governs how quickly Q responds to changes in the model error covariance. This adaptive mechanism allows the filter to balance between stability and responsiveness under varying system dynamics.

As depicted in Fig. 6, the three-dimensional EKF state trajectory illustrates the simultaneous estimation of fundamental harmonic components and a decaying DC offset. The horizontal axes represent the sine and cosine projections of the fundamental harmonic component, while the vertical axis corresponds to the amplitude of the decaying DC component over time. The trajectory reveals the convergence behavior of the EKF, where the state estimates spiral toward a steady orbit as the DC component gradually diminishes. This dynamic response highlights the EKF's effectiveness in accurately tracking both periodic (harmonic) and non-periodic (DC) signal components under noisy and non-ideal conditions.

Figure 7 illustrates the essential components of the EKF framework, emphasizing its two main stages: prediction and correction (update), specifically under the influence of harmonics and DC offset. Figure 7 demonstrates the data flow between the system model, the measurement model, and the EKF algorithm, showcasing how the filter enables accurate state estimation despite the presence of both process and measurement noise. By leveraging its recursive structure, the EKF effectively manages system nonlinearities and inherent uncertainties, making it a robust solution for signal estimation in environments affected by harmonic distortion and DC offset.

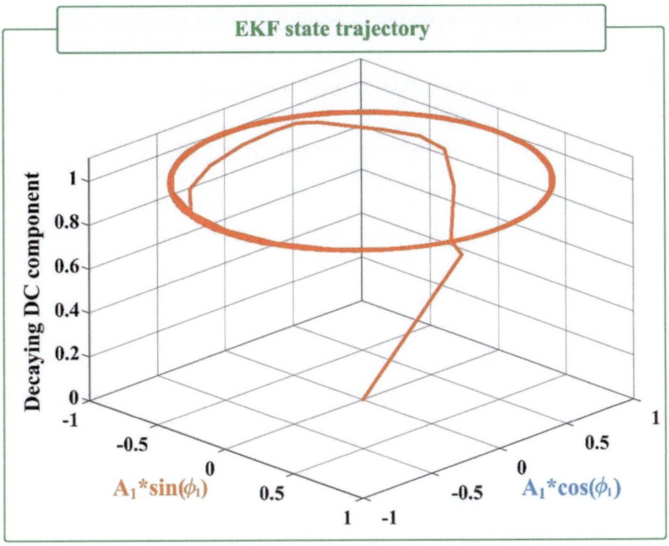

Fig. 6 Tracking fundamental harmonics and decaying DC component

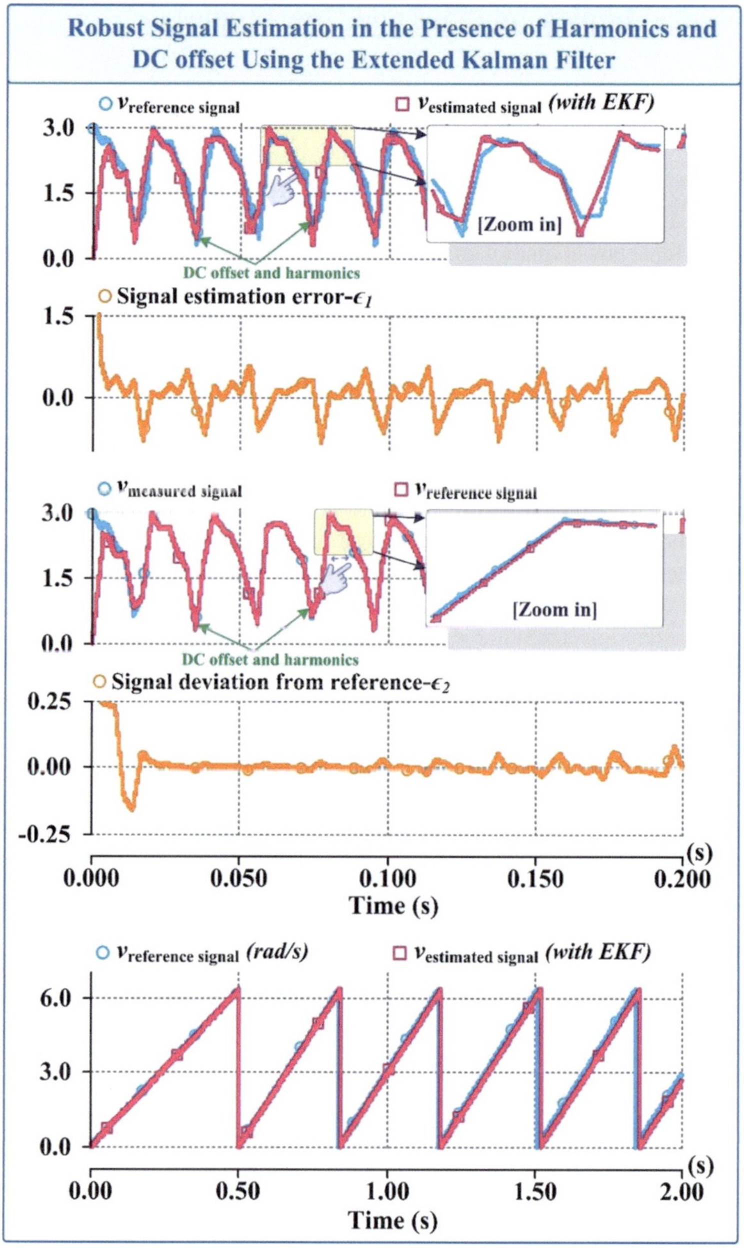

Fig. 7 Signal estimation under harmonics and DC-offset conditions using the EKF

6 Practical Implementations and Case Studies

This section presents practical implementations and case studies demonstrating the application of the EKF for load harmonic extraction and OV voltage signal estimation in a three-phase Shunt Active Power Filter (SAPF) designed for Electric Vehicle (EV) systems. The EKF algorithm is integrated into the control structure of the SAPF to enhance dynamic response, improve harmonic compensation accuracy, and ensure stable power quality under varying load and grid conditions, as illustrated in Fig. 8. To validate the effectiveness and robustness of the proposed EKF-based estimation approach in real-world Electric Vehicle (EV) power electronic environments, case studies are conducted under diverse scenarios. These include highly inductive Nonlinear Loads (NNLs), unbalanced NLL conditions, and grid voltage profiles ranging from normal operation to harmonic distortion and fault-induced disturbances. Simulation results confirm the EKF's capability to accurately extract harmonics and estimate orthogonal vectors, even under complex and dynamic system conditions.

The performance of the EKF-based harmonic extraction and OV estimation has been evaluated in a SAPF used in EV applications, under both harmonic and faulted grid voltage conditions. As illustrated in Fig. 9a and b, at time $t = 0.15s$, harmonic components were injected into the grid voltage, simulating a transition from normal to distorted and faulted grid conditions. The proposed EKF-based algorithm has demonstrated strong performance in accurately extracting load current harmonics and estimating orthogonal voltage components, even in the presence of severe waveform distortions. Furthermore, the load current harmonics, which arise due to nonlinear loads and distorted grid voltages, have been effectively reduced to below 1.0%, highlighting the effectiveness of the EKF in enhancing power quality and filter control robustness under non-ideal operating conditions.

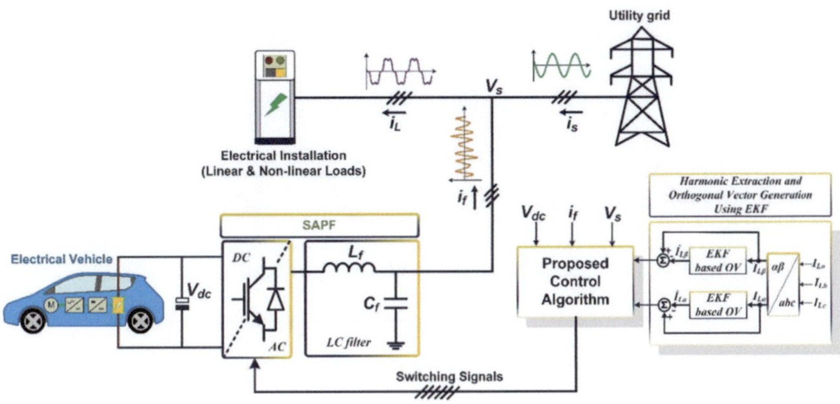

Fig. 8 Application of EKF for load harmonic extraction and OV estimation in a three-phase shunt active power filter used in EVs

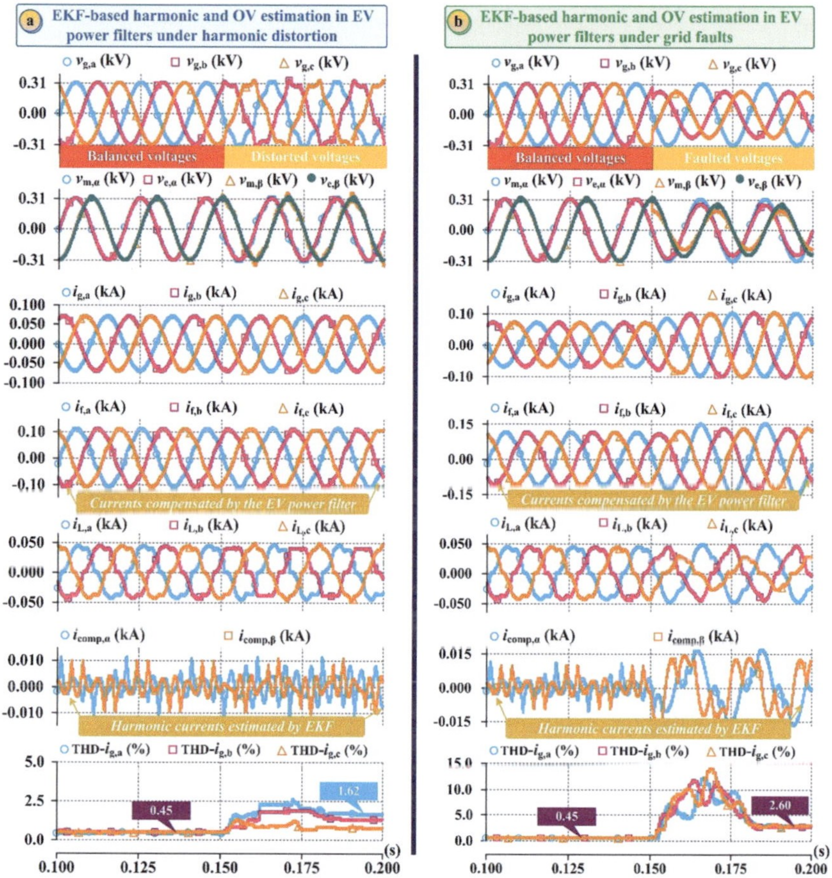

Fig. 9 EKF-based harmonic and OV estimation in EV power filters: **a** distorted grid conditions and **b** grid fault conditions

In this case study, the performance of the EKF-based harmonic extraction and OV estimation is evaluated under highly inductive NNL and unbalanced NLL conditions, as shown in Fig. 10. Initially, up to $t = 0.15s$, only the highly inductive NNL is in operation, introducing substantial current harmonics and phase displacement. After $t = 0.15s$, the system transitions to an unbalanced NLL condition, further challenging the estimation algorithm with asymmetrical load profiles and distorted current waveforms.

Despite these complex loading scenarios, the proposed EKF-based method effectively tracks the harmonic components and estimates accurate orthogonal vectors, demonstrating robustness, adaptability, and suitability for dynamic EV power filtering applications.

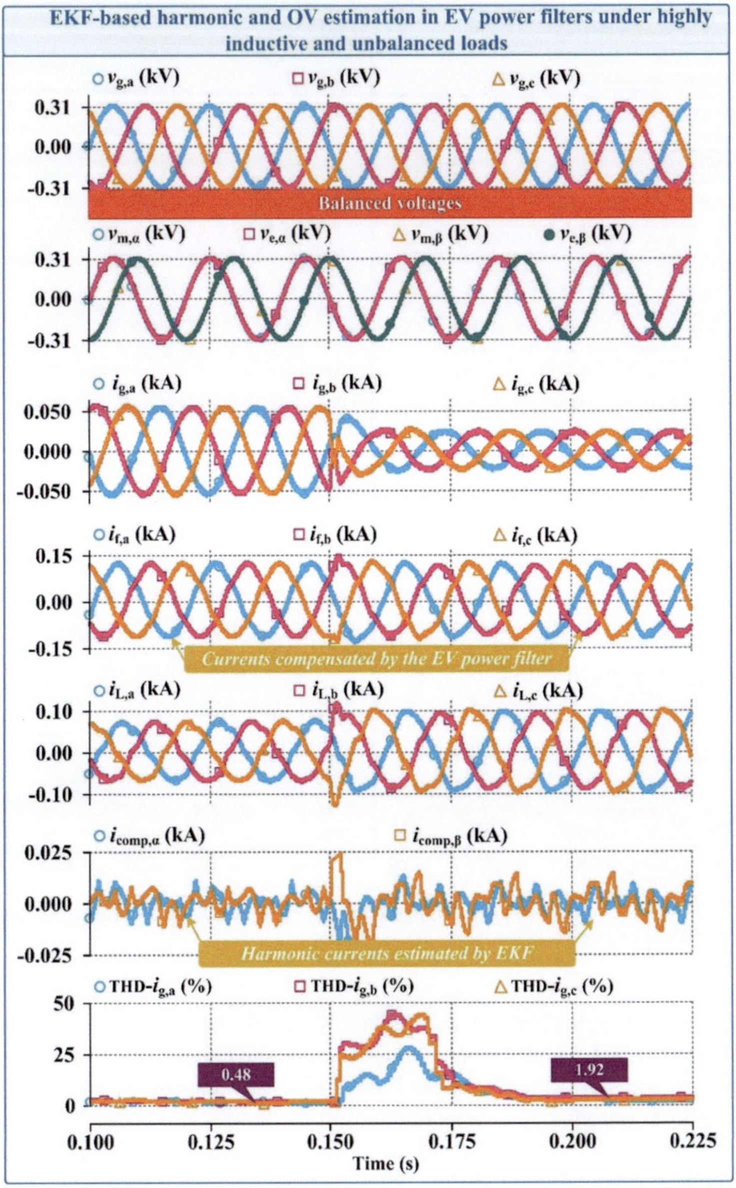

Fig. 10 EKF-based harmonic and orthogonal vector estimation in EV power filters under highly inductive and unbalanced load conditions

7 Conclusion and Chapter Summary

This chapter provided an in-depth exploration of the EKF, highlighting its critical role in nonlinear state estimation within power electronic and EV systems. Starting from the foundational principles of the standard KF, the EKF has been introduced to overcome limitations in systems characterized by nonlinear dynamics and noisy measurements. Through a detailed mathematical formulation, the prediction and correction phases of the EKF have been explained, along with the necessary linearization via Jacobian matrices. Advanced modeling examples demonstrated how the EKF can accurately estimate sinusoidal signal components, including decaying DC offsets and harmonic distortions, under highly dynamic and non-ideal grid conditions. The OV-based EKF approach has been also discussed for extracting fundamental frequency positive and negative sequences, essential for maintaining power quality and effective filtering.

Practical case studies further validated the EKF's effectiveness when applied to SAPFs in EV applications. These studies included scenarios with harmonic-distorted grid voltages, fault-induced voltage disturbances, highly inductive nonlinear loads, and unbalanced load conditions. In all cases, the EKF exhibited strong performance in harmonic extraction, OV estimation, and load compensation, reducing THD to below 1.0%. As a result, the EKF proves to be a robust, adaptive, and efficient estimation framework, suitable for a wide range of nonlinear control and monitoring tasks in power electronics. Its integration into modern EV systems and renewable energy technologies continues to drive innovation in reliable and high-performance power management solutions.

References

Ahmed, M., Harbi, I., Kennel, R., Abdelrahem, M.: Direct Power Control Based on Dead-beat Function and Extended Kalman Filter for PV Systems. J Mod Power Syst Clean Energy 11, 863–872 (2023). https://doi.org/10.35833/MPCE.2021.000793

Biswas SK, Qiao L, Dempster AG (2017) A novel a priori state computation strategy for the unscented kalman filter to improve computational efficiency. IEEE Trans Autom Control 62:1852–1864. https://doi.org/10.1109/TAC.2016.2599291

Celik D, Ahmed H, Meral ME (2023) Kalman Filter-Based Super-Twisting Sliding Mode Control of Shunt Active Power Filter for Electric Vehicle Charging Station Applications. IEEE Trans Power Deliv 38:1097–1107. https://doi.org/10.1109/TPWRD.2022.3206267

Çelik D (2024) An advanced control strategy for power flow management, optimal DC capacitor utilization and compensating harmonics in fuel cell electric vehicle powertrain converters. Int J Hydrog Energy 93:1189–1204. https://doi.org/10.1016/J.IJHYDENE.2024.10.390

Centi, F., Credo, A., Fabri, G., Parasiliti, F., Tursini, M.: Enhancing PWM Rectifier Control through Kalman Filtering. 2024 International Symposium on Power Electronics, Electrical Drives, Automation and Motion, SPEEDAM 2024.1130–1135 (2024). https://doi.org/10.1109/SPEEDAM61530.2024.10609227

Davari SA, Azadi S, Flores-Bahamonde F, Wang F, Wheeler P, Rodriguez J (2024) Compensating the Measurement Error in Model-free Predictive Control of Induction Motor via Kalman Filter-based Ultra-local Model. IEEE Trans Power Electron. https://doi.org/10.1109/TPEL.2024.3443134

Dhaouadi R, Mohan N, Norum L (1991) Design and Implementation of an Extended Kalman Filter for the State Estimation of a Permanent Magnet Synchronous Motor. IEEE Trans Power Electron 6:491–497. https://doi.org/10.1109/63.85891

Ding, L., Wen, C.: High-Order Extended Kalman Filter for State Estimation of Nonlinear Systems Symmetry 2024, Vol. 16, Page 617.16, 617 (2024). https://doi.org/10.3390/SYM16050617

Dou Y (2021) An Improved Prediction Model of IGBT Junction Temperature Based on Backpropagation Neural Network and Kalman Filter. Complexity 2021:5542889. https://doi.org/10.1155/2021/5542889

Ghahremani E, Kamwa I (2011) Dynamic state estimation in power system by applying the extended kalman filter with unknown inputs to phasor measurements. IEEE Trans Power Syst 26:2556–2566. https://doi.org/10.1109/TPWRS.2011.2145396

Golestan S, Guerrero JM, Vasquez JC (2018) Steady-state linear kalman filter-based plls for power applications: A second look. IEEE Trans Ind Electron 65:9795–9800. https://doi.org/10.1109/TIE.2018.2823668

Golestan S, Guerrero JM, Vasquez JC, Abusorrah AM, Al-Turki Y (2019) Single-Phase FLLs Based on Linear Kalman Filter, Limit-Cycle Oscillator, and Complex Bandpass Filter: Analysis and Comparison with a Standard FLL in Grid Applications. IEEE Trans Power Electron 34:11774–11790. https://doi.org/10.1109/TPEL.2019.2906031

Kalman RE (1960) A New Approach to Linear Filtering and Prediction Problems. J Basic Eng 82:35–45. https://doi.org/10.1115/1.3662552

Kazemi Z, Naseri F, Yazdi M, Farjah E (2021) An EKF-SVM machine learning-based approach for fault detection and classification in three-phase power transformers. IET Sci Meas Technol 15:130–142. https://doi.org/10.1049/SMT2.12015

Khodaparast J (2022) A Review of Dynamic Phasor Estimation by Non-Linear Kalman Filters. IEEE Access 10:11090–11109. https://doi.org/10.1109/ACCESS.2022.3146732

Khodarahmi M, Maihami V (2023) A Review on Kalman Filter Models. Archives of Computational Methods in Engineering 30:727–747. https://doi.org/10.1007/S11831-022-09815-7/TABLES/2

Moreno, V.M., Pigazo, Alberto: Kalman filter: recent advances and applications/monograph 584 (2009)

Ofoli, A.R.: Fuzzy-Logic Applications in Electric Drives and Power Electronics. Power Electronics Handbook 1233–1260 (2024). https://doi.org/10.1016/B978-0-323-99216-9.00003-2

Pesend MMS, Taherkhani F (2021) Extended Kalman Filtering Models for Power Harmonics and Frequency Estimation. IEEE Electromagn Compat Mag 10:27–38. https://doi.org/10.1109/MEMC.2021.9477234

Peyman S, Saeid H, Simon H: Nonlinear Filters Theory and Applications. Presented at the (2022)

Reid I Estimation II 1 Discrete-time Kalman filter. (2001)

Rocha KDT, Terra MH (2021) Robust Kalman filter for systems subject to parametric uncertainties. Syst Control Lett 157:105034. https://doi.org/10.1016/J.SYSCONLE.2021.105034

Samantaray SR, Dash PK (2010) High impedance fault detection in distribution feeders using extended kalman filter and support vector machine. Eur T Electr Power 20:382–393. https://doi.org/10.1002/ETEP.321;WGROUP:STRING:PUBLICATION

Simhamed Y, Ykhlef F, Iratni A (2022) A New Classification Scheme Based on Extended Kalman Filter and Support Vector Machine. Electr Power Syst Res 210:108153. https://doi.org/10.1016/J.EPSR.2022.108153

Song S, Lei J, Ma W, Wang Y (2024) Kalman Filter Based Detection Scheme for DC-Link Voltage Measurement Faults in MMC-HVDC Systems. IEEE Trans Power Electron. https://doi.org/10.1109/TPEL.2024.3501253

Tashakor N, Arabsalmanabadi B, Naseri F, Goetz S (2022) Low-Cost Parameter Estimation Approach for Modular Converters and Reconfigurable Battery Systems Using Dual Kalman Filter. IEEE Trans Power Electron 37:6323–6334. https://doi.org/10.1109/TPEL.2021.3137879

Varnoosfaderani HHN, Khorsandi A (2025) State estimation in an islanded hybrid solar-wind DC microgrid using Unscented Kalman Filter. Heliyon 11:42073. https://doi.org/10.1016/J.HELIYON.2025.E42073/ASSET/25E9A4C9-83AC-4829-BA01-1FED86DAF894/MAIN.ASSETS/GR14.JPG

Xu W, Huang C, Golestan S (2022) Design-Oriented Study of the Orthogonal Vector-Based Linear Kalman Filter: Enhancements and Applications. IEEE Trans Ind Electron 69:13359–13372. https://doi.org/10.1109/TIE.2022.3142415

Yu F, Sun Q, Lv C, Ben Y, Fu Y (2014) A SLAM Algorithm Based on Adaptive Cubature Kalman Filter. Math Probl Eng 2014:171958. https://doi.org/10.1155/2014/171958

Zhao Q, Zhang W, Zhang T, Chen J, Ye H (2025) Robust Consensus Based Secondary Control for Microgrids Considering Communication Delay and Noises: A Kalman Filter Based Approach. IEEE Trans Smart Grid. https://doi.org/10.1109/TSG.2025.3568371

Fuzzy Logic-Based Control of Power Converters

Xiaoning Shen, Zhuang Kang, Jiahui Wang, Xinxin Liu, and Jianxing Liu

Abstract Modern industrial systems increasingly rely on efficient power conversion technologies to address energy challenges. To ensure a high-performance operation for power conversion, this chapter introduces a Fuzzy Logic System (FLS)-based Sliding-Mode Control (SMC) strategy tailored for three-phase three-level Neutral-Point-Clamped (NPC) converters. The introduced approach integrates three inter-dependent control loops: a power tracking loop employing Fuzzy Super-Twisting Algorithm (FSTA) to ensure dynamic response and chattering suppression, a voltage regulation loop utilizing fuzzy observer-based SMC for precise DC-link voltage stabilization, and a voltage balancing loop with Proportional-Integral (PI) control to equalize capacitor voltages. Experimental validation under varied operating conditions confirms the advantages of the introduced strategy compared to conventional methods.

Keywords Fuzzy logic · Based control · Sliding · Mode control · Neutral · Point · Clamped converter

Nomenclature

Symbols	Description
C	DC-link capacitance, F
L	Line inductance, H
R_{dc}	DC-link resistance, Ω

X. Shen · Z. Kang · J. Liu (✉)
Department of Control Science and Engineering, Harbin Institute of Technology, Harbin, China
e-mail: jx.liu@hit.edu.cn

J. Wang
School of Automation, Harbin University of Science and Technology, Harbin 150001, China

X. Liu
School of Automation, Nanjing University of Information Science and Technology, Nanjing, China

E. Kabalci (ed.), *Modern Control Methods for Power Electronics*, Energy Systems in Electrical Engineering, https://doi.org/10.1007/978-981-92-1974-2_6

$Vdc1$	Voltage value of the upper DC-link capacitor, V
$Vdc2$	Voltage value of the lower DC-link capacitor, V
$x1{=}Vdc1{+}Vdc2$	The sum of two DC-link capacitor voltages, V
$x1{=}Vdc1{-}Vdc2$	The difference of two DC-link capacitor voltages, V
$v_{\alpha\beta} = [v_\alpha, v_\beta]^{\mathrm{T}}$	The grid voltages in $\alpha\beta$ coordinate, V
$i_{\alpha\beta} = [i_\alpha, i_\beta]^{\mathrm{T}}$	The phase current in $\alpha\beta$ coordinate, A
$u_{\alpha\beta\gamma} = [u_\alpha, u_\beta, u_\gamma]^{\mathrm{T}}$	The average duty cycles in $\alpha\beta\gamma$ coordinate
p	Active power, kW
q	Reactive power, kVar
ω	The grid angular frequency, rad
n	The dimension of the input of fuzzy logic system inputs
$c = [c_1, c_2, ..., c_n]^{\mathrm{T}}$	The input of fuzzy logic system
y	The output of fuzzy logic system
$F_j^l,\ j = 1, 2, ..., n$	The fuzzy sets
$G_c^l,\ j = 1, 2, ..., n$	The fuzzy sets
M_0	The number of fuzzy rules
$\mu_{F_j^l}(\cdot),\ l = 1, 2, ..., M_0$	The membership functions of fuzzy set F_j^l
$\mu_{G_c^l}(\cdot),\ l = 1, 2, ..., M_0$	The membership functions of fuzzy set G_c^l
$\bar{y}_c^l,\ l = 1, 2, ..., M_0$	The point at which $\mu_{G_c^l}(y) = 1$
$\theta = [\bar{y}_c^1, ..., \bar{y}_c^{M_0}]^{\mathrm{T}}$	The ideal constant weight vector
$\hat{\theta}_v$	The estimated weight vector
$\tilde{\theta}_v = \theta_v - \hat{\theta}_v$	The weight vector estimation error
$\xi = [\xi^1, ..., \xi^{M_0}]^{\mathrm{T}}$	The basis function vector
$\omega^{l_0} = \mu_{F^{l_0}}(g)$	The membership value
$Vf{=}0.5g2$	A Lyapuonv function
$p^\star$	The reference of active power
$d_p = \dot{p}^\star$	The derivative of the reference of active power
$d_q = \dot{q}^\star$	The derivative of the reference of reactive power
$e_p = p^\star - p$	The tracking errors of active power
$sp{=}ep$	The active power sliding mode variable
$sq{=}eq$	The reactive power sliding mode variable
$e_q = q^\star - q$	The tracking errors of reactive power
$u_{\alpha\beta}^{eq} = [u_\alpha^{eq}, u_\beta^{eq}]^T$	The equilibrium point
ω_{m1}	The membership value of Rules 1
ω_{m2}	The membership value of Rules 2
ω_{m3}	The membership value of Rules 3
s_{m0}	A positive boundary layer constant
I	The identity matrix
$\lVert\cdot\rVert_2$	The Euclidean norm of vectors
$\lambda_{\max}(P)$	The maximum eigenvalue of the matrix P
$z_1 = \frac{1}{2}x_1^2$	Voltage regulation state
$z_1^\star$	Voltage regulation reference signal
$e_1 = z_1^\star - z_1$	Voltage regulation error
ksv	A positive constant

$sv{=}ksve1$	Voltage regulation sliding mode variable
$d_v = \frac{4z_1}{CR_{dc}}$	Unknown lumped disturbance in voltage regulation loop

1 Introduction

1.1 Recent Advances and Challenges in Advanced Control of Power Converters

Recently, the escalating global concerns over fossil fuel depletion and ecological sustainability have positioned energy-efficient technologies as a critical priority for modern industrial systems (Roy et al. 2022; Asna et al. 2024; Cui et al. 2024). These imperatives drive innovation in transportation electrification, emphasizing demands for enhanced safety, operational reliability, energy efficiency, and cost-effectiveness (Lin et al. 2020; Askari and Farzanehfard 2024). Within this context, power converters serve as pivotal components, enabling efficient energy conversion to meet these multifaceted objectives.

Three-level NPC converters have gained prominence in applications due to their modular architecture, reduced semiconductor stress, and optimal balance between cost-efficiency and operational performance (Leon et al. 2017; Guo et al. 2021; Xuan et al. 2019; Rivera et al. 2015; Zhou and Cheng 2019; Siemens 2019; ABB Switzerland Ltd 2011; Infineon 2021). For instance, NPC converters are employed in aircraft electric starter/generator systems to ensure reliable energy delivery (Guo et al. 2021), while their integration into plug-in electric vehicle charging stations supports rapid DC charging capabilities (Rivera et al. 2015). Industrial adoption is further evidenced by commercial solutions such as SIEMENS' SINAMICS GM150 medium-voltage drives for marine applications (Siemens 2019), ABB's ACS6000 drives for high-power motor systems (ABB Switzerland Ltd 2011), and Infineon's bidirectional charging infrastructure designs for grid interaction (Infineon 2021). As shown in Fig. 1, this chapter focuses on a back-to-back NPC topology where the grid-side converter interfaces with a machine-side converter and motor, modeled as a resistive load R_{dc} to simulate active power disturbance.

Several key challenges in the control of NPC converters are presented below:

(1) Multiloop Complexity: NPC converters demand three interdependent control loops (power tracking, voltage regulation, and voltage balancing), contrasting with the simpler two-loop architecture of traditional two-level converters (Leon et al. 2017). This complexity amplifies controller design challenges due to heightened control objectives and stringent performance requirements.
(2) Power Tracking Trade-Offs: The power tracking loop must synchronize active and reactive power with reference values while reconciling conflicting objectives: transient response and steady-state current quality.

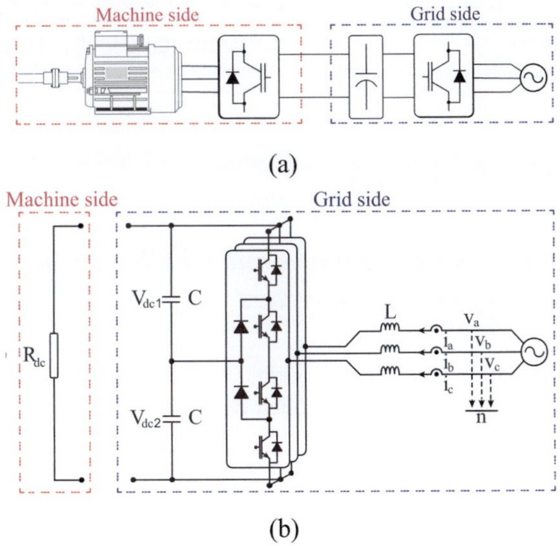

Fig. 1 Topology of converter system: **a** Topology of back-to-back converter system. **b** The simplified topology of back-to-back converter system

(3) Disturbance Sensitivity: In voltage regulation, unknown load dynamics necessitate external disturbance compensation via feedforward mechanisms, as the load resistance R_{dc} remains an unmodeled variable.

(4) Capacitor Voltage Imbalance: Voltage disparities between DC-link capacitors exacerbate harmonic distortion and component stress, mandating dedicated balancing control.

Over the past few years, Proportional-Integral (PI) control, fuzzy logic, Model Predictive Control (MPC), and Sliding-Mode Control (SMC) have been widely applied to NPC systems (Lin et al. 2020; Vazquez et al. 2020; Shen 2025; Yin et al. 2019; Shen et al. 2025; Soliman et al. 2019; Yang et al. 2019; Lu 2018). Among these methods, SMC stands out as a preferred choice due to its robustness to parametric uncertainties and external disturbances, making it particularly suited for NPC applications (Liu et al. 2024 Liu et al. 2024). However, achieving rapid dynamic response necessitates high switching gains, which intensify the undesired chattering phenomenon. These oscillations pose significant risks by potentially exciting unmodeled high-frequency dynamics, thereby degrading system stability (Lin et al. 2024).

To address this challenge, efforts to suppress chattering have centered on higher-order sliding-mode approaches (Ding et al. 2022), with the Super-Twisting Algorithm (STA) (Zhang et al. 2023; Amin et al. 2021; Khoshhava et al. 2021) emerging as a prominent solution. Empirical evidence demonstrates STA's efficacy: it has successfully reduced chattering in two-level converters (Liu et al. 2018) and DC microgrid inverters (Yin et al. 2022) and also used in event-triggered implementations (Kang et al. 2025). Furthermore, its adaptation to NPC inverters for renewable energy integration further validates its versatility (Tabart et al. 2018). Despite these advantages, STA's reliance on large control gains creates a paradoxical situation where dynamic performance improvements inevitably amplify chattering, resulting in a design trade-off.

To overcome these limitations, recent advancements integrate Fuzzy Logic Systems (FLSs) with SMC, giving rise to Fuzzy SMC (FSMC) strategies. This hybrid approach synergistically combines the robustness of SMC with the adaptability of fuzzy logic, achieving dual benefits: (1) chattering suppression through smoothed switching functions while (2) maintaining dynamic responsiveness. Three key applications illustrate this advancement: autonomous vehicle path-tracking systems leverage fuzzy-tuned SMC coefficients to reduce computational overhead (Hwang et al. 2018); fractional-order FSMC schemes enhance precision in permanent magnet motor control via adaptive gain adjustment (Chen et al. 2019); and simplified FSMC implementations achieve voltage regulation in DC motor drives with minimal complexity (Zhang et al. 2016). Notably, FSMC offers three distinct advantages: firstly, chattering suppression via softened switching dynamics; secondly, reduced computational load through streamlined fuzzy rule sets; thirdly, rigorous stability guarantees via Lyapunov theory. However, existing research reveals a critical gap: FSMC applications in NPC converters remain underexplored, particularly regarding hybrid architectures combining FLS with STA, i.e., Fuzzy Super-Twisting Algorithm (FSTA). This research gap directly motivates the introduced FSTA-based control framework for NPC systems, which will be detailed in subsequent sections.

1.2 Fuzzy Logic Control in Power Electronic Systems

Fuzzy logic control has emerged as a pivotal methodology for managing the inherent nonlinearities and uncertainties in power electronic systems. Since the inception of fuzzy set theory by Zadeh in 1965 (Zadeh 1965), and the subsequent development of the first fuzzy controller by Mamdani and Assilian in 1975 (Mamdani and Assilian 1975), fuzzy logic control has been extensively applied across various power electronics domains, including converters, inverters, and power quality enhancement devices.

In the realm of power quality improvement, fuzzy logic controllers have been effectively utilized to address issues such as voltage sags, harmonics, and reactive power compensation. A comprehensive review (Cziker and Beleiu 2024) highlights the application of fuzzy logic controllers in devices like unified power quality conditioners and active power filters, demonstrating their superiority over traditional PI controllers in mitigating power quality disturbances.

Furthermore, the evolution of fuzzy logic control has advanced with Type-2 fuzzy logic controllers, which excel in handling uncertainties and imprecise information. A critical study by (Mendel et al. 2014) provides an extensive overview of Type-2 fuzzy logic controllers, discussing their theoretical foundations, design methodologies, and applications in areas such as robotics, automotive systems, and industrial automation.

In renewable energy systems, fuzzy logic controllers have been employed to optimize the performance of photovoltaic systems, particularly in Maximum Power Point Tracking (MPPT) (Sarvi and Azadian 2022). Studies have confirmed that fuzzy logic control–based MPPT algorithms can adapt more effectively to changing environmental conditions compared to conventional methods, leading to improved energy harvesting efficiency.

Moreover, the integration of fuzzy logic control with other control techniques, such as neural networks (Hou et al. 2021), sliding-mode control (Long et al. 2022), and model predictive control (Liu et al. 2022; Liu et al. 2024), has led to the development of hybrid control strategies. These hybrid controllers have demonstrated improved dynamic performance and robustness in various power electronic applications, including DC-DC converters (Babes et al. 2022) and motor drives (Yang et al. 2022; Ma et al. 2023). Despite the advantages, the design and implementation of fuzzy logic controllers pose challenges, particularly in the tuning of membership functions and rule bases. Recent research has focused on developing systematic design methodologies and optimization techniques (Herrera et al. 1995; Omizegba and Monikang 2009; Shouran and Alsseid 2022; Hamza et al. 2017; Bouallègue et al. 2015; Liu et al. 2021; Yilmaz et al. 2021) to address these challenges, ensuring that fuzzy logic controllers can be effectively tailored to specific applications.

Accordingly, fuzzy logic control serves as a powerful tool in the control of power electronic systems, offering flexibility and robustness against system uncertainties and model nonlinearities. Ongoing research and development are expected to further enhance the capabilities and applications of fuzzy logic control in evolving power electronics landscape.

1.3 Chapter Overview

This chapter examines the integration of fuzzy logic control methodologies in modern power conversion systems, focusing on their application to three-level Neutral-Point-Clamped (NPC) converters. Section 2 introduces the fundamentals of fuzzy logic control. Section 3 presents the design methodology for fuzzy controllers. Section 4 describes the system model of NPC converters and the fundamental principle of fuzzy logic control. Section 5 details the design of a fuzzy logic-based sliding-mode controller for NPC converters, while Sect. 6 validates its efficacy through experimental case studies. The chapter concludes with a summary of key contributions and future research directions. By bridging theoretical insights with experimental validation, this chapter establishes a theoretical and practical framework for addressing key challenges in power converter control, such as dynamic performance optimization, chattering suppression, and disturbance rejection.

2 Fundamentals of Fuzzy Logic Control

2.1 Fuzzy Sets

Fuzzy sets extend the classical set theory by allowing elements to belong to a set with varying degrees of membership, rather than adhering to a strict binary classification. In traditional set theory, an element either completely belongs to a set or not at all, represented by a characteristic function assigning 1 for membership and 0 otherwise. However, such binary delineation often fails to capture the inherent vagueness present in real-world systems.

Fuzzy sets, by contrast, are characterized by membership functions that map each element x in the universe of discourse U to a membership degree in the interval (Roy et al. 2022). This value reflects the extent to which x belongs to the fuzzy set A. For instance, in the fuzzy set "tall people," a height of 180 cm may correspond to a membership degree of 0.9, while 160 cm might correspond to 0.3, depending on the contextually defined notion of "tall."

Mathematically, a fuzzy set A in a universe U is expressed as follows:

$$A = \{(x, \mu_A(x)) | x \in U\}, \tag{1}$$

where $\mu_A(x)$ is the corresponding membership function.

This definition allows smooth transitions between full membership and non-membership, capturing the continuum between absolute truth and falsehood. Consequently, fuzzy sets are particularly suited for representing imprecise phenomena such as human reasoning, linguistic concepts, or control processes in uncertain environments.

The flexibility of fuzzy sets makes them a foundational tool in fuzzy logic, where they support approximate reasoning and enable systems to handle uncertainties in a mathematically rigorous way. Compared with crisp sets, fuzzy sets provide gradual and interpretable transitions, making them highly applicable in control engineering, decision-making, and pattern recognition.

2.2 Membership Functions

Membership functions are fundamental components of fuzzy logic systems, quantifying how strongly a given input value belongs to a fuzzy set (Altas 2017). Various types of membership functions exist, each with distinct properties affecting system sensitivity, interpretability, and computational efficiency.

The triangular membership function is one of the simplest and most widely used forms, characterized by a linear increase and decrease, forming a sharp peak at its center. This sharpness enhances the system's sensitivity to small variations in input. It is mathematically described as follows:

$$\mu_1(x) = \begin{cases} (|x| - |x_1|)/(|x_2| - |x_1|), \ x_1 \le x \le x_2 \\ (|x_3| - |x|)/(|x_3| - |x_2|), \ x_2 \le x \le x_3 \\ 0, \text{otherwise} \end{cases}, \tag{2}$$

where x is the input of membership functions, and x_1, x_2, x_3, x_4 are designed membership function parameters.

In contrast, the trapezoidal membership function features a flat region at the top, introducing a crisp zone where the membership grade remains at its maximum. This design reduces sensitivity near the peak and is often preferred when robustness to input variations is desired. Mathematically, it is represented as follows:

$$\mu_2(x) = \begin{cases} (|x| - |x_1|)/(|x_2| - |x_1|), \ x_1 \le x \le x_2 \\ 1, \ x_2 \le x \le x_3 \\ (|x_4| - |x|)/(|x_4| - |x_3|), \ x_3 \le x \le x_4 \\ 0, \text{otherwise} \end{cases}. \tag{3}$$

For smoother transitions, the Gaussian membership functions provide a bell-shaped curve, reducing sensitivity to sharp input changes while preserving the inherent fuzziness of the system:

$$\mu_3(x) = \mathrm{e}^{-\frac{1}{2}\left(\frac{x - x_p}{w}\right)^2}, \tag{4}$$

where x_p is the center and w controls the width.

The generalized bell-shaped function offers additional flexibility through its slope parameter m:

$$\mu_4(x) = \frac{1}{1 + \left|\frac{x - x_p}{w}\right|^{2m}}. \tag{5}$$

The Cauchy membership function, known for its heavy tails, is expressed as follows:

$$\mu_5(x) = \frac{1}{1 + \left(\frac{x - x_p}{w}\right)^{2m}}. \tag{6}$$

Lastly, sigmoid functions with S-shaped curves are suitable for modeling the lower and upper boundaries of fuzzy regions:

$$\mu_6(x) = \frac{1}{1 + \mathrm{e}^{-a(x-c)}}, \tag{7}$$

where a determines the slope, and c sets the inflection point.

Each membership function type exhibits different trade-offs in adaptability, smoothness, computational efficiency, and modeling accuracy. The selection should align with the system's control objectives, system nonlinearity, and computational resources.

2.3 Linguistic Variables

Linguistic variables form the foundation of fuzzy logic control systems, allowing both inputs and outputs to be described using words or phrases rather than precise numerical values. Unlike conventional variables, which are exact and quantitative, linguistic variables capture subjective concepts such as "high," "low," or "moderate." For example, the variable AGE may take on linguistic values such as YOUNG or OLD. In fuzzy control systems, a commonly used linguistic variable is ERROR, which can take on values such as "POSITIVE" (P), "ZERO" (Z), and "NEGATIVE" (N), thereby facilitating intuitive control rule formulation.

2.4 Fuzzy Logic Operators

Fuzzy logic operators generalize classical Boolean logic to allow reasoning with values between 0 and 1. The main fuzzy operators include negation (NOT), conjunction (AND), disjunction (OR), and implication (Cirstea et al. 2002).

The negation (NOT) operator reflects the degree to which an element does not belong to a fuzzy set, defined as follows:

$$\mu_{\overline{A}}(x) = 1 - \mu_A(x). \tag{8}$$

The conjunction (AND) operator models the intersection of fuzzy sets and is typically implemented using the minimum function:

$$\mu_{A \wedge B}(x) = \min[\mu_A(x), \mu_B(x)]. \tag{9}$$

The disjunction (OR) operator represents the union of fuzzy sets, commonly defined via the maximum function:

$$\mu_{A \vee B}(x) = \max[\mu_A(x), \mu_B(x)]. \tag{10}$$

The implication operator is fundamental to fuzzy rule-based control systems, particularly for encoding IF–THEN rules. Several implication methods have been proposed in the literature. Among them, the most widely used in fuzzy control applications is Mamdani's implication, defined as follows:

$$\mu_R(x, y) = \min[\mu_A(x), \mu_B(y)]. \tag{11}$$

Other alternatives include Zadeh's implication, Gödel's implication, and Łukasiewicz's implication, each differing in how they model the relationship between antecedent and consequent fuzzy sets. The specific representation is as follows:

$$\text{Zadeh's implication}: \mu_R(x, y) = \max(\min[\mu_A(x), \mu_B(y)], 1 - \mu_A(x)), \tag{12}$$

$$\text{Gödel's implication}: \mu_R(x, y) = \begin{cases} 1, & \text{if } \mu_A(x) \le \mu_B(y) \\ \mu_B(y), & \text{otherwise} \end{cases}, \tag{13}$$

$$\text{Lukasiewicz's implication}: \mu_R(x, y) = \min[1, 1 - \mu_A(x) + \mu_B(y)]. \tag{14}$$

The choice of implication operator affects system behavior and is guided by computational constraints and control requirements. Mamdani's approach is favored in many real-time control systems due to its simplicity and ease of implementation.

2.5 *Rule Base*

The rule base is a fundamental component of a Fuzzy Logic Controller (FLC), encoding expert knowledge in the form of IF–THEN statements that link input conditions to control actions. A typical fuzzy rule is written as follows:

$$\text{IF } x_1 \text{ is } A_1 \text{ AND } x_2 \text{ is } A_2 \text{ AND} \ldots \text{THEN } y \text{ is } B, \tag{15}$$

where A_1, A_2, and B are fuzzy sets defined over the inputs x_1, x_2 and output variables y, respectively. The complete rule base comprises multiple such rules:

$$\mathcal{R} = \{R_1, R_2, \ldots, R_N\}, \tag{16}$$

where R_i denotes the ith fuzzy rule in the system.

Each rule is activated depending on the degree of match between current input values and the corresponding fuzzy sets, computed via membership functions. The inference mechanism (e.g., Mamdani or Sugeno method) processes these rules to produce an aggregated fuzzy output for defuzzification. The richness and coverage of the rule base largely determine the flexibility and effectiveness of the fuzzy controller.

2.6 Defuzzification Techniques

After fuzzy inference, the fuzzy system produces a fuzzy output set. However, most real-world control systems require a precise and actionable numerical output. This conversion process is known as defuzzification.

One of the most commonly used methods is the center-average method (centroid or center of gravity) technique, which calculates the center of the area under the aggregated output membership function:

$$y^* = \frac{\int y \cdot \mu(y)\, dy}{\int \mu(y)\, dy}. \tag{17}$$

This method provides a balanced output but is computationally intensive. Another widely adopted approach is the weighted center-average method, especially when output membership functions are symmetrical and their peaks are known. It is defined as follows:

$$y^* = \frac{\sum_{i=1}^{n} \mu_i \cdot y_i}{\sum_{i=1}^{n} \mu_i}, \tag{18}$$

where μ_i is the membership degree of the ith output fuzzy set, and y_i is the corresponding output.

Additional methods include the max membership principle (selecting the output with the highest membership), mean of maximum, and first/last maximum, each suited for different trade-offs between computational simplicity and output smoothness.

2.7 Mamdani Versus Takagi–Sugeno Fuzzy Systems

Fuzzy systems are widely adopted in power electronic control for handling nonlinearities, model uncertainties, and external disturbances. Among them, the Mamdani and Takagi–Sugeno (T–S) models are the most prevalent. Mamdani fuzzy inference systems employ fuzzy sets for both antecedents and consequents, resulting in a fuzzy output that requires defuzzification. This type of system is highly interpretable and aligns well with heuristic design procedures or expert-based knowledge, which is often advantageous in designing intuitive controllers for converters, inverters, and electric drives.

Conversely, Takagi–Sugeno fuzzy inference system combines fuzzy antecedents with functional (often constant or linear) consequents, generating crisp outputs via weighted averaging without defuzzification. This characteristic results in greater suitability for real-time applications such as active power filters, DC-DC converters, and renewable energy interfaces, where computational efficiency and control precision are critical. Furthermore, the mathematical structure of T–S models allows for rigorous control design and stability analysis using Lyapunov-based or linear matrix inequality approaches.

In summary, Mamdani-type systems offer superior interpretability and are ideal for intuitive or human-in-the-loop design. In contrast, T–S systems provide better computational performance and precision, making them more suitable for high-speed control systems. The choice between them depends on application-specific requirements including response speed, control accuracy, and implementation complexity.

3 Design Methodology for Fuzzy Controllers

The design of a fuzzy logic controller follows a systematic methodology that integrates system knowledge with fuzzy logic principles to achieve effective and adaptive control. The main design steps are outlined as follows.

3.1 Selecting Input and Output Variables

The first step is to identify the relevant system signals that will serve as the inputs and outputs of the controller. Inputs typically include error signals, changes in error, or other measurable process variables. Outputs correspond to control actions such as duty cycle, torque, voltage, or frequency, depending on the specific application. The selection of input and output variables should sufficiently characterize the system's dynamic behavior.

3.2 Crafting the Rule Base

The rule base encodes the control logic using a series of linguistic IF–THEN rules. Each rule describes a mapping from fuzzy inputs to fuzzy outputs, for example:

$$\text{IF } x_1 \text{ is } A_1 \text{ AND } x_2 \text{ is } A_2 \text{ THEN } y \text{ is } B, \tag{19}$$

where $A_1, A_2,$ and B are fuzzy sets defined over the inputs x_1, x_2 and output variables y, respectively. These rules are derived from expert knowledge, heuristic insights, or extracted from data via fuzzy identification techniques.

3.3 Tuning Membership Functions

Membership functions define the fuzzy sets associated with each linguistic term. Their shape and parameters have a significant impact on the controller's responsiveness and smoothness.

Common tuning approaches include the following:

Trial-and-Error: Manual adjustment based on observed system performance.

Optimization-Based Tuning: Application of genetic algorithms (Herrera et al. 1995), particle swarm optimization (Omizegba and Monikang 2009; Shouran and Alsseid 2022), or other metaheuristics (Hamza et al. 2017; Bouallègue et al. 2015) to minimize control error.

Adaptive Schemes: Online adjustment based on performance criteria or error feedback (Liu et al. 2021; Yilmaz et al. 2021).

3.4 Fuzzification and Input Normalization

Fuzzification converts crisp input values into degrees of membership in fuzzy sets using the corresponding membership functions. Prior to this, normalization is applied to scale physical inputs into a standardized domain (typically $[-1, 1]$ or (Roy et al. 2022)). If the input is x and its known range is $[x_{\min}, x_{\max}]$, normalization is computed as follows:

$$x_{\text{norm}} = \frac{x - x_{\min}}{x_{\max} - x_{\min}}. \tag{20}$$

3.5 Inference Mechanism

The inference engine evaluates the fuzzy rules and combines their outputs. For Mamdani-type fuzzy logic controllers, the output fuzzy sets generated by all activated rules are aggregated using minimum (AND) and maximum (OR) operators:

$$\mu(y) = \max\left[\min_j \mu_{A_j}(x_j) \wedge \mu_B(y)\right]. \tag{21}$$

3.6 Defuzzification and Output Denormalization

Defuzzification transforms the aggregated fuzzy output into a crisp control value, typically using the center-average method:

$$y^* = \frac{\int y \cdot \mu(y)\,dy}{\int \mu(y)\,dy} \text{ or } y^* = \frac{\sum_{i=1}^{n} \mu_i \cdot y_i}{\sum_{i=1}^{n} \mu_i}. \tag{22}$$

Following defuzzification, denormalization maps the normalized output back to its corresponding value domain:

$$y = y^* \cdot (y_{\max} - y_{\min}) + y_{\min}, \tag{23}$$

where $y_{\max}$ and $y_{\min}$ are the maximum and minimum values of the output, respectively.

4 System Description and Fuzzy Logic Control Fundamental Principle

This section presents the mathematical model of NPC converters, laying the foundation for subsequent control strategy design. Additionally, the section introduces several fundamental principles of fuzzy logic control.

4.1 System Description of Power Converters

The mathematical description of NPC converter in $\alpha\beta\gamma$ coordinate frame is given in (Portillo et al. 2013) by

$$\begin{aligned}
\frac{di_\alpha}{dt} &= \frac{v_\alpha}{L} - \frac{x_1 u_\alpha}{2L} + \frac{x_2}{L}\left(\frac{1}{2\sqrt{6}}\left(u_\beta^2 - u_\alpha^2\right) - \frac{1}{\sqrt{3}} u_\alpha u_\gamma\right), \\
\frac{di_\beta}{dt} &= \frac{v_\beta}{L} - \frac{x_1 u_\beta}{2L} + \frac{x_2}{L}\left(\frac{1}{\sqrt{6}} u_\alpha u_\beta - \frac{1}{\sqrt{3}} u_\beta u_\gamma\right), \\
\dot{x}_1 &= \frac{u_{\alpha\beta}^T i_{\alpha\beta}}{C} - \frac{2x_1}{CR_{dc}}, \\
\dot{x}_2 &= \frac{2\sqrt{2} u_{\alpha\beta}^T i_{\alpha\beta} u_\gamma + \left(u_\alpha^2 - u_\beta^2\right) i_\alpha - 2u_\alpha u_\beta i_\beta}{\sqrt{6}C},
\end{aligned} \tag{24}$$

Using the instantaneous reactive power theory from (Akagi et al. 1984), the active power and reactive powers are defined as $p = v_\alpha i_\alpha + v_\beta i_\beta$ and $q = v_\alpha i_\beta - v_\beta i_\alpha$. Assuming that $x_2 \approx 0$ and balanced grid voltages, the instantaneous power dynamics conform to the orthogonal decomposition:

$$\begin{aligned}\dot{p} &= \frac{1}{L}||v_{\alpha\beta}||^2 - \frac{1}{2L}x_1 v_{\alpha\beta}^T u_{\alpha\beta} + \omega q,\\ \dot{q} &= -\frac{1}{2L}x_1\left(Jv_{\alpha\beta}\right)^T u_{\alpha\beta} - \omega p,\\ J &= \begin{bmatrix} 0 & -1 \\ 1 & 0 \end{bmatrix}.\end{aligned} \tag{25}$$

4.2 Fundamental Principle of Fuzzy Logic Control

A FLS generally comprises four components: the fuzzifier, the fuzzy rule base, the fuzzy inference engine, and the defuzzifier (Wang and Mendel 1992). The fuzzy rule base is constructed using the following IF–THEN statements:

$$R^l : \text{IF } c_1 \text{ is } F_1^l \text{ and ... and } c_n \text{ is } F_n^l, \text{ THEN } y \text{ is } G_c^l, \tag{26}$$

Employing a singleton fuzzifier, center-average defuzzification, and a fuzzy inference engine, the FLS output can be represented as follows:

$$y = \frac{\sum_{l=1}^{M_0} \overline{y}_c^l \prod_{j=1}^n \mu_{F_j^l}(c_j)}{\sum_{l=1}^{M_0} \prod_{j=1}^n \mu_{F_j^l}(c_j)}, \tag{27}$$

where $\overline{y}_c^l$ is the value at which $\mu_{G_c^l}(y) = 1$.

Setting $\xi^l = \frac{\prod_{j=1}^n \mu_{F_j^l}(c_j)}{\sum_{l=1}^{M_0} \prod_{j=1}^n \mu_{F_j^l}(c_j)}$, the FLS can be rewritten in the following compact form:

$$y = \theta^{\mathrm{T}}\xi, \tag{28}$$

This formulation demonstrates that the FLS effectively approximates its output as an unknown ideal weight vector θ multiplied by a known basis function ξ. In practice, an FLS serves two principal roles. First, it can function as a fuzzy observer to estimate disturbances, typically using Gaussian membership functions. Second, it may be integrated with other control strategies to regulate system behavior.

For instance, consider the dynamic system described by $\dot{g} = f + u$, where u is the control input and f represents disturbance. Here, the system variable g and the control u are utilized as the FLS input and output, respectively. The fuzzy sets for g and u are designated as F^{l_0}, $l_0 = 1, 2, ..., 2M + 1$, where M is a positive constant. The fuzzy rules are stated as follows:

$$R^{l_0} : \text{IF } g \text{ is } F^{l_0}, \text{THEN } u \text{ is } u_{l_0},\ l_0 = 1, 2, ..., 2M + 1. \tag{29}$$

Figure 2 illustrates the membership functions with positive constants g_{l_1}, $l_1 = 1, ..., M$. The outputs are chosen such that $u_{l_0} > 0$ when $l_0 < M + 1$, $u_{M+1} = 0$, and $u_{l_0} < 0$ when $l_0 > M + 1$. Consequently, the controller can be expressed as follows:

$$u = \frac{\sum_{l_0=1}^{2M+1} u_{l_0}\omega^{l_0}}{\sum_{l_0=1}^{2M+1} \omega^{l_0}}, \tag{30}$$

To ensure system stability, a Lyapunov function $V_f = 0.5g^2$ is introduced with its derivative as follows:

$$\dot{V}_f = g(u_1\omega^1 + u_2\omega^2 + ... + u_{2M+1}\omega^{2M+1} + f). \tag{31}$$

Stability is further guaranteed by imposing three conditions corresponding to the intervals $g < -g_M$, $-g_M \le g < -g_{M-1}$, and $-g_1 \le g < 0$.

Condition 1: For $g < -g_M$, only $\omega^1 = 1$ is active while all other membership functions vanish. Thus, $u = u_1$ and $\dot{V}_f = g(u_1 + f)$. Choosing u_1 sufficiently larger than the maximum disturbance magnitude $|f|_M$ (the maximum of $|f|$) results in $\dot{V}_f < 0$, ensuring that g converges to $g \ge -g_M$ in finite time.

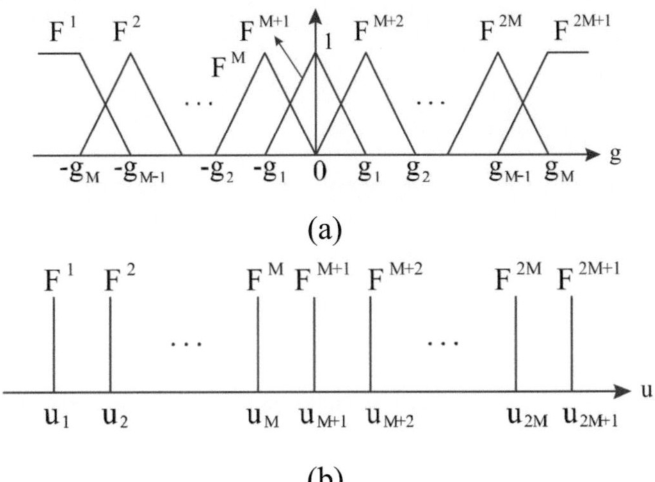

Fig. 2 Membership functions: **a** Membership functions of the input fuzzy sets F^{l_0} for g. **b** Membership functions of the output fuzzy sets F^{l_0} for u

Condition 2: When $-g_M \leq g < -g_{M-1}$, the membership values are computed as follows:

$$\omega^1 = -\frac{g + g_{M-1}}{g_M - g_{M-1}}, \; \omega^2 = \frac{g + g_M}{g_M - g_{M-1}}, \tag{32}$$

with the remaining memberships being zero. Selecting u_1 and u_2 such that both are larger than $|f|_M$ ensures that $\omega^1 u_1 + \omega^2 u_2 > |f|_M$, leading to $\dot{V}_f = g(\omega^1 u_1 + \omega^2 u_2 + f) < 0$ and guaranteeing finite-time convergence of g to $g \geq -g_{M-1}$.

Condition 3: Finally, within the interval $-g_1 \leq g < 0$, the active memberships are as follows:

$$\omega^M = -\frac{g}{g_1}, \; \omega^{M+1} = \frac{g + g_1}{g_1}, \tag{33}$$

with all other memberships equal to zero, yielding $u = \omega^M u_M$. Under these conditions, g converges to the bounded region $|g|<|f|_M g_1/u_M$. A similar analysis applies for g positive.

In this chapter, the singleton fuzzifier and center-average defuzzification approach is adopted because of its reliability, computational simplicity, and extensive use in the literature. Moreover, a three-function membership design is selected to strike an effective balance between performance and complexity.

5 Fuzzy Logic Control Scheme Design of Power Converter

To address these challenges, this section develops a FLS-based SMC framework for NPC converters. In the power tracking loop, a FSTA optimized via FLS is investigated to ensure bounded convergence of active and reactive power tracking errors. In the DC-link voltage regulation loop, a fuzzy observer-integrated SMC strategy is employed for robust voltage stabilization. And the DC-link voltage balancing loop utilizes the conventional Proportional-Integral (PI) control for balancing the DC-link voltages of two capacitors. The key innovations and contributions of the scheme are as follows:

(1) An innovative integration of FLS with both STA and SMC architectures is established for simultaneous power regulation and DC-link voltage management in NPC converters. Lyapunov stability theory rigorously validates the convergence of both FSTA and Fuzzy SMC (FSMC) controllers. Practical implementation is facilitated through systematic parameter tuning guidelines, while comprehensive comparative experiments and theoretical evaluations collectively confirm the framework's enhanced efficacy across diverse operational scenarios.

(2) The introduced FSTA architecture in the power tracking loop addresses inherent limitations of conventional STA through FLS augmentation. This hybridization achieves dual improvements: accelerated transient response during power

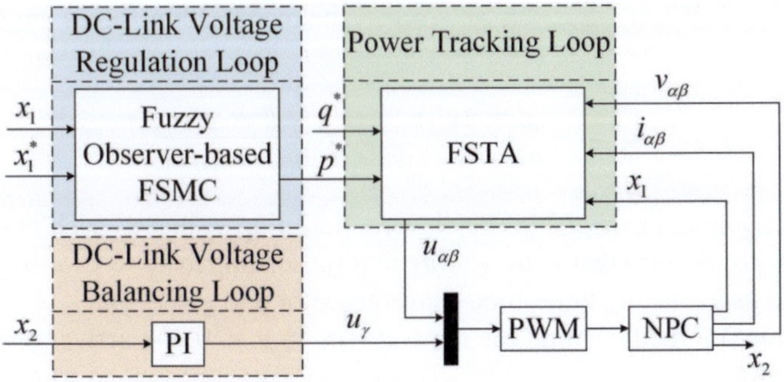

Fig. 3 Control structure of the introduced strategy

reference transitions and superior steady-state performance characterized by mitigated chattering effects and enhanced current waveform quality.

(3) Within the DC-link voltage regulation loop, the FSMC strategy retains the inherent robustness of classical SMC while alleviating chattering via continuous approximation of switching functions. To counteract performance degradation from unmodeled disturbances, a fuzzy disturbance observer is integrated, with its real-time estimations dynamically compensating the control law. This synergistic design ensures exceptional disturbance rejection without compromising computational efficiency.

The comprehensive control architecture is illustrated in Fig. 3. Subsequent subsections elaborate on the controller's design methodology, stability verification, and operational analysis.

5.1 Power Tracking Loop

In this section, a finite-time sliding-mode control with fuzzy logic enhancement (FSTA) is employed to rapidly drive the tracking errors of active and reactive power into predefined bounded regions while simultaneously suppressing chattering and preserving current quality during steady-state operation.

By solving $\dot{p} = 0$ and $\dot{q} = 0$ (Portillo et al. 2013), the equilibrium point $u_{\alpha\beta}^{eq} = [u_\alpha^{eq}, u_\beta^{eq}]^{\mathrm{T}}$ is obtained. Subsequently, the controller is specified as follows:

$$\begin{aligned} u_{\alpha\beta}^{eq} &= \frac{2}{x_1 \left\| v_{\alpha\beta} \right\|^2} \left[\left(\left\| v_{\alpha\beta} \right\|^2 + L\omega q \right) v_{\alpha\beta} - L\omega p J v_{\alpha\beta} \right], \\ u_{\alpha\beta} &= u_{\alpha\beta}^{eq} - \mu(e_p) v_{\alpha\beta} - \mu(e_q) J v_{\alpha\beta}, \end{aligned} \tag{34}$$

where the functions $\mu(e_p)$ and $\mu(e_q)$ will be detailed later. The dynamics of the tracking errors are given by

$$
\begin{aligned}
\dot{e}_p &= -\frac{1}{L}\left\|v_{\alpha\beta}\right\|^2 + \frac{1}{2L}x_1 v_{\alpha\beta}^T u_{\alpha\beta} - \omega q + d_p,\\
\dot{e}_q &= \frac{1}{2L}x_1\left(Jv_{\alpha\beta}\right)^T u_{\alpha\beta} + \omega p + d_q,
\end{aligned}
\tag{35}
$$

In practical applications, since the power converter's rated power is limited by the maximum current of its power semiconductors, both $d_p = \dot{p}^*$ and $d_p = \dot{q}^*$ are bounded, i.e., $|d_p| \le d_{p0}$ and $|d_q| \le d_{q0}$. By substituting (34) into (35), the error dynamics can be recast as follows:

$$
\begin{aligned}
\dot{e}_p &= B\mu(e_p) + d_p,\\
\dot{e}_q &= B\mu(e_q) + d_q,
\end{aligned}
\tag{36}
$$

where $B = -\frac{x_1}{2L}\left\|v_{\alpha\beta}\right\|^2$.

For further analysis, the sliding-mode variable is defined as $s_m = e_m$ for $m \in \{p, q\}$. The variable s_m along with the auxiliary function u_{s_m} for $m \in \{p, q\}$ serve as the input and output of the fuzzy logic system. Three fuzzy sets, i.e., P (positive), N (negative), and Z (zero), are selected. And the corresponding fuzzy rule base is given by.

Rule 1: If s_m is P, then u_{s_m} is P.
Rule 2: If s_m is Z, then u_{s_m} is Z.
Rule 3: If s_m is N, then u_{s_m} is N.

Utilizing the membership functions depicted in Fig. 4, a singleton fuzzifier and center-average defuzzification yield the expression for u_{s_m} as follows:

$$
u_{s_m} = \omega_{m1} + \omega_{m2} \times 0 - \omega_{m3},
\tag{37}
$$

where the membership values $\omega_{m1}, \omega_{m2}, \omega_{m3} \in [0, 1]$ satisfy $\omega_{m1} + \omega_{m2} + \omega_{m3} = 1$. Consequently, the tracking errors for active and reactive power are driven into bounded regions in finite time by employing the FSTA defined below:

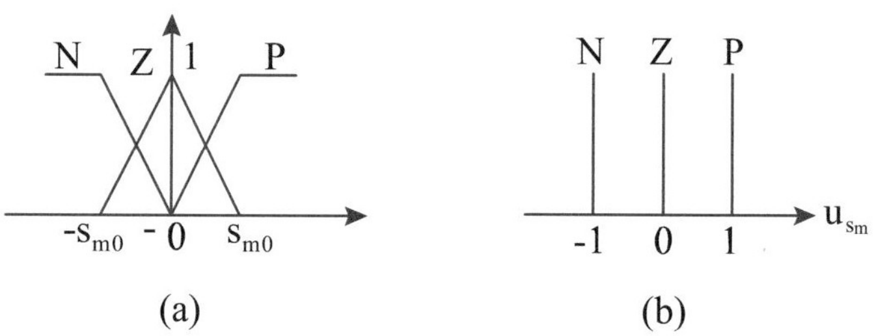

Fig. 4 Membership functions of fuzzy sets for power tracking loop: **a** Membership functions of the input fuzzy sets for s_m. **b** Membership functions of the output fuzzy sets for u_{s_m}

$$\mu(e_m) = -\frac{1}{B}\left(k_{m1}\phi_{m1} + \int_0^t k_{m2}\phi_{m2}d\tau\right),\ \phi_{m1} = |s_m|^{\frac{1}{2}}u_{s_m}, \tag{38}$$

$$u_{s_m} = \begin{cases} \text{sign}(s_m) & |s_m| > s_{m0} \\ s_m/s_{m0} & |s_m| \le s_{m0} \end{cases},\ \phi_{m2} = \begin{cases} \frac{1}{2}\text{sign}(s_m) & |s_m| > s_{m0} \\ \frac{3|s_m|s_m}{2s_{m0}^2} & |s_m| \le s_{m0} \end{cases}, \tag{39}$$

Notice that the output u_{s_m} of the introduced FLS is equivalent to a saturation function with a boundary layer defined by s_{m0} (Wai and Su 2006). Selecting alternative membership functions would adjust u_{s_m} accordingly to meet specific performance requirements.

Since the active power and reactive power controllers exhibit similar properties, only the stability proof for the active power controller is presented here; the reactive power analysis follows similarly.

For clarity in the subsequent theoretical analysis, the active power controller (38) is re-expressed as follows:

$$\mu(e_p) = -\frac{1}{B}\left(k_{p11}\overline{\phi}_{p1} + \int_0^t k_{p21}\overline{\phi}_{p2}d\tau\right), \tag{40}$$

where $\overline{\phi}_{p1} = k_{p3}\phi_{p1}$, $k_{p11} = k_{p1}/k_{p3}$, $k_{p21} = k_{p2}/k_{p3}^2$, $\overline{\phi}_{p2} = k_{p3}^2\phi_{p2}$, and k_{p3} is a positive constant.

Introducing an auxiliary variable $z_p = -\int_0^t k_{p21}\overline{\phi}_{p2}d\tau$, the time derivatives of $\overline{\phi}_{p1}$ and z_p become

$$\begin{aligned} \dot{\overline{\phi}}_{p1} &= \overline{\phi}_{p1'}[B\mu(e_p) + d_p] = \overline{\phi}_{p1'}[-k_{p11}\overline{\phi}_{p1} + z_p + d_p], \\ \dot{z}_p &= -k_{p21}\overline{\phi}_{p2} = -k_{p21}\overline{\phi}_{p1'}\overline{\phi}_{p1}, \end{aligned} \tag{41}$$

where $\overline{\phi}_{p1'}$ is the partial derivative of $\overline{\phi}_{p1}$ described below

$$\overline{\phi}_{p1'} = \frac{\partial\overline{\phi}_{p1}}{\partial s_p} = \begin{cases} \frac{k_{p3}}{2|s_p|^{\frac{1}{2}}} & |s_p| > s_{p0} \\ \frac{3k_{p3}|s_p|^{\frac{1}{2}}}{2s_{p0}} & s_{p0} \ge s_p \ge -s_{p0} \end{cases}. \tag{42}$$

A Lyapunov function is chosen as $V_p = \zeta^T P\zeta$, where $\zeta = [\overline{\phi}_{p1}, z_p]^{\mathrm{T}}$, and $P = \begin{bmatrix} \beta + 4\varepsilon^2 & -2\varepsilon \\ -2\varepsilon & 1 \end{bmatrix}$ is a positive definite matrix with positive constants β and ε.

Then, the time derivatives of ζ and $V_p = \zeta^T P\zeta$ are given by

$$\begin{aligned} \dot{\zeta} &= \overline{\phi}_{p1'}\begin{bmatrix} -k_{p11} & 1 \\ -k_{p21} & 0 \end{bmatrix}\zeta + \begin{bmatrix} \overline{\phi}_{p1'}d_p \\ 0 \end{bmatrix} = \overline{\phi}_{p1'}A\zeta + D, \\ \dot{V}_p &= \dot{\zeta}^T P\zeta + \zeta^T P\dot{\zeta} = \overline{\phi}_{p1'}\zeta^T(A^T P + PA)\zeta + 2D^T P\zeta. \end{aligned} \tag{43}$$

With the relation $k_{p21} = 2\varepsilon k_{p11} + \beta + 4\varepsilon^2$, define $Q = -(A^T P + PA) - 2\varepsilon I$ which simplifies to

$$Q = \begin{bmatrix} 2\beta k_{p11} - 2\varepsilon - 4\varepsilon(\beta + 4\varepsilon^2) & 0 \\ 0 & 2\varepsilon \end{bmatrix}, \tag{44}$$

Then $Q > 0$ provided that $k_{p11} > \frac{1}{\beta}\left(\varepsilon + 2\varepsilon\left(\beta + 4\varepsilon^2\right)\right)$ is established. Thus, it follows that

$$\begin{aligned} \dot{V}_p &\le -2\varepsilon\overline{\phi}_{p1'}\|\zeta\|_2^2 + 2D^T P\zeta \\ &\le -2\varepsilon\overline{\phi}_{p1'}\left|\overline{\phi}_{p1}\right|\|\zeta\|_2 + 2\overline{\phi}_{p1'}d_{p0}\max\{\beta + 4\varepsilon^2, 2\varepsilon\}\|\zeta\|_2 \\ &= -2\overline{\phi}_{p1'}\|\zeta\|_2\left(\varepsilon\left|\overline{\phi}_{p1}\right| - d_{p0}\max\{\beta + 4\varepsilon^2, 2\varepsilon\}\right), \end{aligned} \tag{45}$$

This establishes that $\overline{\phi}_{p1}$ and z_p remain bounded and that the sliding-mode variable s_p is driven into a bounded set in finite time. The convergence properties of s_p are analyzed in two cases:

Case 1: For $|s_p|>s_{p0}$, one has $\left|\overline{\phi}_{p1}\right| = k_{p3}\left|s_p\right|^{\frac{1}{2}} \ge k_{p3}\left|s_{p0}\right|^{\frac{1}{2}}$. Choosing k_{p3} as specified in (46) with a small positive constant ε_0, then (45) can be rewritten as follows (47).

$$k_{p3} > \frac{d_{p0}\max\{\beta + 4\varepsilon^2, 2\varepsilon\}}{\varepsilon\left|s_{p0}\right|^{\frac{1}{2}}} + \varepsilon_0, \tag{46}$$

$$\dot{V}_p \le -2\varepsilon\varepsilon_0\left|s_p\right|^{\frac{1}{2}}\overline{\phi}_{p1'}\|\zeta\|_2 \le -\varepsilon\varepsilon_0 k_{p3}\left(\frac{V_p}{\lambda_{\max}(P)}\right)^{\frac{1}{2}}. \tag{47}$$

Integrate (47) with time, the convergence time of s_p from the initial value of sliding-mode variable $s_p(0)$ to s_{p0} (if $s_p(0) > s_{p0}$) or $-s_{p0}$ (if $s_p(0) < -s_{p0}$) can be derived as

$$T_c \le \left(V_p(s_p(0))^{\frac{1}{2}} - V_p(s_{p0})^{\frac{1}{2}}\right)\frac{2(\lambda_{\max}(P))^{\frac{1}{2}}}{\varepsilon\varepsilon_0 k_{p3}}, \tag{48}$$

where $V_p(s_p(0))$ and $V_p(s_{p0})$ denote the initial function value of V_p and its value at the boundary, respectively.

Therefore, s_p reaches the region $|s_p| \le s_{p0}$ in finite time.

Case 2: For $|s_p| \le s_{p0}$, we have $\left|\overline{\phi}_{p1}\right| = k_{p3}\left|s_p\right|^{\frac{3}{2}}/s_{p0}$. Consequently, s_p ultimately converges to a bounded region satisfying

$$\left|s_p\right|^{\frac{3}{2}} \le \frac{s_{p0}d_{p0}\max\{\beta + 4\varepsilon^2, 2\varepsilon\}}{\varepsilon k_{p3}}. \tag{49}$$

5.2 DC-Link Voltage Regulation Loop

In this subsection, a FSMC combined with a fuzzy observer is designed based on Sect. 4.2 to enforce the DC-link voltage regulation error to a specified bound.

Defining $z_1 = x_1^2/2$, the dynamic Eq. (24) for the DC-link voltage can be rewritten as follows:

$$\dot{z}_1 = \frac{2}{C}p^\star - d_v, \tag{50}$$

where $p^\star = \frac{1}{2}x_1 u_{\alpha\beta}^{\mathrm{T}} i_{\alpha\beta}^\star$ is the reference active power to be designed later (Portillo et al. 2013). And the unknown lumped disturbance $d_v = 4z_1/CR_{dc}$ is represented using the fuzzy observer with a prescribed accuracy δ_v (Wang and Mendel 1992).

$$d_v = \theta_v^{\mathrm{T}} \xi(z_1) + \delta_v. \tag{51}$$

Let the regulation error be given by $e_1 = z_1^* - z_1$ and define the sliding-mode variable as $s_v = k_{sv} e_1$. Calculating the derivative of s_v yields

$$\dot{e}_1 = -\frac{2}{C}p^\star + d_v. \tag{52}$$

Let s_v and auxiliary function u_{vsmc} be the input and output linguistic variables of FLS with the same fuzzy sets and rules described in power tracking loop. Replace s_{m0}, s_m, and u_{s_m} in Fig. 4 with s_{v0}, s_v, and u_{vsmc}, then u_{vsmc} is determined as

$$u_{vsmc} = \omega_{v1} - \omega_{v3}. \tag{53}$$

The DC-link voltage regulation loop is stabilized via the following FSMC strategy and fuzzy observer:

$$\begin{aligned} p^\star &= \frac{C}{2}\left(\hat{d}_v + k_v u_{vsmc}\right), \\ \hat{d}_v &= \hat{\theta}_v^T \xi(z_1), \\ \dot{\hat{\theta}}_v &= k_{sv} s_v \xi(z_1), \end{aligned} \tag{54}$$

where

$$\begin{aligned} \xi(z_1) &= \frac{1}{\sum_{l=1}^{3} \mu_l(z_1)} [\mu_1(z_1), \mu_2(z_1), \mu_3(z_1)]^{\mathrm{T}}, \\ \mu_l(z_1) &= \exp\left(-\frac{(z_1/\delta_{v1} + (1-l) \times 0.5)^2}{\delta_{v2}^2}\right), \quad l = 1, 2, 3. \end{aligned} \tag{55}$$

with δ_{v1} and δ_{v2} as positive constants.

Stability of the DC-link voltage regulation loop is assessed using the Lyapunov function

$$V_v = \frac{1}{2}s_v^2 + \frac{1}{2}\tilde{\theta}_v^{\mathrm{T}}\tilde{\theta}_v, \tag{56}$$

Differentiating V_v with respect to time, one obtains

$$\begin{aligned} \dot{V}_v &= k_{sv}s_v\left(-k_v u_{vsmc} + d_v - \hat{d}_v\right) - \tilde{\theta}_v^{\mathrm{T}} \\ \dot{\hat{\theta}}_v &= -k_{sv}s_v(k_v(\omega_{v1} - \omega_{v3}) - \delta_v), \end{aligned} \tag{57}$$

Since $s_v(\omega_{v1} - \omega_{v3}) > 0$, the sliding-mode variable s_v converges to a bounded region $|s_v| \leq |\delta_v| s_{v0}/k_v$ provided that $k_v > |\delta_v|$. This confirms the stability of the DC-link voltage regulation loop.

5.3 DC-Link Voltage Balancing Loop

In the final control layer, a simple Proportional-Integral (PI) strategy is implemented to equalize the voltage difference between the two DC-link capacitors (Leon et al. 2017; Lin et al. 2020).

$$u_y = -k_{pb}x_2 - k_{ib}\int_0^t x_2 d\tau, \tag{58}$$

Detailed stability analysis for this loop can be found in (Lin et al. 2020).

5.4 Tuning Procedure of Control Parameters

This subsection introduces a systematic parameter calibration process to achieve optimal controller performance, comprising two phases: simulated parameter tuning and experimental fine-tuning.

(1) Simulation Tuning

To ensure operational safety and efficiency, preliminary parameter tuning is conducted on a simulated NPC converter system. A sequential parameter adjustment approach is adopted to reduce optimization complexity:

Step I: Conventional PI control is initially deployed across all control loops. Parameters for the voltage balancing loop are derived through transfer function analysis and Bode diagram interpretation following the methodology in (Lin et al. 2020), establishing a foundational configuration for subsequent adjustments.

Step II: The PI controller in the power tracking loop is progressively replaced through two sub-stages. In the first stage, the PI control strategy in the power tracking loop is replaced with STA (34) and (38) with ϕ_{m1} and ϕ_{m2} described as follows, while maintaining the other two control loops unchange.

$$\phi_{m1} = |s_m|^{\frac{1}{2}} \operatorname{sign}(s_m), \quad \phi_{m2} = \frac{1}{2} \operatorname{sign}(s_m), \tag{59}$$

where $m \in \{p, q\}$. Normally, symmetrical parameters $k_{p1} = k_{q1}$ and $k_{p2} = k_{q2}$ are calculated via error dynamics principles (Tabart et al. 2018).

Then, in the second stage, modify ϕ_{m1} and ϕ_{m2} in (59) to implement FSTA via (38) and (39). The threshold s_{m0} for $m \in \{p, q\}$ satisfies $s_{p0} = s_{q0}$ and is calibrated slightly above steady-state error magnitudes observed in Step I. At this time, the introduced FSTA has better chattering suppression ability than STA. While increasing k_{m1} and k_{m2} enhances transient response, careful balance is maintained to prevent excessive chattering. In this chapter, the parameters k_{m1} and k_{m2} are chosen within 3 times of STA parameters to balance dynamic performance and chattering phenomenon.

Step III: The PI voltage regulator is replaced with fuzzy observer-based FSMC, which is also divided into two stages. In the first stage, the following fuzzy observer-based SMC strategy is employed:

$$p^\star = \frac{C}{2}\big(\hat{d}_v + k_v \operatorname{sign}(s_v)\big). \tag{60}$$

with the estimated disturbance $\hat{d}_v$ obtained based on the fuzzy observer (54) and (55).

Membership function parameter δ_{v1} in (55) is selected as the maximum value of $|z_1|$ to avoid singularity of ξ, which can be tested in Step I. And membership function parameter δ_{v2} in (55) is the width of the function μ_l, which can be tuned per (Angelov and Filev 2004). Then, to ensure the dynamic performance, the parameter k_v is usually chosen very large. Initial $k_v = 10^5$ ensures robust disturbance rejection, later refined with k_{sv} to balance estimation convergence and ripple.

In the second stage, replace $\operatorname{sign}(s_v)$ in (60) with softened switching function u_{vsmc} in (54), where s_{v0} slightly exceeds the steady-state maximum value of $|s_v|$ measured in Step I. This preserves SMC's robustness while attenuating chattering. The dynamic performance can be improved with the increase of k_v, and the tuning principle of k_v is similar to that of k_{m1} and k_{m2} at the end of Step II.

(2) Experimental Fine-Tuning

Field-testing reveals amplified voltage/power fluctuations compared to simulations. Accordingly, the parameters s_{m0} and s_{v0} are proportionally increased, with minor adjustments to other parameters for optimal real-world performance.

5.5 Discussion

The introduced FSTA and FSMC architectures address inherent limitations of traditional STA/SMC through fuzzy-logic enhancements, as analyzed below.

(1) FSTA Superiority

First, the enhanced dynamic performance of the introduced FSTA compared to conventional STA in the power tracking loop is analyzed. By comparing Eqs. (38), (39), and (59), it can be observed that FSTA and STA exhibit different ϕ_{m1} and ϕ_{m2}. During power transients $|s_{m0}|>s_{m0}$, both FSTA and STA follow identical convergence laws. Assuming the same initial conditions s_{m0} and identical parameter settings for ε, β, and ε_0, the finite-time convergence to the region $|s_{m0}|\leq s_{m0}$ is governed solely by parameter k_{p3}, as seen in (48). As stated in Sect. 5.4, the introduced FSTA adopts larger values for k_{m1} and k_{m2} than STA, which leads to an increased k_{p3} and thus improved transient response.

Next, the chattering suppression capabilities of both strategies are examined. The parameter s_{m0} is chosen slightly greater than the maximum value of $|e_m|$ in steady-state, implying that the power tracking loop remains stable once the state enters region $|s_{m0}|\leq s_{m0}$. In (38), the term $k_{m1}\phi_{m1}$ impose a greater impact on chattering than the continuous integral term $\int_0^t k_{m2}\phi_{m2}d\tau$. In steady-state, STA yields $|\phi_{m1}| = |s_m|^{\frac{1}{2}}$, while FSTA results in $|\phi_{m1}| = |s_m|^{\frac{3}{2}}/s_{m0} < |s_m|^{\frac{1}{2}}$ due to fuzzy tuning. By appropriately selecting control parameters, FSTA achieves a smaller control effect induced by $k_{m1}\phi_{m1}$, thereby reducing chattering. Consequently, the introduced FSTA simultaneously enhances dynamic performance during power variations and suppresses steady-state chattering more effectively than STA, with only a modest increase in computational complexity.

(2) FSMC Superiority

When the DC-link voltage varies, $|s_v|>s_{v0}$, both FSMC and SMC can be represented as (60). According to the tuning strategy outlined in Sect. 5.4, the parameter k_v is typically larger than $\left|d_v - \hat{d}_v\right|_M$ (the maximum value of $\left|d_v - \hat{d}_v\right|$). A new Lyapunov function $V_{vs} = 0.5s_v^2$ is constructed, and the convergence time of the state s_v from an initial point $s_v(0)$ back to the region $|s_v|\leq s_{v0}$ can be derived as follows:

$$T_{vs} \leq \frac{\sqrt{2}}{k_{sv}k_{vo}}\left(V_{vs}(s_v(0))^{\frac{1}{2}} - V_{vs}(s_{v0})^{\frac{1}{2}}\right), \tag{61}$$

where $k_{vo} = k_v - \left|d_v - \hat{d}_v\right|_M$ is a positive constant. Evidently, a larger k_v shortens the convergence time, explaining why FSMC achieves faster convergence. This conclusion can also be intuitively understood from Eq. (60): a higher k_v provides greater control effort, accelerating system stabilization.

Chattering behavior is also compared. Once the system enters region $|s_v|\leq s_{v0}$ is arrived, it operates in a stable state. In Eq. (60), the control law for SMC meets that

$|k_v \text{sign}(s_v)|=k_v$, whereas in the first equation of (54), the control action of FSMC satisfies $k_v|u_{vsmc}|= k_v|s_v|/s_{v0} < k_v$. Therefore, a proper choice of k_v will result in a reduced control gain and mitigated chattering in steady-state. In summary, FSMC enhances the control effort during transients for rapid voltage regulation and adaptively reduces it in steady-state to alleviate chattering, effectively addressing the limitations of traditional SMC. Similar to FSTA, the marginal computational overhead introduced by fuzzy logic components is outweighed by substantial improvements in dynamic response and operational smoothness.

6 Application Examples

This section validates the effectiveness of the introduced fuzzy logic system-based sliding-mode control methodology through comprehensive experimental evaluations on a three-level NPC converter system. Experimental validation spans three operational scenarios: DC-link voltage variation, active power transients, and reactive power tracking, with performance metrics encompassing voltage stabilization, disturbance rejection, power regulation, and chattering suppression.

The prototype of NPC converter system governed by a TMS320F28377D DSP is shown in Fig. 5. Key system parameters include: DC-link voltage reference (750 V), phase inductance (2 mH), DC-link capacitance (6 mF), grid specifications (400 V line voltage, 50 Hz frequency), and 10 kHz sampling frequency. Comparative analysis involves four control schemes: conventional PI, observer-based STA, Disturbance Observer-Based SMC (DOB-SMC) (Shen et al. 2025), and the introduced strategy. Data acquisition utilized oscilloscope measurements and DSP-recorded power trajectories.

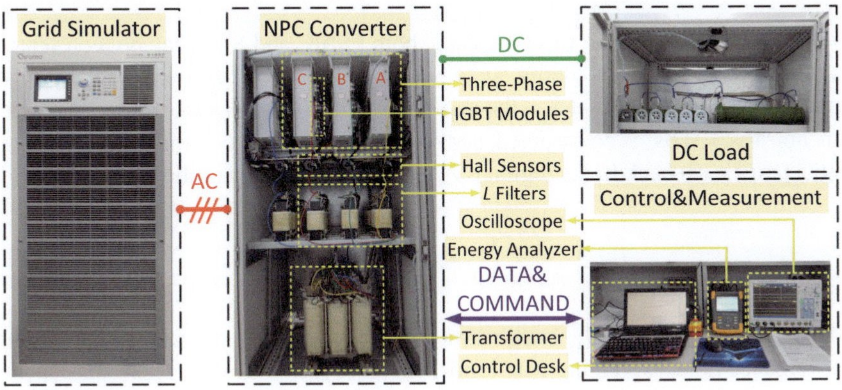

Fig. 5 Prototype of NPC converter system

Table 1 Control parameters

	PI	Observer-based STA	DOB-SMC	FLS-based SMC
Power tracking loop	$kpp = 0.00000009$, $kip = 1.3e\text{-}6$	$kp1 = kq1 = 50000$, $kp2 = kq2 = 35000000$	$kpp = 0.00000009$, $kip = 1.3e\text{-}6$	$kp1 = kq1 = 50000$, $sp0 = sq0 = 400$, $kp2 = kq2 = 80000000$
DC-link voltage regulation loop	$kpv = 0.13$, $kiv = 2$	$ksv = 37$, $\delta v1 = 320000$, $kv1 = 384$, $\delta v2 = 0.15$, $kv2 = 1700000$	$kpv = 1$, $kiv = 0.01$, $ks = 260$, $l = 115$, $\varepsilon = 3000$	$ksv = 37$, $\delta v1 = 320000$, $svo = 200000$, $\delta v1 = 0.15$, $kv = 525000$
DC-link voltage balancing loop	$kpb = 0.005$, $kib = 0.00001$	$kpb = 0.005$, $kib = 0.00001$	$kpb = 0.005$, $kib = 0.00001$	$kpb = 0.005$, $kib = 0.00001$

Additionally, control parameters presented in Table 1 were optimized using the tuning methodology described in Sect. 5.4. Specifically, k_{pp} and k_{ip} are the proportional and integral gains of the power tracking loop; k_{pv} and k_{iv} correspond to the proportional and integral gains of the DC link voltage regulation loop, k_{pb} and k_{ib} are the proportional and integral gains for the DC-link voltage balancing loop. Furthermore, $k_{p1,}$ k_{q1} and $k_{p2,}$ k_{q2} represent the proportional and integral gains of the *p*/*q*-axis power tracking loop. Definitions of other parameters can be found in the Nomenclature or in Sect. 5.

6.1 DC-Link Voltage Variation

As demonstrated in Fig. 6, 40 V step reduction in DC-link voltage (750 V → 710 V) at 0.2 s demonstrates voltage regulation capabilities. While all controllers achieve reference tracking, the FLS-based SMC strategy exhibits superior transient characteristics. Specifically, there is no overshoot in the voltage curve of the introduced strategy, and the settling time of PI, observer-based STA, DOB-SMC, and the introduced strategy is 0.173 s, 0.050 s, 0.081 s, and 0.032 s respectively. Accordingly, these results confirm the introduced scheme's enhanced voltage stabilization efficacy through intelligent switching surface adaptation.

6.2 Active Power Variation

Load resistance reduction from 360 Ω to 106 Ω (1.56 kW → 5.31 kW active power) at 0.2 s under unity power factor operation reveals dynamic performance, and the dynamic responses of DC-link voltage, active power, and reactive power of four strategies are depicted in Fig. 7. Considering that voltage balancing loop of all the strategies utilize PI controllers, the steady-state voltage difference comparison in Fig. 8 focuses specifically on the introduced strategy. As illustrated in Fig. 8, all

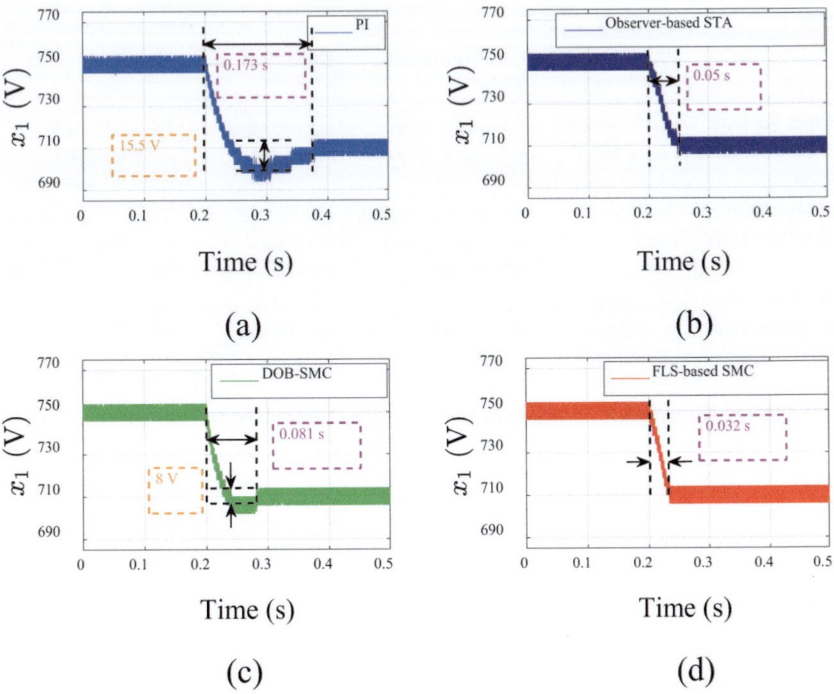

Fig. 6 Experimental results under DC-link voltage variation: **a** PI. **b** Observer-based STA. **c** DOB-SMC. **d** The introduced FLS-based SMC strategy

control methods successfully achieve the stable operation of the system. The DC-link voltage maintains precise regulation at its 750 V reference value while simultaneously tracking both active and reactive power references. While the reactive power ripple (peak-to-peak measurement) measures 500 VAr for PI, DOB-SMC, and the introduced strategy, marginally higher than the 365 VAr observed in observer-based STA, all values remain within acceptable operational limits. Notably, the PI-controlled voltage balancing loop effectively maintains DC-link voltage differential fluctuations centered around zero.

The disturbance rejection capability manifests primarily through two quantifiable parameters: DC-link voltage recovery duration and transient voltage depression. Comparative analysis reveals the introduced controller demonstrates superior performance with a reduced voltage depression of 20 V (equivalent to observer-based STA and 4 V lower than conventional PI and DOB-SMC implementations). Concurrently, it achieves faster transient recovery in 0.088 s, obtaining a 45.7% improvement over PI (0.162 s), 40.1% over observer-based STA (0.147 s), and 37.6% over DOB-SMC (0.141 s), respectively.

Current harmonic characteristics during steady-state, depicted in Fig. 9, were quantified through spectral analysis. Total Harmonic Distortion (THD) serves as a critical metric for both current quality and chattering intensity. Lower THD values

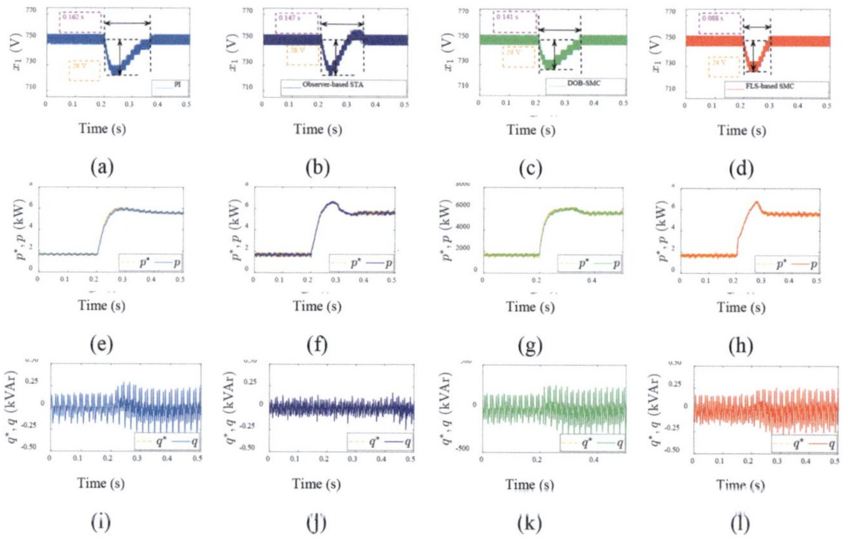

Fig. 7 Experimental results under active power variation: **a**, **e**, **i** DC-link voltage, active power, and reactive power of PI. **b**, **f**, **j** DC-link voltage, active power, and reactive power of observer-based STA. **c**, **g**, **k** DC-link voltage, active power, and reactive power of DOB-SMC. **d**, **h**, **l** DC-link voltage, active power, and reactive power of the introduced FLS-based SMC strategy

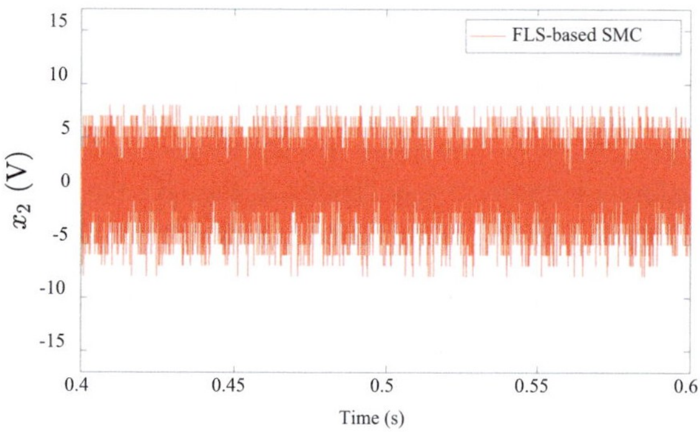

Fig. 8 DC-link voltage difference of the introduced strategy in steady-state

correspond to superior current waveforms and effective chattering suppression. Notably, all SMC variants (observer-based STA, DOB-SMC, and the introduced strategy) demonstrate THD levels comparable to or exceeding PI performance. This empirical evidence confirms that while maintaining equivalent current quality and chattering suppression characteristics, the introduced strategy outperforms comparative controllers in dynamic response and disturbance rejection capabilities.

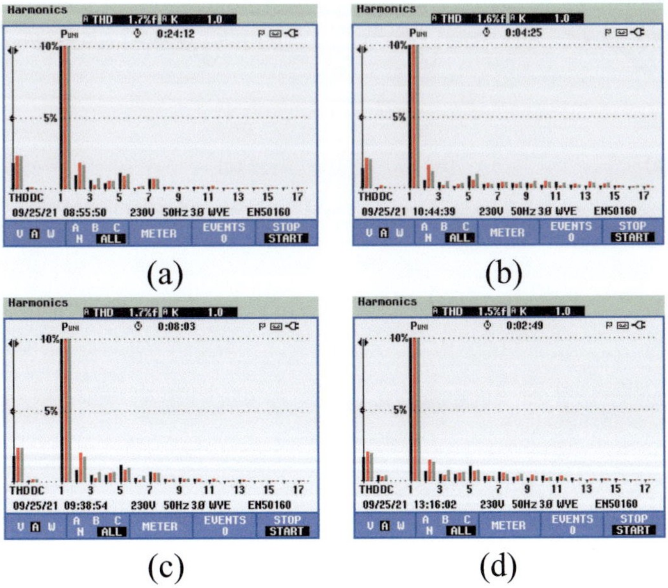

Fig. 9 Current harmonic spectrum with $p = 5.31$ kW and $q = 0$ VAr: **a** PI. **b** Observer-based STA. **c** DOB-SMC. **d** The FLS-based SMC strategy

6.3 Reactive Power Variation

To evaluate power tracking capabilities, a step disturbance was introduced by shifting the reactive power reference from 0VAr to -5VAr at 0.2 s under no-load grid compensation conditions. As evidenced in Fig. 10, the DC-link voltage and active power maintain satisfactory reference tracking with minimal perturbation during reactive power transients. Detailed examination of reactive power trajectories in Fig. 10i–l reveals the introduced controller achieves a convergence duration of 0.027 s, corresponding to 39.71%, 51.92%, and 38.57% of the settling times required by PI, observer-based STA, and DOB-SMC controllers, respectively. This empirically confirms the superior transient convergence characteristics of the introduced method.

Notably, both PI and DOB-SMC implementations exhibit transient overshoot in active/reactive power responses, that is, a consequence of inherent compromises between tracking rapidity and overshoot suppression during controller parameter optimization. In contrast, the introduced strategy completely eliminates such overshoot phenomena while maintaining tracking precision.

Harmonic analysis in Fig. 11 reaffirms comparable THD metrics across all controllers, verifying effective chattering attenuation in FLS-based sliding-mode implementations. These findings effectively establish the introduced controller's dual advantage: achieving accelerated power tracking while maintaining competitive harmonic performance relative to conventional approaches.

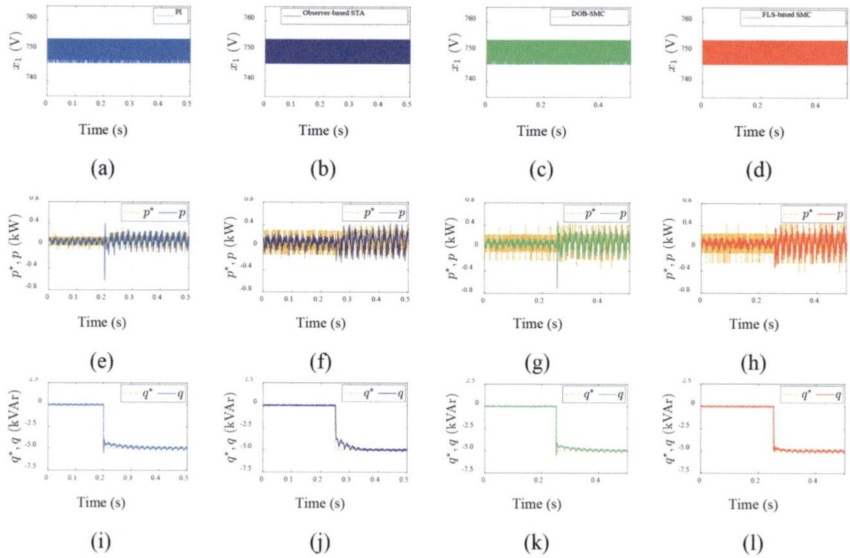

Fig. 10 Experimental results under reactive power variation: **a**, **e**, **i** DC-link voltage, active power, and reactive power of PI. **b**, **f**, **j** DC-link voltage, active power, and reactive power of observer-based STA. **c**, **g**, **k** DC-link voltage, active power, and reactive power of DOB-SMC. **d**, **h**, **l** DC-link voltage, active power, and reactive power of the introduced FLS-based SMC strategy

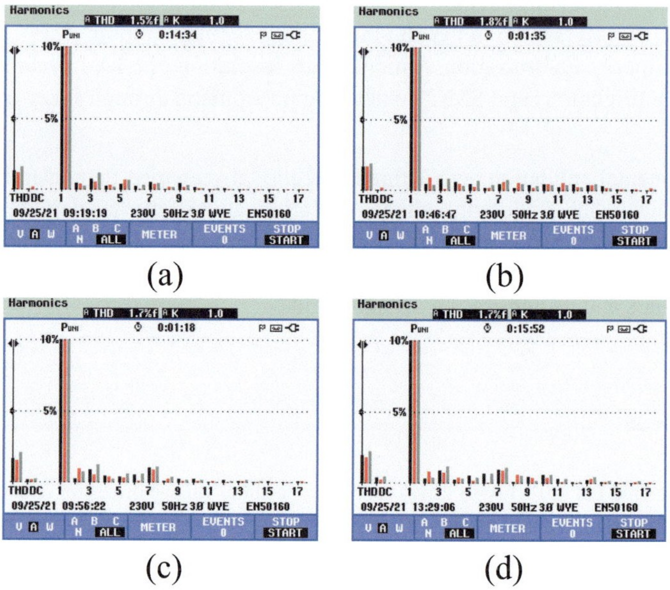

Fig. 11 Current harmonic spectrum with $p = 0$ kW and $q = -5$ VAr: **a** PI. **b** Observer-based STA. **c** DOB-SMC. **d** The FLS-based SMC strategy

6.4 Computational Complexity Comparison

A systematic comparison of computational demands was conducted through real-time execution profiling. Controller complexity was quantified via maximum execution cycle durations, yielding measured values of 79.01 μs (PI), 86.76 μs (observer-based STA), 79.46 μs (DOB-SMC), and 85.71 μs (introduced FLS-SMC). These results demonstrate that the advanced control architecture introduces negligible computational overhead while delivering enhanced dynamic performance, confirming practical feasibility for real-time implementation in power electronic systems.

7 Conclusion and Chapter Summary

This chapter presents an innovative fuzzy logic-based sliding-mode control framework for three-phase three-level NPC power converters. The presented methodology realizes multi-objective optimization capability: concurrent realization of voltage stabilization, power tracking, and harmonic suppression. Meanwhile, the introduced strategy synergistically integrates fuzzy logic systems with advanced control architectures through two key innovations:

(1) Hybrid Control Architecture: FLS-STA integration in the power tracking loop and FLS-SMC fusion in the DC-link voltage regulation loop;
(2) Performance Optimization: Simultaneous resolution of STA's dynamic performance limitations and SMC's chattering phenomena through fuzzy compensation.

Experimental validation across three operational scenarios demonstrates significant advancements, including voltage regulation superiority, accelerated stabilization, dynamic disturbance rejection, steady-state chattering suppression and computational practicality. Future research directions will focus on real-time optimization of fuzzy rules and its application in more fields.

References

ABB Switzerland Ltd (2011) ABB drives for marine medium voltage drives for reliable and efficient operations at sea. Accessed 2011. https://new.abb.com/drives/medium-voltage-ac-drives/acs6000

Akagi H, Kanazawa Y, Nabae A (1984) Instantaneous reactive power compensators comprising switching devices without energy storage components. IEEE Trans Ind Appl IA–20(3):625–630

Altas IH (2017) Fuzzy logic control in energy systems with design applications in MatLab/Simulink. The Institution of Engineering and Technology, United Kingdom

Amin MM, El-Sousy FFM, Mohammed OA, Aziz GAA, Gaber K (2021) MRAS-based super-twisting sliding-mode estimator combined with block control and DTC of six-phase induction motor for ship propulsion application. IEEE Trans Ind Appl 57(6):6646–6658

Angelov PP, Filev DP (2004) An approach to online identification of Takagi-Sugeno fuzzy models. IEEE Trans Syst Man Cybern B Cybern 34(1):484–498

Askari S, Farzanehfard H (2024) Fixed frequency highly efficient resonant converter with low component count for CC and CV charges of electric vehicles batteries. IEEE Trans Intell Transp Syst 25(10):13494–13500

Asna M, Shareef H, Prasanthi A, Errouissi R, Wahyudie A (2024) A novel multi-level charging strategy for electric vehicles to enhance customer charging experience and station utilization. IEEE Trans Intell Transp Syst 25(9):11497–11508

Babes B, Mekhilef S, Boutaghane A, Rahmani L (2022) Fuzzy approximation-based fractional-order nonsingular terminal sliding mode controller for DC-DC buck converters. IEEE Trans Power Electron 37(3):2749–2760

Bouallègue S, Toumi F, Haggège J, Siarry P (2015) Advanced metaheuristics-based approach for fuzzy control systems tuning. In: Zhu Q, Azar AT (eds) Complex system modelling and control through intelligent soft computations. Springer International Publishing, Cham, Heidelberg, New York, Dordrecht, London. Studies in Fuzziness and Soft Computing, vol 319, pp 627–653

Chen S-Y, Chiang H-H, Liu T-S, Chang C-H (2019) Precision motion control of permanent magnet linear synchronous motors using adaptive fuzzy fractional-order sliding-mode control. IEEE/ASME Trans Mechtron 24(2):741–752

Cirstea MN, Dinu A, Khor J, McCormick M (2002) Neural and fuzzy logic control of drives and power systems. Elsevier, U.K., Oxford

Cui D et al (2024) Coordinated charging scheme for electric vehicle fast-charging station with demand-based priority. IEEE Trans Transport Electrific 10(3):6449–6459

Cziker AC, Beleiu HG (2024) Fuzzy control systems for power quality improvement—a systematic review exploring their efficacy and efficiency. Appl Sci 14(11):4468

Ding S, Mei K, Yu X (2022) Adaptive second-order sliding mode control: a lyapunov approach. IEEE Trans Autom Control 67(10):5392–5399

Guo F, Yang T, Diab A, Yeoh S, Bozhko S, Wheeler P (2021) An enhanced virtual space vector modulation scheme of three-level NPC converters for more-electric-aircraft applications. IEEE Trans Ind Appl 57(5):5239–5251

Hamza MF, Yap HJ, Choudhury IA (2017) Recent advances on the use of meta-heuristic optimization algorithms to optimize the type-2 fuzzy logic systems in intelligent control. Neural Comput Appl 28(5):979–999

Herrera F, Lozano M, Verdegay JL (1995) Tuning fuzzy logic controllers by genetic algorithms. Int J Approx Reason 12(34):299–315

Hou S, Chu Y, Fei J (2021) Robust intelligent control for a class of power-electronic converters using neuro-fuzzy learning mechanism. IEEE Trans Power Electron 36(8):9441–9452

Hwang C-L, Yang C-C, Hung JY (2018) Path tracking of an autonomous ground vehicle with different payloads by hierarchical improved fuzzy dynamic sliding-mode control. IEEE Trans Fuzzy Syst 26(2):899–914

Infineon (2021) Gate-driver ICs for EV-charging stations and wallboxes. Accessed 2021. https://www.infineon.com/dgdl/Infineon-Gate_driver_ICs_for_EV_charging_and_wall_boxes-ProductPresentation-v01_00-EN.pdf?fileId=5546d4627aa5d4f5017b1601946e60d7

Kang Z, Lin X, Shen X, Liu Z, Gao Y, Liu J (2025) Event-triggered generalized super-twisting sliding mode control for position tracking of PMSMs. IEEE Trans Ind Informat. https://doi.org/10.1109/TII.2025.3556086

Khoshhava MA, Zarchi HA, Markadeh G (2021) Sensor-less speed and flux control of dual stator winding induction motors based on super twisting sliding mode control. IEEE Trans Energy Convers 36(4):3231–3240

Leon JI, Vazquez S, Franquelo LG (2017) Multilevel converters: control and modulation techniques for their operation and industrial applications. Proc IEEE 105(11):2066–2081

Lin H, Liu J, Leon JI, Marquez A, Luo W, Franquelo LG (2020) Observer-based sliding-mode control of a DC/DC buck converter for railway systems. IET Renew Power Gener 14(18):3579–3588
Lin H et al (2020) Integral sliding-mode control-based direct power control for three-level NPC converters. Energies 13(1):227
Lin H et al (2024) Sliding mode control for NPC converters via a dual layer nested adaptive tuning technique. IEEE Trans Ind Informat 20(9):11418–11428
Liu Z, Mohammadzadeh A, Turabieh H, Mafarja M, Band SS, Mosavi A (2021) A new online learned interval type-3 fuzzy control system for solar energy management systems. IEEE Access 9:10498–10508
Liu X, Qiu L, Fang YT, Wang K, Li YD, Rodríguez J (2022) A fuzzy approximation for FCS-MPC in power converters. IEEE Trans Power Electron 37(8):9153–9163
Liu X, Qiu L, Fang YT, Wang K, Li YD, Rodríguez J (2024) Fuzzy approximation ARX model-based intelligent two-horizon robust FCS-MPC for power converter. IEEE Trans Transport Electrific 10(3):5783–5793
Liu L et al (2024) Multidisturbances compensation for three-level NPC converters in microgrids: a robust adaptive sliding mode control approach. IEEE Trans Ind Informat 20(6):8917–8931
Liu L, Zhang Z, Yin Y, Vazquez S, Zhao Y, Kennel R (2024) An efficient robust power-voltage control for three-level NPC converters in microgrids. IEEE Trans Ind Informat 20(4):5849–5863
Liu J, Yin Y, Luo W, Vazquez S, Franquelo LG, Wu L (2018) Sliding mode control of a three-phase AC/DC voltage source converter under unknown load conditions: industry applications. IEEE Trans Syst Man Cybern Syst 48(10):1771–1780
Long B, Lu PJ, Chong KT, Rodriguez J, Guerrero JM (2022) Robust fuzzy-fractional-order nonsingular terminal sliding-mode control of LCL-type grid-connected converters. IEEE Trans Ind Electron 69(6):5854–5866
Lu Y (2018) Adaptive-fuzzy control compensation design for direct adaptive fuzzy control. IEEE Trans Fuzzy Syst 26(6):3222–3231
Ma P, Yu J, Wang Q-G, Liu J (2023) Filter- and observer-based finite-time adaptive fuzzy control for induction motors systems considering stochastic disturbance and load variation. IEEE Trans Power Electron 38(2):1599–1609
Mamdani EH, Assilian S (1975) An experiment in linguistic synthesis with a fuzzy logic controller. Int J Man Mach Stud 7(1):1–13
Mendel J, Hagras H, Tan W-W, Melek WW, Ying H (2014) Introduction to type-2 fuzzy logic control: theory and applications. Wiley-IEEE Press
Omizegba EE, Monikang FA (2009) Fuzzy logic controller parameter tuning using particle swarm algorithm. Int J Model Identificat Control 6(1):26–31
Portillo R, Vazquez S, Leon JI, Prats MM, Franquelo LG (2013) Model based adaptive direct power control for three-level NPC converters. IEEE Trans Ind Informat 9(2):1148–1157
Rivera S, Wu B, Kouro S, Yaramasu V, Wang J (2015) Electric vehicle charging station using a neutral point clamped converter with bipolar DC bus. IEEE Trans Ind Electron 62(4):1999–2009
Roy P, He J, Zhao T, Singh YV (2022) Recent advances of wind-solar hybrid renewable energy systems for power generation: a review. IEEE Open J Ind Electron Soc 3:81–104
Sarvi M, Azadian A (2022) A comprehensive review and classified comparison of MPPT algorithms in PV systems. Energy Syst 13(2):281–320
Shen X et al (2025) Fixed-time sliding mode control for NPC converters with improved disturbance rejection performance. IEEE Trans Ind Informat. https://doi.org/10.1109/TII.2025.3540481
Shen X et al (2025) Finite-time sliding mode control for NPC converters with enhanced disturbance compensation. IEEE Trans Circuits Syst I Reg Papers 72(4):1822–1831
Shouran M, Alsseid A (2022) Particle swarm optimization algorithm-tuned fuzzy cascade fractional order PI-fractional order PD for frequency regulation of dual-area power system. Processes 10(3):477
Siemens (2019) Drives for every demand: the SINAMICS family of medium voltage drives. Accessed 2019. https://assets.new.siemens.com/siemens/assets/api/uuid:dc24aadc-96b8-40aa-98f9-12c53fbbb774/poster-sinamics-mvdrives-en.pdf

Soliman MA, Hasanien HM, Azazi HZ, El-Kholy EE, Mahmoud SA (2019) An adaptive fuzzy logic control strategy for performance enhancement of a grid-connected PMSG-based wind turbine. IEEE Trans Ind Informat 15(6):3163–3173

Tabart Q, Vechiu I, Etxeberria A, Bacha S (2018) Hybrid energy storage system microgrids integration for power quality improvement using four-leg three-level NPC inverter and second-order sliding mode control. IEEE Trans Ind Electron 65(1):424–435

Vazquez S, Acuna P, Aguilera RP, Pou J, Leon JI, Franquelo LG (2020) DC-link voltage-balancing strategy based on optimal switching sequence model predictive control for single-phase H-NPC converters. IEEE Trans Ind Electron 67(9):7410–7420

Wai R-J, Su K-H (2006) Adaptive enhanced fuzzy sliding-mode control for electrical servo drive. IEEE Trans Ind Electron 53(2):569–580

Wang L-X, Mendel JM (1992) Fuzzy basis functions, universal approximation, and orthogonal least-squares learning. IEEE Trans Neural Netw 3(5):807–814

Xuan Y, Yang X, Chen W, Liu T, Hao X (2019) A novel NPC dual-active-bridge converter with blocking capacitor for energy storage system. IEEE Trans Power Electron 34(11):10635–10649

Yang C, Jiang Y, Na J, Li Z, Cheng L, Su C-Y (2019) Finite-time convergence adaptive fuzzy control for dual-arm robot with unknown kinematics and dynamics. IEEE Trans Fuzzy Syst 27(3):574–588

Yang T, Sun N, Fang Y (2022) Adaptive fuzzy control for uncertain mechatronic systems with state estimation and input nonlinearities. IEEE Trans Ind Informat 18(3):1770–1780

Yilmaz BM, Tatlicioglu E, Savran A, Alci M (2021) Adaptive fuzzy logic with self-tuned membership functions based repetitive learning control of robotic manipulators. Appl Soft Comput 104:107183

Yin Y et al (2019) Observer-based adaptive sliding mode control of NPC converters: an RBF neural network approach. IEEE Trans Power Electron 34(4):3831–3841

Yin Y et al (2022) Observer-based sliding-mode control for grid-connected power converters under unbalanced grid conditions. IEEE Trans Ind Electron 69(1):517–527

Zadeh LA (1965) Fuzzy sets. Inf Control 8(3):338–353

Zhang X, Wang Y, Liu G, Yuan X (2016) Robust regenerative charging control based on T-S fuzzy sliding-mode approach for advanced electric vehicle. IEEE Trans Transport Electrific 2(1):52–65

Zhang L, Chen Z, Yu X, Yang J, Li S (2023) Sliding-mode-based robust output regulation and its application in PMSM servo systems. IEEE Trans Ind Electron 70(2):1852–1860

Zhou J, Cheng P-T (2019) Modulation methods for 3L-NPC converter power loss management in STATCOM application. IEEE Trans Ind Appl 55(5):4965–4973

Artificial Neural Network-Based Control in Power Electronics

Ayse Kocalmis Bilhan

Abstract Humanity has lived together with nature and learned many solutions from nature that make their lives easier. They have tried to understand the events that occur in nature with their thoughts, minds, and feelings. Artificial neural networks (ANNs) also represent a branch of artificial intelligence modeled according to the structure and function of the human brain. Artificial neural networks (ANNs) have emerged as a significant technology in power electronics, providing intelligent, adaptive, and data-driven control solutions. This chapter explores the integration of ANN-based control strategies in various power electronic applications, including converters, inverters, motor drives, and renewable energy sources. By leveraging the learning capabilities of ANNs, power electronic systems can overcome traditional control limitations, enhancing efficiency, stability, and fault tolerance. The chapter discusses the fundamental principles of ANN architecture, training methodologies, and activation functions, followed by practical implementations in DC-DC, DC-AC, AC-AC, and AC-DC converter systems. Furthermore, the role of ANNs in renewable energy systems, such as solar and wind energy, is examined, highlighting their impact on forecasting, Maximum Power Point Tracking (MPPT), and predictive maintenance. Challenges such as computational complexity, data dependency, and integration issues are also addressed, along with future trends in ANN-based control for power electronics. The findings underscore the potential of ANNs in advancing intelligent and self-optimizing power systems, paving the way for smarter and more resilient energy solutions.

Keywords Artificial neural network · Converters · Power electronics · Inverters · Control

A. Kocalmis Bilhan (✉)
Electrical and Electronics Engineering Department, Faculty of Engineering and Architecture, Nevsehir Haci Bektas Veli University, Nevşehir, Turkey
e-mail: akbilhan@nevsehir.edu.tr

E. Kabalci (ed.), *Modern Control Methods for Power Electronics*, Energy Systems in Electrical Engineering, https://doi.org/10.1007/978-981-92-1974-2_7

Nomenclature

NN	Neural Network
ANN	Artificial Neural Network
FFNN	Feedforward Neural Network
DC	Direct Current
AC	Alternative Current
PWM	Pulse-Width Modulation
MSE	Mean Square Error
MRE	Mean Relative Error
MAE	Mean Absolute Error
RMSE	Root Mean Square Error
COV	Coefficient of Variation
CC	Correlation Coefficient
COD	Coefficient of Determination
EC	Efficiency Coefficient
OIMP	Overall Index of Model Performance
CRM	Coefficient of Residual Mass

1 Introduction to Artificial Neural Networks

The search to resolve the complexities of the human brain has fascinated humanity since the beginning. Observing the complex systems of nature, they discovered various solutions that helped enhance their daily lives, whether it was finding food, shelter, or developing social structures. Driven by curiosity, humans have continually sought to make sense of the phenomena they encounter in nature, leaning on their capacity for thought, reasoning, and emotional understanding. This deep connection between humanity and the natural world has fueled a pursuit of knowledge. The emergence of the computer, a remarkable innovation of modern technology, was a significant milestone born from this quest for understanding. By closely studying the workings of the human brain (its neural networks, thought processes, and problem-solving abilities), scientists and engineers tried to design machines that could mimic these functions. This effort not only reflects our fascination with the brain but also highlights the profound influence nature has had on technological advancements (Hassabis et al. 2017).

The human brain is a complicated organ composed of billions of specialized cells known as neurons. Unlike most cells in the human body, which have a defined lifespan and are constantly being replaced, neurons possess a remarkable ability to persist throughout a person's life without undergoing natural death. This unique feature may contribute to the enduring nature of the information stored in our brains. It is estimated that the human brain contains roughly 100 billion neurons, each playing a crucial role in transmitting and processing information. Research has shown that individual differences in people's brains can significantly influence how they process and interpret information. This variability among brains points to the existence of diverse information processing systems, each tailored to the specific experiences and abilities of the individual. One of the most fascinating insights gained from neuroscience is that the nerves in the human brain operate in parallel. This means that multiple processes can occur simultaneously, allowing for efficient communication and rapid responses to stimuli. This parallel processing capability is essential for complex tasks such as decision-making, problem-solving, and the overall functioning of cognitive processes (Maass 2016).

In 1942, two influential figures, neurobiologist Warren McCulloch and mathematician-statistician Walter Pitts came together and proposed a ground-breaking idea: Brain functions could be replicated through mathematical analogs rather than relying solely on biological neurons. This innovative approach leads the way for the exploration of neural networks (NNs) and ANN. In an important paper published in 1943, McCulloch and Pitts demonstrated that a biological neuron could be conceptualized as a threshold function (McCulloch and Pitts 1943). This model suggested that neurons conduct only when the input they receive exceeds a certain threshold, thereby introducing a mathematical basis for understanding neural activity. Building on this concept, Donald Hebb released a crucial book in 1949 titled "Organization of Behavior," which provided insights into how learning occurs at the cellular level (Hebb 1949). Hebb's theory of neuronal connectivity and reinforcement can be seen as a pioneer in modern learning algorithms, including the widely used backpropagation algorithm in neural networks (Rumelhart et al. 1986). The 1950s marked the beginning of accelerated learning research, driven by substantial advancements in computational technology and the increasing capability of emerging computers to perform rapid numerical calculations. In this transformative period, Frank Rosenblatt brought out the perceptron, the first single-layer artificial neural network model (Rosenblatt 1958). This model was based on a simplified abstraction of neuron structure, capable of performing basic pattern recognition tasks. During the following decade, in the 1960s, Bernard Widrow and Marcian Hoff contributed further to the field by formulating the delta rule. This innovative learning rule allows for the adjustment of connection weights within a neural network by assessing the error defined as the difference between the actual output generated by the network and the desired output. This feedback mechanism represented a significant step toward improving the learning capabilities of artificial systems. The theoretical underpinnings of artificial neural networks were largely established between 1969 and 1982, during which time researchers developed various architectures and algorithms that would shape the future of the field. This period laid the groundwork for the current era of artificial neural networks, where advanced deep learning techniques are utilized across diverse applications (from computer vision to natural language processing) transforming how machines learn and interact with the world (Schmidhuber 2015).

ANNs are inspired based on the structure and functionality of the human brain, drawing inspiration from its remarkable complexity and interconnection. Nevertheless, it is important to recognize that the brain's highly interconnected architectural structure is a unique quality that has not yet been replicated by current technology or digital computing systems. The basic elements of ANNs are artificial neurons (ANs). These artificial neurons that comprise these neural networks can be seen as relatively simplistic in comparison to the sophisticated neurons found in the biological brain. These artificial neurons vary significantly in shape and size, reflecting the diverse roles they play within the neural network. By forming connections among themselves, artificial neurons establish intricate networks organized in distinct layers, which contribute to their overall functionality.

It is important to understand the structure of brain neurons to recognize the structure of the artificial neuron. Each neuron consists of a cell body known as the soma, which houses the nucleus and vital cellular components. Dendrites are located at one end of a neuron. These dendrites are essential for receiving incoming signals from other neurons or external sources, acting as the primary channel for information entry. The axon is a long, thin part of a neuron that looks like a single fiber. It sends electrical signals away from the cell body to communicate with other neurons or the outside environment. The axon further divides into numerous branches, establishing connections with multiple target cells. When the dendrites receive signals, they convey this information to the soma, where the neuron integrates the incoming data to determine whether to generate an output signal. If the decision is made to transmit a signal, an electrical impulse, known as an action potential, is initiated within the neuron. This action potential travels down the axon and is relayed through its branching structure to connect with other neurons, thereby facilitating communication within the neural network. A basic example of this structure is illustrated in Fig. 1.

In biological systems, the process of learning is fundamentally linked to the dynamic adjustments of synaptic connections between neurons. The human journey of "learning by doing" involves a continuous brain evolution shaped by countless experiences and interactions with the environment. As the life goes on, the synaptic connections that facilitate communication between neurons are not just modified; they undergo intricate restructuring, allowing for both the strengthening of existing connections and the creation of entirely new ones. This complex interplay is what underlies the phenomenon of learning. Similarly, in ANNs, learning operates on

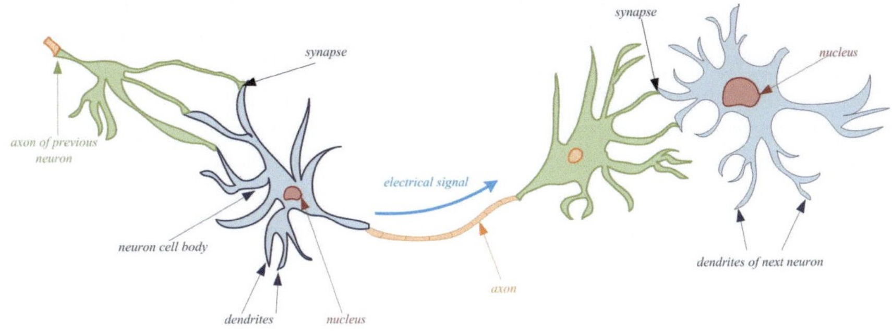

Fig. 1 General structure of a biological neuron

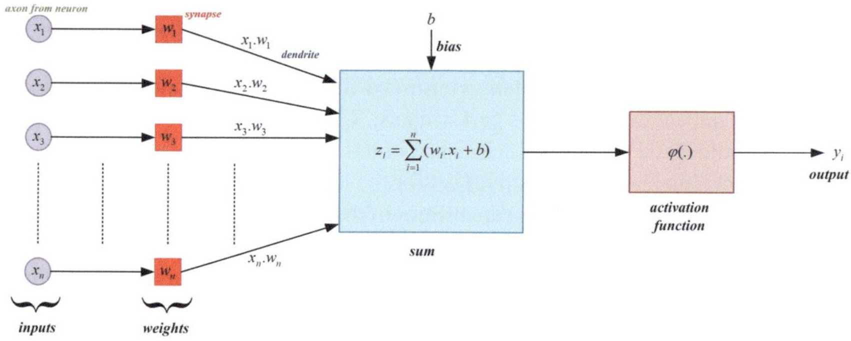

Fig. 2 Structure of ANN

a comparable principle. These computational models mimic the way human brains learn by utilizing a training process that involves a wide array of examples. During this training phase, input data and corresponding output data are processed systematically. The training algorithm plays a crucial role in this process, as it iteratively adjusts the connection weights between the artificial neurons based on the data provided (Maier et al. 2023). This adjustment continues until the network achieves a stable outcome, known as convergence. Through this method, ANNs learn to recognize patterns and make predictions, effectively mirroring the fundamental learning mechanisms observed in biological systems. Figure 2 illustrates the structure of ANNs.

In Fig. 2, the input values, represented by the symbol x, refer to the individual data points. Each input value is multiplied by a specific connection weight, denoted as w, which determines the significance of that input in the overall calculation. In the simplest ANN structure, these weighted signals undergo a straightforward addition process. This sum is then sent to an activation function ($\varphi(.)$), which processes the combined signal to generate the final output. This output can take various forms, depending on the application. The design of this system allows for flexibility, enabling the implementation of various additional functions and diverse network structures tailored to different processing needs and scenarios.

The activation function $\varphi(.)$ plays a crucial role in the dynamics of ANNs by facilitating changes in the output of the sum function over time. This function, often referred to as the transfer function, takes the net input generated from the sum process and transforms it into the output. Notably, activation functions are typically nonlinear, allowing for more complex relationships between inputs and outputs. In the context of neural cell models, various activation functions can be utilized. For instance, some functions may be better suited for binary classification, while others might excel in regression tasks or multi-class recognition scenarios. By modeling the output of neurons, activation functions effectively constrain the output to specified ranges, which are generally either [0, 1] for functions like the sigmoid or [−1, 1] for others like the hyperbolic tangent. It is important to ensure that the activation functions chosen for use in an ANN possess properties of differentiability and continuity.

These characteristics are vital for the training process of the network, enabling effective backpropagation of errors during the learning phase. The selection of the most suitable activation function is often determined through experimentation, as it can significantly impact the network's performance. Ultimately, the choice of activation function is influenced by the characteristics of the data that the ANN is handling, as well as the specific learning objectives set for the network. A well-chosen activation function can enhance the learning capabilities of the network, allowing it to adapt and generalize effectively based on the input it receives. Figure 3 illustrates the various types of activation functions (Szandała 2021), and mathematical equations of these functions are given in Table 1.

ANNs can be primarily classified into two distinct categories: feedforward networks and recurrent (or feedback) networks. In feedforward networks, information processing is linear and occurs in a single direction—from input to output. This structure means that once the data flows through the network layers, it does not return to the previous layers. This unidirectional flow is particularly effective for tasks that require static input–output mapping, where the relationship between inputs and outputs does not change over time. Conversely, recurrent neural networks (RNNs) possess a more complex architecture that allows signals to flow both forward and backward, as well as laterally across connections. This feature enables RNNs to maintain a form of memory, making them highly suitable for dynamic or temporal mapping tasks, where the context of previous inputs can influence the current output.

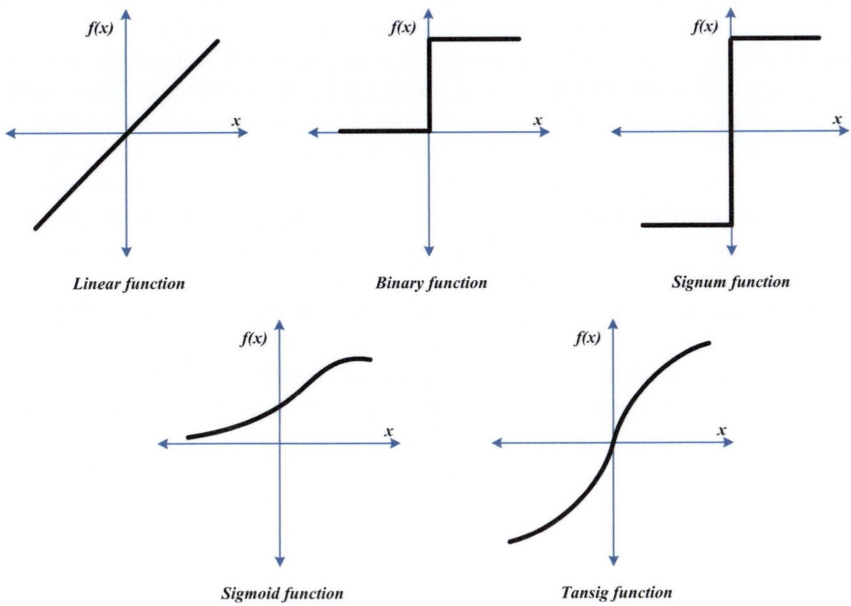

Fig. 3 Activation functions of ANN

Table 1 Mathematical equations of some the activation functions

Name	Formula	
Linear function	$f(x) = x$	(1)
Binary function	$f(x) = \begin{cases} 0 \; for \;\; x < 0 \\ 1 \; for \; x >= 0 \end{cases}$	(2)
Signum function	$f(x) = \begin{cases} -1 \; for \; x < 0 \\ 0 \;\; for \; x = 0 \\ 1 \;\; for \; x > 0 \end{cases}$	(3)
Sigmoid function	$f(x) = \frac{1}{1+e^{-x}}$	(4)
Tansig function	$\tanh(x) = \frac{2}{1+e^{-2x}} - 1$	(5)

This characteristic is particularly valuable in applications such as time series analysis, where understanding the sequence and timing of inputs is crucial. Further distinguishing ANNs, we can classify them as static or dynamic based on whether they model steady-state or time-variant systems. A static ANN, such as a feedforward network, computes consistent outputs for given inputs, while a dynamic ANN, such as an RNN, adjusts its outputs based on the temporal sequence of inputs over time.

Over the years, various ANN models have been developed for different learning tasks and complexities. Some notable models include: Perceptron (the simplest type of neural network, usually with a single layer of neurons, used for simple classification tasks), Multilayer perceptrons (MLPs—These consist of an input layer, one or more hidden layers, and an output layer which are used to model more complex functions and solve a variety of tasks), Convolutional neural networks (CNNs—these networks specialize in image processing by utilizing convolutional layers that autonomously and efficiently learn spatial hierarchies of features.), Recurrent neural networks (RNNs—re well-suited for processing sequential data, such as speech or text, due to their ability to retain information from previous inputs through an inherent memory mechanism), Adaline and Madaline, Backpropagation (BP) Network, Radial Basis Function Network (RBFN), Modular Neural Network (MNN), Learning Vector Quantization (LVQ) Network, Fuzzy Neural Network (FNN), Kohonen's Self-Organizing Feature Map (SOFM), Adaptive Resonance Theory (ART) Network, Real-Time Recurrent Network, Elman Network, Hopfield Network, Boltzmann Machine, etc. (Skansi 2018). The schematic diagrams of a single-layer feedforward network, a multilayer feedforward network (FFNN), an RBFN, and a discrete-time recurrent network architecture are shown in Fig. 4.

ANNs can be represented by the Eq. 6 (Mount et al. 2016);

$$y = f(\theta, x) + \varepsilon \tag{6}$$

where y is the model output, x is model input, θ is a set of model parameters, and ε is the random error associated with the model prediction.

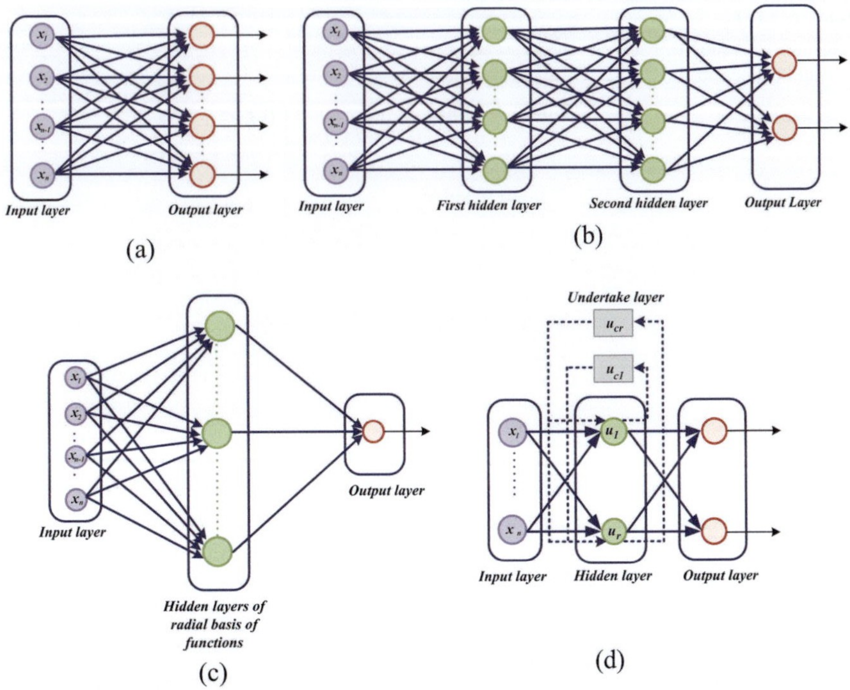

Fig. 4 Schematic diagram of **a** A single-layer feedforward network, **b** A multilayer feedforward network, **c** A Radial Basis Function Network, **d** An Elman Neural Network

In multilayer artificial neural networks (ANNs), the functional relationship that transforms model inputs into outputs is defined by the connectivity between input-layer nodes and output-layer nodes, which is mediated by hidden nodes and transfer functions. Relation for input and output nodes are given Eqs. 7 and 8.

$$I_j = \sum_{i=0}^{n} w_{ji}.x_i, \tag{7}$$

$$y_j = g(I_j). \tag{8}$$

2 Integration of Artificial Neural Networks in Power Electronics

The process of implementing neural network-based control in power electronic systems involves a systematic approach to designing and applying ANNs for the regulation of various components within these systems. The process follows several key steps each contributing to the overall effectiveness and efficiency of the control strategy. The first step in implementing neural network control is to identify the system's dynamics. This involves collecting input–output data under various operating conditions. This dataset serves as the foundation for training and testing the ANN. Following system identification, the next step is the design of an appropriate ANN architecture to the specific application. This includes determining the optimal number of network layers, the number of neurons within each layer, and the selection of suitable activation functions. These design choices are influenced by the system's complexity and the desired control objectives.

Once the architecture is defined, the ANN proceeds to the training phase, where it utilizes the previously gathered input–output data. This design phase includes determining the optimal number of layers within the network, the appropriate number of neurons for each layer, and selecting the right activation functions. These architectural choices are vital, as they directly impact the network's ability to learn and adapt to the complexities of the system. Factors such as the nonlinearity of the system, the types of disturbances it may encounter, and the desired precision of control objectives must be carefully considered during this stage. After the training is complete, it is essential to validate the ANN's performance. This is accomplished using a separate validation dataset that was not part of the training phase. This crucial step assesses the network's ability to generalize its learning to new and unseen data, which is vital for real-world applications where systems may encounter unpredictable conditions. The final phase involves the real-time deployment of the trained ANN controller within the power electronic system. In this stage, the ANN generates control signals based on the system's current state, dynamically adapting to changes and disturbances. Given the complexity of power electronic applications and the stringent real-time constraints they impose, the successful implementation of an ANN-based controller demands not only high computational efficiency but also robust hardware capabilities to ensure reliable performance. This combination of intelligent control and advanced hardware sets the foundation for achieving optimal operation in a variety of power electronic applications.

The integration of ANNs into power electronics has led to significant advancements, unlocking a range of innovative solutions for various applications, including control systems, parameter estimation, and optimization of power systems. ANNs perform well in modeling complex nonlinear relationships that are often encountered in power electronic devices, such as converters and inverters. By taking advantage of their learning capabilities, ANNs continuously adapt to changing conditions and input data, allowing for real-time adjustments and improvements. This adaptability not only enhances the operational performance of power electronic systems but also increases their overall efficiency, contributing to more reliable and effective energy management solutions in various industries (Bose 2009; Karanayil and Rahman 2011; Zhao et al. 2021).

2.1 Control Systems

One of the critical applications of ANNs in power electronics is in excitation control for synchronous machines. Excitation systems play a vital role in regulating the terminal voltage and reactive power output of synchronous generators, which is essential for maintaining power system stability. Traditional control methods, such as Proportional-Integral-Derivative (PID) controllers, often face challenges due to nonlinearities and variations in the system. ANNs provide an effective alternative by learning the nonlinear dynamics of the machine and adjusting control signals as needed (Zhang et al. 2020a). Voltage and reactive power (VAR) optimization is crucial for power distribution networks, as it helps ensure voltage stability while minimizing power losses. ANN-based predictive control models can forecast voltage fluctuations and optimize reactive power compensation. Accurate demand forecasting is also vital for the efficient operation of smart grids, enabling better resource allocation and load management. ANNs have been extensively applied in load forecasting due to their ability to identify complex patterns in electricity consumption data (Allahmoradi et al. 2024).

Power electronic systems are inherently nonlinear, necessitating adaptive control strategies to maintain stability and performance. ANN-based controllers have been effectively utilized in various power electronic applications, including DC-DC converters, inverters, and motor drives. By continuously learning from real-time system behavior, ANNs can dynamically adjust control parameters to accommodate disturbances, variations, and changes in load. This adaptability enhances robustness compared to conventional fixed-gain controllers, making ANN-based adaptive control a promising approach for complex power electronic applications (Farhangi 2010).

In addition to traditional neural network implementations, biologically inspired models, such as the Brain Emotional Learning Based Intelligent Controller (BELBIC), have also been developed for power electronics applications. BELBIC mimics the emotional learning processes of the human brain to create intelligent control strategies for nonlinear and dynamic systems. Its ability to adapt to changing conditions makes BELBIC particularly useful in applications requiring real-time decision-making and high reliability (Albalawi et al. 2002).

2.2 Estimation and Prediction

ANNs have become significant in the estimation and prediction fields of power electronics, providing advanced solutions for modeling complex, nonlinear systems. Their capacity to learn from data and adapt to changing conditions has resulted in significant improvements across various applications, including demand forecasting and system state prediction. The integration of ANNs into these tasks showcases the synergy between artificial intelligence and power engineering, leading to more efficient, reliable, and intelligent power management solutions (Fakhryza et al. 2024; Yarikkaya and Vardar 2023).

In particular, ANNs have been utilized for system state forecasting, which is vital in power electronics. By analyzing sensor data such as temperature, pressure, and vibration, ANNs can predict the future state of components. This capability facilitates proactive maintenance and minimizes downtime. This data-driven approach proves especially valuable when traditional modeling is hindered by system complexity. By learning from historical data and identifying patterns, ANNs can effectively forecast the remaining useful life of components, ultimately enhancing system reliability and performance. Additionally, ANNs have been employed to predict electricity consumption by examining patterns influenced by factors such as weather conditions, time of day, and consumer behavior. This ability is essential for the efficient operation of smart grids, as accurate demand forecasts enable better resource allocation and load management. The adaptability of ANNs allows them to manage the nonlinear and dynamic nature of power consumption data, resulting in more reliable and economical grid management (Manno et al. 2024; Syed et al. 2021).

2.3 *Fault Diagnosis and Condition Monitoring*

Artificial neural networks (ANNs) provide advanced methods for detecting anomalies and predicting system health. Their ability to learn from complex datasets and identify complex patterns has significantly increased system reliability and operational efficiency over traditional maintenance strategies (Ma et al. 1999). In fault diagnosis, ANNs can identify and classify faults in power electronic components by analyzing operational data. For example, ANNs have been used to detect bearing faults in machines by processing vibration signals, which leads to timely interventions and reduced downtime. Their capacity to handle nonlinear relationships and adapt to new fault patterns makes them superior to conventional diagnostic methods (Zarei et al. 2014). For condition monitoring, ANNs enable the continuous assessment of equipment health by predicting potential failures before they occur. By learning from historical performance data, ANNs can forecast the remaining useful life of components, allowing for proactive maintenance scheduling. This predictive capability is particularly valuable in complex systems where traditional modeling poses challenges, thereby enhancing overall system reliability and performance (Mohagheghi et al. 2009). The integration of ANNs into fault diagnosis and condition monitoring represents a significant advancement in power electronics, promoting a shift from reactive to predictive maintenance approaches. This transition not only improves operational efficiency but also extends the lifespan of critical components, demonstrating the transformative impact of artificial intelligence in modern industrial practices.

3 Artificial Neural Networks Based Controllers in Specific Applications

The growing demand for energy-efficient power conversion systems across various sectors, including renewable energy, industrial automation, and consumer electronics, has driven the development of advanced and high-speed power converters. As it is shifted toward cleaner and more sustainable energy sources, the need for converters that can handle the complex and variable requirements of modern power systems is becoming increasingly critical. Especially, DC-DC (buck, boost, etc.), DC-AC (inversion), AC-AC (chopper), AC-DC (rectification) circuits, and their control techniques have gained more significance. However, these converters exhibit highly nonlinear behaviors due to switching operations, magnetic saturation, and varying load conditions. Conventional controllers require complex mathematical model to approximate system behavior. ANNs, on the other hand, can learn system dynamics through training, eliminating the need for precise mathematical models. ANN-based controllers offer rapid response times due to their parallel processing capability. This feature is particularly crucial in high-frequency switching applications such as active power filters, motor drives, and grid-connected inverters. Unlike conventional controllers that rely on sequential computations, ANNs can process multiple inputs simultaneously, ensuring faster control action and reduced transient response times. This capability makes them highly suitable for applications such as renewable energy integration and electric vehicle power management systems.

3.1 Artificial Neural Networks Based Controllers in DC-DC Converter Applications

DC-DC converters play a significant role in efficiently regulating output DC voltage and power derived from various distributed energy sources, such as solar panels, batteries, and other alternative renewable energy sources. Despite their widespread use, the design of effective control systems for these converters faces several significant challenges. One primary issue is the need to maintain precise and rapid adjustments to the desired output voltages. This is complicated by factors like low switching frequencies that are typically required for high-power applications, fluctuations in load conditions, disturbances in the DC input voltage supply, deviations in component parameters, and the constraints posed by current and pulse-width modulation (PWM) saturation within the converters (Dong et al. 2021).

Traditional methods for controlling DC-DC converters face several significant challenges. One major issue is their inability to effectively mitigate distortions that occur due to variations in load, which can result in unstable and fluctuating output voltage levels. Additionally, these conventional approaches often involve complex loop architectures that can complicate system design and implementation. Moreover, they tend to exhibit slower response times, making it difficult to adapt quickly to changing conditions.

To overcome these limitations, it has been increasingly focused on the development of advanced control techniques for DC-DC converters. These new methodologies aim to enhance the stability and responsiveness of converters, enabling them to maintain constant voltage levels despite dynamic load changes. A noteworthy innovation in this area is the integration of artificial neural networks (ANNs). These intelligent systems utilize learning algorithms to dynamically adjust converter parameters, thereby improving performance. Compared to traditional control strategies, ANN-based controllers offer faster response times and superior disturbance rejection capabilities, making them a promising alternative for optimizing the operation of DC-DC converters (Khan et al. 2021). The schematic diagram of ANN-based control for DC-DC converter is shown in Fig. 5.

In the diagram shown in Fig. 5, the load voltage, the load current, or other variable values can be measured with "measurement block" and used as the input information of the ANN. The ANN used here can have more than one hidden layer. The output of the ANN determines the switching signal of the DC-DC converter. Here, a buck, boost, or buck-boost DC-DC converter circuit can be preferred according to the purpose of use.

Also, circuit parameters for DC-DC converters can be modeled or determined based on artificial intelligence (AI). If the steps to be followed for this are listed:

I— A switching loss database is created.

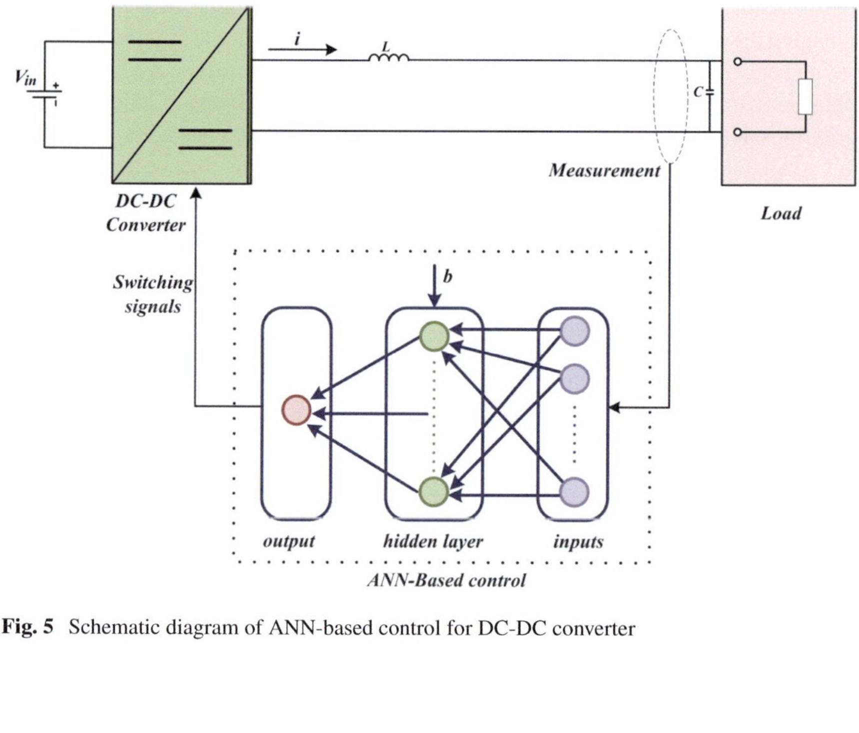

Fig. 5 Schematic diagram of ANN-based control for DC-DC converter

II— Then an artificial neural network (ANN) is trained from the database.

III— Other ANNs are trained for other semiconductor switches requiring less training data using transfer learning (TL).

IV— Finally, a heuristic optimization algorithm is used to obtain the most efficient and optimum device design. Similarly, neural networks are used to evaluate various output parameters of power electronic devices, thus achieving optimum loading of power elements.

3.2 Artificial Neural Networks Based Controllers in DC-AC Converter Applications

ANNs have an important usage area of DC-AC power electronic converters, commonly known as ***inverters***. These converters transform direct current (DC) from sources like photovoltaic panels and batteries into alternating current (AC) suitable for grid integration and various AC loads (Bilhan and Sunter 2020). Traditional inverter control strategies often struggle with challenges such as system nonlinearities, parameter variations, and dynamic load conditions. ANN-based controllers offer adaptive and intelligent solutions to these challenges, enhancing inverter performance and reliability. The basic scheme of ANN-based control of DC-AC inverter is illustrated in Fig. 6. The ANN takes as inputs the measured variables of the filter current i_f, the output voltage V_o, the output current i_o, the reference voltage, etc. The output of the ANN is used to switch the switches of the DC-AC converter.

The ability of ANNs to learn and adapt makes them particularly suited for managing the complexities of DC-AC conversion. They can handle system nonlinearities and uncertainties more effectively than conventional controllers. An ANN-based controller can be trained to predict and compensate for variations in load and input conditions, leading to improved voltage regulation and reduced harmonic distortion. This adaptability enables the inverter to maintain optimum performance across a wide range of operating conditions. One of the critical performance metrics for inverters is the Total Harmonic Distortion (THD) of the output waveform. High THD can lead to inefficiencies. ANN-based controllers have demonstrated significant potential in minimizing THD. Additionally, inverters often operate under varying load conditions and may encounter disturbances that can affect performance. ANN-based controllers can adapt to these changes in real time, maintaining stable operation without the need for manual retuning. This adaptability is particularly beneficial in applications such as renewable energy systems, where input conditions can be highly variable.

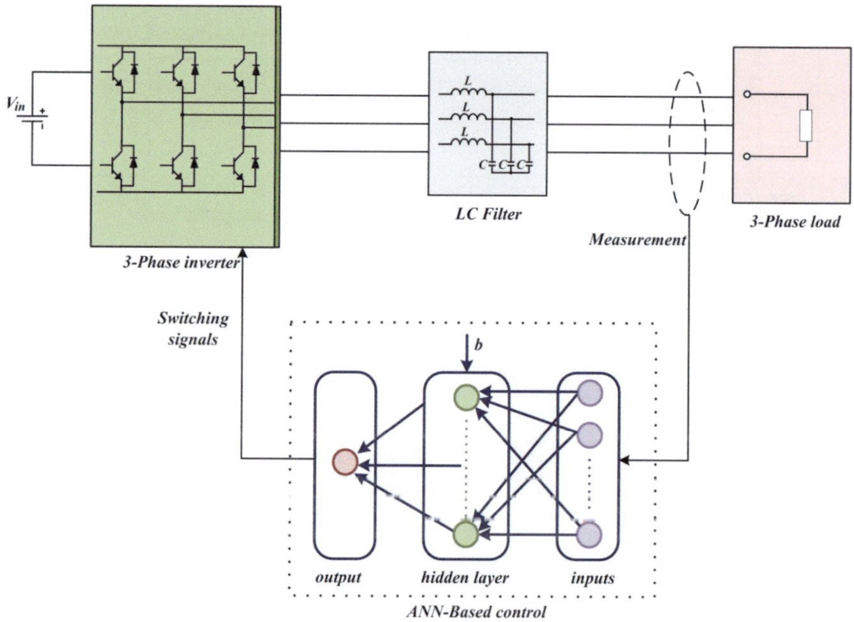

Fig. 6 Schematic diagram of ANN-based control for DC-AC converter

3.3 Artificial Neural Networks Based Controllers in AC-AC Converter Applications

AC-AC converters are essential devices that convert alternative current (AC) from one frequency or amplitude to another without undergoing an intermediate conversion to direct current (DC). This direct conversion process is vital in various applications, including motor drives that control electric motors, voltage regulators that maintain stable output voltages, and frequency changers that adjust the frequency of electrical power for specific needs. One of the primary challenges associated with traditional control methods for AC-AC converters is their struggle to manage the nonlinear behaviors and dynamic responses that are characteristic of these systems. This can lead to inefficiencies and difficulties in achieving precise control. In contrast, controllers based on ANNs demonstrate significant advantages in handling these complexities. By utilizing learning algorithms and adaptive mechanisms, ANN-based controllers can adjust their parameters in real time, effectively improving the performance and reliability of AC-AC converter systems in various industrial and commercial applications. Figure 7 illustrates a single-phase AC-AC converter with an artificial neural network (ANN)-based control system.

In Fig. 7, while the AC input voltage is converted to a specific output voltage and frequency, this process is managed by the switching signals generated by the ANN. This process is managed by the switching signals generated by the ANN.

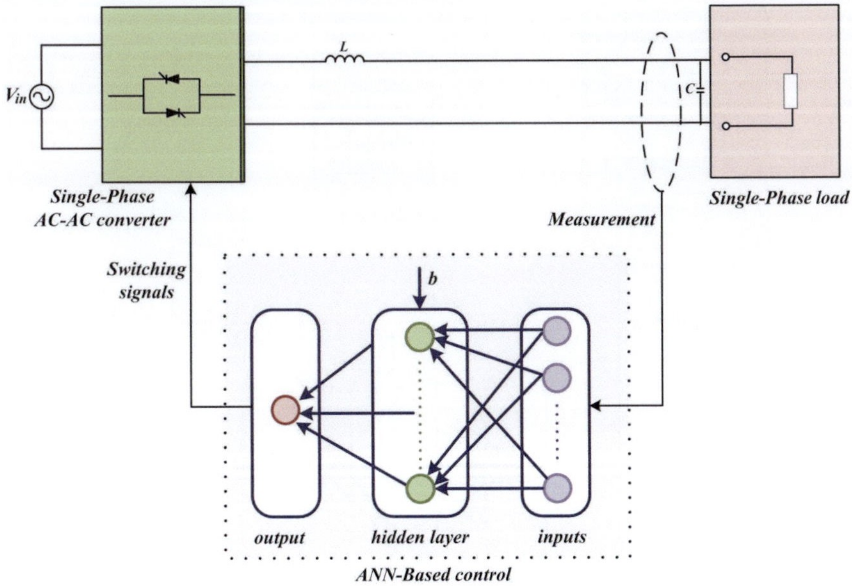

Fig. 7 Schematic diagram of the ANN-based control for the AC-AC converter

The inductance and capacitance at the converter's output enhance the quality of the waveform by reducing harmonic distortions, while various electrical devices are powered on the load side. The measurement and feedback mechanism allows the system to adapt to dynamic changes by transmitting the output voltage and current values to the ANN-based control system. In this setup, the input layer of the ANN receives measurement data, the hidden layer learns the system's behavior and makes control decisions, and the output layer regulates the converter's operation by generating appropriate switching signals.

AC-AC converters display nonlinear characteristics due to their switching operations and varying load conditions. Artificial neural networks (ANNs) can effectively model these nonlinearities by learning from data to predict system behavior under different scenarios. This ability enhances the accuracy and reliability of the control system, leading to improved performance in applications such as voltage regulation and harmonic reduction. Furthermore, the operating conditions of AC-AC converters can change rapidly, which requires controllers that can adapt in real time. ANN-based controllers can adjust to these variations, maintaining optimal performance without the need for manual retuning. This adaptability is particularly advantageous in industrial environments where load demands fluctuate, ensuring consistent operation and extending equipment lifespan. Additionally, ANNs excel in pattern recognition, making them well-suited for fault detection in AC-AC converters. By analyzing electrical signals, ANN-based systems can identify anomalies that indicate faults, enabling proactive maintenance and reducing downtime. This capability is critical in applications where reliability is essential.

3.4 Artificial Neural Networks Based Controllers in AC-DC Converter Applications

AC-DC converters play a crucial role in modern electrical systems by converting AC, which fluctuates in polarity and magnitude, into DC, which maintains a constant polarity. These converters are integral to a wide range of applications, including renewable energy systems like solar and wind energy setups, electric vehicle charging infrastructures, and various industrial power supply systems that require stable and reliable DC output. Figure 8 illustrates the fundamental architecture of an ANN-based control system for an AC-DC converter.

As seen in Fig. 8, the input AC supply is provided and converted to DC using switching elements. The ANN-controlled system, on the other hand, produces appropriate output values that rely on real-time measurements from various input variables, including voltage, current, and load conditions, by the designed system and using the trained neural network. This control structure increases system performance by offering a more flexible and adaptive structure compared to classical control methods. Complementing this setup is a meticulously designed output filter circuit, composed of an inductor (L) and a capacitor (C). This filter circuit is essential for minimizing voltage fluctuations, thereby ensuring that a more stable and consistent DC voltage is delivered to the load.

One of the challenges faced by AC-DC converters is their tendency to exhibit nonlinear characteristics, especially when traditional control methods are applied. This nonlinearity arises from the inherent switching operations within the converters

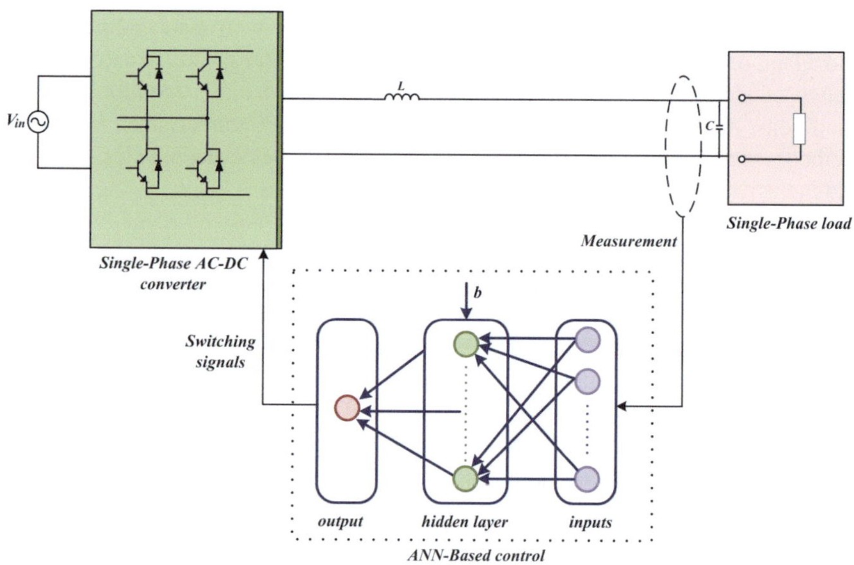

Fig. 8 Fundamental structure of an ANN-based control system for an AC-DC converter

and changes in load conditions that can occur frequently in real-world scenarios. Artificial neural networks (ANNs) are particularly capable of modeling these complexities. By analyzing large datasets and learning from historical performance, ANNs can predict how the system will behave under different conditions. This predictive capability significantly enhances the accuracy and reliability of control systems, leading to improved performance in critical applications such as power factor correction, where the goal is to optimize energy consumption, and harmonic reduction, which helps minimize electrical noise and ensure consistent power quality.

Another noteworthy aspect of AC-DC converters is that their operating conditions can shift rapidly due to varying input voltage levels or fluctuating load demands. To address this issue, it is imperative to use advanced controllers capable of real-time adaptation. ANN-based controllers excel in this regard, as they can dynamically adjust operational parameters in response to changing conditions. This means that the converters can maintain optimal performance levels without requiring time-consuming manual retuning, which is particularly beneficial in renewable energy contexts where input power, such as sunlight or wind strength, can be irregular and unpredictable.

4 Artificial Neural Networks Based Controllers in Motor Drives

Conventional control techniques, such as Proportional-Integral-Derivative (PID) controllers and field-oriented control (FOC), often struggle to maintain precision in highly dynamic and nonlinear motor systems. ANNs, with their ability to learn and adapt, offer superior control strategies that enhance the speed, torque, and efficiency of electric motors. ANN-based controllers can handle variations in load conditions, disturbances, and model uncertainties more effectively than traditional methods, making them suitable for applications such as electric vehicles, robotics, and industrial automation (Qiu et al. 2024).

One of the primary applications of ANNs in motor drives is speed and torque control. Conventional controllers rely on mathematical models that may not fully capture the complexities of motor behavior, leading to performance degradation under fluctuating conditions. In conventional control methods of induction motors, rotor speed is commonly calculated as difference between synchronous angular velocity (ω_s) and slip angular velocity (ω_k) which are given in Eq. (9) (Boyar et al. 2024a). These methods depend on parameters and are particularly inaccurate at low speeds because the effect of voltage drop decreases and the variability of stator resistance is evident.

$$\left.\begin{array}{c}\hat{\psi}_s = \int (u_s - i_s R_s) dt \\ \hat{\psi}_s = \left[\psi_{sd}, \psi_{sq}\right]^T \\ \theta = \tan^{-1}\left(\frac{\hat{\psi}_{sq}}{\hat{\psi}_{sd}}\right) \\ \omega_s = \dot{\theta} = \frac{\hat{\psi}_{sd}.\dot{\hat{\psi}}_{sq} - \hat{\psi}_{sq}.\dot{\hat{\psi}}_{sd}}{\hat{\psi}_{sd}^2 + \hat{\psi}_{sq}^2} \\ \omega_k = \frac{1}{\tau_r}\frac{i_{sq}}{i_{sd}} \\ \omega_r = \omega_s - \omega_k\end{array}\right\} \tag{9}$$

ANNs, however, can be trained with real-time and historical data to approximate optimal control actions, leading to better dynamic response and efficiency. It has been demonstrated that ANN-based controllers for induction motors provide faster transient response, reduced overshoot, and improved steady-state accuracy compared to traditional PID controllers (Lodi et al. 2024). Implementing the ANNs' speed and flux estimators involves the training, testing, and validation stages.

5 Collecting Training Data

The training data is collected by simulating conventional motor control methods, such as the Direct Torque Control, direct field-oriented control, indirect field-oriented control, etc., as shown in Fig. 9.

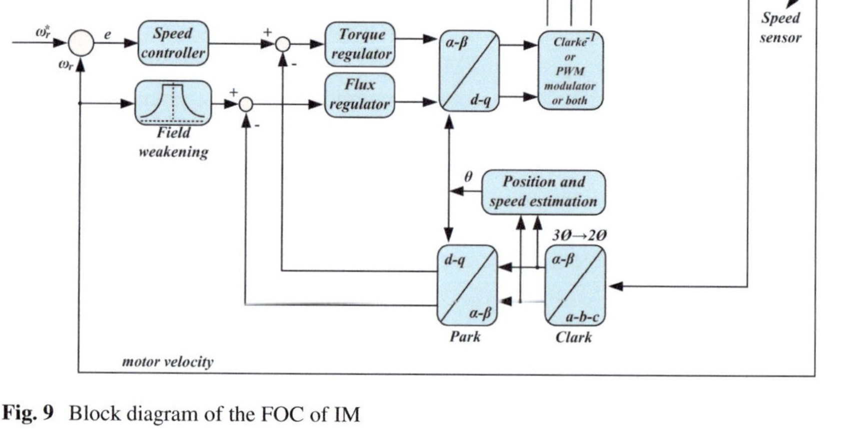

Fig. 9 Block diagram of the FOC of IM

The Direct Torque Control (DTC) method directly regulates flux and torque production without explicitly controlling the stator currents. This regulation technique employs a simple control structure that utilizes two hysteresis controllers and predefined look-up tables. In contrast, the field-oriented control (FOC) method transforms machine parameters into a synchronously rotating coordinate system aligned with the rotor flux, enabling the decoupling of flux and torque while mitigating the effects of machine nonlinearity. The FOC approach decomposes the stator current into two components: torque and flux. The required current measurements for coordinate transformation are performed in the stationary d-q reference frame, and the current vectors I_{ds} and I_{qs} are subsequently converted into the rotating d-q reference frame.

Data is collected by randomly selecting load torque disturbances to cover a wide range of operating conditions. The collected data is the measured voltages and currents at $\alpha\beta$. In addition, the actual rotor speed is also recorded. In addition, noise is added to increase the robustness of the feedforward neural network against noise, and the data is recorded. All the recorded data is exported to train the ANN. A dataset is created with the data collected for a certain period in a small step range.

6 Designing, Training, and Validation of Artificial Neural Network Structure

It is an important issue to select the type of ANN for the accurate estimation of the flux and speed. The ANN type can be a feedforward NN (FFNN), RBFN, BPN, etc. For training, testing, and validation purposes, the MATLAB/Toolbox can be used, which is given in Fig. 10. One can apply another program. The training function is selected as one of the techniques (trainlm, step, linear, etc.). To evaluate the quality of predictions made by neural networks, several evaluation criteria can be employed, including mean square error (MSE), mean relative error (MRE), mean absolute error (MAE), root mean square error (RMSE), coefficient of variation (COV), correlation coefficient (CC), coefficient of determination (COD), efficiency coefficient (EC), overall index of model performance (OIMP), and coefficient of residual mass (CRM). These definitions and metrics are thoroughly detailed in Table 2 with Eq. (10–19), where n_s indicates the number of observations, while d and y represent the desired (observed) values and the corresponding predicted outputs, respectively. Such a suite of evaluation criteria is crucial for understanding and improving the performance of neural network models in predictive tasks (Elsheikh et al. 2019). Also, it can be tuned how many hidden layers can be used to estimate the rotor speed in MATLAB/Toolbox accurately.

Each ANN training is done offline to save implementation time. After determining the efficiency and accuracy of the ANN, the process of exporting the ANN as a Simulink block is achieved using the neural network (NN) MATLAB/Toolbox as shown in Fig. 11. Then, this block is used where it is necessary.

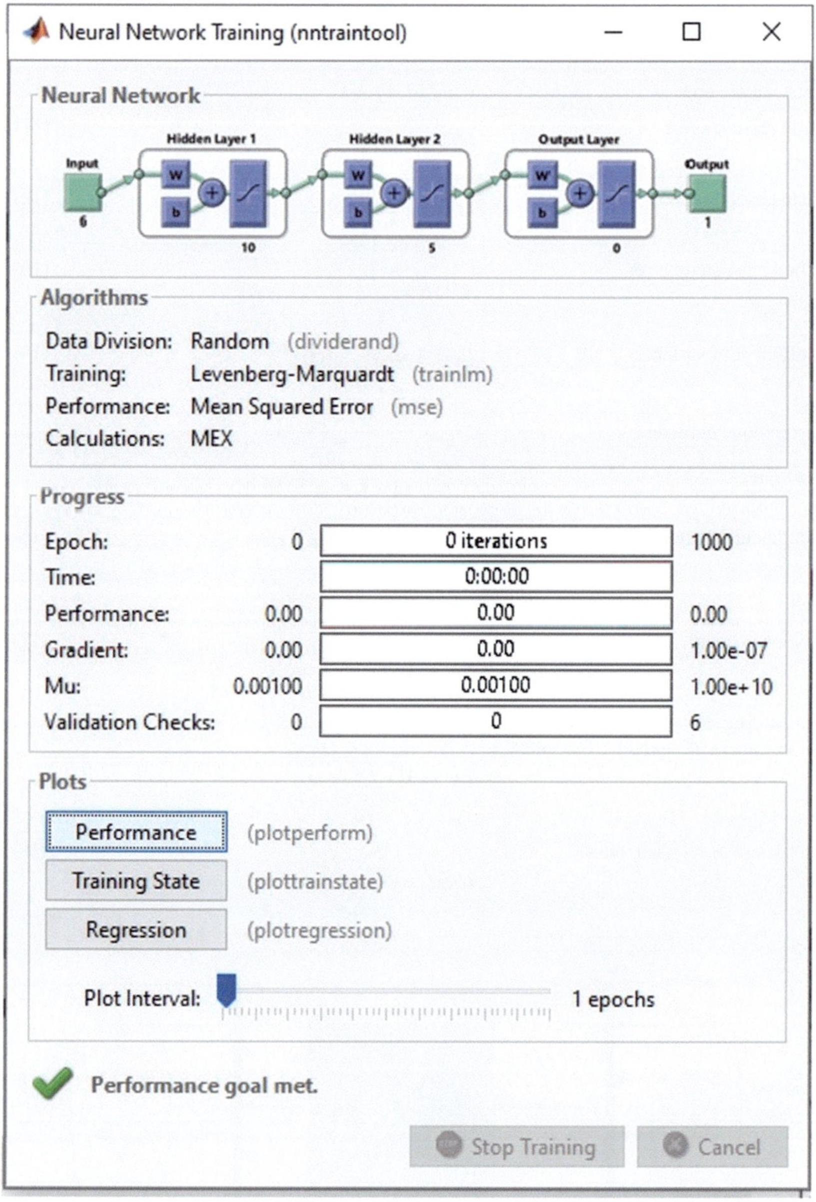

Fig. 10 Example of the neural network training

Table 2 Evaluation criteria

Method	Formula	
Mean square error	$y = f(\theta, x) + \varepsilon$	(10)
Mean relative error	$MRE = \frac{1}{n_s}\sum_{i=1}^{n_s}\frac{d_i - y_i}{d_i}$	(11)
Mean absolute error	$MAE = \frac{1}{n_s}\sum_{i=1}^{n_s}\lvert d_i - y_i\rvert$	(12)
Root mean square error	$RMSE = \sqrt{\frac{1}{n_s}\sum_{i=1}^{n_s}(d_i - y_i)^2}$	(13)
Coefficient of variance	$Cov = \frac{RMSE}{\frac{1}{n_s}\sum_{i=1}^{n_s} y_i} x100$	(14)
Correlation coefficient	$Corr = \frac{n_s\sum_{i=1}^{n_s}(d_i . y_i) - \left(\sum_{i=1}^{n_s} d_i - \sum_{i=1}^{n_s} y_i\right)}{\left[n_s\sum_{i=1}^{n_s} d_i^2 - (\sum_{i=1}^{n_s} d_i)^2\right]\left[\left[n_s\sum_{i=1}^{n_s} y_i^2 - (\sum_{i=1}^{n_s} y_i)^2\right]\right]}$	(15)
Coefficient of determination	$R^2 = \frac{(\sum_{i=1}^{n_s}(d_i - \bar{d}) - (y_i - \bar{y}))^2}{\sum_{i=1}^{n_s}(d_i - \bar{d})^2 x \sum_{i=1}^{n_s}(y_i - \bar{y})^2}$	(16)
Efficiency coefficient	$EC = 1 - \frac{\sum_{i=1}^{n_s}(d_i - y_i)^2}{\sum_{i=1}^{n_s}(d_i - \bar{d})^2}$	(17)
Overall index of model performance	$EC = 1 - \frac{\sum_{i=1}^{n_s}(d_i - y_i)^2}{\sum_{i=1}^{n_s}(d_i - \bar{d})^2}$	(18)
Coefficient of residual mass	$CRM = \frac{\sum_{i=1}^{n_s} y_i - \sum_{i=1}^{n_s} d_i}{\sum_{i=1}^{n_s} d_i}$	(19)

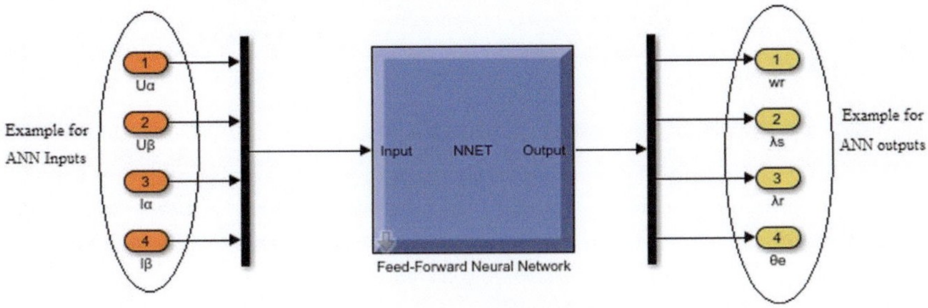

Fig. 11 Example of the neural network Simulink block

By continuously learning from input–output relationships, ANN controllers can adapt to parameter variations, eliminating the need for frequent manual tuning.

Fault detection and diagnosis in motor drives have also been significantly enhanced through ANN-based methods. Electric motors are prone to faults such as bearing wear, rotor misalignment, and stator winding failures, which can lead to severe performance degradation and increased maintenance costs. Traditional fault diagnosis techniques rely on predefined thresholds and expert knowledge, which may not always detect subtle fault signatures. ANN-based fault diagnosis systems, on the other hand, use pattern recognition and classification algorithms to detect anomalies in motor operation. By analyzing vibration signals, current waveforms, and temperature data, ANN models can identify faults at an early stage, allowing for predictive maintenance and reduced downtime. Studies have shown that ANN-based fault diagnosis achieves high accuracy in identifying both incipient and severe faults, outperforming conventional diagnostic methods (Shao et al. 2017).

Another area where ANNs have revolutionized motor drives is in sensorless control. Many high-performance motor drive systems require position and speed sensors, which add complexity, cost, and potential reliability issues. ANN-based sensorless control techniques estimate rotor position and speed using motor terminal voltages and currents, eliminating the need for physical sensors. By training an ANN with data from a motor's electrical and mechanical characteristics, sensorless control algorithms can accurately predict motor states in real time. Research has shown that ANN-based sensorless control provides better noise immunity and robustness against parameter variations compared to traditional model-based estimators (Ichikawa et al. 2006). This advancement is particularly beneficial in applications such as electric vehicles, where reducing sensor dependency enhances reliability and cost-effectiveness.

Energy efficiency optimization is another key application of ANNs in motor drives. Optimizing energy consumption in electric motors is critical, especially in industrial settings where motors account for a significant portion of energy usage. ANN-based optimization techniques dynamically adjust operating parameters such as switching frequencies, duty cycles, and modulation strategies to minimize losses and improve efficiency. A notable example is the use of ANN-driven space vector pulse-width modulation (SVPWM) in inverters, which improves the power factor and reduces harmonics, leading to better overall efficiency. Studies indicate that ANN-based energy optimization strategies can achieve energy savings of up to 10–15% compared to conventional approaches (Langer et al. 2014). In addition to control and optimization, ANNs are also being integrated with other artificial intelligence techniques to further enhance motor drive performance. Hybrid approaches combining ANNs with fuzzy logic, genetic algorithms, and reinforcement learning have shown promising results in adaptive motor control. These hybrid models leverage the strengths of each technique (ANNs for learning complex patterns, fuzzy logic for handling uncertainties, and genetic algorithms for optimization), resulting in superior performance in nonlinear and uncertain environments (Ibrahim et al. 2022). It was emphasized that less energy consumption can be achieved with different ANN-based optimization techniques, such as Lung performance-based optimization, in the calculation of energy consumption demands of hybrid energy storage systems used in electric vehicles (Boyar et al. 2024b). The growing availability of high-performance embedded processors and edge computing has also facilitated the real-time implementation of ANN-based motor drive systems, making them increasingly viable for practical applications.

7 Artificial Neural Networks Based Controllers in Renewable Energy Systems

ANNs have become essential in renewable energy systems within power electronics due to their ability to manage nonlinear, complex problems. Renewable energy sources like solar, wind, hydro, and hybrid systems inherently exhibit fluctuations due to environmental conditions, making their integration into power systems difficult using traditional control methods. ANNs, with their ability to learn and adapt, have provided intelligent solutions for forecasting, control, fault detection, and energy management, thereby improving efficiency, stability, and reliability in power electronic-based renewable energy systems (Zhang et al. 2022b).

One of the most important applications of ANNs in renewable energy is in solar photovoltaic (PV) systems. Since solar energy output depends heavily on environmental factors such as sunlight intensity, temperature, and cloud cover, which are shown in Fig. 12, accurately predicting power generation is crucial for grid stability.

ANN models are commonly employed to forecast solar radiation and estimate PV power output. By processing large datasets, including historical weather patterns and real-time atmospheric conditions, ANNs can predict energy production with high accuracy. Compared to conventional statistical models, ANN-based forecasting methods achieve superior precision, reducing uncertainties in energy management. In Fig. 13, the structure of the artificial neural network (ANN) used for modeling the solar collector, thermal performance, and air heater is shown (Dikmen et al. 2014; Farkas and Géczy-Víg 2003; Ghritlahre and Prasad 2018).

The ANN structure shown in Fig. 13a shows an input layer with three input parameters (global solar radiation, ambient temperature, and working fluid inlet temperature) and an output layer with a single output neuron (working fluid outlet temperature). All available step functions (log-sigmoid, tan-sigmoid, hard limit, and linear functions) for the neurons of the layers are analyzed and selected (Farkas and Géczy-Víg 2003). In the ANN structure shown in Fig. 13b, the inputs of the model are solar radiation, ambient temperature, average storage tank temperature, and slope angle. Again, in this model, all available step functions for the neurons of the layers are analyzed and the most suitable one is selected. The performance of the ANN model can be evaluated using various standard statistical performance evaluation criteria such as RMSE, COV and multiple determination coefficient (Dikmen et al. 2014). In Fig. 13c, an ANN model developed to determine the thermal performance of a unidirectional flow solar air heater is shown. The input parameters of

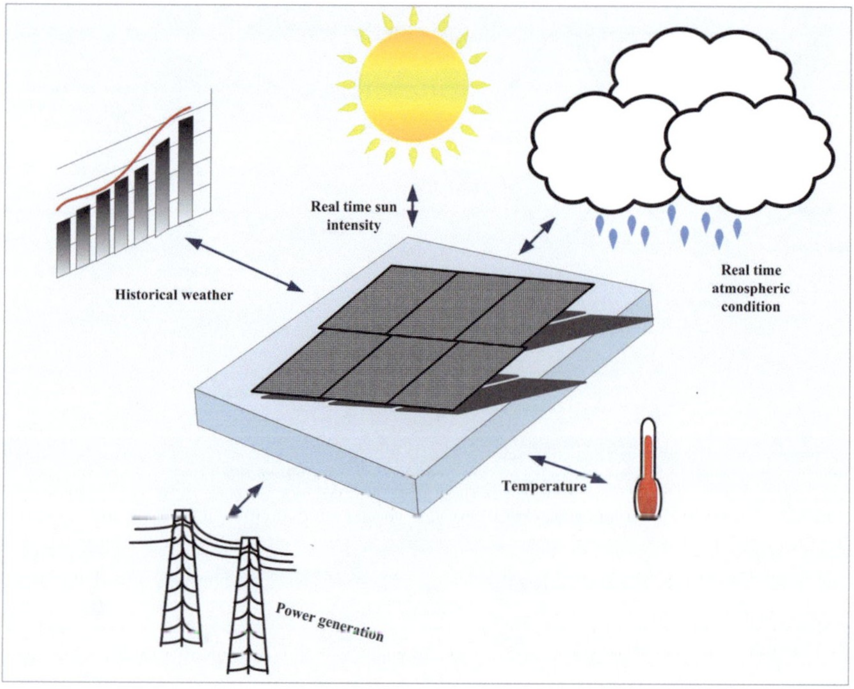

Fig. 12 Environmental factors for solar panels

ANN are ambient temperature, air inlet temperature, mass flow rate, and solar radiation intensity. The output parameters are air temperature difference, heat gained by air and thermal efficiency. During the training of this ANN model, four different training functions, namely SCG backpropagation, single-step secant backpropagation, Levenberg–Marquardt backpropagation, and conjugate gradient BP with Polak-Ribiére updates, were compared using MSE and coefficient of multiple predictors to obtain the best performance. As a result of these comparisons, it was determined that Levenberg–Marquardt BP training function showed the best performance among other training functions (Ghritlahre and Prasad 2018).

In addition to forecasting, ANNs play a significant role in Maximum Power Point Tracking (MPPT), which is critical for optimizing power extraction from solar panels. Traditional MPPT methods, such as Perturb and Observe (P&O) and Incremental Conductance (IC), often struggle with rapidly changing irradiance conditions, leading to inefficiencies (Zhao et al. 2020). ANN-based MPPT algorithms, however, continuously learn and adjust to changing environmental conditions, ensuring optimal power delivery. It has been shown that ANN-based MPPT techniques can significantly improve tracking speed and accuracy, especially in dynamic weather conditions, outperforming traditional controllers. Fault detection and diagnosis in PV systems also benefit from ANN integration. Problems such as partial shading, short circuits, and deterioration over time can be encountered in solar panels.

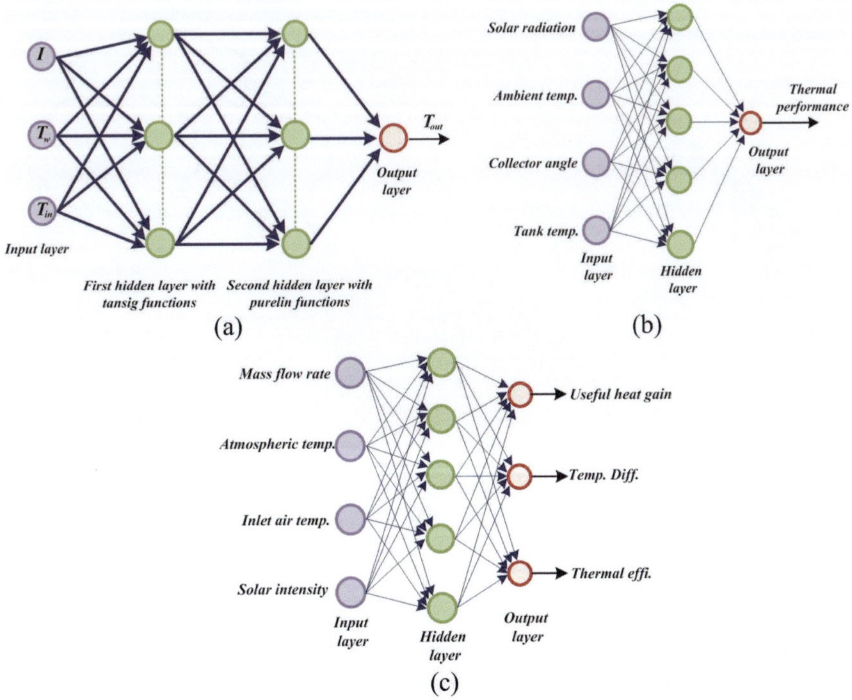

Fig. 13 ANN model for **a** solar collector, **b** thermal performance, **c** solar air heater

As shown in Fig. 14, ANNs can be trained to detect anomalies by analyzing electrical parameters like voltage, current, and power output. By identifying faults early, these intelligent models help prevent energy losses and reduce maintenance costs. Research indicates that ANN-based fault detection systems exhibit higher sensitivity and faster response times than conventional diagnostic techniques.

Wind energy systems, like solar power, exhibit inherent variability due to fluctuating wind speeds and atmospheric conditions. Accurate forecasting of wind energy production is essential for ensuring grid stability and efficient energy distribution. ANNs have proven effective in predicting short-term and long-term wind power output by analyzing historical wind speed data, temperature, and air pressure variations. A hybrid ANN model incorporating wavelet transform techniques has demonstrated a 20% improvement in wind speed forecasting accuracy compared to standalone ANN models. Such advancements allow energy providers to better manage wind farm operations and optimize power dispatch strategies (Arslan Tuncar et al. 2024). Beyond forecasting, ANNs also contribute to the intelligent control of wind turbines. Conventional control methods may struggle to adapt to sudden wind speed changes, leading to mechanical stress and inefficiencies. ANN-based turbine control

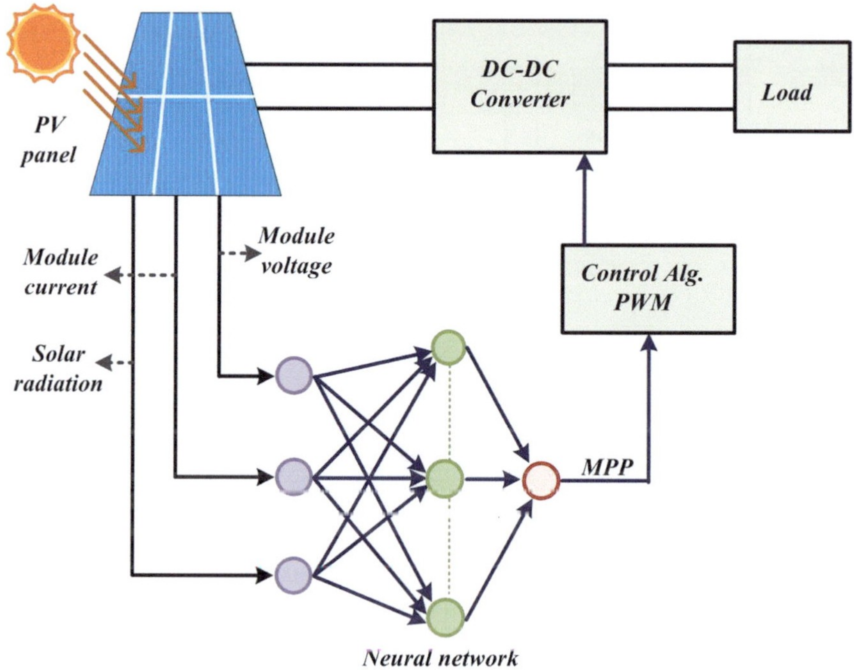

Fig. 14 Schematic diagram of the MPPT for the PV system using ANNs

systems can dynamically adjust blade pitch angles and generator torque based on real-time input data, enhancing power capture while reducing wear and tear. In Fig. 15, a scheme of ANN-controlled wind turbine-driven PMSG is shown.

In this model, the output mechanical power for a variable speed wind turbine is calculated as in Eq. 20.

$$P_m = \frac{1}{2}\rho A C_T(\lambda, \beta) V_w^3 \tag{20}$$

where C_T (λ) is the torque coefficient, ρ is the air density, A is the wind turbine rotor swept area, V_w is the wind speed (m/s), β is the blade pitch angle, and λ is the tip speed ratio and it is defined as Eq. 21. Moreover, the maximum power of the wind turbine and the corresponding rotor speed are given in Eq. 22.

$$\lambda = \frac{\omega_r . r}{V_w} \tag{21}$$

$$\left.\begin{array}{ll} P_{t\max} = K_1 . V_w^3 & K_1 = 0.5\rho \Lambda C_{p\max} \\ \omega_r^* = K_2 V_w & K_2 = \frac{\lambda_{opt}}{r_m} \end{array}\right\} \tag{22}$$

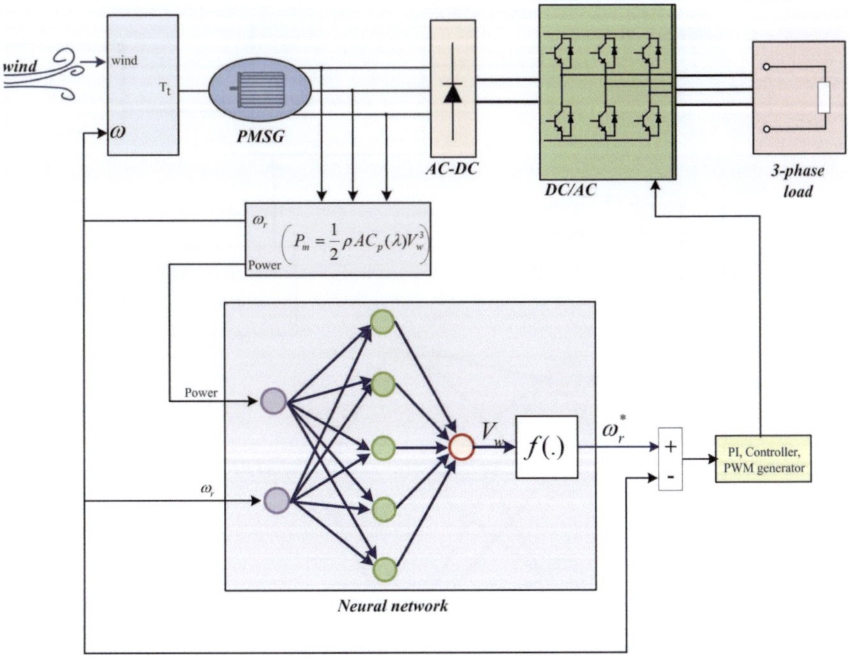

Fig. 15 Schematic diagram of ANN-controlled wind turbine-driven PMSG

where P_{tmax} is the turbine maximum power at a certain wind velocity, C_p is the power coefficient, and $\omega_r{}^*$ is the desired rotor speed when the turbine operates at maximum power point, and r_m is the turbine radius. As Eq. 21 shows, if the wind speed is available, the wind turbine generator can be controlled to vary the rotor speed proportionally to the wind speed to obtain the optimum power. The dynamics of turbine power and generator power can be described as Eq. 23.

$$\frac{dP_e}{d\omega_r} = \frac{dP_m}{d\omega_r} - J\frac{d\omega_r}{dt} - J\omega_r\frac{\frac{d^2\omega_r}{dt^2}}{\frac{d\omega_r}{dt}} \tag{23}$$

where ω_r is turbine rotor speed, J is system inertia, P_m is turbine power, and P_e is generator power.

In Fig. 15, an ANN is used to implement wind velocity estimation. In this figure, sample data P_m is generated based on the turbine power equation, utilizing preselected samples of rotor speed and wind velocity. Specifically, pairs of rotor speed and power data are systematically compiled, resulting in a comprehensive dataset that consists of 10,000 data pairs in the format $(\omega_r(i), P_m(i, j))$ where i ranges from 1 to 100, representing different rotor speeds, and j also ranges from 1 to 100, reflecting various wind velocities. This extensive collection of data pairs is utilized as the input matrix for the neural network. Furthermore, the wind velocity samples are assigned as target outputs for the training process. A three-layer neural network architecture is employed to facilitate effective learning. The input layer consists of two linear neurons that capture essential input features. The hidden layer comprises five tan-sigmoid neurons, which enhance the network's capability to model nonlinear relationships. Finally, the output layer includes a single linear neuron that produces the predicted wind velocity based on the provided inputs. This structured approach enables the neural network to accurately learn the underlying patterns between turbine performance and wind conditions (Li et al. 2005).

Another crucial application of ANNs in wind energy is fault detection and predictive maintenance. Wind turbines are subject to mechanical failures, such as gearbox malfunctions and generator overheating. ANN-based fault diagnosis systems analyze vibration signals, acoustic emissions, and electrical outputs to detect early signs of wear and tear. By implementing predictive maintenance strategies, wind farm operators can prevent unexpected breakdowns and optimize maintenance schedules, ultimately reducing operational costs.

Hybrid renewable energy systems, which integrate multiple sources like solar and wind, as shown in Fig. 16, present additional challenges in energy management and optimization. The variability of each energy source requires intelligent decision-making to balance power generation and storage effectively. ANN-based energy management systems can optimize energy distribution between sources, predict demand fluctuations, and improve battery storage efficiency. These systems enhance grid stability by ensuring a smooth energy supply despite fluctuations in individual power sources.

Figure 16 shows the implementation of a three-layer Radial Basis Function Network (RBFN) NN controller for the boost converter of a PV array. In RBFN, the input layer contains three nodes: V_{pv}, I_{pv}, and *temperature*, the hidden layer contains nine nodes, and the output layer contains one output node. The RBFN structure is given in Fig. 17. The PWM module connected to the output of this module is used to generate PWM pulses to control the duty cycle of the DC-DC boost converter. The nodes in this layer are used to directly transmit the numerical inputs to the next layer. Within the hidden layer, each neuron employs a Gaussian function, which is a specialized form of radial basis function. This function serves as a membership function, determining the degree to which an input belongs to a particular category based on its distance from the center of the Gaussian curve. In the output layer, the single node computes the overall output as the summation of all incoming signals.

Once the RBFN has been initialized, a supervised learning of the gradient is used to train this system. The derivation is the same as that of the backpropagation algorithm. It is employed to adjust the parameters of the RBFN by using the training patterns. By recursive application of the chain rule, the error term for each layer is calculated and updated. The adjustment of the parameters for learning and the weight of links enhances the performance of solar systems. After initializing the RBFN, the training process begins through a supervised learning approach that utilizes gradient descent. This technique mirrors the principles of the backpropagation algorithm, which is fundamental in neural network training. During this process, the provided training datasets are used to fine-tune the RBFN parameters. As it is proceeding, it

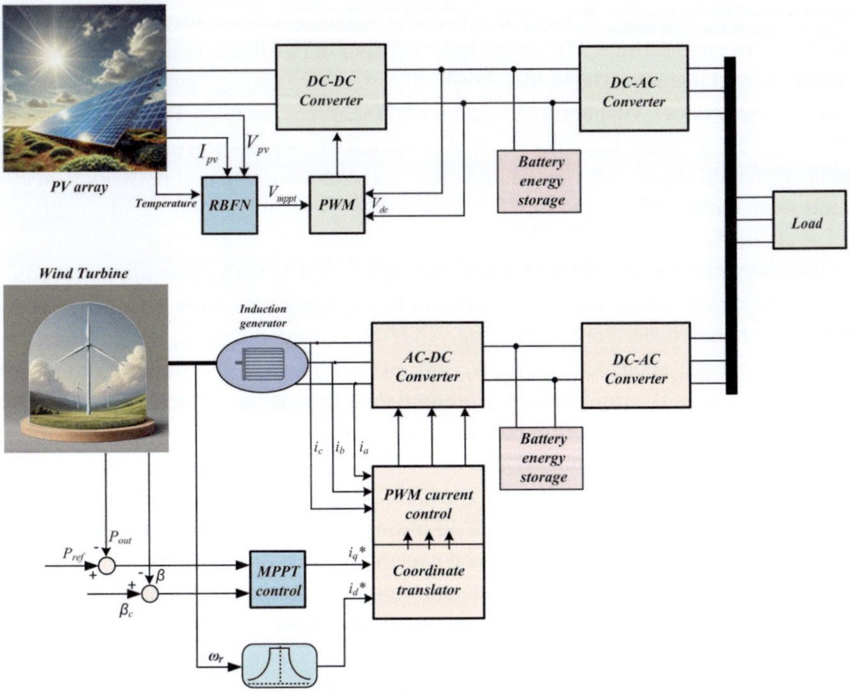

Fig. 16 ANN-based controlled hybrid renewable energy system block scheme

Fig. 17 Three-layer RBFN

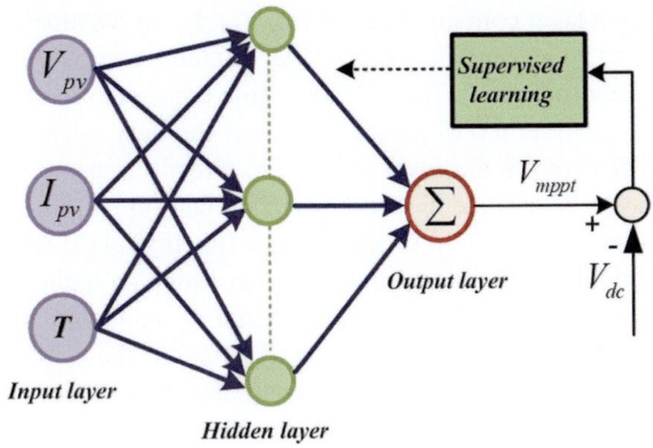

is applied the chain rule in a recursive manner to diligently compute the error term associated with each layer of the network. This error calculation enables us to make precise updates to the weights and parameters, thereby improving the network's ability to learn from the training patterns efficiently. By systematically adjusting the learning parameters and the interconnection weights, it can significantly enhance the RBFN's performance. This optimization process is crucial for improving the efficacy and accuracy of solar energy systems, ensuring they operate more effectively in real-world applications (Lin et al. 2011).

The wind energy system illustrated in Fig. 16 consists of an induction generator, a current control AC-DC converter, a field orientation mechanism that includes a coordinate translator, a current-controlled DC-AC inverter, and an MPPT controller, which utilizes both Proportional-Integral (PI) control and an Elman Neural Network (ENN). The architecture of the ENN for the control of wind energy, illustrated in Fig. 18, includes the input layer, hidden layer, context layer, and output layer with two inputs and one output.

The proposed ENN takes the feedback into account, and a better learning efficiency can be obtained. To elevate the sensitivity of the neurons toward the historical patterns of input data, self-connections are integrated into both the context nodes and the output feedback nodes. This strategic enhancement empowers the ENN to handle complex, nonlinear problems more effectively, leading to improved convergence precision and a notable reduction in overall learning time. Upon initialization, the ENN employs a supervised learning methodology, utilizing gradient descent as the primary optimization technique to fine-tune its parameters. The derivation is the same as that of the backpropagation algorithm. It is employed to adjust the parameters w_{jo}, w_{rj}, and w_{ij} of the ENN by using the training patterns. This involves systematically applying the chain rule to compute the error terms for each layer within the

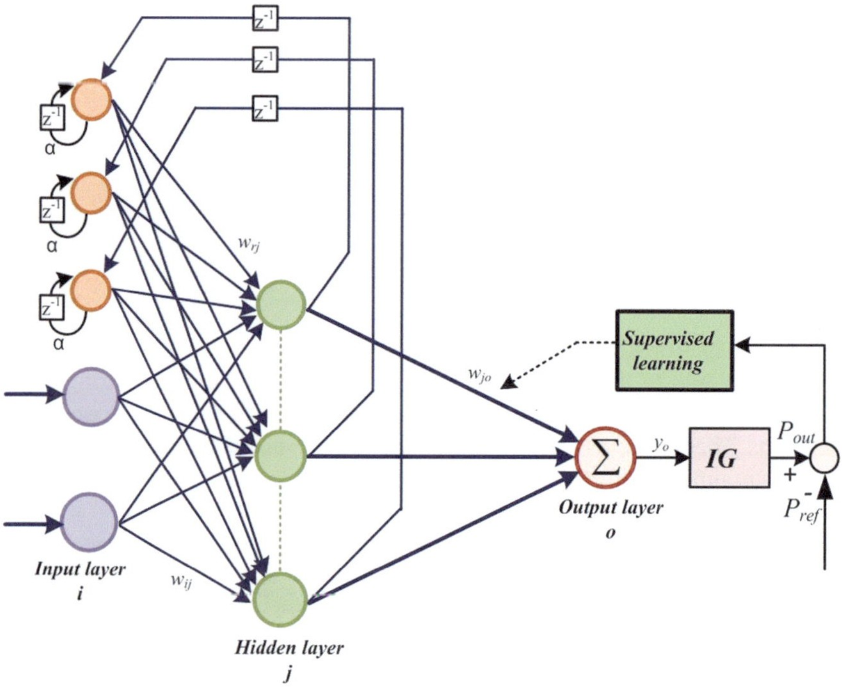

Fig. 18 Three-layer ENN

network. As the training progresses, the ENN meticulously updates these error terms to refine its understanding of the data patterns. This process facilitates an adaptive learning environment, ensuring that the network not only improves its accuracy over time but also becomes adept at recognizing and adjusting to the variations in input data efficiently (Lin et al. 2011).

8 Challenges and Limitations

Integrating artificial neural networks (ANNs) into power electronics presents numerous advantages, but it also comes with specific challenges that need to be addressed.

8.1 Computational Complexity

The training process for ANNs, especially those involving deep learning architectures, demands considerable computational resources. This is due to the intricate calculations required for processing vast amounts of data across multiple layers of neurons. To enable these models to function in real-time applications within power electronic systems, such as inverter control or energy management, there must be a focus on developing highly efficient algorithms. Additionally, specialized hardware, like Graphics Processing Units (GPUs) or Field Programmable Gate Arrays (FPGAs), may be necessary to meet the computational intensity and ensure rapid response times.

8.2 Data Requirements

The performance of ANNs is intrinsically tied to the quality and quantity of the data used for training. In the field of power electronics, this involves gathering extensive datasets that are both high-quality and representative of real-world operational conditions. These datasets must cover many scenarios, including various load conditions, environmental influences, and potential fault situations that the systems may encounter during operation. The challenge lies not only in the collection of such rich datasets, potentially requiring sophisticated data acquisition systems and sensors, but also in ensuring that the data accurately reflects the dynamics of the power electronic systems. If the data is insufficiently diverse or fails to encompass critical operational extremes, the resulting ANN models may perform poorly in real-world applications, leading to inefficiencies or system failures.

8.3 Integration with Existing Systems

The application of ANNs in current power electronic frameworks requires careful strategic planning and consideration. This integration involves assessing the compatibility of ANN models with current hardware and software configurations, as well as ensuring that the overall system maintains reliability and stability. This may require upgrading old systems, developing interfaces for communication between ANN algorithms and conventional controllers, and thoroughly testing the integrated system under various operational scenarios to confirm that it meets performance standards and safety regulations.

9 Application Examples

ANNs have found extensive applications in power electronics, enhancing control, optimization, fault detection, and energy management. One of the most significant applications of ANNs in power electronics is in the control of power converters and inverters, where they offer superior dynamic response, adaptability, and robustness against parameter variations compared to conventional controllers (Wang et al. 2022). ANNs are also widely used in fault detection and condition monitoring, enabling real-time diagnostics and predictive maintenance by analyzing operational parameters and identifying anomalies before failures occur (Kumawat et al. 2023). Moreover, ANN-based techniques have been implemented in energy management systems, particularly in renewable energy applications such as solar photovoltaic (PV) and wind energy conversion systems, where they optimize power generation, improve grid integration, and enhance stability (Saadi et al. 2024). In electric drives, ANN-based controllers are employed to improve motor control, enhancing speed regulation, torque control, and efficiency in induction motors, permanent magnet synchronous motors (PMSMs), and brushless DC motors (Qiu et al. 2024). Additionally, ANNs contribute to power quality improvement by mitigating harmonics, compensating reactive power, and optimizing power factor correction in smart grid applications (Mohsen et al. 2023).

Another crucial area where ANNs play a vital role is in load forecasting and demand-side management, where they analyze historical data to predict future power consumption patterns with high accuracy, assisting utilities in grid planning and operation. In battery energy storage systems (BESS), ANN-based models optimize charging and discharging strategies, extend battery life, and enhance the efficiency of energy storage solutions (Raoofi and Yildiz 2023). Furthermore, ANNs are used in fuel cell electric vehicle (EV) power management, where they optimize energy distribution between the battery, regenerative braking system, and drivetrain to maximize efficiency and driving range (Liu et al. 2024). Another emerging application is in smart transformers and solid-state transformers (SSTs), where ANN-based controllers enable real-time voltage regulation, fault detection, and adaptive control for improved reliability and efficiency (Vandoorn et al. 2013).

10 Future Trends

The future of artificial neural network (ANN)-based control systems in power electronics is expected to concentrate on several key advancements: efficiency, adaptability, and real-time optimization. As power electronic systems evolve in complexity, ANN-based controllers are poised to leverage sophisticated techniques like deep learning and reinforcement learning to significantly enhance dynamic performance. This improvement will not only increase the systems' responsiveness to changing conditions but also bolster their fault tolerance, allowing for more reliable operation.

In addition to these capabilities, predictive maintenance will become a crucial component of ANN-based control, enabling systems to anticipate failures and schedule repairs before problems arise. Integrating computing and cloud-based artificial intelligence will play an essential role in this transformation by facilitating rapid decision-making and enabling remote diagnostics. This means that operators can monitor systems from afar, receive real-time insights, and address issues without needing to be physically present. Furthermore, hybrid ANN approaches are expected to be popular. These methods will mix traditional control strategies, which have been tried and tested over decades, with advanced machine learning techniques. This combination will enhance stability and robustness, making these systems particularly suited for critical applications such as renewable energy generation (like solar and wind), electric vehicles, and smart grid technologies.

As the computational power of hardware continues to increase and artificial intelligence algorithms become more refined, ANN-based control systems are expected to revolutionize the field of power electronics. These advancements will lead to the development of intelligent, self-optimizing systems capable of achieving unprecedented levels of efficiency and reliability, paving the way for a more sustainable and technologically advanced future.

11 Conclusion and Chapter Summary

In this chapter, the integration of artificial neural network (ANN)-based control in power electronics is investigated and its importance in improving system performance, adaptability, and efficiency is highlighted. ANN-based control systems are rapidly transforming power electronics by offering intelligent, self-learning, and adaptive control mechanisms. Their ability to handle nonlinearities, predict system behavior, and optimize performance in real time makes them invaluable in modern power electronic applications. The implementation of ANN-based controllers has shown significant improvements in system stability, fault tolerance, and efficiency across various applications, including energy conversion, motor drives, and renewable energy systems.

Despite challenges such as computational complexity and data dependency, advancements in hardware acceleration, computing, and cloud-based AI solutions are expected to mitigate these limitations, further expanding ANN applications in power electronics. Future research will likely focus on enhancing ANN models through deep learning and reinforcement learning techniques, leading to even more robust, adaptive, and efficient control systems.

As power electronic systems continue to evolve, the role of ANN-based control will become increasingly vital, paving the way for intelligent, self-optimizing power systems that support sustainable energy solutions and the next generation of smart grids.

References

Albalawi H, El-Shimy ME, Abdel Meguid H, Kassem AM, Zaid SA (2002) Analysis of a hybrid wind/photovoltaic energy system controlled by brain emotional learning-based intelligent controller. Sustainability 14(8) https://doi.org/10.3390/su14084775

Allahmoradi S, Afrasiabi S, Liang X, Zhao J, Shahidehpour M (2024) Data-driven Volt/VAR optimization for modern distribution networks: a review. IEEE Access 12:71184–71204. https://doi.org/10.1109/ACCESS.2024.3403035

Arslan Tuncar E, Sağlam Ş, Oral B (2024) A review of short-term wind power generation forecasting methods in recent technological trends. Energy Rep 12:197–209. https://doi.org/10.1016/j.egyr.2024.06.006

Bilhan AK, Sunter S (2020) Simulation and design of three-level cascaded inverter based on soft computing method. Tehnički Vjesnik 27(2):489–496. https://doi.org/10.17559/TV-20180503115747

Bose BK (2009) The past, present, and future of power electronics [guest introduction]. IEEE Ind Electron Mag 3(2):7–11. https://doi.org/10.1109/MIE.2009.932709

Boyar A, Kabalci E, Kabalci Y (2024) Sensorless speed controller of an induction motor with MRAS-based model predictive control. Comput Electr Eng 118:109350. https://doi.org/10.1016/j.compeleceng.2024.109350

Boyar A, Kabalcı Y, Kabalcı E (2024) Elektrikli araçların hibrit enerji depolama sistemleri için akciğer performansına dayalı optimizasyon algoritması temelli kontrolör geliştirilmesi. NÖHÜ Müh Bilim Derg 13(3):3. https://doi.org/10.28948/ngumuh.1449613

Dikmen E, Ayaz M, Ezen HH, Küçüksille EU, Şahin AŞ (2014) Estimation and optimization of thermal performance of evacuated tube solar collector system. Heat Mass Transf 50(5):711–719. https://doi.org/10.1007/s00231-013-1282-0

Dong W, Li S, Fu X, Li Z, Fairbank M, Gao Y (2021) Control of a buck DC/DC converter using approximate dynamic programming and artificial neural networks. IEEE Trans Circuits Syst I Regul Pap 68(4):1760–1768. https://doi.org/10.1109/TCSI.2021.3053468

Elsheikh AH, Sharshir SW, Abd Elaziz M, Kabeel AE, Guilan W,Haiou Z (2019) Modeling of solar energy systems using artificial neural network: A comprehensive review. Sol Energy 180:622–639. https://doi.org/10.1016/j.solener.2019.01.037

Fakhryza KA, Budisusila EN, Nugroho AA (2024) Application of artificial neural network for peak load forecasting in 150 kV Semarang power system. J Electr Syst Inf Technol 11(1):40. https://doi.org/10.1186/s43067-024-00165-x

Farhangi H (2010) The path of the smart grid. IEEE Power Energy Mag 8(1):18–28. https://doi.org/10.1109/MPE.2009.934876

Farkas I, Géczy-Víg P (2003) Neural network modelling of flat-plate solar collectors. Comput Electron Agric 40(1):87–102. https://doi.org/10.1016/S0168-1699(03)00013-9

Ghritlahre HK, Prasad RK (2018) Application of ANN technique to predict the performance of solar collector systems–a review. Renew Sust Energ Rev 84:75–88. https://doi.org/10.1016/j.rser.2018.01.001

Hassabis D, Kumaran D, Summerfield C, Botvinick M (2017) Neuroscience-inspired artificial intelligence. Neuron 95(2):245–258. https://doi.org/10.1016/j.neuron.2017.06.011

Hebb DO (1949) The organization of behavior: a neuropsychological theory. John Wiley and Sons, Inc., New York, p. 335, https://doi.org/10.1002/sce.37303405110.

Ibrahim KSMH, Huang YF, Ahmed AN, Koo CH, El-Shafie A (2022) A review of the hybrid artificial intelligence and optimization modelling of hydrological streamflow forecasting. Alex Eng J 61(1):279–303. https://doi.org/10.1016/j.aej.2021.04.100

Ichikawa S, Tomita M, Doki S, Okuma S (2006) Sensorless control of permanent-magnet synchronous motors using online parameter identification based on system identification theory. IEEE Trans Ind Electron 53(2):363–372. https://doi.org/10.1109/TIE.2006.870875

Karanayil B, Rahman MF (2011) 38–artificial neural network applications in power electronics and electrical drives. In: Rashid MH (ed) Power Electronics Handbook, 3rd edn. Butterworth-Heinemann, Boston, pp 1139–1154. https://doi.org/10.1016/B978-0-12-382036-5.00038-0

Khan HS, Mohamed IS, Kauhaniemi K, Liu L (2021) Artificial neural network-based voltage control of DC/DC converter for DC microgrid applications. In: 2021 6th IEEE workshop on the electronic grid (eGRID), pp 1–6. https://doi.org/10.1109/eGRID52793.2021.9662132

Kumawat H, Tundwal P, Dave V (2023) Artificial intelligence-enabled fault detection and diagnosis for improved power quality in hybrid renewable energy systems. In: 2023 3rd international conference on advancement in electronics and communication engineering (AECE), pp 480–485. https://doi.org/10.1109/AECE59614.2023.10428200

Langer N, Bhat AH, Agarwal P (2014) Neural-network-based space-vector pulse-width modulation for capacitor voltage balancing of three-phase three-level improved power quality converter. IET Power Electron 7(4):973–983. https://doi.org/10.1049/iet-pel.2013.0216

Li H, Shi KL, McLaren PG (2005) Neural-network-based sensorless maximum wind energy capture with compensated power coefficient. IEEE Trans Ind Appl 41(6):1548–1556. https://doi.org/10.1109/TIA.2005.858282

Lin W-M, Hong C-M, Chen C-H (2011) Neural-network-based MPPT control of a stand-alone hybrid power generation system. IEEE Trans Power Electron 26(12):3571–3581. https://doi.org/10.1109/TPEL.2011.2161775

Liu B, Wei X, Sun C, Wang B, Huo W (2024) A controllable neural network-based method for optimal energy management of fuel cell hybrid electric vehicles. Int J Hydrog Energy 55:1371–1382. https://doi.org/10.1016/j.ijhydene.2023.10.215

Lodi KA, Beig AR, Al Jaafari KA, Aung Z (2024) ANN-based improved direct torque control of open-end winding induction motor. IEEE Trans Ind Electron 71(10):12030–12040. https://doi.org/10.1109/TIE.2024.3357865

Ma H, Xu D, Lee Y-S (1999) Fault diagnosis of power electronic circuits based on neural network and waveform analysis. In: Proceedings of the IEEE 1999 international conference on power electronics and drive systems, PEDS'99 (Cat. No.99TH8475), vol 1, pp 234–237. https://doi.org/10.1109/PEDS.1999.794566

Maass W (2016) Searching for principles of brain computation. Curr Opin Behav Sci 11:81–92. https://doi.org/10.1016/j.cobeha.2016.06.003

Maier HR et al (2023) Exploding the myths: an introduction to artificial neural networks for prediction and forecasting. Environ Model Softw 167:105776. https://doi.org/10.1016/j.envsoft.2023.105776

Manno A, Intini M, Jabali O, Malucelli F, Rando D (2024) An ensemble of artificial neural network models to forecast hourly energy demand. Optim Eng 25(4):2315–2343. https://doi.org/10.1007/s11081-024-09883-7

McCulloch WS, Pitts W (1943) A logical calculus of the ideas immanent in nervous activity. Bull Math Biophys 5(4):115–133. https://doi.org/10.1007/BF02478259

Mohagheghi S, Harley RG, Habetler TG, Divan D (2009) Condition monitoring of power electronic circuits using artificial neural networks. IEEE Trans Power Electron 24(10):2363–2367. https://doi.org/10.1109/TPEL.2009.2017806

Mohsen S, Bajaj M, Kotb H, Pushkarna M, Alphonse S, Ghoneim SSM (2023) Efficient artificial neural network for smart grid stability prediction. Int Trans Electr Energy Syst 2023(1):9974409. https://doi.org/10.1155/2023/9974409

Mount NJ et al (2016) Data-driven modelling approaches for socio-hydrology: opportunities and challenges within the Panta Rhei Science Plan. Hydrol Sci J 61(7):1192–1208. https://doi.org/10.1080/02626667.2016.1159683

Qiu W, Zhao X, Tyrrell A, Perinpanayagam S, Niu S, Wen G (2024) Application of artificial intelligence-based technique in electric motors: a review. IEEE Trans Power Electron 39(10):13543–13568. https://doi.org/10.1109/TPEL.2024.3410958

Raoofi T, Yildiz M (2023) Comprehensive review of battery state estimation strategies using machine learning for battery management systems of aircraft propulsion batteries. J Energy Storage 59:106486. https://doi.org/10.1016/j.est.2022.106486

Rosenblatt F (1958) The perceptron: a probabilistic model for information storage and organization in the brain. Psychol Rev 65(6):386–408. https://doi.org/10.1037/h0042519

Rumelhart DE, Hinton GE, Williams RJ (1986) Learning representations by back-propagating errors. Nature 323(6088):533–536. https://doi.org/10.1038/323533a0

Saadi M, Djalel D, Meghni B, Rekioua D (2024) Intelligent energy management strategy and sizing methodology for hybrid systems in isolated regions. Chin J Electr Eng 10(3):50–62. https://doi.org/10.23919/CJEE.2024.000091

Schmidhuber J (2015) Deep learning in neural networks: an overview. Neural Netw 61:85–117. https://doi.org/10.1016/j.neunet.2014.09.003

Shao S-Y, Sun W-J, Yan R-Q, Wang P, Gao RX (2017) A deep learning approach for fault diagnosis of induction motors in manufacturing. Chin J Mech Eng 30(6):1347–1356. https://doi.org/10.1007/s10033-017-0189-y

S. Skansi (2018) An overview of different neural network architectures. In: Skansi S (ed) Introduction to deep learning: from logical calculus to artificial intelligence. Springer International Publishing, Cham, pp 175–183. https://doi.org/10.1007/978-3-319-73004-2_10

Syed D et al (2021) Deep learning-based short-term load forecasting approach in smart grid with clustering and consumption pattern recognition. IEEE Access 9:54992–55008. https://doi.org/10.1109/ACCESS.2021.3071654

Szandała T (2021) Review and comparison of commonly used activation functions for deep neural networks. In: Bhoi AK, Mallick PK, Liu C-M, Balas VE (eds) Bio-inspired Neurocomputing. Springer, Singapore, pp 203–224. https://doi.org/10.1007/978-981-15-5495-7_11

Vandoorn TL, De Kooning JDM, Meersman B, Guerrero JM, Vandevelde L (2013) Voltage-based control of a smart transformer in a microgrid. IEEE Trans Ind Electron 60(4):1291–1305. https://doi.org/10.1109/TIE.2011.2165463

Wang D et al (2022) Model predictive control using artificial neural network for power converters. IEEE Trans Ind Electron 69(4):3689–3699. https://doi.org/10.1109/TIE.2021.3076721

Yarikkaya S, Vardar K (2023) Neural network based predictive current controllers for three phase inverter. IEEE Access 11:27155–27167. https://doi.org/10.1109/ACCESS.2023.3258679

Zarei J, Tajeddini MA, Karimi HR (2014) Vibration analysis for bearing fault detection and classification using an intelligent filter. Mechatronics 24(2):151–157. https://doi.org/10.1016/j.mechatronics.2014.01.003

Zhang L, Ling J, Lin M (2022) Artificial intelligence in renewable energy: a comprehensive bibliometric analysis. Energy Rep 8:14072–14088. https://doi.org/10.1016/j.egyr.2022.10.347

Zhang Z, Wang G, Wang K, Wang Z (2022) Application of artificial neural network in power electronic control. In: He J, Li Y, Yang Q, Liang X (eds) The proceedings of the 16th annual conference of China electrotechnical society. Springer Nature, Singapore, pp 977–989. https://doi.org/10.1007/978-981-19-1532-1_104

Zhao Z et al (2020) A dynamic particles MPPT method for photovoltaic systems under partial shading conditions. Energy Convers Manag 220:113070. https://doi.org/10.1016/j.enconman.2020.113070

Zhao S, Blaabjerg F, Wang H (2021) An overview of artificial intelligence applications for power electronics. IEEE Trans Power Electron 36(4):4633–4658. https://doi.org/10.1109/TPEL.2020.3024914

Adaptive and Robust Control Techniques in Power Electronics

Xiaoning Shen, Cheng Li, Zhuang Liu, Yabin Gao, and Jianxing Liu

Abstract Adaptive and robust control methods are essential for managing variations and disturbances in power converters. This chapter introduces a two-stage adaptive robust control framework for the grid-connected three-level neutral-point-clamped (NPC) converters. The control structure is constructed by an adaptive-gain second-order sliding mode (SOSM) controller and a switched high-gain observer (HGO). To achieve DC-link voltage regulation and instantaneous power tracking, the adaptive-gain SOSM method is deployed in two control loops, without the requirement of known bounds of the disturbance derivative. Compared with a fixed-gain SOSM approach, the adaptive gain version achieves faster transient response and improved steady-state performance. On the other hand, classical observers-based disturbance compensation technique is widely used in converter systems to reject disturbances. However, the disturbance estimation performance of the conventional disturbance observer often degrades significantly under measurement noise due to the trade-off between the immunity to measurement noise and the speed of state reconstruction. To overcome this limitation, a switched gain HGO is embed into the voltage regulation loop alongside the adaptive gain SOSM controller. By dynamically switching the observer gain, the HGO effectively suppresses performance deterioration induced by noise. Finally, a set of comparative experiments was conducted and the effectiveness of the introduced scheme is confirmed by experimental results.

Keywords Robust control · Adaptive gain · High · Gain observer · Neutral · Point · Clamped converter

Nomenclature

v_a, v_b, v_c	Grid voltages in *abc* coordinate.
i_a, i_b, i_c	Grid current in *abc* coordinate.

X. Shen · C. Li · Z. Liu · Y. Gao · J. Liu (✉)
Department of Control Science and Engineering, Harbin Institute of Technology, Harbin, P. R. China
e-mail: jx.liu@hit.edu.cn

E. Kabalci (ed.), *Modern Control Methods for Power Electronics*, Energy Systems in Electrical Engineering, https://doi.org/10.1007/978-981-92-1974-2_8

$\delta_a, \delta_b, \delta_c$	Average duty cycles in natural *abc* reference frame.
$v_{\alpha\beta}$	$\left[v_\alpha, v_\beta\right]^T$, Grid voltages in $\alpha\beta$ coordinate.
$i_{\alpha\beta}$	$\left[i_\alpha, i_\beta\right]^T$, Grid current in $\alpha\beta$ coordinate.
$\delta_{\alpha\beta}$	$\left[\delta_\alpha, \delta_\beta\right]^T$, Average duty cycles in $\alpha\beta$ coordinate.
v_{dc}	$v_{c1} + v_{c2}$, DC-link capacitor voltage.
C, R_{dc}, L	DC-link capacitor, DC-link resistance and line inductor
ω	The grid angular frequency.
SMC	Sliding mode control.
SOSM	Second-order sliding mode.
ESO	Extended state observer.
HGO	High-gain observer

Chapter Overview

This chapter examines the integration of robust and adaptive control methodologies in modern power conversion systems, focusing on their application to three-level neutral-point-clamped (NPC) converters. Section 2 discusses the current challenges and advancements in advanced control technique in power electronics. Section 3 introduces the fundamental principle of adaptive and robust control. Section 4 outlines the operational principles and topology of NPC converters. Section 5 details the design of a two-stage adaptive robust control framework, comprising of a second-order sliding mode (SOSM) controller with adaptive gain and a switched high-gain observer (HGO), while Sect. 6 validates its efficacy through experimental case studies. This chapter unites theoretical analysis with experimental verification to formulate an adaptive and robust control framework in power electronics, for tackling principal challenges in the power converter control, including dynamic performance enhancement, measurement noise tolerance, and disturbance rejection.

1 Introduction

In recent years, active front end (AFE) converters have emerged as a viable solution among various industrial applications (Jean-Pierre et al. 2023), including motor drive systems (Bharatiraja et al. 2018; Abuelnaga et al. 2023) and DC microgrids (MGs) (Vuyyuru et al. 2019; Li et al. 2022; Shen et al. 2025). Compared to the traditional uncontrollable rectifier, the AFE offers dependable DC bus voltage regulation, lower current harmonic distortion, and bi-directional power flow (Matijevic et al. 2023). As shown in Fig. 1, the grid-connected AFE converter applied in a typical dc MG is illustrated, the photovotaic systems, battery management systems, wind power and lumped dc loads are involved (Wang et al. 2021). Serving as the interface between the AC grid and the DC microgrids, the grid connected AFE rectifier plays

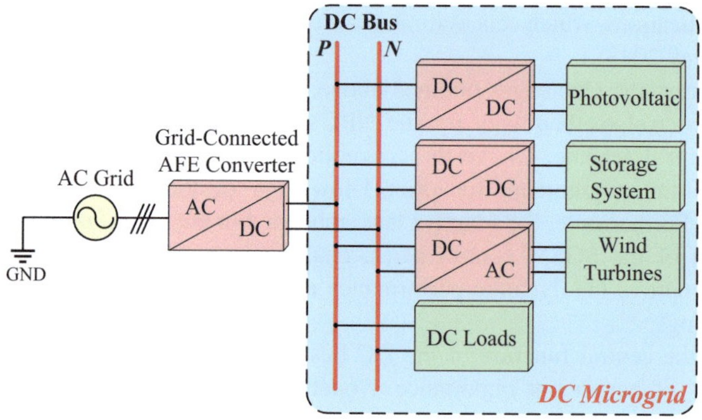

Fig. 1 Grid-connected AFE converter utilized for typical direct current MG

a critical role in overall system operation. Among the available topologies, three-level NPC converters are increasingly implemented in DC microgrids owing to its higher voltage capacity and improved waveform quality compared to a conventional two-level converter. (Li et al. 2024; Xu et al. 2022).

Control system development of three-level NPC converters remains difficult for both the researchers and industry practitioners, because the system model is highly nonlinear, and the parameter uncertainty and unknown disturbances are not easy to address. For tackling these challenges, a number of effective nonlinear control methods have been presented, including model predictive control (Zafra et al. 2024), passivity-based control (Wang et al. 2017), neural control (Liu et al. 2025), and sliding mode control (SMC) (Liu et al. 2022; Shen et al. 2024). Among these control strategies, in particular, SMC methods show promise for regulating NPC converters, owing to their advantages such as finite convergence time and strong insensitivity against parameter uncertainties (Liu et al. 2023; Yang et al. 2025). However, traditional SMC approaches commonly cause the chattering phenomenon, which negatively impacts the performance of power electronic devices (Levant 2010). An appealing approach to mitigate chattering is the super-twisting second-order sliding mode control algorithm (Shen et al. 2023), where the discontinuous function utilized in conventional SMC is smoothed through integration, thereby substantially reducing the chattering effect. In (Gong et al. 2021), a new SOSM current controller is developed for LC-coupling hybrid active power filters, achieving favorable transient response and steady-state operation. Reference (Shen et al. 2020) demonstrates the application of a SOSM controller in both voltage and current loops of the NPC converter. Additionally, SOSM controllers have been implemented in various fields, including grid-connected converters (Yang et al. 2025; Xia et al. 2019), DC-DC converters (Rakhtala and Casavola 2022), and motor drives (Lascu et al. 2020). Although SOSM controllers offer several benefits, their implementation typically requires prior knowledge of the disturbance derivative's upper bound. Moreover, it is essential to avoid excessively high gains in SOSM controllers to prevent significant amplification of chattering in actual applications, which acts as limitations on the controller's performance (Utkin and Poznyak 2013).

To address the challenges outlined above, this chapter introduces a robust and adaptive control methodology for the NPC converters, that is, the adaptive-gain SOSM control scheme. Unlike the conventional SOSM controller, this approach eliminates the requirement for prior knowledge of the disturbance derivative boundary. Furthermore, the adaptive law embedded in the algorithm prevents over-estimation of the SOSM control gains, enabling automatic adjustment of these gains to improve the dynamic performance and robustness without magnification of chattering.

Given the central function of the DC bus in DC microgrids, ensuring voltage stability is of paramount importance. Nonetheless, actions like load connection/disconnection and variations in power generation can lead to voltage fluctuation, which may adversely affect the MG's performance. To address this, incorporating the disturbance compensation term is necessary to enhance the power electronic system's robustness. For instance, reference (Lu et al. 2018) integrates an extended state observer (ESO) into the voltage loop to mitigate DC bus voltage fluctuations induced by load power variations. In (Liu et al. 2018), an ESO-based H-infinity control strategy is developed for regulating the DC-link voltage. Reference (Khalil 2017) reviews the HGO and its use for disturbance rejection in permanent magnet synchronous motors. Additionally, (Wang et al. 2019) utilizes a HGO to estimate disturbances in alternating current machines.

Although the ESO and HGO exhibit satisfactory disturbance rejection capabilities, they are susceptible to the detrimental impacts of measurement noise. Fundamentally, there is a compromise between fast disturbance estimation and increased susceptibility to measurement noise (Kwakernaak and Sivan 1972). In practical applications, due to the inherent presence of measurement noise, the observer gains cannot be excessively increased, as this would exacerbate noise amplification. Consequently, the performance of such observers is limited by the requirement to balance estimation speed and noise sensitivity.

To mitigate the impact of measurement noise, a switched HGO is incorporated into the DC-link voltage regulation loop for the three-level NPC converter to reject external disturbances. Considering the inherent noise present in the converter's measurements, the switched HGO leverages varying-gain mechanism to improve robustness against such noise. Specifically, a high gain is employed during the dynamic phase to achieve effective disturbance compensation, while a lower gain is used in the steady-state phase to suppress the adverse influence of noise.

The key contributions and innovations of this chapter are outlined below:

1. The adaptive-gain SOSM controllers are deployed for both the voltage regulation and the instantaneous power tracking. Unlike traditional SOSM approaches, this method removes the dependency on knowing the boundaries of the disturbance derivative beforehand. In addition, the adaptive algorithm enhances converter performance by dynamically adjusting gains, that is, increasing them during

transients to improve dynamic response and decreasing them in steady-state to suppress chattering.

2. A switched HGO is integrated with the adaptive-gain SOSM controller in the DC-link voltage regulation to counteract disturbances. Unlike classical linear observers, this observer utilizes distinct gains for transient and steady-state phases. High gains during transients ensure rapid disturbance estimation, while low gains in steady-state mitigate the impact of measurement noise, achieving both estimation speed and noise robustness.
3. Detailed tuning guidelines are provided for the entire control strategy, facilitating their implementation in practical applications.
4. Comparative experiments are conducted on a three-level NPC converter setup, contrasting the introduced control scheme with a representative ESO-based SOSM method. The results validate the superiority of the introduced scheme, demonstrating improved disturbance rejection capability, reduced chattering, and enhanced dynamic/steady-state performance.

2 Fundamental Principle of Robust and Adaptive Control

2.1 Robust Control

2.1.1 Sliding Mode Control

As a widely used methodology in robust control, SMC offers significant advantages. Consider the nonlinear system as follows:

$$\dot{x}(t) = f(t, x) + g(t, x)u \tag{1}$$

where $\boldsymbol{x(t)} \in \boldsymbol{R^n}$ denotes the system state vector, $\boldsymbol{u} \in \boldsymbol{R^m}$ represents the control input, $\boldsymbol{f(t, x)} \in \boldsymbol{R^n}$ and $\mathbf{g}(\boldsymbol{t}, \boldsymbol{x}) \in \boldsymbol{R^{n\times m}}$. The control input $\boldsymbol{u}$ itches according to the following logic:

$$u_i = \begin{cases} u_i^+(t, \mathbf{x}), \ s_i(t, \mathbf{x}) > 0 \\ u_i^-(t, \mathbf{x}), \ s_i(t, \mathbf{x}) < 0 \end{cases}, \quad i = 1, 2, \cdots, m \tag{2}$$

where $u_i^+(t, \boldsymbol{x})$, $u_i^-(t, \boldsymbol{x})$, and $s_i(t, \boldsymbol{x})$ are appropriate smooth continuous functions, and $s_i(t, \boldsymbol{x}) = 0$ is the i-th sliding mode variable, and

$$\mathbf{S}(\mathbf{x}) = [s_1(\mathbf{x}), s_2(\mathbf{x}), \cdots, s_m(\mathbf{x})]^T = 0 \tag{3}$$

is an (n-m)-dimensional sliding manifold.

The control task at this time is to construct continuous control functions $u_i^+(t, \boldsymbol{x})$, $u_i^-(t, \boldsymbol{x})$ and a sliding manifold $\boldsymbol{S}(\boldsymbol{x}) = 0$ such that the closed-loop system (1)–(2) exhibits a sliding motion on the (n-m)-dimensional sliding manifold $\boldsymbol{S}(\boldsymbol{x}) = 0$.

The state trajectory of a sliding mode control system consists of two key stages:

1. The reaching phase, where the system state converges toward the sliding surface. In this process, the reaching condition guarantees that, starting from any initial state, the system will arrive the sliding manifold in finite-time.
2. The sliding phase, in which the system slides along the sliding surface driven by the designed control effect.

These two phases are illustrated in Fig. 2.

Corresponding to the two phases of the motion of the SMC system, the SMC law design mainly consists of two steps:

1. Construct appropriate sliding mode variables to ensure that the system dynamics restricted to the sliding surface have the desired dynamic characteristics.
2. Design the control law so that even in the presence of uncertainties and disturbances, the sliding surface is still attractive to the system state, and the system trajectory can reach the sliding surface within a finite-time. The control law can be local or global, depending on the specific control requirements.

In the sliding mode control, a customized equation, called the sliding mode variable, is first presented. When the designed sliding mode variable equals zero, the position where the sliding mode variable is located is called the sliding surface. By skillfully designing the sliding mode variable, the system trajectory can achieve the desired closed loop performance when it reaches the sliding surface. Sliding mode

Fig. 2 Schematic diagram of sliding mode control

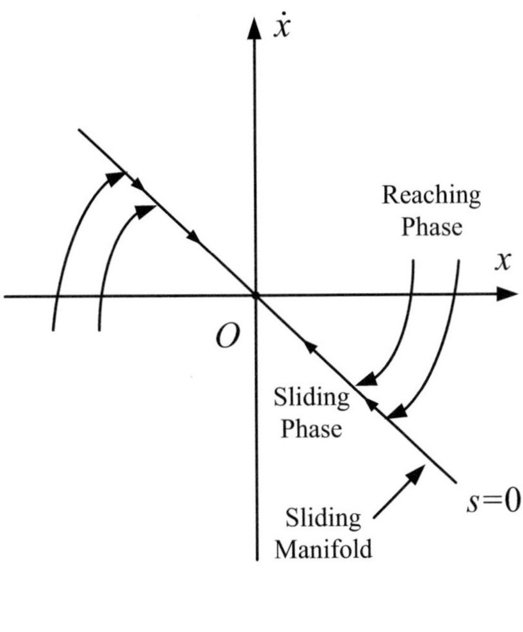

control relies on a control input that switches continuously, ensuring the trajectory converges to the sliding surface and undergoes high-frequency, low-amplitude oscillations known as "sliding mode motion." Through this control method, sliding mode control has the following advantages: (1) Insensitivity to disturbances and uncertainties; (2) Fast dynamic response; (3) The sliding mode variable can finite-time reach zero. Generally, the SMC design consists of two stages. The initial stage involves constructing a switching equation that enables the system mode to perform "sliding mode motion" on the sliding surface that meets the design requirements. The subsequent stage focuses on selecting the control law capable of driving the system state to the sliding surface in the presence of uncertainties and external disturbances.

However, this rapid switching behavior in SMC can lead to a chattering phenomenon in the system, as shown in Fig. 3. Chattering is the main drawback of SMC. On the one hand, it may damage the switching devices. On the other hand, it may also activate the high-frequency unmodeled dynamics in the system, potentially leading to instability. In response to this issue, researchers worldwide have conducted extensive studies and proposed various solutions.

(1) **Quasi-Sliding Mode Method**

This method replaces the sign function employed in SMC with a saturation function, and adopts different control methods inside and outside the boundary layer, effectively reducing the chattering of SMC. Although the quasi-sliding mode method reduces chattering, it sacrifices the control precision. Instead of converging exactly to the sliding surface, the sliding mode variable converges only to a small vicinity around zero. To overcome this problem, some scholars proposed the idea of adaptive adjustment of the boundary layer.

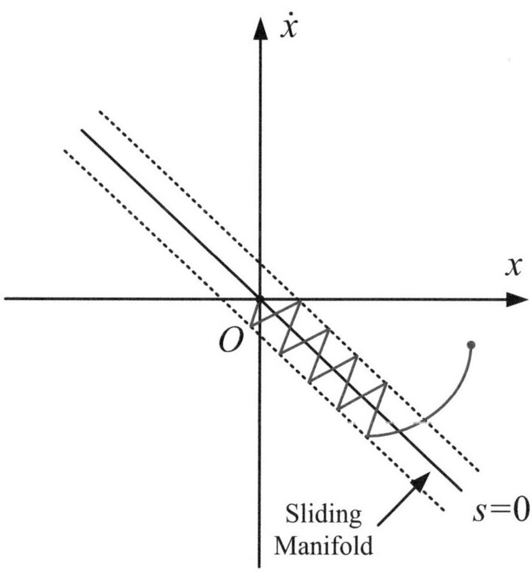

Fig. 3 System trajectory under sliding mode control

(2) **Reaching-Law Method**

Currently, the mainstream reaching-law approaches primarily consist of constant, exponential, and power-rate reaching laws. In addition, in recent years, scholars from various countries have also proposed various reaching-law variants, which have been widely applied in the industrial field and achieved good results.

(3) **Observer-Based Method**

When designing SMC for uncertain systems, to ensure robustness, the switching gains in SMC methods are typically set higher than the upper bound of the disturbance. However, the larger the switching gain, the greater the chattering. Therefore, if an observer is used to observe the disturbances and uncertainties, the gain of the SMC merely requires to exceed the upper limit of the estimation error, and thus the chattering can be significantly reduced.

(4) **Adaptive-Gain Method**

In recent years, for systems where the bounds of disturbances exist but are unknown, some scholars have proposed the adaptive-gain SMC method. The method designs the control gains of SMC in a time-varying form, which are automatically tuned based on the magnitude of the disturbance, avoiding over-estimating the value of the control gain. That is to say, the control gains are minimized to the extent that they remain adequate to reject disturbances and guarantee the establishment of the sliding motion. Currently, the existing adaptive-gain sliding mode control methods are mainly divided into three forms: (1) The first category is the method of adjusting the gain using the equivalent control value. This kind of method takes the filtered value of the equivalent control as the estimated magnitude of the disturbance. Once the sliding motion is established, the high-frequency part of the control action is suppressed via a low-pass filter and exploited for disturbance estimation in control gain design. Eventually, the value of the SMC gain consists of the filtered signal and some constants used to compensate for the possible error between the actual disturbance and its estimated value by the filter. (2) The second category is the method of designing a monotonically increasing controller gain. This kind of method keeps increasing the controller gain until the sliding motion occurs, and then fixes the gains at that value. If the disturbance increases, the SMC may become unstable. In this scenario, the control gains will continue to increase until a new sliding mode is re-established. The drawback of this method is that the control gain does not decrease. When the disturbance decreases, the control gain does not drop, so the chattering may still be significant. To overcome this drawback, some scholars have proposed the third-type method. (3) The third category is the method of designing a control gain that can both increase and decrease. This kind of method guarantees that the sliding mode variable finite-time reaches a vicinity near zero, while avoiding excessive control gain estimation, thereby mitigating chattering.

(5) **High-Order SMC Method**

In Filippov's sense, high-order sliding mode refers to a motion that evolves on a particular integral manifold of a discontinuous system. It is characterized by simultaneous convergence of the sliding variable and its successive derivatives (up to a certain order) to 0. This property can be described by introducing a new manifold, namely the sliding set.

The r-th order sliding set for the sliding surface $s(t, x) = 0$ is represented as follows:

$$s = \dot{s} = \ddot{s} = \cdots = s^{(r-1)} = 0 \tag{4}$$

The preceding equation establishes an r-dimensional constraint condition that limits the system's admissible state trajectories.

According to the Levant's definition of high-order sliding mode in 1996, the r-th order sliding set (4) must be non-empty, and supposed as a local integral set under Filippov's framework. (That is to say, it is composed of Filippov solutions of a discontinuous system). Then, the corresponding motion that satisfies (4) is known as the "r-th order sliding mode" for the sliding surface $s(t, x) = 0$.

The system has a SOSM dynamics if and only if the system trajectory is located at the intersection of $s = 0$ and $\dot{s} = 0$ in the state-space, as shown in Fig. 4.

Adopting high-order SMC presents the primary challenge of requiring increased prior information. Generally, the design of an r-th order SMC approach on the sliding surface $s = 0$ requires the information of $s, \dot{s}, \cdots, s^{(r-1)}$ (it is known that the super-twisting algorithm only requires the information of s.) Theoretically, the values of $s, \dot{s}, \cdots, s^{(r-1)}$ can be obtained through a precise robust differentiator with finite-time convergence.

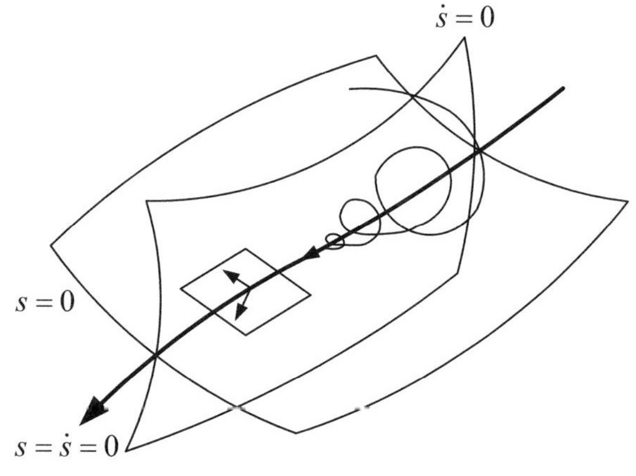

Fig. 4 Second-order sliding mode trajectory

2.1.2 H-Infinity Control

H-infinity control constitutes one of the effective methodologies in robust control. In 1981, the Canadian scholar Zames first proposed and investigated the H-infinity control problem, which immediately garnered widespread attention within the control community and experienced rapid development. To date, a relatively comprehensive theoretical framework has been established.

Originally, H-infinity control emerged from frequency domain considerations. It was observed that many design challenges in control systems can be reformulated as minimizing the H-infinity norm of a certain closed loop transfer function matrix, or ensuring that this norm remains below a prescribed positive threshold. After more than a decade of concerted effort, H-infinity control has matured into a fully developed set of frequency domain design theories and methods. Concurrently, state space approaches have elucidated the internal dynamics and structural properties of systems, offering the advantages of clear physical interpretation and suitability for computer design.

In optimal control theory, the most effective performance metrics are the H_2 norm and the H_∞ norm, both of which are formulated and defined in the frequency domain with respect to the closed loop transfer function matrix.

For a stable transfer function matrix *G(s)*, its H_2 norm is defined as follows:

$$\|\mathbf{G}\|_2 \overset{\text{def}}{=} \left(\frac{1}{2\pi}\int_{-\infty}^{\infty} \mathrm{tr}\left[\mathbf{G}^{\mathrm{H}}(\mathrm{j}\omega)\mathbf{G}(\mathrm{j}\omega)\right]\mathrm{d}\omega\right)^{1/2} \tag{5}$$

while the H_∞ norm is defined as

$$\|\mathbf{G}\|_\infty \overset{\text{def}}{=} \sup\overline{\sigma}[\mathbf{G}(\mathrm{j}\omega)] \tag{6}$$

The problem of designing a control system that both stabilizes the closed loop dynamics and minimizes the effect of disturbances on the controlled outputs can be cast as minimizing the H_∞ norm of the transfer function matrix *G(s)*. In this context, the H_∞ norm of *G(s)* serves as a performance index for system design, and the controller synthesis that seeks to minimize this index is known as H_∞ optimal control.

2.2 Adaptive Control

2.2.1 Model Reference Adaptive Control

Model reference adaptive control system comprises an ideal model (reference model) whose output specifies the desired dynamic response of the plant, thereby embodying the performance requirements. Whenever a discrepancy exists between the reference model's output and the actual system output, the adaptive mechanism (i.e., the adaptation algorithm) makes the necessary decisions, and the actuator adjusts the tunable controller parameters so as to eliminate the error. In this way, the actual system (i.e., the adjustable plant) is driven to match the dynamic response of the reference model.

Model reference adaptive control system consists of two loops. The inner loop comprises the controller and the plant, which together form the tunable system. The outer loop comprises the reference model and the adaptive mechanism. When disturbances cause the plant's behavior to deviate from the optimal trajectory specified by the reference model, a generalized error arises by comparing the reference model output with the plant output. This generalized error generates feedback according to a prescribed adaptation law and adjusts the controller parameters or produces an auxiliary control input. As a result, the tunable system's output is driven to coincide with the reference model's output, causing the generalized error to diminish toward zero.

2.2.2 Self-Tuning Control

Self-tuning control system is organized into two loops. The inner loop comprises the plant and the controller and functions analogously to a conventional feedback control system. The outer loop consists of a recursive parameter estimator and a controller parameter calculator. In this architecture, online parameter identification is seamlessly integrated with controller design. During operation, the plant's input output data are fed into the recursive parameter estimator to perform real-time plant identification. The estimated model parameters, together with a prescribed performance indicator, are then used to compute the controller parameters online, thereby synthesizing the control action so that the control objectives are ultimately achieved. In practice, the influence of stochastic disturbances and measurement noise must be taken into account, ensuring that even when the plant undergoes large variations, its dynamic behavior still meets the prescribed quality requirements.

Regarding control laws, self-tuning control systems can employ conventional strategies (including minimum-variance, PI, and PID control), minimum-variance control, quadratic optimal control, pole zero configuration, and so forth. As for parameter estimation methods, they may utilize stochastic-approximation techniques, various least-squares algorithms, auxiliary variable methods, or maximum-likelihood estimation. By combining different control laws with distinct parameter estimation techniques, a wide variety of self-tuning control algorithms can be constructed.

2.2.3 Dynamic Gain Control

The idea of dynamic-gain-based adaptive methods is to introduce a time-varying high gain (generally required to be monotonically increasing) based on the available state/output information, in order to compensate for system uncertainties and thus directly construct a satisfactory continuous-feedback controller. Consider the state-feedback stabilization of an uncertain nonlinear system:

$$\dot{x} = u + Ax^2 \tag{7}$$

where $x \in \boldsymbol{R}$ and $u \in \boldsymbol{R}$ are states and input of system, $A \in \boldsymbol{R}$ are unknown constants. On constructing the following controller

$$u = -x\left(Bx^2 + B + 1\right), \quad \dot{B} = x^2\left(1 + x^2\right), \quad B(0) > 0 \tag{8}$$

It can be observed that B is positive and monotonically increasing. This monotonicity guarantees that the dynamic gain grows sufficiently large to directly compensate for the system uncertainty, without the need to estimate the unknown parameter. Choose the Lyapunov function $V = \frac{1}{2}x^2$

$$\dot{V} =\leq -x^2 - (B - |A|)\left(x^2 + x^4\right). \tag{9}$$

By a proof-by-contradiction argument, one can show that for any initial condition, both $x(t)$ and $B(t)$ remain bounded on $[0, +\infty)$, and moreover

$$\int_0^t (x^2(s) + x^4(s))\mathrm{d}s < +\infty \tag{10}$$

Combining this result with the boundedness of $\dot{x}$ and invoking Barbalat's lemma, it follows that

$$\lim_{t \to +\infty} x(t) = 0 \tag{11}$$

3 System Description of Power Converters

The circuit configuration of the three-phase three-level NPC converter is illustrated in Fig. 5. On the alternating-current side, the utility grid is interfaced with the NPC converter through three-phase line inductors denoted as L. The direct-current side is connected to a DC load. Functioning as the interface between the AC grid and DC load, NPC converters aim to facilitate bidirectional power transfer while maintaining a stable DC-link voltage for the load side.

Assuming the AC grid operates in a balanced state and the DC-link capacitor voltages are approximately equal $v_{c1} \cong v_{c2}$, the dynamic models for AC currents and DC-link voltage in the stationary $\alpha\beta$ coordinate system can be derived based on the circuit setup shown in Fig. 5, as reported in Yin et al. (2021):

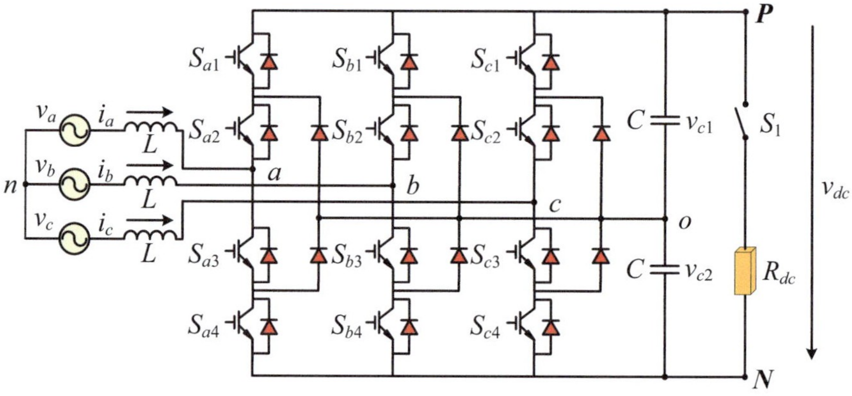

Fig. 5 Topology of the three-phase three-level NPC converter

$$L\begin{bmatrix} \frac{\mathrm{d}i_\alpha}{\mathrm{d}t} \\ \frac{\mathrm{d}i_\beta}{\mathrm{d}t} \end{bmatrix} = \begin{bmatrix} v_\alpha \\ v_\beta \end{bmatrix} - \frac{v_{dc}}{2}\begin{bmatrix} \delta_\alpha \\ \delta_\beta \end{bmatrix}, \tag{12}$$

$$C\frac{\mathrm{d}v_{dc}}{\mathrm{d}t} = \delta_\alpha i_\alpha + \delta_\beta i_\beta - \frac{2v_{dc}}{R_{dc}}. \tag{13}$$

On the other hand, the instantaneous active and reactive powers are deduced refer to Akagi and Aredes (2017):

$$p = i_{\alpha\beta}^T v_{\alpha\beta}, \quad q = i_{\alpha\beta}^T J v_{\alpha\beta}, \tag{14}$$

where $J = \begin{bmatrix} 0 & -1 \\ 1 & 0 \end{bmatrix}$.

By substituting (12) and (13)–(14), the derivatives of active/reactive powers relative to time are obtained as follows:

$$\begin{aligned} \dot{p} &= \tfrac{1}{L}\left(||v_{\alpha\beta}||^2 - \tfrac{v_{dc}}{2} v_{\alpha\beta}^T \delta_{\alpha\beta}\right) + \omega q, \\ \dot{q} &= -\tfrac{v_{dc}}{2L} \delta_{\alpha\beta}^T J v_{\alpha\beta} - \omega p, \end{aligned} \tag{15}$$

where $\|v_{\alpha\beta}\|^2 = v_\alpha^2 + v_\beta^2$.

For the power tracking control in this chapter, the direct power control technique founded on output regulation subspaces (ORS-DPC), as put forward in Vazquez et al. (2008), is applied to separately govern the active/reactive powers. When running in the steady state, the active/reactive powers maintain unchanging values. In other words, their first-order time derivatives, $\dot{p}$ and $\dot{q}$, are both zero, from which the following equations can be obtained:

$$\delta_{\alpha\beta}^* = \left(\frac{2}{v_{dc}} + \frac{2L\omega q}{v_{dc}||v_{\alpha\beta}||^2}\right) v_{\alpha\beta} - \left(\frac{2L\omega p}{v_{dc}||v_{\alpha\beta}||^2}\right) J v_{\alpha\beta}, \tag{16}$$

where $\delta^*_{\alpha\beta}$ represents the average duty cycles under a stable state, which is also referred to as the equilibrium point.

Evidently, the original main control goal has been replaced by the requirement of steering the active and reactive powers towards the equilibrium point. As a result, the control input of the power tracking loop can be formulated by:

$$\delta_{\alpha\beta} = \delta^*_{\alpha\beta} - u_p v_{\alpha\beta} - u_q J v_{\alpha\beta}, \tag{17}$$

where u_p functions as the active power control input while u_q serves as the reactive power control input. The detailed design procedures for these two controllers will be presented in the subsequent section.

4 Adaptive Robust Controller Design of Power Converters

4.1 Switched High-Gain Observer

To address the problem that the traditional observer is adversely affected by the existing measurement noise, a switched HGO is incorporated into the voltage regulation loop.

To resolve the issue of the conventional observer being subject to a negative effect with respect to the existing measurement noise, a switched HGO is integrated into the voltage regulation loop.

(1) *Observer Design:* Based on the power balancing principle, the DC-link voltage dynamics described in (13) could be reformulated as follows:

$$\frac{C}{2}\frac{\mathrm{d}v_{dc}}{\mathrm{d}t} = \frac{p - p_d}{v_{dc}}, \tag{18}$$

where p_d denotes the active power of the DC load. It is important to emphasize that the losses in the NPC converter and the line inductor are not taken into account in this formulation.

Assumption 1 The inner loop exhibits a much wider bandwidth compared to the outer loop. This indicates that when designing the voltage regulation loop (outer loop), the dynamics of the instantaneous power tracking loop (inner loop) are not factored in. It also suggests that the actual active power p is approximately equal to its reference value p^*.

Given the presence of measurement noise, Eq. (18) could be expressed as:

$$\frac{C}{2}\dot{x}_1 = u_v - x_2, \quad \dot{x}_2 = \phi(t), \quad y = x_1 + m(t) \tag{19}$$

where y denotes the output signal and $m(t)$ denotes the measurement noise.

$$x_1 = v_{dc}^2/2, \quad u_v = p^*, \quad x_2 = p_d, \tag{20}$$

Assumption 2 According to Khalil (2017), both x_2 and $\dot{x}_2$ are bounded, and $\phi(t)$ satisfies the local Lipschitz condition. The measurement noise is limited by a bound M, such that $|m(t)| \leq M$.

In light of these characteristics, a high-gain observer architecture emerges as a viable solution for estimating and counteracting the disturbance effect caused by x_2. The construction details of this observer configuration are elaborated in the following expressions:

$$\begin{aligned} \frac{C}{2}\dot{\hat{x}}_1 &= u_v - \hat{x}_2 + \frac{a_1}{\varepsilon}(x_1 - \hat{x}_1 + m(t)), \\ \dot{\hat{x}}_2 &= -\frac{a_2}{\varepsilon^2}(x_1 - \hat{x}_1 + m(t)) \end{aligned} \tag{21}$$

where a_1, a_2, and ε are the parameters of observer.

Reference (Khalil 2017) has provided the proof that, by leveraging the second-order HGO formulated in (21), one can obtain the ultimate bound of the estimation errors, which is presented as:

$$\|\tilde{x}\| \leq \frac{cM}{\varepsilon}, \tag{22}$$

where $\tilde{x} = [\tilde{x}_1, \tilde{x}_2]^T$ and $\tilde{x}_1 = x_1 - \hat{x}_1$, $\tilde{x}_2 = x_2 - \hat{x}_2$ are observation errors, c is a positive constant.

Remark 1 With the high-gain observer in place, the estimated disturbance $\hat{x}_2$ is designed to track the disturbance x_2 Nevertheless, the amplitude of $\hat{x}_2$ fluctuations during steady-state operation is tied to the ultimate bound $\|\tilde{x}\|$. According to Eq. (22), this ultimate bound $\|\tilde{x}\|$ will increase due to the measurement noise. This increase is problematic as it negatively impacts the steady-state behavior of the NPC converter. One might consider reducing this issue by increasing the parameter ε. While this does reduce the amplification, it also has an unintended consequence: the speed at which disturbances are estimated during transient periods slows down. This creates a clear trade-off that restricts the effectiveness of the HGO when employed in NPC converters. On one hand, you have the system's ability to handle measurement noise. On the other hand, you achieve the rapidness of disturbance estimation.

To tackle this problem, this chapter explores the use of a HGO based on a switched-gain approach that restricts the effectiveness of the HGO when employed in NPC converters (Ahrens and Khalil 2009). Drawing on the system dynamics presented in (19), the switched HGO is developed as follows:

$$\begin{aligned} \frac{C}{2}\dot{\hat{x}}_1 &= u_v - \hat{x}_2 + h_1(y - \hat{x}_1), \\ \dot{\hat{x}}_2 &= -h_2(y - \hat{x}_1). \end{aligned} \tag{23}$$

The switched function of parameters is developed as:

$$h_1 = \frac{a_1}{\varepsilon_1}, h_2 = \frac{a_2}{\varepsilon_1^2}, \quad \text{if } |y - \hat{x}_1| > d_s,$$
$$h_1 = \frac{a_1}{\varepsilon_2}, h_2 = \frac{a_2}{\varepsilon_2^2}, \quad \text{if } |y - \hat{x}_1| \le d_s, \tag{24}$$

where d_s, ε_1, ε_2 are three positive constants, $\varepsilon_2 > \varepsilon_1$.

The parameter d_s serves as a threshold detector to trigger the transition between the observer gain values. According to the switching criterion given in (24), two operational modes can be distinguished:

$|y - \hat{x}_1| > d_s$, the system is in a transient state. Here, a small gain parameter $\varepsilon = \varepsilon_1$ is selected to enable rapid convergence of the disturbance estimate, ensuring fast dynamic response.

$|y - \hat{x}_1| \le d_s$, the system has reached steady-state operation. A larger parameter $\varepsilon = \varepsilon_2$ is then applied to reduce the observer's sensitivity to measurement noise, enhancing steady-state accuracy.

The behavior of the switched gain mechanism is visually depicted in Fig. 6. Note that abrupt transients may cause $|y - \hat{x}_1|$ to temporarily overshoot the switching boundary $\delta_s = [-d_s, d_s]$, leading to the so-called "peaking phenomenon." As discussed in Ahrens and Khalil (2009), this can induce chattering in the ε parameter until the transient subsides. To mitigate this issue, a time-delay T_d is incorporated into the switching logic. Specifically, once $|y - \hat{x}_1|$ enters the switching zone δ_s, the gain transition is delayed by T_d to guarantee that estimation errors settle within a positively invariant set before switching occurs.

The steps for implementing the switched high-gain observer in the NPC converter are as follows:

(1) When the condition $\left|y - \hat{x}_1\right| > d_s$ is met, the observer gains are set as$h_1 = a_1/\varepsilon_1$, $h_2 = a_2/\varepsilon_1^2$, and simultaneously, the delay timer is reset.
(2) Once $\left|y - \hat{x}_1\right|$ enters the switching zone δ_s, the delay timer starts running.
(3) If$\left|y - \hat{x}_1\right| \in [-d_s, d_s]$,, after a time duration of T_d elapses, the observer gains are changed to $h_1 = a_1/\varepsilon_2$ and$h_2 = a_2/\varepsilon_2^2$.

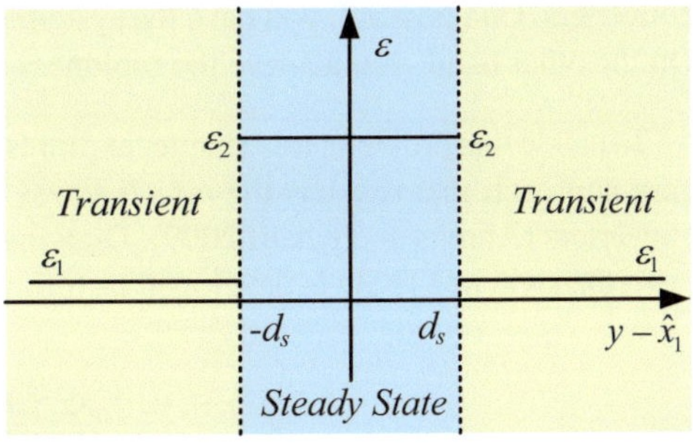

Fig. 6 Graphical representation of the switching mechanism

Lemma 1 Ahrens and Khalil 2009) There exists some scalars $\zeta_n > 0$ and $T_n > 0$ such that the trajectories of system (23) satisfy the following condition:

$$||\tilde{x}|| \leq \zeta_n, \quad \forall t \geq T_n. \tag{25}$$

(1) *Tuning Guideline of the Switched HGO:* Here are the detailed tuning guidelines for the switched High-Gain Observer, which are implemented in a two-step manner:

The first tuning stage is dedicated to calibrating a conventional HGO. Through the utilization of Eqs. (19) and (21), one can derive:

$$\begin{bmatrix} \dot{\tilde{x}}_1 \\ \dot{\tilde{x}}_2 \end{bmatrix} = \underbrace{\begin{bmatrix} -\frac{2a_1}{\varepsilon C} & -\frac{2}{C} \\ \frac{a_2}{\varepsilon^2} & 0 \end{bmatrix}}_{A} \begin{bmatrix} x_1 \\ x_2 \end{bmatrix} + \begin{bmatrix} 0 \\ 1 \end{bmatrix} \phi(t) \tag{26}$$

where matrix A possesses the Hurwitz property, and we can represent its characteristic polynomial in the following way:

$$\lambda^2 + \frac{2a_1}{\varepsilon C}\lambda + \frac{2a_2}{\varepsilon^2 C} \tag{27}$$

Following that, the calculation of the damping coefficient ξ_v and natural frequency ω_v are carried out, and the results are presented as:

$$\xi_v = a_1\sqrt{\frac{1}{2a_2C}}, \quad \omega_v = \sqrt{\frac{2a_2}{\varepsilon^2 C}} \tag{28}$$

Consequently, based on the bandwidth of the HGO specified in Gao (2003), the parameters a_1, a_2 and ε could be ascertained. In other words,

$$\omega_b = \omega_v\sqrt{(4\xi_v^4 - 4\xi_v^2 + 2)} \tag{29}$$

Typically, when the bandwidth is larger, disturbances can be reconstructed more swiftly. Nevertheless, this also results in a decrease in the system's ability to resist measurement noise.

The next stage involves tuning the switched HGO. Initially, to determine optimal values for ε_1 and ε_2, a standard HGO is implemented on the NPC converter. This standard HGO is adjusted following the procedure detailed in the first step. Through conducting experiments with various ε_1 values and assessing the corresponding dynamic performance of the observer, one can pick an appropriate ε_1 that enables a rapid transient response (for example, ε_1 is set to 0.01 in this chapter). In a similar fashion, an acceptable ε_2 value can be identified using the same method, but with the aim of minimizing the negative impact of measurement noise (for instance, ε_2 is selected as 0.5 in this chapter). It should be emphasized that the choices of ε_1 and ε_2 parameters must meet the lower-bound requirements stated in Ahrens and Khalil (2009).

Next, the delay time T_d and the parameter d_s need to be determined. Usually, the delay time T_d ought to be a relatively small number, as this ensures that ε can switch quickly during operation. However, choosing an extremely small value is not recommended. According to Ahrens and Khalil (2009), T_d must meet a certain lower-bound requirement; otherwise,ε will switch repeatedly during the peaking phenomenon.

As for the parameter d_s, its selection can be based on the final upper bound of $|y - \hat{x}_1|$ in the steady state. The initial value of the parameter d_s can be directly configured as:

$$d_s = \Gamma \tag{30}$$

where Γ represents the ultimate upper bound of $|y - \hat{x}_1|$ in the steady-state. Subsequently, the switched HGO for the NPC converter is set up based on the parameters chosen previously, and its performance as an observer is evaluated through experiments. Finally, considering the forward voltages and losses present in the actual power converter, experimental fine-tuning is necessary to ultimately attain the desired dynamic and steady-state performance.

4.2 Adaptive-Gain Second-Order Sliding Mode Control

To endow the NPC converter with excellent transient- and steady-state performance, an adaptive-gain SOSM control scheme is utilized in both the DC-link voltage regulation loop and the power tracking loop.

(1) *Adaptive-Gain SOSM Controller in Voltage Regulation Loop:* To streamline the design procedure, the sliding variable is defined as:

$$s = x_1^* - x_1 \tag{31}$$

where $x_1^* = \frac{v_{dc}^{*\,2}}{2}$ is the reference value of x_1.

In order to facilitate the controller design, system (19) could be reformulated as:

$$\dot{x}_1 = -a_v(t) + b_v(t)u_v, \tag{32}$$

where $a_v(t) = 2x_2/C$ and $b_v(t) = 2/C$. Following this, computing the time derivative of the sliding variable yields:

$$\dot{s}_v = a_v(t) - b_v(t)u_v, \tag{33}$$

By applying the principles of equivalent control theory, the control input is designed by:

$$\begin{aligned} u_v &= \alpha_v |s_v|^{1/2}\mathrm{sign}(s_v) + \nu_v + \frac{\hat{a}_v(t)}{b_v(t)}, \\ \dot{\nu}_v &= \beta_v \mathrm{sign}(s_v), \end{aligned} \tag{34}$$

where α_v and β_v are two positive constants in the fixed-gain SOSM, $\hat{a}_v(t) = 2\hat{x}_2/C$ could be derived using the switched HGO presented in Sect. 5.1. By inserting Eq. (34) into (33), $\dot{s}_v$ can be rewritten as:

$$\begin{aligned} \dot{s}_v &= -\alpha_{vb}|s_v|^{1/2}\mathrm{sign}(s_v) + \chi_v, \\ \dot{\chi}_v &= -\beta_{vb}\mathrm{sign}(s_v) + \dot{\tilde{a}}_v, \end{aligned} \tag{35}$$

Given Lemma 1, we can conclude that $\dot{\tilde{a}}_v$ is bounded and adheres to the following condition:

$$|\dot{\tilde{a}}_v| < \xi_1 \quad \forall t \geq T_n, \tag{36}$$

where $\xi_1 > 0$ is a positive constant.

Taking condition (36) into account, if the following condition is met (Shen et al. Aug. 2020), then within a finite time $t \geq T_n$, the states of system (35) will finite-time converge to the sliding mode manifold:

$$\frac{\lambda_{\min}\{G\}}{||f||_2} > \xi_1, \tag{37}$$

where

$$G = \frac{1}{2}\begin{bmatrix} \alpha_{vb}^3 + 2\alpha_{vb}\beta_{vb} & -\alpha_{vb}^2 \\ -\alpha_{vb}^2 & \alpha_{vb} \end{bmatrix}, \quad f = \begin{bmatrix} -\alpha_{vb} \\ 2 \end{bmatrix}. \tag{38}$$

While the fixed-gain SOSM controller can achieve global finite-time convergence through careful gain selection, it faces a critical dilemma: higher gains enhance dynamic responsiveness but exacerbate chattering, whereas lower gains mitigate chattering at the cost of slower transient performance. This trade-off inherently limits the controller's effectiveness in NPC converters. To overcome this limitation, this chapter presents an adaptive-gain SOSM control scheme for the converter's voltage regulation. By dynamically adjusting the controller gains based on real-time system behavior, the introduced approach effectively reconciles the competing objectives of rapid disturbance rejection and minimized steady-state oscillations, eliminating the need for static gain tuning and significantly improving overall system performance.

Unlike conventional SOSM controllers that rely on prior knowledge of the estimated disturbance error gradient boundary ξ_1, the adaptive-gain SOSM method presented here eliminates this requirement by dynamically adjusting gains based on real-time system behavior.

The design process for the adaptive-gain SOSM voltage regulator follows a similar framework to its fixed-gain counterpart; however, in this approach, α_v and β_v are redefined as time-varying gains governed by the following adaptive law (Shtessel et al. 2012):

$$\dot{\alpha}_v = \begin{cases} \varpi_1 \sqrt{\frac{\tau_1}{2}} \text{sign}(|s_v| - F_v), & \text{if } \alpha_v > \alpha_{vt} \\ \kappa_v, & \text{if } \alpha_v \leq \alpha_{vt} \end{cases}$$
$$\beta_v = m_v \alpha_v, \tag{39}$$

where ϖ_1, τ_1, F_v, κ_v, m_v, and α_{vt} are positive constants.

Lemma 2 Shtessel et al. 2012) The control gains α_v and β_v are bounded with α_v^* and β_v^* as:

$$\alpha_v < \alpha_v^*, \ \beta_v < \beta_v^*, \quad \forall t \geq 0. \tag{40}$$

where $\alpha_v^* > 0$ and $\beta_v^* > 0$ are positive constants.

Theorem 1 After finite time $t \geq T_n$, the system dynamics (35) will reach the sliding mode manifold, i.e., $|s_v| \leq \eta_1$ and $|\dot{s}_v| \leq \eta_2$, if the adaptive gains α_v and β_v are employed according to (39) and (40), where $s_v(0) > F_v$, $\alpha_v(0) > \alpha_{vt} > 0$, $\eta_1 \geq F_v$ and $\eta_2 > 0$.

Proof The complete proof of Theorem 1 could be found in reference (Shtessel et al. 2012).

Remark 2 Essentially, the gains of the adaptive-gain SOSM scheme are automatically adjusted to prevent overestimation of the control gains α_v and β_v. When implemented in the NPC converter, this allows satisfactory features to be achieved, namely rapid dynamic response and mitigated steady-state chattering. Initially, the control gains are driven to increase rapidly according to the adaptive mechanism until the SOSM is attained. Thereafter, the gains are reduced, and during this reduction, the system dynamics gradually diverge from the sliding mode manifold. Thus, a critical detector F_v is incorporated into the adaptive law to avoid further divergence from the established SOSM. Should the sliding mode variable exceed the detector's threshold $|s_v| \geq F_v$, the control gains are caused to increase reversely, compelling the system trajectories to reconverge to the sliding mode manifold and initiate the next cycle.

By implementing the advanced adaptive-gain SOSM scheme, a fast dynamic response is enabled through the use of sufficiently high adaptive gains. Simultaneously, adopting relatively low control gains effectively mitigates steady-state chattering. Consequently, this control approach leads to a significant enhancement in the NPC converter's operational performance.

An in-depth analysis is conducted on the chattering suppression of the adaptive-gain SOSM control scheme. Assume that the system states are initially distant from the sliding surface, i.e., $|s_v| > F_v$, then, in accordance with the adaptive law defined in (39), the control gains undergo a dynamic increase until s_v enters F_v. Under this condition, the gain α_v can be mathematically expressed as:

$$\alpha_v = \alpha_v(0) + \varpi_1\sqrt{\frac{\tau_1}{2}} \cdot t \tag{41}$$

Evidently, through the selection of a sufficiently large $\varpi_1\sqrt{\tau_1/2}$, substantial values are rapidly attained by the control gains, thereby enabling powerful control authority during transients and ensuring a fast dynamic response. Once $|s_v| < F_v$, the gains are decreased with the identical gradient $\varpi_1\sqrt{\tau_1/2}$ until the condition $\alpha_v \leq \alpha_{vt}$ is satisfied. Subsequently, the gains increase with a smaller gradient κ_v, and the gain α_v is expressed as:

$$\alpha_v = \alpha_{vt} + \kappa_v \cdot t \tag{42}$$

Then, the time derivative of α_v will change to the following form if the value of α_v satisfies $\alpha_v > \alpha_{vt}$

$$\dot{\alpha}_v = -\varpi_1\sqrt{\frac{\tau_1}{2}} \tag{43}$$

which indicates that the value of α_v will be diminished until $\alpha_v \leq \alpha_{vt}$ is achieved. As a result, in the steady-state, the value of α_v dynamically fluctuates within a small range around α_{vt} (the control gain β_v undergoes the same variation pattern because $\beta_v = m_v\alpha_v$).

It is important to emphasize that the parameter α_{vt} is typically calibrated to a minimal magnitude—just sufficient to preserve the finite-time convergence characteristic. This calibration ensures that during steady-state operation, the control gains α_v and β_v remain at low values. As a result, the control effort is minimized, and chattering is significantly reduced (Utkin and Poznyak 2013; Shtessel et al. 2012).

(2) *Adaptive-Gain SOSM Controller in Power Tracking Loop:* The tracking errors of power are directly utilized to construct the sliding mode variables:

$$\begin{aligned} s_p &= p^* - p, \\ s_q &= q^* - q, \end{aligned} \tag{44}$$

where p^* and q^* represent the reference values for active and reactive powers, respectively.

The adaptive-gain SOSM controllers are also employed for power tracking. Subsequently, utilizing the ORS-DPC approach, the controllers presented in (17) are formulated by:

$$\begin{aligned} u_i &= \alpha_i |s_i|^{1/2} \mathrm{sign}(s_i) + \nu_i, \\ \dot{\nu}_i &= \beta_i \mathrm{sign}(s_i), \end{aligned} \tag{45}$$

where $i \in \{p, q\}$, and the adaptive gains $\alpha_i = \alpha_i(s_i, \dot{s}_i, t)$ and $\beta_i = \beta_i(s_i, \dot{s}_i, t)$ are designed as (Shtessel et al. 2012):

$$\dot{\alpha}_i = \begin{cases} \varpi_3 \sqrt{\frac{\tau_3}{2}} \mathrm{sign}(|s_i| - F_i), & \text{if } \alpha_i > \alpha_{it} \\ \kappa_i, & \text{if } \alpha_i \leq \alpha_{it} \end{cases} \tag{46}$$

$$\beta_i = m_i \alpha_i,$$

where ϖ_3, τ_3, F_i

, κ_i, m_i, and α_{it} represent six positive constants.

By substituting (17) into (44), the derivatives of s_p and s_q are derived by:

$$\begin{aligned} \dot{s}_p &= a_p(t) - b_{pq}(t) u_p, \\ \dot{s}_q &= a_q(t) - b_{pq}(t) u_q, \end{aligned} \tag{47}$$

where $a_p(t) = \dot{p}^*$, $a_q(t) = \dot{q}^*$ and

$$b_{pq}(t) = \frac{v_{dc} \left\| v_{\alpha\beta} \right\|^2}{2L}. \tag{48}$$

By directly substituting the adaptive-gain SOSM controller from (47), the following expression can be derived:

$$\begin{aligned} \dot{s}_p &= -\alpha_p b_{pq}(t) \left| s_p \right|^{\frac{1}{2}} \mathrm{sign}(s_p) + \chi_p, \\ \dot{\chi}_p &= -\beta_p b_{pq}(t) \mathrm{sign}(s_p) + \dot{a}_p(t), \end{aligned} \tag{49}$$

and

$$\begin{aligned} \dot{s}_q &= -\alpha_q b_{pq}(t) \left| s_q \right|^{\frac{1}{2}} \mathrm{sign}(s_q) + \chi_q, \\ \dot{\chi}_q &= -\beta_q b_{pq}(t) \mathrm{sign}(s_q) + \dot{a}_q(t). \end{aligned} \tag{50}$$

Assumption 3 The functions $\dot{a}_p(t)$ and $\dot{a}_q(t)$ are bounded with.

$$\left|\dot{a}_p(t)\right| \leq \xi_p, \quad \left|\dot{a}_q(t)\right| \leq \xi_q, \tag{51}$$

where ξ_p and ξ_q are the unknown constants.

The validity of Assumption 3 is examined here. Given that the power tracking loop possesses a sufficiently higher bandwidth compared to the outer loop, it follows that $p \approx p^*$ and $q \approx q^*$. Under this approximation, one can readily obtain $\dot{a}_p(t) = \ddot{p}^*$ and $\dot{a}_q(t) = \ddot{q}^*$. Based on the system dynamics described in Eqs. (12) through (15), the second-order derivatives of the power reference signals $\ddot{p}^*$ and $\ddot{q}^*$ can be derived as:

$$\begin{aligned}
\ddot{p}^* &= \frac{-1}{2L}\left[\frac{v_{\alpha\beta}^T\delta_{\alpha\beta}}{C}\left(\delta_{\alpha\beta}^T i_{\alpha\beta} - \frac{2v_{dc}}{R_{dc}}\right) + v_{dc}v_\alpha(\dot{\delta}_\alpha + \omega\delta_\beta) + \right. \\
&\quad \left. v_{dc}v_\beta(\dot{\delta}_\beta - \omega\delta_\alpha)\right] - \frac{\omega v_{dc}\delta_{\alpha\beta}^T J v_{\alpha\beta}}{2L} - \omega^2 i_{\alpha\beta}^T v_{\alpha\beta} \\
\ddot{q}^* &= \frac{1}{2L}\left[\frac{\delta_{\alpha\beta}^T J v_{\alpha\beta}}{C}\left(\frac{2v_{dc}}{R_{dc}} - \delta_{\alpha\beta}^T i_{\alpha\beta}\right) + v_{dc}v_\alpha(\omega\delta_\alpha - \dot{\delta}_\beta) + \right. \\
&\quad \left. v_{dc}v_\beta(\omega\delta_\beta + \dot{\delta}_\alpha)\right] - \frac{\omega v_{\alpha\beta}^T}{L}\left(v_{\alpha\beta} - \frac{v_{dc}}{2}\delta_{\alpha\beta}\right) - \omega^2 i_{\alpha\beta}^T J v_{\alpha\beta}
\end{aligned} \tag{52}$$

It is evident that, when operating as expected, the functions $\dot{a}_p(t) = \ddot{p}^*$ and $\dot{a}_q(t) = \ddot{q}^*$ remain bounded, thereby validating the reasonableness of Assumption 3.

It is important to note that the finite-time convergence of the power tracking loop can also be guaranteed according to the findings in Shtessel et al. (2012).

(3) *Tuning Methodology of the Adaptive-Gain SOSM Controller:* The tuning procedure for the gains in the adaptive-gain SOSM controller is presented as follows. As an illustration, the voltage regulation loop's gain tuning can be split into two main stages.

The initial stage involves tuning the fixed-gain SOSM outlined in Tabart et al. (2018).

Based on Eq. (35), $\ddot{s}_v$ are derived as:

$$\ddot{s}_v = -\frac{1}{2}\alpha_{vb}|s_v|^{-1/2}\dot{s}_v - \beta_{vb}\frac{s_v}{|s_v|} + \dot{\tilde{a}}_v, \tag{53}$$

Assuming reaching the sliding surface, it becomes evident that $|s_v| < \Lambda$ with $\Lambda > 0$. Considering the worst scenario, where $|s_v| = \Lambda$, Eq. (53) can then be expressed as:

$$\ddot{s}_v = -\frac{1}{2}\alpha_{vb}\Lambda^{-1/2}\dot{s}_v - \beta_{vb}\frac{s_v}{\Lambda} + \dot{\tilde{a}}_v, \tag{54}$$

where the variable $\dot{\tilde{a}}_v$ is bounded and remains a small quantity when the system is in steady-state operation. Two new variables, W_1 and W_2 are introduced and defined as follows:

$$W_1 = \frac{1}{2}\alpha_{vb}\Lambda^{-1/2}, \quad W_2 = \frac{\beta_{vb}}{\Lambda} \tag{55}$$

Therefore, Eq. (54) can be simplified to:

$$\ddot{s}_v + W_1\dot{s}_v + W_2 s_v = \dot{\tilde{a}}_v, \tag{56}$$

The characteristic function can thus be expressed as:

$$\lambda^2 + W_1\lambda + W_2 = 0, \tag{57}$$

which can be rewritten as:

$$\lambda^2 + 2\xi_n\omega_n\lambda + \omega_n^2 = 0 \tag{58}$$

where ξ_n denotes the damping ratio, and ω_n denotes the natural frequency. The following tuning functions are adopted with $W_1 = 2\xi_n\omega_n$ and $W_2 = \omega_n^2$:

$$\alpha_{vb} = 4\xi_n\omega_n\Lambda^{1/2}, \quad \beta_{vb} = \omega_n^2\Lambda \tag{59}$$

and the gains of the SOSM controller could be calculated by:

$$\alpha_v = 2C\xi_n\omega_n\Lambda^{1/2}, \quad \beta_v = \frac{1}{2}C\omega_n^2\Lambda \tag{60}$$

It can be seen that the control gains α_v and β_v could be chosen based on ξ_n and ω_n. Moreover, due to practical factors such as power losses and forward voltage drops in the real converter, experimental fine-tuning is also necessary.

In the subsequent stage, the tuning process of the gain adaption law is carried out, which is further divided into three components:

(1) Set a relatively small gain α_v based on the previously discussed error dynamics approach to minimize chattering in the steady state. Subsequently, one can directly assign the parameter α_{vt} within the adaptive mechanism as $\alpha_{vt} = \alpha_v$, thereby maintaining low chattering in steady conditions (e.g., $\alpha_{vt} = 3\ \alpha_{pt,qt} = 1.41e - 6$ in this chapter).
(2) The term $\varpi_1\sqrt{\tau_1/2}$ reflects the dynamic adjustment rate of gain α_v. If $\alpha_v > \alpha_{vt}$, this adaptation process can be described as follows:

$$\alpha_v = \alpha_{vt} \pm \varpi_1\sqrt{\tau_1/2} \cdot t \tag{61}$$

To guarantee rapid dynamic response during transients, this parameter should be set to a comparatively large value. Following that the coefficient κ_v determines the growth rate of the control gain when $\alpha_v \le \alpha_{vt}$, which can be expressed as:

$$\alpha_v = \alpha_{vt} + \kappa_v \cdot t \tag{62}$$

The coefficient can be selected to be close to α_{vt}, i.e., $\kappa_v \approx \alpha_{vt}$ (for instance, $\kappa_v = 5$, $\kappa_{p,q} = 1.41 \times 10^{-6}$ in this chapter). Following this, one can directly determine the parameter m_v based on the settings provided in the proof of Theorem 1 in Sect. 5.2 (for instance, m_v= 325, $m_{p,q}$= 160 in this chapter), as described below:

$$m_v = \frac{\varpi_2}{\varpi_1}\sqrt{\frac{\tau_2}{\tau_1}} \tag{63}$$

(3) The final aspect involves determining the detector parameter F_v. To enhance the likelihood of activating a rapidly increasing control gain during transients, one can directly set the parameter as $F_v = \Lambda$. Furthermore, considering the presence of losses and forward voltage drops, an experimental trail-and-error is necessary to ensure desirable performance both in dynamic response and steady-state operation.

4.3 PI Control

To ensure the voltage balance between the DC-link capacitors, a PI controller is employed as follows:

$$\delta_{ba} = k_{pb}\left(e_{dc}^* - e_{dc}\right) + k_{ib}\int_0^t \left(e_{dc}^* - e_{dc}\right)dt \tag{64}$$

where δ_{ba} is the control input of the voltage balancing loop, k_{pb} and k_{ib} are the gains of the PI controller, and $e_{dc} = v_{c1} - v_{c2}$, e_{dc}^* is the reference value of e_{dc} which is selected as $e_{dc}^* = 0$ in this chapter.

The introduced control scheme for the NPC converter has been fully developed and the complete control block diagram is illustrated in Fig. 7.

5 Experimental Verification

This section details the execution of a series of experiments conducted on an actual three-level NPC converter setup, depicted in Fig. 8. To mimic the three-phase AC grid, a Chroma 61,830 grid simulator was employed. DC loads were attached to the DC side of the NPC converter, acting as unknown power perturbations. The control strategies were implemented using a DSP TMS320F28377D on the control board, with a sampling frequency set at 6.4 kHz. The specific parameters are listed in Table 1.

To assess the performance, a multitude of comparative experiments are performed among the ESO-based SOSM control scheme (ESO-SOSM), the ESO-based

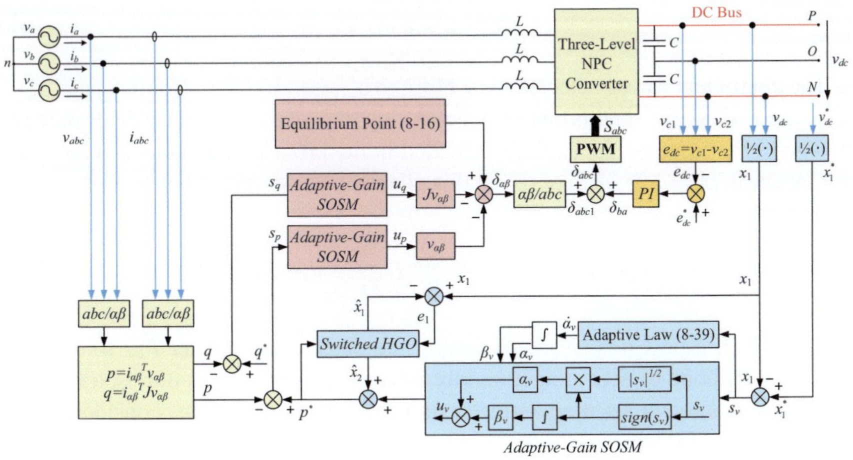

Fig. 7 Introduced switched HGO-based adaptive-gain SOSM control scheme for the three-level NPC converter

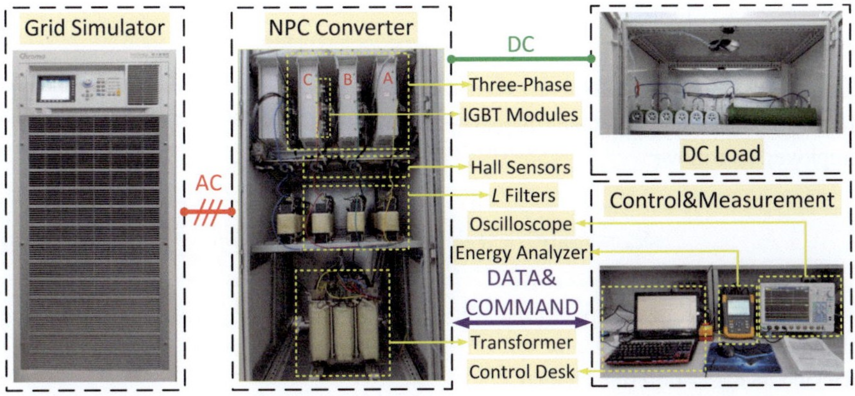

Fig. 8 Laboratory prototype of three-phase NPC converter

Table 1 Plant parameters of experiment

Description	Parameters	Values	Unites
AC voltage	v_{abc}	230	V_{rms}
Line inductor	L	2	mH
DC-link capacitor	C	6000	μF
Reference of DC voltage	v_{dc}^*	750	V
Reference of reactive power	q^*	0	kvar
Grid frequency	f	50	Hz
Switching frequency	f_{sw}	6.4	kHz
Sampling frequency	f_s	6.4	kHz

adaptive-gain SOSM scheme (ESO-AGSOSM), and the introduced control scheme. In this section, the introduced method is referred to as SHGO-AGSOSM for short. Moreover, to more thoroughly demonstrate the advantages of the SHGO-AGSOSM control strategy, it is also contrasted with high-gain versions of the ESO-SOSM method. These include ESO#5-SOSM (high-gain ESO), ESO-SOSM#5 (high-gain SOSM in the voltage regulation loop), and ESO#5-SOSM#5 (both high-gain ESO and high-gain SOSM in the voltage regulation loop). The parameters of comparative control methods are presented in Table 2.

To guarantee an equitable comparison among these control methods, each controller listed in Table 2 has undergone meticulous tuning to optimize performance. Specifically, the ESO parameters were adjusted according to the observer bandwidth as described in Gao (2003). The fixed-gain SOSM controllers were tuned following the guidelines detailed in Sect. 5.2. Meanwhile, the introduced SHGO-AGSOSM control approach was fine-tuned based on the tuning principles outlined

Table 2 Parameter setting table

Controller	Power tracking loop	Voltage regulation loop	Voltage balancing loop
ESO-SOSM	$\alpha_p = \alpha_q = 1.41e-6, \beta_p = \beta_q = 2.27e-5$, $h_p = h_q = 13$	$\alpha_v = 7, \beta_v = 5120, h_v = 16 a_1 = 6.6, a_2 = 500$ (ESO)	k_{pb} =5e -3k_{ib} =1e -5
ESO-AGSOSM	$\varpi_3 = 2.67e-3$, $\tau_3 = 0.08 F_{p,q} = 200, h_{p,q} = 13$	$\varpi_1 = 2500, \tau_1 = 0.08$, $F_v = 800, h_v = 16 a_1 = 6.6, a_2 = 500$ (ESO)	k_{pb} =5e -3 k_{ib} =1e -5
ESO#5-SOSM	$\alpha_p = \alpha_q = 1.41e-6, \beta_p = \beta_q = 2.27e-5$, $h_p = h_q = 13$	$\alpha_v = 7, \beta_v = 5120, h_v = 16 a_1 = 11.2, a_2 = 2000$ (ESO)	k_{pb} =5e -3k_{ib} =1e -5
ESO-SOSM#5	$\alpha_p = \alpha_q = 1.41e-6, \beta_p = \beta_q = 2.27e-5$, $h_p = h_q = 13$	$\alpha_v = 15, \beta_v = 6200, h_v = 16 a_1 = 6.6, a_2 = 500$ (ESO)	k_{pb} =5e -3k_{ib} =1e -5
ESO#5-SOSM#5	$\alpha_p = \alpha_q = 1.41e-6, \beta_p = \beta_q = 2.27e-5$, $h_p = h_q = 13$	$\alpha_v = 15, \beta_v = 6200, h_v = 16 a_1 = 11.2, a_2 = 2000$ (ESO)	k_{pb} =5e -3k_{ib} =1e -5
SHGO-AGSOSM	$\varpi_3 = 2.67e-3$, $\tau_3 = 0.08 F_{p,q} = 200, h_{p,q} = 13$	$\varpi_1 = 2500, \tau_1 = 0.08$, $F_v = 800, h_v = 16 a_1 = 0.28, a_2 = 1, d_s = 1850$, T_d =1e -3	k_{pb} =5e -3k_{ib} =1e -5

in Sect. 5. Notably, during experimental implementation, the "sat" function specified in Eq. (65) replaced the traditional "sign" function typically used in sliding mode control. This substitution serves to mitigate chattering effects and enhance the stability and smoothness of system operation.

$$\mathrm{sat}(s) = \begin{cases} s/h, & \text{if } |s| \le h \\ \mathrm{sign}(s), & \text{if } |s| > h \end{cases} \tag{65}$$

where $h > 0$ represents the width of the boundary layer. It's important to note that, as a general rule, a small h confers a stronger control influence and enables higher control precision. Nevertheless, this comes at the cost of increased chattering. Conversely, a wide boundary layer reduces chattering but compromises tracking accuracy. Consequently, the selection of the boundary layer width demands careful consideration, tailored to the specific requirements of each system. In this chapter, the determination of the width h is based on the tracking precision.

5.1 Comparative Studies Among SHGO-AGSOSM, ESO-AGSOSM, and ESO-SOSM Control Schemes

Initially, comparative experiments were carried out among the ESO-SOSM scheme, the ESO-AGSOSM scheme, and the introduced SHGO-AGSOSM control scheme. The behavior of the DC-link voltage was examined. Figure 9 illustrates the DC-link voltage's transient response when the reference value shifts from 750 to 650 V. Evidently, compared to the ESO-SOSM control scheme, both the ESO-AGSOSM scheme and the introduced scheme demonstrate superior transient performance, characterized by lower voltage overshoot and shorter settling times. This enhanced performance can be attributed to the rapid increase in gains during the transient phase. Additionally, a voltage step-down experiment from 650 to 750 V, depicted in Fig. 10, yields comparable results.

The robustness of the introduced control scheme was evaluated. At $t = 0.15$ s, a 150 Ω DC load was linked to the NPC converter, serving as an external disturbance. The waveforms of the DC-link voltage, three-phase AC current, and active/reactive powers are presented in Figs. 11, 12 and 13 respectively. It is evident that the introduced scheme surpasses the other two schemes in terms of disturbance rejection capacity. This superiority is well-founded. In the case of the ESO, the gains cannot be set at a sufficiently high level because of the presence of measurement noise. However, the SHGO manages to overcome this limitation.

Moreover, the steady-state control of comparative methods with a 5.3 kW power transmission case was tested (where two DC loads, $R_1 = 150$ Ω and $R_2 = 360$ Ω, were concurrently connected to the DC bus). Figure 14 displays the three-phase AC current performance. The harmonic spectra of the AC currents are illustrated

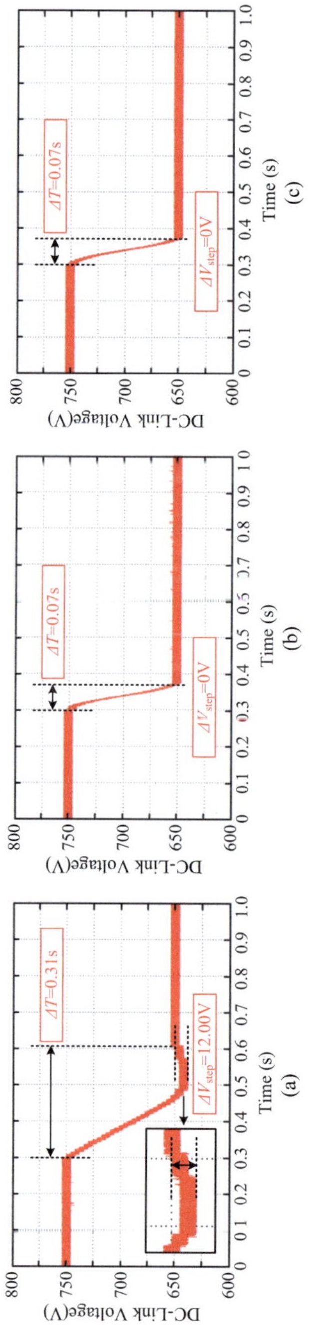

Fig. 9 DC-link voltage step comparison from 750 to 650 V **a** ESO-SOSM **b** ESO-AGSOSM **c** Introcuced SHGO-AGSOSM

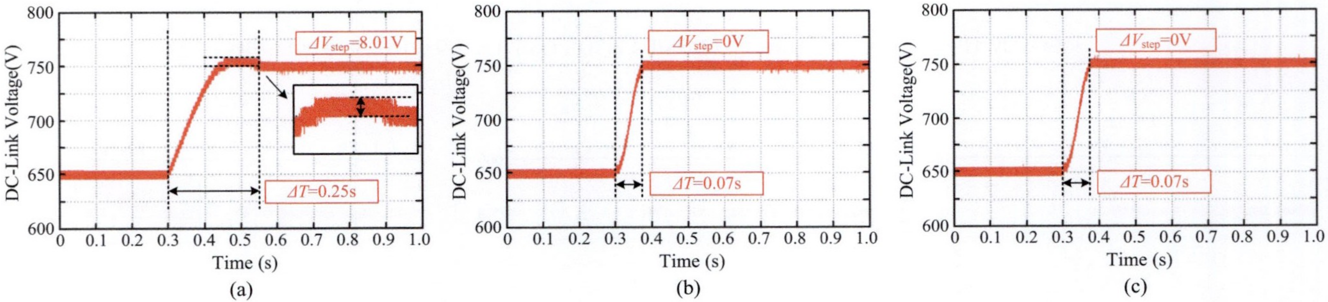

Fig. 10 DC-link voltage step comparison from 650 to 750 V **a** ESO-SOSM **b** ESO-AGSOSM **c** Introduced SHGO-AGSOSM

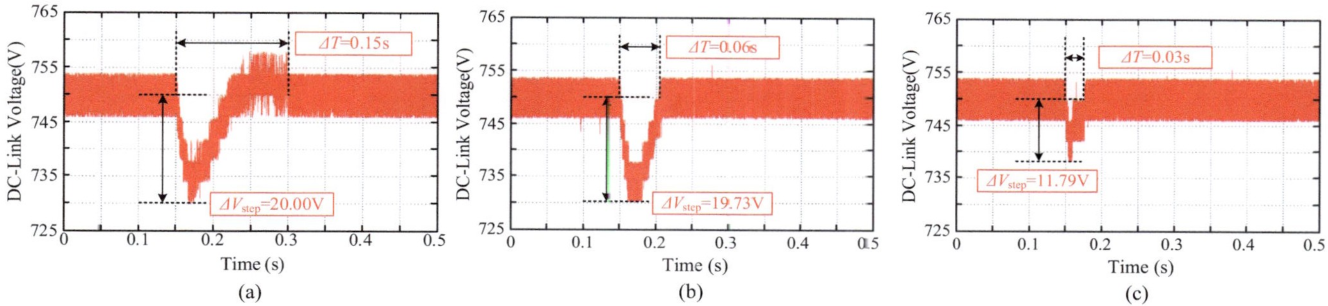

Fig. 11 The DC-link voltage comparison under a 150Ω load step **a** ESO-SOSM **b** ESO-AGSOSM **c** Introduced SHGO-AGSOSM

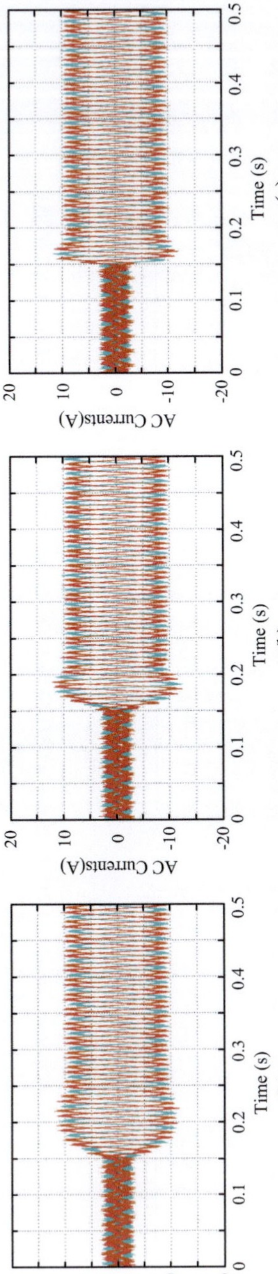

Fig. 12 Three-phase AC current waveform comparison under a 150 Ω load step **a** ESO-SOSM **b** ESO-AGSOSM **c** Introduced SHGO-AGSOSM

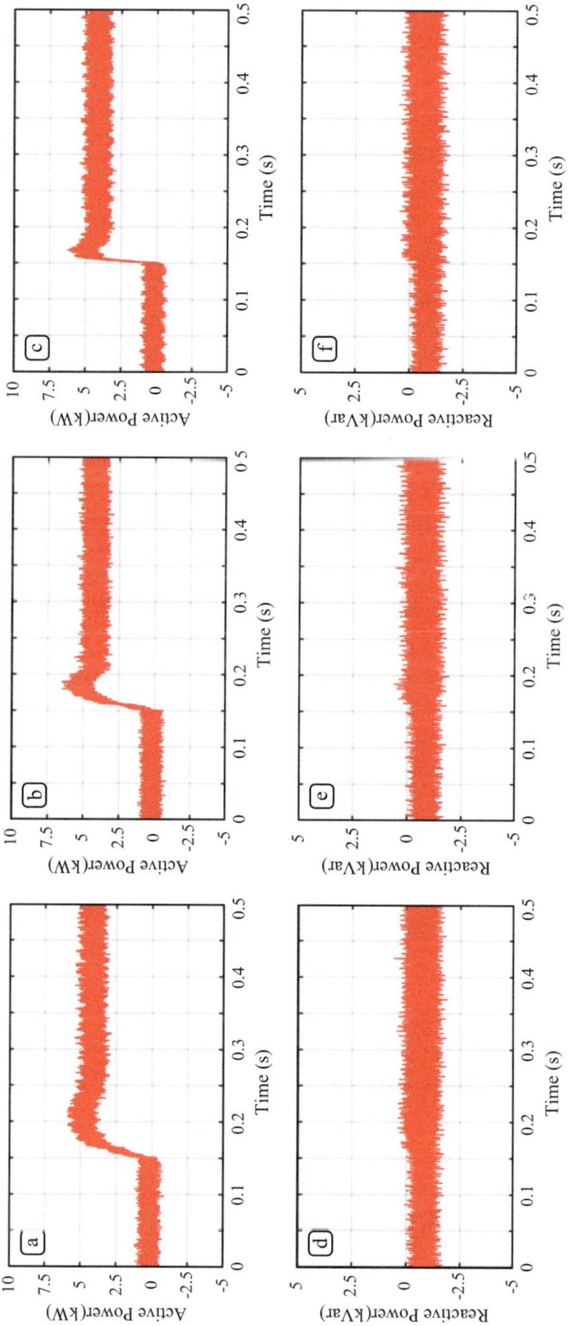

Fig. 13 The active and reactive power comparison under a 150Ω load step. Active power **a** ESO-SOSM **b** ESO-AGSOSM **c** Introduced SHGO-AGSOSM. Reactive power **d** ESO-SOSM **e** ESO-AGSOSM **f** Introduced SHGO-AGSOSM

in Fig. 15. Notably, the controller introduced in this chapter demonstrated superior steady-state performance compared to the other control approaches.

5.2 Comparative Study of SHGO-AGSOSM and ESO-SOSM Control Schemes with High Gains

In this subsection, the introduced method is compared with the high-gain ESO-SOSM schemes. Figures 16 and 17 depict the DC-link voltage stepping performance, which confirms that the employment of high gains in the SOSM controller accelerates the transient response speed. As a result, the transient performance of the ESO-SOSM#5 and ESO#5-SOSM#5 schemes is comparable to that of the introduced control scheme. Furthermore, the robustness of these approaches are also assessed. Analogously to previous experiments, a 150Ω DC-link load is attached to the converter at $t = 0.15$ s. The waveforms of DC-link voltage, three-phase AC current, and active/reactive powers are illustrated in Figs. 18, 19, and 20, respectively. Evidently, due to the high gains within the ESO-SOSM-based methods, they eventually attain a robust performance on par with the introduced method.

Conversely, the 5.3 kW power transmission steady-state behaviors of these control scheme are also showcased. Figure 21 illustrates the three-phase AC current waveforms and Fig. 22 displays the corresponding harmonic spectra. Evidently, even though the dynamic response and robustness of the high-gain ESO-SOSM controllers have been enhanced to match those of the introduced control scheme, their steady-state performance has suffered. Specifically, the total harmonic distortion (THD) values for the ESO#5-SOSM, ESO-SOSM#5, and ESO#5-SOSM#5 control approaches are 2.8%, 2.7%, and 3.4% respectively. These values are significantly higher than that of the introduced control scheme. In contrast to these fixed-gain ESO-SOSM approaches, the introduced controller leverages its adaptation law. This enables it to simultaneously achieve superior performance in disturbance rejection, as well as in both transient- and steady-state operations.

5.3 Comparative Study of SHGO-AGSOSM and ESO-SOSM Control Schemes with Increasing Gains

From the above-mentioned discussion, it is obvious that the ESO-SOSM scheme exhibits an inherent trade-off between transient- and steady-state operation. However, the introduced strategy manages to circumvent this limitation, simultaneously delivering excellent disturbance rejection capability, along with satisfactory transient- and steady-state operation. To further validate this finding, the introduced scheme is compared with ESO-SOSM controllers that feature increasing gains, rather than just

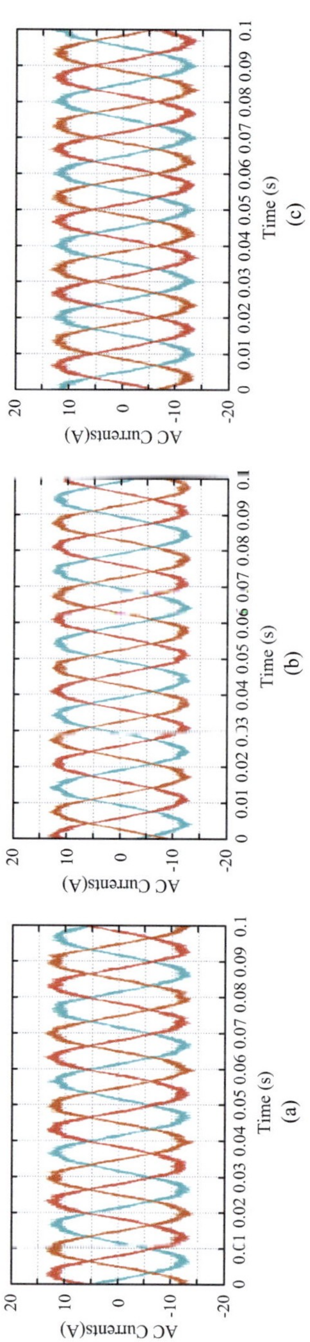

Fig. 14 Three-phase AC current waveform comparison for a 5.3 kW power transmission **a** ESO-SOSM **b** ESO-AGSOSM **c** Introduced SHGO-AGSOSM

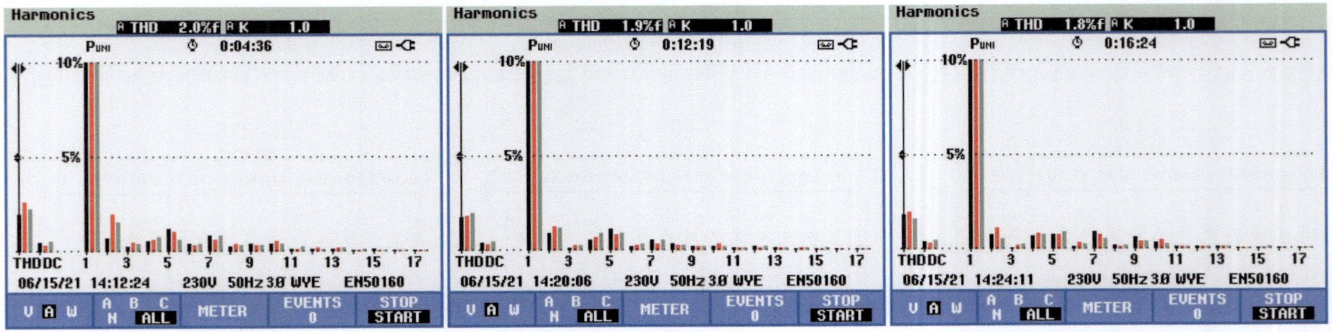

(a) (b) (c)

Fig. 15 Three-phase ac current harmonic spectrum comparison for a 5.3 kW power transmission **a** ESO-SOSM **b** ESO-AGSOSM **c** Introduced SHGO-AGSOSM

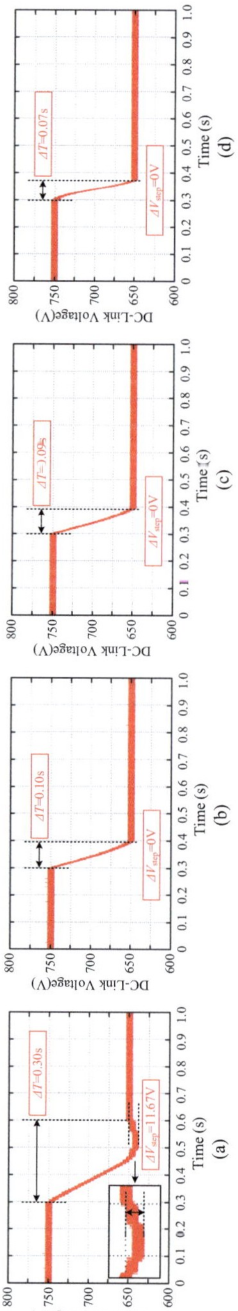

Fig. 16 DC-link voltage step comparison from 750 to 650 V **a** ESO#5-SOSM **b** ESO-SOSM#5 **c** ESO#5-SOSM#5 **d** Introduced SHGO-AGSOSM

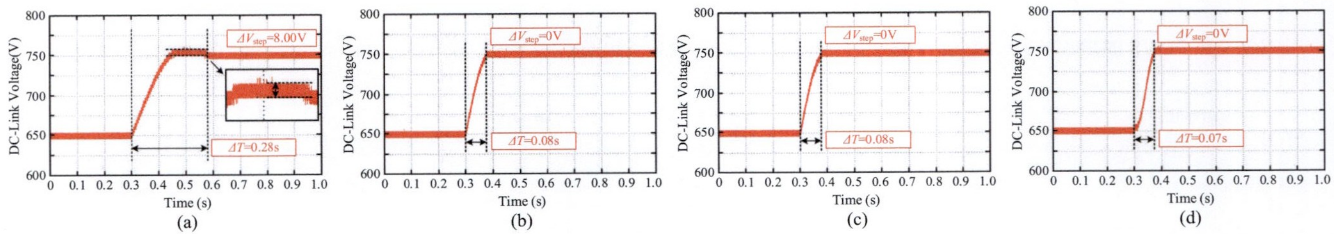

Fig. 17 DC-link voltage step comparison from 650 to 750 V **a** ESO#5-SOSM **b** ESO-SOSM#5 **c** ESO#5-SOSM#5 **d** Introduced SHGO-AGSOSM

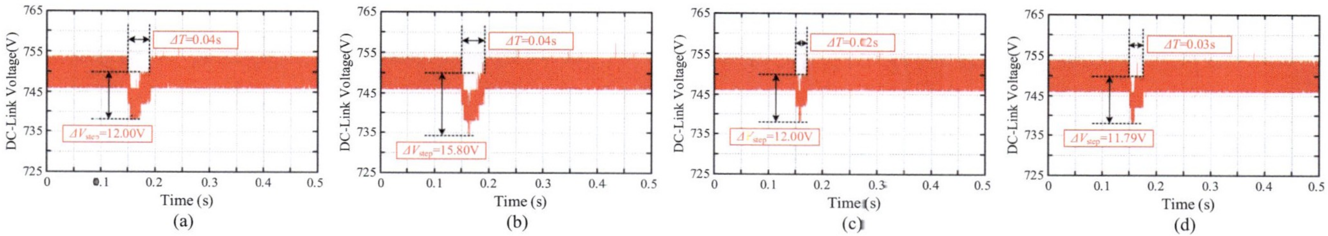

Fig. 18 DC-link voltage comparison under a 150Ω load step **a** ESO#5-SOSM **b** ESC-SOSM#5 **c** ESO#5-SOSM#5 **d** Introduced SHGO-AGSOSM

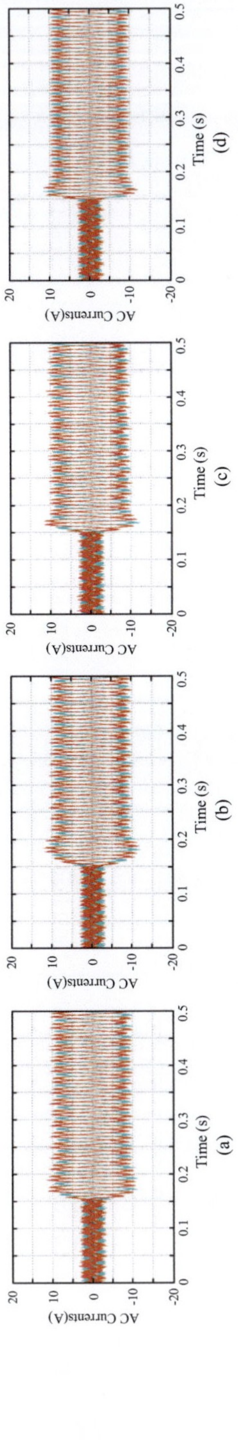

Fig. 19 Three-phase AC current waveform comparison under a 150Ω load step **a** ESO#5-SOSM **b** ESO-SOSM#5 **c** ESO#5-SOSM#5 **d** Introduced SHGO-AGSOSM

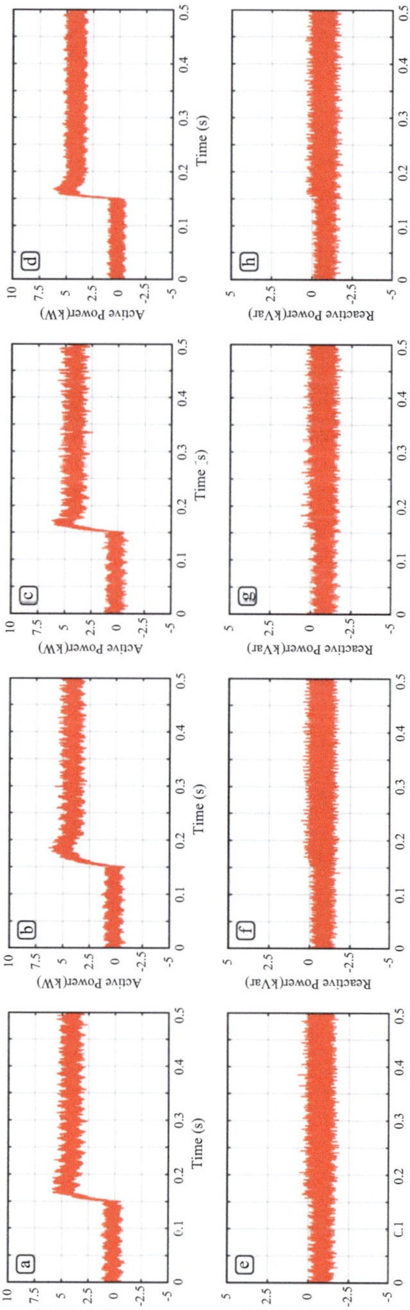

Fig. 20 The active and reactive power comparison under a 150Ω load step. Active power **a** ESO#5-SOSM **b** ESO-SOSM#5 **c** ESO#5-SOSM#5 **d** Introduced SHGO-AGSOSM. Reactive power **e** ESO#5-SOSM **f** ESO-SOSM#5 **g** ESO#5-SOSM#5 **h** Introduced SHGO-AGSOSM

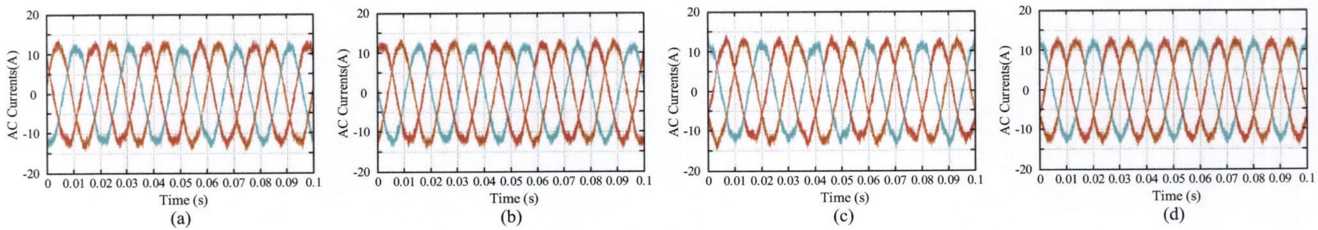

Fig. 21 Three-phase AC current waveforms for a 5.3 kW power transmission **a** ESO#5-SOSM **b** ESO-SOSM#5 **c** ESO#5-SOSM#5 **d** Introduced SHGO-AGSOSM

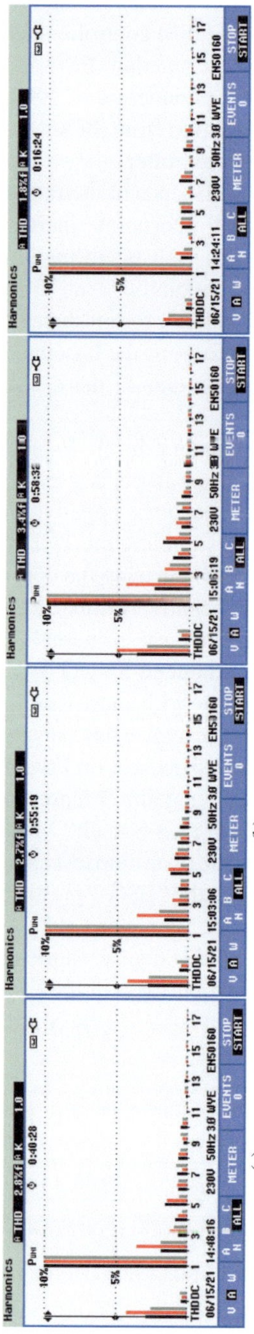

Fig. 22 Three-phase AC current harmonic spectrums for a 5.3 kW power transmission **a** ESO#5-SOSM **b** ESO-SOSM#5 **c** ESO#5-SOSM#5 **d** Introduced SHGO-AGSOSM

a single fixed-gain. To clearly demonstrate the effect of increased gains on system behavior, all experimental data have been reanalyzed and replotted as line graphs.

Table 3 lists the parameters of ESO-SOSM controllers featuring varying observer gains (where the sole difference from the standard ESO-SOSM lies in the observer gains). Likewise, Table 4 presents the parameters of ESO-SOSM controllers that vary in control gains (here, the sole variation from the standard ESO-SOSM is in the gains of the voltage regulation SOSM controller).

Figure 23 assesses the transient performance of the introduced SHGO-AGSOSM and ESO-SOSM control strategies that incorporate incrementally higher observer gains. This evaluation focuses on metrics such as settling time and overshoot voltage when the DC-link voltage reference transitions between 750 and 650 V. For simplicity in the figures, the "-SOSM" suffix is omitted; for instance, "ESO" represents "ESO-SOSM," and the observer gains escalate from the baseline ESO-SOSM to ESO#5-SOSM. Notably, as the observer gains increase, the dynamic performance shows minimal change.

Figure 24 illustrates the settling time and droop voltage of the DC-link voltage when a 150 Ω load step is applied. This graph is used to gauge the disturbance rejection capabilities of the systems. It becomes apparent that as the observer gains rise, so does the robustness. However, examining the THD values during the steady-state, as depicted in Fig. 25, reveals that increasing the observer gains also amplifies measurement noise. This amplification negatively impacts the steady-state behaviors.

Taking into account the comprehensive experimental data presented in Fig. 23 through Fig. 25, it is clear that the introduced SHGO-AGSOSM approach outperforms the ESO-SOSM control strategies with increasing observer gains. It excels simultaneously in robustness, transient response, and steady-state smoothness.

Concurrently, equivalent tests were carried out on ESO-SOSM control strategies featuring incrementally higher controller gains. Figure 26 showcases the experimental outcomes of the dynamic behavior assessment. Meanwhile, the results of the robustness evaluation and the steady-state operation test are respectively illustrated in Figs. 27 and 28. For clarity in the figures, the "ESO-" prefix is omitted; thus, "SOSM" stands for "ESO-SOSM," and the controller gains progress from the base ESO-SOSM configuration to ESO-SOSM#5. Evidently, as the controller gains increase, both the transient performance and the robustness of the ESO- SOSM-based systems improve. Nevertheless, this enhancement comes at the cost of reduced current quality

Table 3 The ESO-SOSM control scheme with different observer gains

	ESO#1-SOSM	ESO#2-SOSM	ESO#3-SOSM	ESO#4-SOSM
(a_1, a_2)	(7.52, 800)	(8.44, 1100)	(9.36, 1400)	(10.28, 1700)

Table 4 The ESO-SOSM control scheme with different controller gains

	ESO-SOSM#1	ESO-SOSM#2	ESO-SOSM#3	ESO-SOSM#4
(α_v, β_v)	(8.6, 5336)	(10.2, 5552)	(11.8, 5768)	(13.4, 5984)

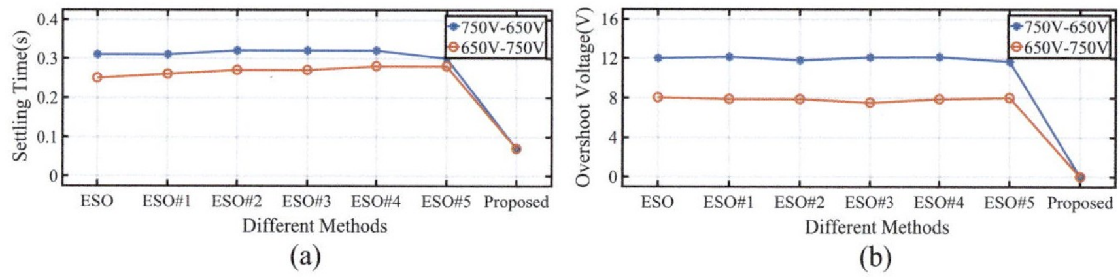

Fig. 23 DC-link voltage performance comparison **a** Settling time **b** Overshoot voltage

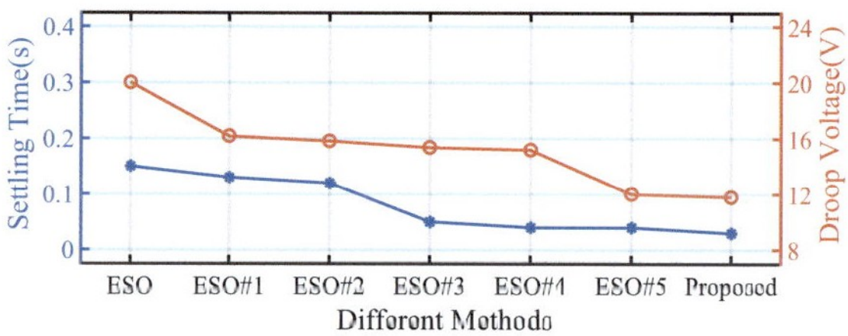

Fig. 24 The settling time and droop voltage comparisons of the DC-link voltage under a 150Ω load step

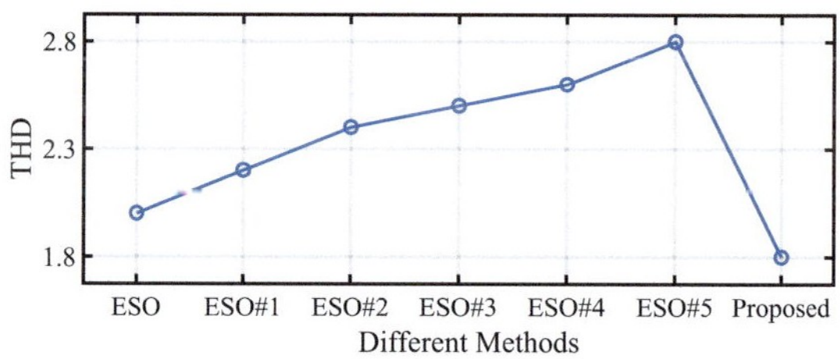

Fig. 25 The three-phase AC current THD comparison for a 5.3 kW power transmission

during steady-state operation. In contrast, the introduced SHGO-AGSOSM approach uniquely manages to achieve superior performance across all aspects: exceptional disturbance rejection capability, fast dynamic response, and outstanding steady-state operation, all at the same time.

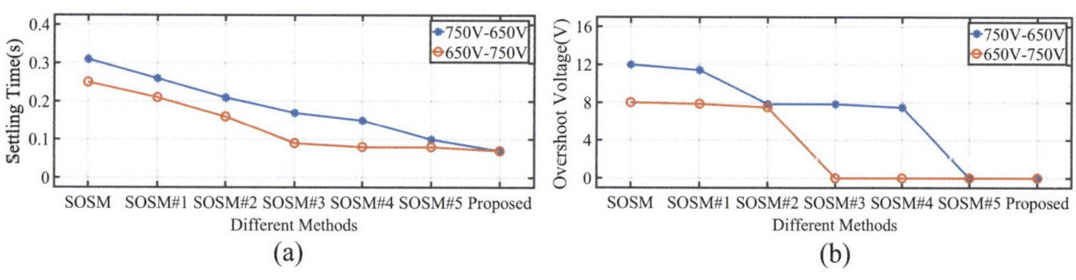

Fig. 26 DC-link voltage performance comparison **a** Settling time **b** Overshoot voltage

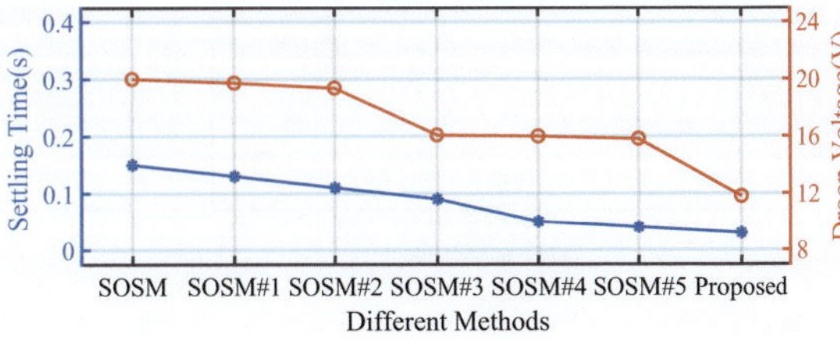

Fig. 27 The settling time and droop voltage comparisons of the DC-link voltage under a 150Ω load step

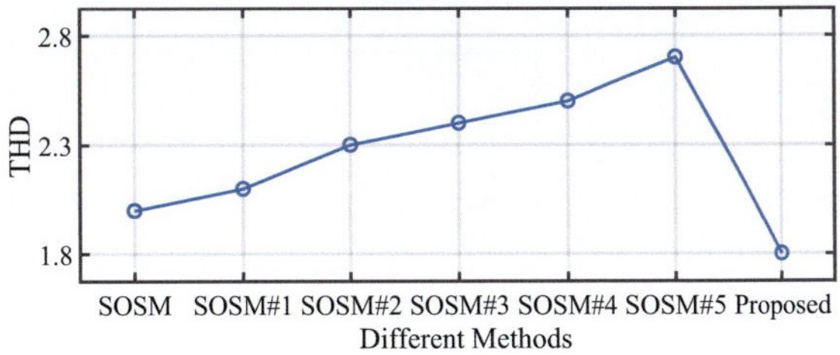

Fig. 28 The three-phase AC current THD comparison for a 5.3 kW power transmission

6 Conclusion and Chapter Summary

This chapter introduces an adaptive robust control scheme designed specifically for NPC converters. In the voltage regulation and power tracking loops, three adaptive-gain SOSM controllers are implemented. Their aims are to drive the DC-link voltage and instantaneous power towards respective reference values. Notably, this control approach does not rely on estimating the upper bound of the disturbance derivative, a parameter that is notoriously difficult to determine accurately in NPC converters. Thanks to the dynamic adaptation of the gains, this algorithm offers the NPC converter a faster dynamic response and more desirable steady-state performance compared to traditional SOSM methods. Furthermore, to mitigate the adverse impacts of external disturbances on control performance, a switched HGO is integrated into the voltage regulation loop. Unlike classical linear observers, this switched HGO leverages gain-varying mechanism, which effectively resolves the trade-off between the rapidness of disturbance estimation and tolerance to measurement noise. Consequently, it not only boosts the system's immunity against noise but maintains a strong anti-disturbance capability. The superiority of the introduced control approach has been rigorously evaluated through comparative analysis with the ESO-SOSM control strategy. Experimental results demonstrate that the introduced control scheme excels in delivering both remarkable dynamic and steady-state performance. It also exhibits outstanding disturbance rejection characteristics, making it a highly suitable option for grid-connected NPC converters.

References

Abuelnaga AH, Ni Z, Badawi S, Narimani M, Sayed-Ahmed A, Zargari NR (2023) A new active front-end control for regenerative cascaded h-bridge motor drives with filter-less interfacing capability. IEEE Trans Power Electron 38(7):8559–8570

Ahrens JH, Khalil HK (2009) High-gain observers in the presence of measurement noise: a switched-gain approach. Automatica 45(4):936–943

Akagi H, Aredes M (2017) Instantaneous power theory and applications to power conditioning. Wiley, Hoboken, NJ, USA

Bharatiraja C, Selvaraj R, Chelliah TR, Munda JL, Tariq M, Maswood AI (2018) Design and implementation of fourth arm for elimination of bearing current in NPC-MLI-fed induction motor drive. IEEE Trans Ind Appl 54(1):745–754

Gao Z (2003) Scaling and bandwidth-parameterization based controller tuning. Proc Amer Control Conf 4989–4996

Gong C, Sou W, Lam C (2021) Second-order sliding-mode current controller for LC-coupling hybrid active power filter. IEEE Trans Ind Electron 68(3):1883–1894

Jean-Pierre G, Altin N, Shafei AE, Nasiri A (2023) An expanded Lyapunov-function based control strategy for cascaded h-bridge multilevel active front-end converters. IEEE Open J. Power Electron. 4:117–127

Khalil HK (2017) High-gain observers in feedback control: application to permanent magnet synchronous motors. IEEE Control Syst Mag 37(3):25–41

Kwakernaak H, Sivan R (1972) Linear Optimal Control Systems. Wiley, Hoboken, NJ, USA

Lascu C, Argeseanu A, Blaabjerg F (2020) Super-twisting sliding-mode direct torque and flux control of induction machine drives. IEEE Trans Power Electron 35(5):5057–5065

Levant A (2010) Chattering analysis. IEEE Trans Autom Control 55(6):1380–1389

Li K, Formentini A, Dewar D, Zanchetta P (2022) Controller design of an active front-end converter keeping in consideration grid dynamic interaction. IEEE Trans Ind Electron 69(5):5195–5206

Li B, Wu X, Kish GJ, Li YR (2024) Dual 3L-NPC converter system with improved power quality for bipolar DC distribution. IEEE Trans Power Electron 39(12):16553–16565

Liu J et al (2022) Sliding mode control of grid-connected neutral-point-clamped converters via high-gain observer. IEEE Trans Ind Electron 69(4):4010–4021

Liu Z et al (2023) Fixed-time sliding mode control for DC/DC Buck converters with mismatched uncertainties. IEEE Trans. Circuits Syst. I-Regul. Pap. 70(1):472–480

Liu X, Qiu L, Fang Y, Wang K, Li Y, Rodríguez J (2025) Event-driven based reinforcement learning predictive controller design for three-phase NPC converters using online approximators. IEEE Trans Power Electron 40(4):4914–4926

Liu J, Yin Y, Luo W, Vazquez S, Franquelo LG, Wu L (2018) Sliding mode control of a three-phase AC/DC voltage source converter under unknown load conditions: industry applications. IEEE Trans Syst Man Cybern Syst **48**(10):1771–1780

Lu J, Golestan S, Savaghebi M, Vasquez JC, Guerrero JM, Marzabal A (2018) An enhanced state observer for DC-link voltage control of three-phase AC/DC converters. IEEE Trans Power Electron 33(2):936–942

Matijevic E, Sharma R, Zare F, Kumar D (2023) Extremum seeking as a tool for active damping of active front-end converters. IEEE Trans Ind Electron 70(4):3404–3413

Rakhtala SM, Casavola A (2022) Real-time voltage control based on a cascaded super twisting algorithm structure for DC–DC converters. IEEE Trans Ind Electron 69(1):633–641

Shen X et al (2020) High-performance second-order sliding mode control for NPC converters. IEEE Trans. Ind. Inform. 16(8):5345–5356

Shen X et al (2023) Adaptive-gain second-order sliding-mode control of NPC converters via super-twisting technique. IEEE Trans Power Electron 38(12):15406–15418

Shen X et al (2024) Sliding-mode control of neutral-point-clamped power converters with gain adaptation. IEEE Trans Power Electron 39(8):9189–9201

Shen X et al (2025) Fixed-time sliding mode control for NPC converters with improved disturbance rejection performance. IEEE Trans Ind Inform. https://doi.org/10.1109/TII.2025.3540481

Shtessel Y, Taleb M, Plestan F (2012) A novel adaptive-gain super-twisting sliding mode controller: methodology and application. Automatica 48(5):759–769

Tabart Q, Vechiu I, Etxeberria A, Bacha S (2018) Hybrid Energy storage system microgrids integration for power quality improvement using four-leg three-level NPC inverter and second-order sliding mode control. IEEE Trans Ind Electron 65(1):424–435

Utkin VI, Poznyak AS (2013) Adaptive sliding mode control with application to super-twist algorithm: equivalent control method. Automatica 49(1):39–47

Vazquez S et al (2008) A model-based direct power control for three-phase power converters. IEEE Trans Ind Electron 55(4):1647–1657

Vuyyuru U, Maiti S, Chakraborty C (2019) Active power flow control between DC microgrids. IEEE Trans. Smart Grid 10(5):5712–5723

Wang J, Mu X, Li Q-K (2017) Study of passivity-based decoupling control of T-NPC PV grid-connected inverter. IEEE Trans Ind Electron 64(9):7542–7551

Wang F, Wang J, Kennel RM, Rodríguez J (2019) Fast speed control of ac machines without the proportional-integral controller: using an extended high-gain state observer. IEEE Trans Power Electron 34(9):9006–9015

Wang R, Sun Q, Hu W, Li Y, Ma D, Wang P (2021) SoC-based droop coefficients stability region analysis of the battery for stand-alone supply systems with constant power loads. IEEE Trans Power Electron 36(7):7866–7879

Xia J, Guo Y, Zhang X, Jatskevich J, Amiri N (2019) Robust control strategy design for single-phase grid-connected converters under system perturbations. IEEE Trans Ind Electron 66(11):8892–8901

Xu X, Zheng Z, Wang K, Yang B, Li Y (2022) A carrier-based common-mode voltage elimination method for back-to-back three-level NPC converters. IEEE Trans Power Electron 37(3):3040–3052

Yang R, Gong Y, Paszke W (2025) ILC-based tracking control for linear systems with external disturbances via an SMC scheme. IEEE Trans Autom Sci Eng 22:9698–9707

Yin Y et al (2021) Observer-based sliding-mode control for grid-connected power converters under unbalanced grid conditions. IEEE Trans Ind Electron 69(1):517–527

Zafra E et al (2024) Computationally efficient sphere decoding algorithm based on artificial neural networks for long-horizon FCS-MPC. IEEE Trans Ind Electron 71(1):39–48

Sensorless Control of Power Converters

Aydin Boyar, Yasin Kabalci, and Ersan Kabalci

Abstract This chapter provides a comprehensive overview of sensorless control techniques for power converters by focusing on electric machine drive applications. The chapter covers the fundamental principles and implementation methods of high-performance motor drive systems that are critical in modern industrial applications. Various control strategies including Scalar Control, Direct Torque Control (DTC), Field-oriented control (FOC), and Model Predictive Control (MPC) are analytically discussed. The control principles and dynamic performances are analyzed in detail. The importance of independent control solutions over traditional mechanical sensor-based approaches is growing in response to rising efficiency expectations and the need to reduce operational costs in industrial automation systems. It is evident that sensorless control systems are preferable in industrial applications when considering the reliability problems, maintenance expenses, environmental durability constraints, and failure risk factors associated with mechanical sensors. In addition, widely used sensorless prediction methods such as Model Reference Adaptive System (MRAS) and Sliding Mode Observer (SMO) are investigated. These methods are evaluated in terms of their robustness, prediction accuracy, and computational requirements. Additionally, induction motor drive application studies designed with sensorless control methods are examined and detailed analyses are included in the chapter.

Keywords Sensorless control · Induction motor · Electric motor control

A. Boyar (✉) · E. Kabalci
Department of Electrical and Electronics Engineering, Faculty of Engineering and Architecture, Nevsehir Haci Bektas Veli University, Nevsehir, Turkey
e-mail: aydinboyar@nevsehir.edu.tr

E. Kabalci
e-mail: kabalci@nevsehir.edu.tr

Y. Kabalci
Department of Electrical and Electronics Engineering, Faculty of Engineering, Nigde Ömer Halisdemir University, Nigde, Turkey
e-mail: yasinkabalci@ohu.edu.tr

E. Kabalci (ed.), *Modern Control Methods for Power Electronics*, Energy Systems in Electrical Engineering, https://doi.org/10.1007/978-981-92-1974-2_9

Nomenclature

$\hat{i}_{s\alpha}, \hat{i}_{s\beta}$	α, β Components of the estimated stator current
$\widehat{\Psi}_{r\alpha}, \widehat{\Psi}_{r\beta}$	α And β components of the estimated rotor flux
$\widehat{\omega}_{MRAS^{CC}}$	Estimated speed of the induction motor with $MRAS^{CC}$
$\widehat{\omega}_{MRAS^{EMF}}$	Estimated speed of the induction motor with $MRAS^{EMF}$
$\widehat{\omega}_{MRAS^{RF}}$	Estimated speed of the induction motor with $MRAS^{RF}$
$\widehat{\omega}_{MRAS^{RP}}$	Estimated speed of the induction motor with $MRAS^{RP}$
$\widehat{\omega}_m$	Predicted speed
$\widehat{\omega}_{SMO}$	Estimated speed of the induction motor with SMO
$\widehat{\omega}^e_{SMO}$	Speed estimate before the low-pass filter with the SMO
$e_m, \hat{e}_m$	Estimated electromotive forces
$i^*_{sd,q}$	Reference stator current components
i^p_s	Estimated of the stator current
$V^*_{sd,q}$	Reference stator voltage components
$\widehat{\Psi}_r$	Estimated the rotor flow
$\widehat{\Psi}_s$	Estimated the stator flux
Ψ^p_s	Next stator flux value
ω^*_m	Reference speed
$\Delta V_{c1}, \Delta V_{c2}$	Voltage difference of each capacitor
μ	Auxiliary variable
ANPC	Active neutral-point clamped
C	Constant coefficient
CCS	Continuous control set
DTC	Direct torque control
EMF	Electromotive force
FCS	Finite control set
$f_{nominal}$	Nominal frequency
FOC	Field-oriented control
i_a, i_b, i_c	Phase currents
$i_{m\alpha}, i_{m\beta}$	Components of the magnetizing current
i_s, i_r	Stator and rotor current
i_{sd}, i_{sq}	Dq components of stator current
$i_{s\alpha}, i_{s\beta}$	α,β Components of the stator current
KF	Kalman filter
K_p, K_i	Proportional and integral coefficients of the PI
K_v	Coefficient
LO	Luenberger observer
LPF	Low-pass filter
L_s, L_r, L_m	Stator, rotor and magnetizing inductances
MPC	Model predictive control
MPCC	Model predictive current control
MPTC	Model predictive torque control
MRAS	Model reference adaptive system

$MRAS^{CC}$	Stator current-based MRAS
$MRAS^{EMF}$	Electromotive force MRAS
$MRAS^{RF}$	Rotor flux-based MRAS
$MRAS^{RP}$	Reactive power-based MRAS
P-, Q-, X-, Y-MRAS	Active power, reactive power and virtual variable-based MRAS
$r, \hat{r}$	Components of the motor such as flux
RMSE	Root mean square error
R_s, R_r	Stator and rotor resistances
S_a, S_b, S_c	Each phase switching states
SMO	Sliding mode observer
S_ω, S_μ	Speed and auxiliary variable switching functions
T, p_p, ω_m	Electromagnetic torque, number of poles and speed
$T_{nominal}$	Nominal torque
T_p	Estimated torque value
V/f	Voltage/frequency
V_{ma}	Adaptive model in Q-,P-,X-,Y-MRAS
V_{mr}	Reference model in Q-,P-,X-,Y-MRAS
V_{rd}, V_{rq}	Dq components of rotor voltage
V_s	Stator voltage
$V_{s(nominal)}$	Nominal stator voltage
V_{sd}, V_{sq}	Dq components of stator voltage
$V_{s\alpha}, V_{s\beta}$	α And β components of the stator voltage
z	Stator voltage and/or stator current
ε_ω	Error rate
$\varepsilon_{\omega CC}$	Error rate in $MRAS^{CC}$
$\varepsilon_{\omega EMF}$	Error rate in $MRAS^{EMF}$
$\varepsilon_{\omega RF}$	Error rate of the fluxes
$\varepsilon_{\omega RP}$	Error rate in $MRAS^{RP}$
θ_r	Angle between the d axis and the α axis
θ_{sr}	Angle between the stator and rotor flow vectors
λ_c, λ_T	Weighting factor
Ψ_m	Flux in the air gap
Ψ_{rd}, Ψ_{rq}	Dq components of the rotor flux
$\Psi_{r\alpha}, \Psi_{r\beta}$	α And β components of the rotor flux
Ψ_s, Ψ_r	Stator and rotor fluxes
ω_0, μ_0	Constant coefficients
$\omega_{m(nominal)}$	Nominal speed
ω_{sl}	Slip speed of the motor

1 Electric Motor Control Methods

The control of AC motors involves generating appropriate control signals through the inverter, which functions as a motor drive system, to regulate the desired torque or speed. The control strategies implemented in motor drives can be broadly classified into three categories: scalar control, vector control, and Model Predictive Control (MPC), as illustrated in Fig. 1. Scalar control primarily includes techniques such as Voltage-to-Frequency (V/f) control and basic torque control, while vector control encompasses more advanced approaches like field-oriented control (FOC) and direct torque control (DTC). Although scalar methods are simpler and easier to implement, they are generally less efficient and less reliable compared to vector control strategies. Consequently, scalar control is not the preferred choice in high-performance industrial applications where precision and dynamic response are critical (Kabalcı and Boyar 2021; Poorfakhraei et al. 2021).

The equivalent value of the induction motor is mathematically expressed by Eqs. 1–5 where V_s is stator voltage; R_s and R_r represent the resistances of the stator and rotor; L_s, L_r,L_m represent stator, rotor, and magnetizing inductances; Ψ_s, Ψ_r denote stator and rotor fluxes; i_s, i_r define stator and rotor current values; T,p_p, ω_m represent electromagnetic torque, number of poles, and speed, respectively (Merrassi et al. 2021; Rodriguez et al. 2022).

$$V_s = R_s i_s + \frac{d\psi_s}{dt} \tag{1}$$

$$0 = R_r i_r + \frac{d\psi_r}{dt} + j\omega_r \psi_r \tag{2}$$

$$\psi_s = L_s i_s + L_m i_r \tag{3}$$

$$\psi_r = L_m i_s + L_r i_r \tag{4}$$

$$T = \frac{3}{2} p_p Im\{\overline{\psi}_s i_s\} \tag{5}$$

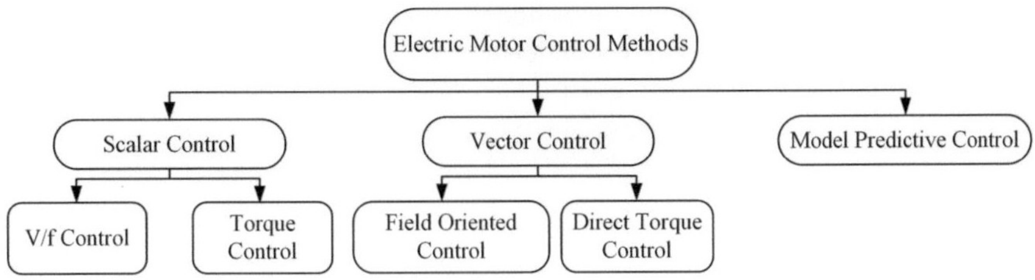

Fig. 1 Electric motor control methods

In solving induction motor drive problems and developing estimation or control algorithms, it is more appropriate to express the induction motor as a state space model. Thus, any system expressed by higher order differential equations can be converted to first-order differential equations. In addition, the system behavior can be easily defined, and it is an effective method used to convert it to a simulation model in a computer environment. For the steady and transient conditions in continuous time, the induction motor can be stated as in Eq. 7 according to the general expression of the state space model given in Eq. 6 where V_{sd} and V_{sq} represent the d and q components of stator voltage, V_{rd} and V_{rq} are the rotor voltages, i_{sd} and i_{sq} represent the stator current, Ψ_{rd} and Ψ_{rq} denote the rotor flux, and $L_\sigma = L_s\text{-}L_m{}^2/L_r$ (Alsofyani and Idris 2013; Boyar 2024).

$$\dot{x} = Ax + Bu \tag{6}$$

$$\underbrace{\begin{bmatrix} \dot{i}_{sd} \\ \dot{i}_{sq} \\ \dot{\psi}_{rd} \\ \dot{\psi}_{rq} \\ \dot{\omega}_m \end{bmatrix}}_{\dot{x}} = \underbrace{\begin{bmatrix} -\left(\frac{R_s}{L_\sigma} + \frac{L_m^2 R_r}{L_\sigma L_r^2}\right) & 0 & \frac{L_m R_r}{L_\sigma L_r^2} & \frac{\omega_m L_m}{L_\sigma L_r} & 0 \\ 0 & -\left(\frac{R_s}{L_\sigma} + \frac{L_m^2 R_r}{L_\sigma L_r^2}\right) & -\frac{\omega_m L_m}{L_\sigma L_r} & \frac{L_m R_r}{L_\sigma L_r^2} & 0 \\ \frac{R_r}{L_r} L_m & 0 & -\frac{R_r}{L_r} & -\omega_m & 0 \\ 0 & \frac{R_r}{L_r} L_m & \omega_m & -\frac{R_r}{L_r} & 0 \\ 0 & 0 & 0 & 0 & 0 \end{bmatrix}}_{A}$$

$$\underbrace{\begin{bmatrix} i_{sd} \\ i_{sq} \\ \psi_{rd} \\ \psi_{rq} \\ \omega_m \end{bmatrix}}_{x} + \underbrace{\begin{bmatrix} \frac{1}{L_\sigma} & 0 \\ 0 & \frac{1}{L_\sigma} \\ 0 & 0 \\ 0 & 0 \\ 0 & 0 \end{bmatrix}}_{B} \underbrace{\begin{bmatrix} V_{sd} \\ V_{sq} \end{bmatrix}}_{u} \tag{7}$$

1.1 Scalar Control

V/f control, which is included in the category of scalar control techniques, was first used in the 1960s to control induction motor speed in both open-loop and closed-loop setups. This control technique is appropriate for open-loop applications such as ventilation systems and fans, where accurate speed tracking is not required. On the other hand, closed-loop systems control the speed and torque of the motor by using both voltage and frequency modulation. The V/f approach, which is widely used in induction motor drives, improves system performance as well as motor dynamic response. The advantages include a simple design, ease of implementation, minimal steady-state inaccuracy, and a low total cost. Furthermore, this control technique necessitates a low initial current and enables frequency and voltage modifications to regulate acceleration and deceleration of the motor. One of the significant advantages of this strategy is that it does not require an extensive knowledge of the motor specifications (Boyar 2024; Hannan et al. 2018; Hughes and Drury 2013; Manias 2017; Trzynadlowski 2001).

The basic principle of V/f control is to acquire the magnetic flux in the air gap at the maximum level by keeping the voltage/frequency ratio constant. Although there is no connection between voltage and frequency, the connection of magnetic flux with voltage and frequency is seen in Eq. 8 where Ψ_m represents the flux in the air gap, and K_v represents the coefficient between V_s and f. Since the speed of the motor is dependent on the stator voltage, adjusting the speed to get maximum flux requires figuring out the stator voltage frequency while maintaining a consistent V/f ratio. To obtain a flux that provides maximum torque, the V/f ratio must be constant throughout all frequencies. The voltage-frequency characteristic of the induction motor when the V/f ratio is constant is seen in Fig. 2a where $V_{s(nominal)}$ denotes the nominal stator voltage and $f_{nominal}$ represents the nominal frequency. When the V/f ratio is constant, the speed of the induction motor can be changed in a wide range except for speeds close to zero. When the V/f ratio is constant, the torque-speed characteristic of the induction motor is seen in Fig. 2b where $T_{nominal}$ represents the nominal torque and $\omega_{m(nominal)}$ represents the nominal speed. It is seen that the induction motor can generate almost the same maximum torque at all speeds up to the nominal speed, except for very low speed values. This region is known as constant torque, and the maximum torque can be obtained from the motor regardless of the speed up to the nominal speed. Above the nominal speed, the flux is inversely proportional to the frequency and decreases. Since the stator current is limited, the torque to be produced decreases inversely proportional to the speed. Consequently, this area is referred to as the constant power area (Boyar 2024; Hannan et al. 2018; Hughes and Drury 2013; Manias 2017; Trzynadlowski 2001).

$$\psi_m \cong \frac{V_s}{f} \cong K_v \tag{8}$$

The general structure of V/f control applied to induction motors is seen in Fig. 3. The error ratio between the motor speed and the reference speed (ω_m^*) is applied to

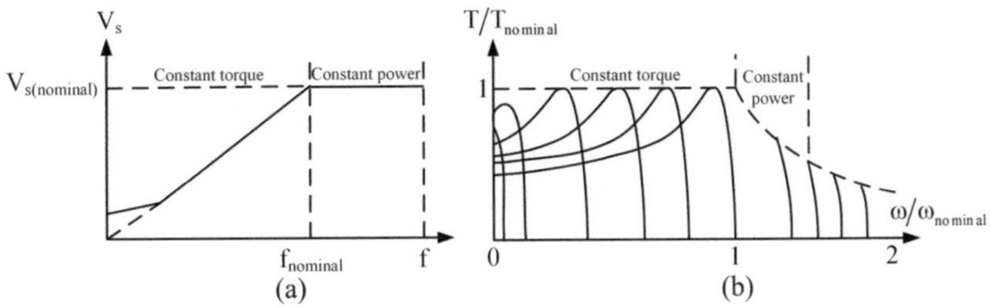

Fig. 2 **a** Voltage-frequency and **b** Torque-speed characteristics of the induction motor when the V/f ratio is constant

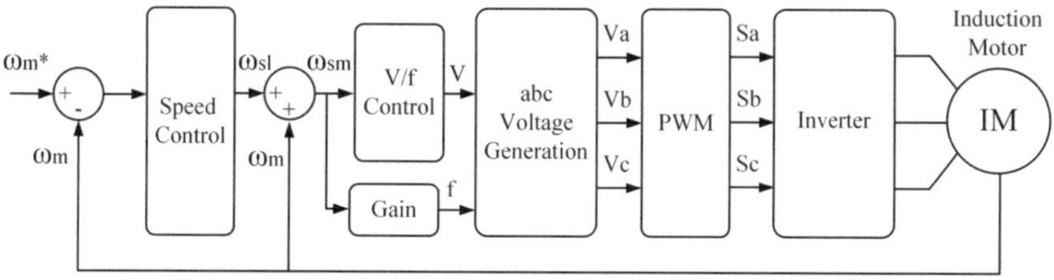

Fig. 3 General structure of the V/f control

the speed control unit, and the slip speed of the motor (ω_{sl}) is obtained. The ω_{sm} is acquired by adding the motor speed to ω_{sl}. Torque control can be performed by utilizing the connection between slip and torque in V/f control. In the V/f control method applied under constant torque, slip varies according to the changes in frequency. Maximum performance is obtained from the motor at the optimum slip value. The disadvantage of this control method is that low performance is obtained at low speeds. In addition, the increase in slip causes the motor current to increase and losses in the system. This situation can cause high costs and shorten the motor life (Hannan et al. 2018; Jannati et al. 2017).

1.2 *Field-oriented control (FOC)*

Field-oriented control, also known as vector control, was initially introduced by Hasse and Blaschke in the late 1960s. The main objective of this method was to address the limitations inherent in traditional scalar control approaches by leveraging the dynamic model of the induction motor, which incorporates voltage, current, and magnetic flux variables (Kabalcı and Boyar 2021). FOC has become a widely accepted technique, particularly for controlling squirrel-cage induction motors, due to its ability to independently regulate torque and flux. Despite this decoupled control capability, the method's performance is sensitive to variations in motor parameters and operating speed, which may adversely affect system robustness. The principle of FOC involves converting a system that is inherently time- and speed-dependent into a time-invariant structure by mapping the variables onto a rotating reference frame defined by the *d* and *q* axes. This conversion allows the induction motor to be controlled similarly to a DC motor. When the a-axis of stator reference frame is aligned with the α-axis, the position of the stator current vector across the abc, αβ, and dq coordinate systems can be visualized, as illustrated in Fig. 4. The initial abc-to-αβ transformation is performed using Eqs. 9 and 10, followed by the αβ-to-dq transformation using Eqs. 11 and 12. In this framework, the q-axis component of the stator current is responsible for producing electromagnetic torque, while the d-axis component governs the magnetic flux of the motor (Boyar 2024; Manias 2017; Ramamoorthy et al. 2021).

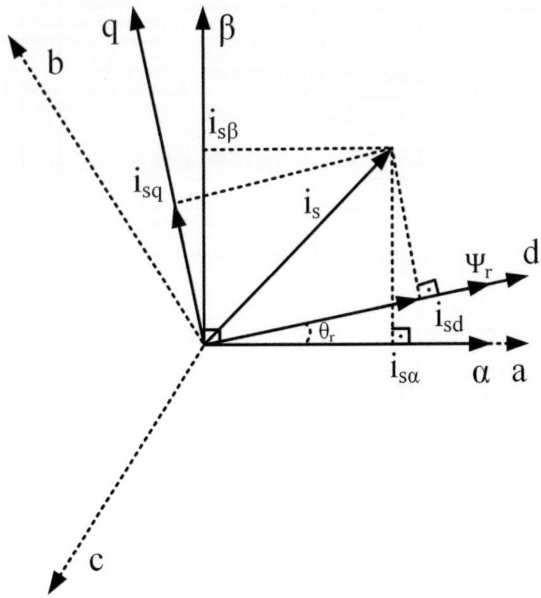

Fig. 4 Vectors of the stator current in the frame abc, αβ, and dq

$$i_{s\alpha} = i_a \tag{9}$$

$$i_{s\beta} = \frac{1}{\sqrt{3}} i_a + \frac{2}{\sqrt{3}} i_b \tag{10}$$

$$i_{sd} = i_{s\alpha} \cos\theta_r + i_{s\beta} \sin\theta_r \tag{11}$$

$$i_{sq} = -i_{s\alpha} \sin\theta_r + i_{s\beta} \cos\theta_r \tag{12}$$

where i_a, i_b represent the a and b phases of the stator current, $i_{s\alpha}$ and $i_{s\beta}$ represent the α,β components of the stator current, and θ_r represents the angle between the d-axis and the α-axis.

According to the approach used to obtain the field orientation, FOC is categorized into two methods as direct FOC and indirect FOC. The structure of direct FOC is shown in Fig. 5. In the direct FOC, the rotor flux can be obtained in two ways. The first technique uses a flux sensor to measure the air gap flux and then calculates the rotor flux computationally. The drawback of this method is the requirement for specialized flux sensors. The second method involves employing sensors to collect data on stator voltage and current in order to compute the stator flux and rotor flux (Panchade et al. 2013).

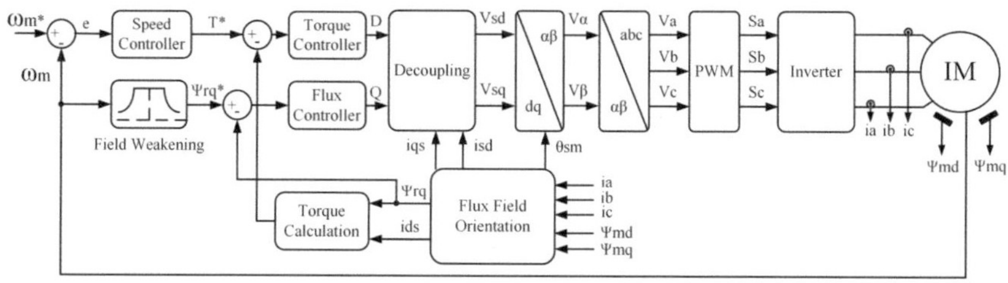

Fig. 5 Direct field-oriented control structure

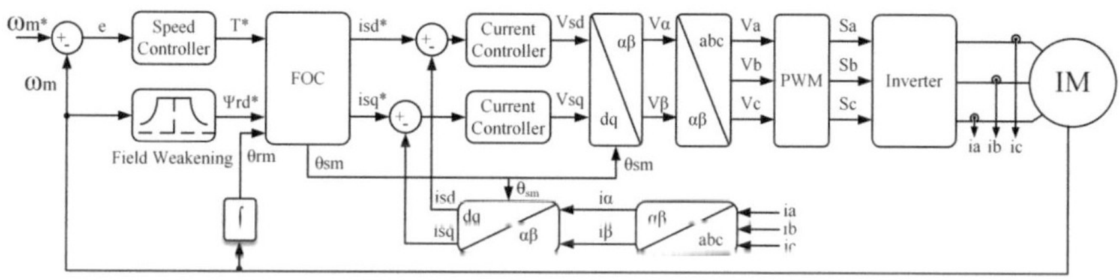

Fig. 6 Indirect field-oriented control structure

The indirect field oriented control approach has been suggested as an alternative to direct field-oriented control since it is not desirable to place sensors in the air gap of electric motors. Figure 6 depicts the overall structure, and in this method, the rotor flux is predicted using current speed and slip estimations without the use of flux sensors based on the current information collected. It provides torque and flux information for the induction motor, which may be accessed using the motor equations. The drawbacks of the indirect field-oriented control method include the need for coordinate transformation and the sensitivity to motor parameters (Kabalcı and Boyar 2021; Manias 2017).

1.3 Direct Torque Control (DTC)

I. Takahashi proposed direct torque control (DTC) in 1986 to regulate induction motors with quick torque response and great efficiency. DTC was created as an alternate control algorithm to FOC approaches due to its simpler control structure and superior torque response. The purpose of this strategy is to maximize the flux and torque generating capabilities of the induction motor using inverter control. Advantages of DTC control over FOC include eliminating the coordinate transformation, obtaining the stator flux in the sinusoidal waveform, and improving the accuracy of flux and speed estimation. The fundamental principle behind the DTC technique is to calculate the stator flux, often known as the stator electromotive force (EMF) integral. The DTC controllers result in cost savings in motor driving applications due to the absence of an encoder. In the FOC control technique, torque and flux are controlled in accordance with the *d*- and *q*-axis components of the stator currents, but in the DTC approach, they are independently controlled by applying voltage vectors (Kabalcı and Boyar 2021; Alsofyani and Idris 2013; Jannati et al. 2017; Sutikno et al. 2014). Figure 7 shows the control structure where the stator flux and electromagnetic torque are applied to the comparator and controlled independently. Flux and torque comparators can be hysteresis comparators. The general structure of this method consists of flux and torque estimators, hysteresis comparators, voltage vector selection, and voltage and current sensors. The purpose of determining the voltage vector is to correct the flux trajectory required to obtain the desired torque response by using the stator flux and torque estimation. Therefore, the stator currents and voltages of the induction motor are required for flux and torque prediction. The stator flux vector is calculated with Eq. 13. If the stator voltage vector is constant in a certain range, the stator flux can be stated with Eq. 14 by using Eq. 13. This shows that the stator flux depends on the change in the stator voltage vector (Kabalcı and Boyar 2021; Poorfakhraei et al. 2021; Jannati et al. 2017; Ammar et al. 2018).

$$\frac{d\psi_s(t)}{dt} = v_s(t) - R_s i_s(t) \tag{13}$$

$$\Delta\psi_s(t) = \psi_s(t) - \psi_s(t - \Delta t) = \int_{t-\Delta t}^{t} v_s \Delta t \tag{14}$$

The relationship between the stator and rotor flux vectors shows that a stable stator flux vector produces a stable rotor flux vector. The torque of an induction motor can be stated as in Eq. 15 where $\sigma = 1\text{-}(L_m^2/L_s L_r)$. The generated torque is affected by the shift in the angle (θ_{sr}) between the stator and rotor flow vectors. The inverter switching frequency depends on the speed and torque of the controlled motor in addition to the change in the stator flux vector. Despite having a simple structure, the hysteresis comparator-based DTC structure has significant disadvantages such as having a variable switching frequency, high torque ripple, and needing a high sampling requirement during the application phase. Variable-frequency switching

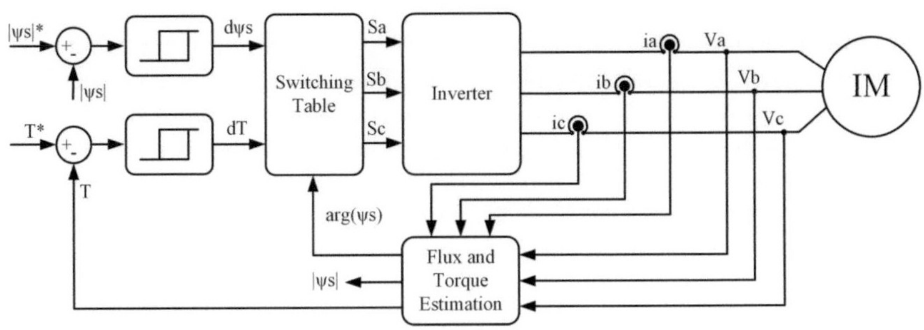

Fig. 7 Direct torque control structure

methods can cause torque deviations and distortions in the operation of the motor. However, thanks to the advantages of multi-level inverter topologies, the DTC control method can be suitable for reducing harmonics and distortions that may occur in the motor (Kabalcı and Boyar 2021; Poorfakhraei et al. 2021; Sutikno et al. 2014).

$$T = \frac{3}{2} p_p \frac{L_m}{\sigma L_s L_r} \psi_s \psi_r \sin \theta_{sr} \tag{15}$$

While estimating the stator flux, the estimation unit in DTC control eliminates the need for rotor parameters and the effect of stator resistance. The simple control structure of DTC reduces the cost and reduces the calculation requirements by eliminating the coordinate transformation. Since stator flux and torque are precisely controlled with the estimated parameters, the need for speed and position sensors is eliminated in the DTC method (Kabalcı and Boyar 2021; Boyar 2024).

1.4 Model Predictive Control (MPC)

Model Predictive Control (MPC) relies on a mathematical model of the system to forecast future behavior and select optimal control actions. Its straightforward implementation, fast dynamic response, and ability to manage multiple control objectives through a cost function make it a highly suitable choice for the control of power inverters and converters. Additionally, MPC eliminates the need for traditional cascade control structures by simplifying overall system design (Zhang et al. 2021; Boyar et al. 2023). In general, there are two primary forms of MPC: Continuous Control Set (CCS) MPC and Finite Control Set (FCS) MPC. In FCS-MPC, the switching states are treated as discrete control variables, and the optimal state is selected by directly minimizing a predefined cost function without the need for any modulation technique. In contrast, CCS-MPC requires continuous control signals, which necessitate a modulation stage and involve more intensive computations. Because of its simplicity, ease of digital implementation, and high dynamic performance, FCS-MPC is increasingly being adopted for the control of power electronic converters and motor drive systems (Boyar and Kabalci 2024). The FOC and DTC are among the most commonly utilized techniques for industrial induction motor drive applications. FOC relies on a PI-regulated current control loop, while DTC uses a predefined switching table that maps torque and flux errors to stator flux vector positions. On the other hand, Model Predictive Current Control (MPCC) is implemented to regulate the stator currents of electric machines with high precision, whereas Model Predictive Torque Control (MPTC) is specifically designed to achieve real-time control of both torque and stator flux components simultaneously. (Rodriguez et al. 2022; Krušelj and Sumina 2019).

The FCS-MPC approach uses the mathematical equivalent of motor to control the induction motor. Equation 16 is utilized to estimate the stator flux ($\widehat{\Psi}_s$), while Eq. 17 is used to estimate the rotor flow ($\widehat{\Psi}_r$). Equation 18 yields the next stator flux value (Ψ_s^p) (Boyar 2024; Wang et al. 2014; Boyar et al. 2024).

$$\hat{\psi}_s(k) = \hat{\psi}_s(k-1) + T_s V_s(k) - R_s T_s i_s(k) \tag{16}$$

$$\hat{\psi}_r(k) = \frac{L_r}{L_m}\hat{\psi}_s(k) + i_s(k)\left(L_m - \frac{L_r L_s}{L_m}\right) \tag{17}$$

$$\psi_s^p(k+1) = \hat{\psi}_s(k) + T_s V_s(k) - R_s T_s i_s(k) \tag{18}$$

The estimate of the stator current (i_s^p) required to predict the torque produced by the induction motor is obtained with Eq. 19 where $k_r = L_m/L_r$, $R_\sigma = R_s + {k_r}^2 R_r$, $\tau_r = L_r/R_r$ ve $\sigma = 1\text{-}(L_m{}^2/L_s L_r)$, $\tau_\sigma = L_\sigma/R_\sigma$, $L_\sigma = \sigma L_s$ (Boyar et al. 2024).

$$i_s^p(k+1) = \left(1 + \frac{T_s}{\tau_\sigma}\right) i_s(k) + \frac{T_s}{\tau_\sigma + T_s}\left\{\frac{1}{R_\sigma}\left[\left(\frac{k_r}{\tau_r} - k_r j\omega_m\right)\hat{\psi}_r(k) + V_s(k)\right]\right\} \tag{19}$$

Equation 20 uses the predicted stator current and flux to estimate the torque value of induction motor (T_p) (Boyar et al. 2024).

$$T^p(k+1) = \frac{3}{2} p_p \mathrm{Re}\left\{\overline{\psi}_s^p(k+1) i_s^p(k+1)\right\} \tag{20}$$

The general block diagram of the MPTC method for induction motor control, which is one of the FCS-MPC methods, is shown in Fig. 8. Each capacitor in an Active Neutral-Point Clamped (ANPC) or Neutral-Point Clamped (NPC) inverter topology needs to have a voltage drop that is half the input voltage for the capacitors to be balanced. In the balanced condition, the difference of these capacitor voltages must be $\Delta V_c = 0$. The voltage difference of each capacitor (ΔV_{c1}, ΔV_{c2}) is acquired with Eqs. 21 and 22 where S_a, S_b, S_c are the switching states of each phase and i_a, i_b, i_c denote the phase currents (Boyar 2024; Boyar et al. 2024; Jun et al. 2020).

$$\Delta v_{c1} = \frac{T_s}{C_1} i_{c1} = \frac{T_s}{C_1}\left(-\frac{S_a(S_a+1)}{2} i_a - \frac{S_b(S_b+1)}{2} i_b - \frac{S_c(S_c+1)}{2} i_c\right) \tag{21}$$

$$\Delta v_{c2} = \frac{T_s}{C_2} i_{c2} = \frac{T_s}{C_2}\left(\frac{S_a(S_a-1)}{2} i_a + \frac{S_b(S_b-1)}{2} i_b + \frac{S_c(S_c-1)}{2} i_c\right) \tag{22}$$

$$\Delta v_c = \Delta v_{c1} - \Delta v_{c2} = -\frac{T_s}{C_{1,2}}\left(S_a^2 i_a + S_b^2 i_b + S_c^2 i_c\right) \tag{23}$$

Equation 24 provides the cost function for balancing the inverter input capacitor voltages and controlling the torque of the induction motor. The difference between the predicted values and the reference torque (T^*) and stator (Ψ_s^*) values is assessed in the cost function. The switching components of inverters are subjected to the

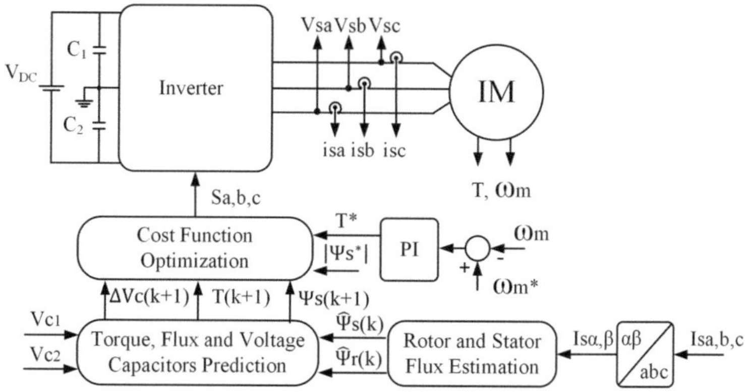

Fig. 8 FCS-MPC block structure for induction motor control

switching state that minimizes the cost function. The weighting factor is represented by λ_c and λ_T (Boyar et al. 2024).

$$g = \left|T * -T^p(k+1)\right| + \lambda_T\left|\psi_s^* - \psi_s^p(k+1)\right| + \lambda_c|\Delta v_c(k+1)| \qquad (24)$$

The comparison of commonly used electric motor control methods is given in Table 1. While scalar control has a high error rate, the FOC method, which was developed as an alternative to this method, is complex and slow in response compared to DTC. The DTC control method has high current and torque ripple compared to FOC. FCS-MPC control is becoming more widespread today due to its fast dynamic response, easy application, low torque, and current ripple (Poorfakhraei et al. 2021; Wang et al. 2018; Yaramasu and Bin 2017).

2 Sensorless Speed Estimation Methods

Speed sensors are used for speed control in electric motor drives. In case of using speed sensors, there are disadvantages such as increasing system cost, covering area, and problems in integrating with the motor. The formation of noise-related distortions in the signal obtained in speed measurement causes negativities in the performance of the motor control system. There are studies on obtaining speed information without sensors in order to enhance the performance of control systems. Other benefits of these systems include lower costs, less maintenance, and fewer pieces of hardware. Induction motor speed is estimated using a variety of techniques. The most widely used of these are the model reference adaptive system (MRAS), Kalman filter (KF), Luenberger observer (LO), and sliding mode observer (SMO) techniques. Compared to SMO and MRAS, the KF approach has more mathematical functions. In addition,

Table 1 Comparison of control methods (Boyar 2024)

Control method	Advantages	Disadvantages
Scalar control	• Simple and low • No feedback needed	• Low dynamic response • Error rate is high • High torque ripple • No torque control
DTC	• Fast dynamic response • Simple structure	• Variable switching frequency • Difficult to control at low speed • High current and torque ripple compared to FOC • Starting current is high • Sensitive to motor parameters
FOC	• Superior dynamic performance • Small torque and current ripple • Constant switching frequency • Full torque generation at zero speed • Low parameter sensitivity	• Feedback needed • High cost • Slow response according to DTC • Modulation method required • According to DTC, it has a complex structure
FCS-MPC	• Easy to apply • Simple structure • No modulation required • Fast dynamic response • Small torque and current ripple	• Sensitive to motor parameters • High calculation • Variable switching frequency

SMO has a disadvantage due to the discontinuous switching function. Among sensorless control techniques, the MRAS is commonly favored due to its relatively simple structure, ease of implementation, and reduced computational complexity compared to alternative methods (Benlaloui et al. 2015; Boyar et al. 2022; Reddy et al. 2020; Mossa et al. 2020).

2.1 *Model Reference Adaptive System (MRAS)*

When speed sensors are included to the motor control system, the cost and space rise. Today, speed estimation algorithms can be easily applied with the development of microprocessor technologies. The speed of an induction motor can be estimated by the MRAS method which has been first proposed by S. Tamai in 1987 (Tamai et al. 1987) and developed by C. Schauder in 1992 (Schauder 1992). MRAS estimation methods consist of three main parts as reference model, adaptive model, and adaptation mechanism as seen in Fig. 9 where z represents the stator voltage and/or stator current, r and $\hat{r}$ represent the components of the motor such as flux. The speed value determined using the adaptive model is not the foundation of the reference model. The error function derived from the two models' outputs is used to predict the speed. The speed of the motor is obtained from the output of the adaptation mechanism and is utilized simultaneously in the adaptive model through a closed loop. The MRAS

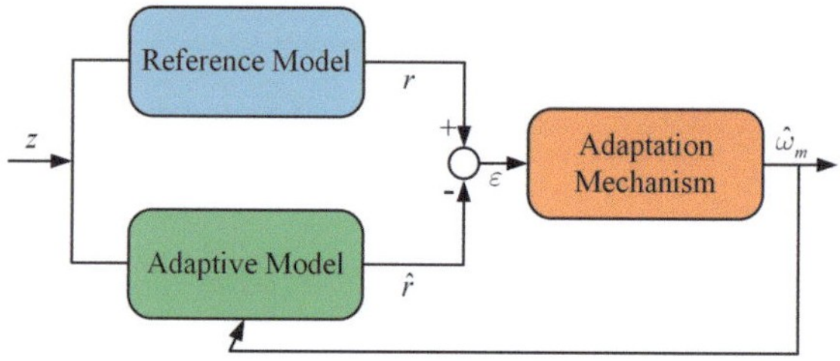

Fig. 9 General structure of MRAS speed estimation

method has several different speed estimating techniques, including rotor flux-based MRAS (MRAS^{RF}), stator current-based MRAS (MRAS^{CC}), electromotive force and reactive power-based MRAS (MRAS^{EMF}, MRAS^{RP}), active power, reactive power, and virtual variable-based model reference adaptive systems (P-, Q-, X-, Y-MRAS) (Korzonek et al. 2019; Bednarz and Dybkowski 2019; Kumar et al. 2016). Stator voltage and current measurements are adequate for predicting the speed of an induction motor. Due to the ease of application of the MRAS method, it is extensively preferred in the speed estimation of induction motors (Boyar 2024; Boyar et al. 2022).

2.1.1 Rotor Flux-Based Model Reference Adaptive System (MRAS^{RF})

The first proposed MRAS method is the rotor flux-based MRAS estimation method. The general structure of the speed estimation system with the MRAS^{RF} method is given in Fig. 10 and includes a reference model, adaptive model, and adaptation mechanism (Boyar et al. 2022; Abu-Rub et al. 2021; Bao and Hou 2017).

The reference model consists of the stator equations of the induction motor and is also known as the voltage model. The adaptive model, referred to as the current model, includes the rotor equations. With these two models, the rotor flux components of the induction motor are estimated and the speed value that allows the error to reach the minimum value is obtained by calculating the error rate in the adaptation mechanism (Boyar et al. 2024; Boyar et al. 2022; Korzonek et al. 2019). The reference model defined as the voltage model is stated by Eqs. 25 and 26 where $V_{s\alpha}$, $V_{s\beta}$ represent the α and β components of the stator voltage. In this model, the α and β components of the rotor flux ($\Psi_{r\alpha}$, $\Psi_{r\beta}$) of the induction motor are obtained by utilizing the voltage and current equations of the stator (Boyar et al. 2024; Abu-Rub et al. 2021; Bao and Hou 2017).

$$\frac{d}{dt}\psi_{r\alpha} = (L_r/L_m)V_{s\alpha} - (L_r/L_m)\left(R_s + \sigma L_s\frac{d}{dt}\right)i_{s\alpha} \tag{25}$$

$$\frac{d}{dt}\psi_{r\beta} = (L_r/L_m)V_{s\beta} - (L_r/L_m)\left(R_s + \sigma L_s\frac{d}{dt}\right)i_{s\beta} \tag{26}$$

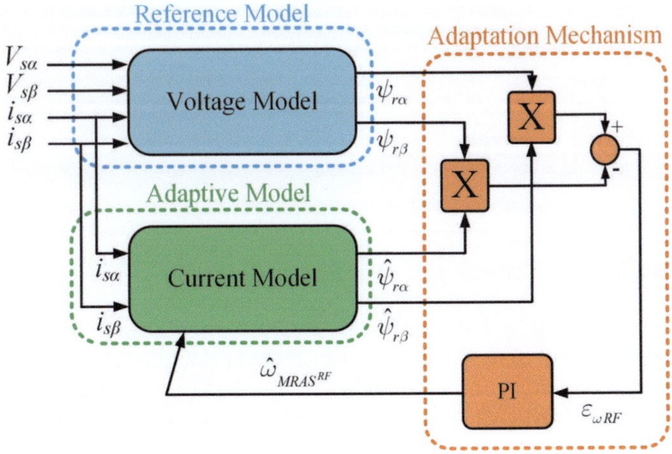

Fig. 10 General structure of MRAS^{RF} speed estimation

The adaptive model referred to as the current model is stated by Eqs. 27 and 28 where $\tau_r = L_r/R_r$. This model contains the stator current and speed information of the induction motor. In the current model, the components of the rotor flux axis ($\widehat{\Psi}_{r\alpha}$, $\widehat{\Psi}_{r\beta}$) are obtained. The error rate of the fluxes obtained with the voltage and current models ($\varepsilon_{\omega RF}$) is calculated by Eq. 29 and applied to the PI controller input. The instantaneous speed estimate of the induction motor ($\widehat{\omega}_{MRAS^{RF}}$) is calculated by Eq. 30. The speed value is obtained by minimizing the error rate of the fluxes estimated from the PI controller output. K_p and K_i represent the proportional and integral coefficients of the PI controller (Boyar 2024; Boyar et al. 2024; Korzonek et al. 2019; Abu-Rub et al. 2021).

$$\frac{d}{dt}\hat{\psi}_{r\alpha} = -(1/\tau_r)\hat{\psi}_{r\alpha} - \hat{\omega}_{MRAS^{RF}}\hat{\psi}_{r\beta} + (L_m/\tau_r)i_{s\alpha} \tag{27}$$

$$\frac{d}{dt}\hat{\psi}_{r\beta} = \hat{\omega}_{MRAS^{RF}}\hat{\psi}_{r\alpha} - (1/\tau_r)\hat{\psi}_{r\beta} + (L_m/\tau_r)i_{s\beta} \tag{28}$$

$$\varepsilon_{\omega RF} = \psi_{r\alpha}\hat{\psi}_{r\beta} - \psi_{r\beta}\hat{\psi}_{r\alpha} \tag{29}$$

$$\hat{\omega}_{MRAS^{RF}} = K_p\varepsilon_{\omega RF} + K_i\int \varepsilon_{\omega RF}dt \tag{30}$$

2.1.2 Stator Current-Based Model Reference Adaptive System (MRAS^{CC})

The basic principle of the stator current-based MRAS speed estimation method depends on the error rate between the estimated stator current value and the actual current value. Its general structure is seen in Fig. 11 and includes reference, adaptive, and adaptation mechanisms (Korzonek et al. 2019).

The induction motor serves as the reference model, while the current model and current estimation make up the adaptive model. The rotor flux estimation is obtained by employing the current model used in the MRAS^{RF} method. The current estimation model is a combination of the voltage and current models and is expressed by Eqs. 31 and 32. $\hat{i}_{s\alpha}$, $\hat{i}_{s\beta}$ represent the α,β components of the estimated stator current. In contrast to MRAS^{RF}, the adaptation mechanism relies on a difference between the induction motor's predicted and measured stator current. The error rate ($\varepsilon_{\omega CC}$) is calculated by Eq. 33 where $e_{is} = i_s - \hat{i}_s$ and applied to the PI controller input. The speed estimation of the induction motor ($\hat{\omega}_{MRAS^{CC}}$) from the PI controller output is carried out by the MRAS^{CC} method with Eq. 34 (Boyar 2024; Korzonek et al. 2020; Orlowska-Kowalska and Dybkowski 2010).

$$\frac{d}{dt}\hat{i}_{s\alpha} = \frac{1}{\sigma L_s}\left(-R_s\hat{i}_{s\alpha} - \frac{L_m^2}{L_r\tau_r}\hat{i}_{s\alpha} + \frac{L_m}{L_r\tau_r}\hat{\psi}_{r\alpha} + \frac{L_m}{L_r}\hat{\omega}_{MRAS^{CC}}\hat{\psi}_{r\beta} + V_{s\alpha}\right) \tag{31}$$

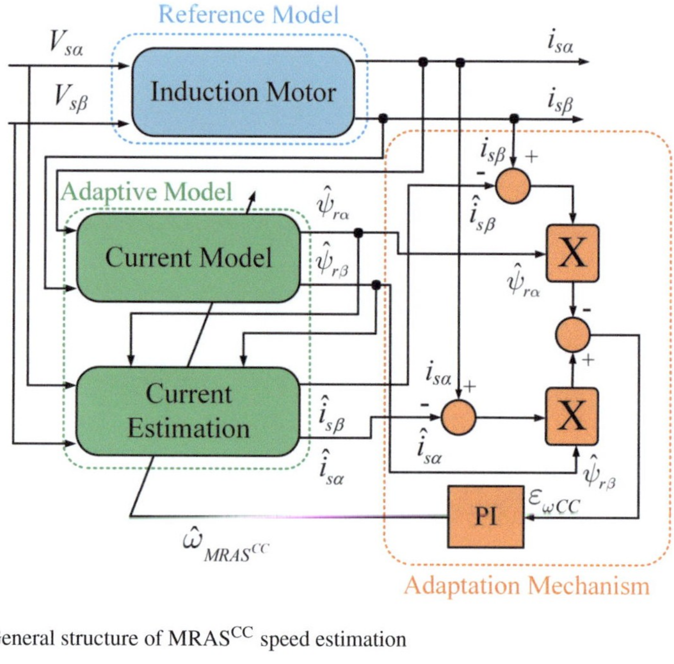

Fig. 11 General structure of MRAS^{CC} speed estimation

$$\frac{d}{dt}\hat{i}_{s\beta} = \frac{1}{\sigma L_s}\left(-R_s\hat{i}_{s\beta} - \frac{L_m^2}{L_r\tau_r}\hat{i}_{s\beta} + \frac{L_m}{L_r\tau_r}\hat{\psi}_{r\beta} + \frac{L_m}{L_r}\hat{\omega}_{MRAS^{CC}}\hat{\psi}_{r\alpha} + V_{s\alpha}\right) \tag{32}$$

$$\varepsilon_{\omega CC} = e_{is\alpha}\hat{\psi}_{r\beta} - e_{is\beta}\hat{\psi}_{r\alpha} \tag{33}$$

$$\hat{\omega}_{MRAS^{CC}} = K_p\varepsilon_{\omega CC} + K_i\int \varepsilon_{\omega CC}dt \tag{34}$$

2.1.3 Electromotive Force and Reactive Power-Based Model Reference Adaptive Systems (MRAS^{EMF}, MRAS^{RP})

With the increasing interest in speed estimation methods, electromotive force (Peng and Fukao 1994) and reactive power dependent model reference adaptive systems have been proposed (Peng and Fukao 1994; Peng et al. 1994). The speed estimation of the induction motor is performed by the MRAS^{EMF} method using the electromotive force. The general structure of MRAS^{EMF} is seen in Fig. 12, and the reference model is stated by Eqs. 35 and 36, while the adaptive model is stated by Eqs. 37 and 38 (Korzonek et al. 2019; Peng and Fukao 1994; Kumar et al. 2015; Marwali and Keyhani 1997).

Instead of using rotor fluxes in the reference and adaptive models, speed estimation is performed by comparing electromotive force estimates (Boyar 2024).

$$e_{m\alpha} = V_{s\alpha} - R_s i_{s\alpha} - \sigma L_s\frac{d}{dt}i_{s\alpha} \tag{35}$$

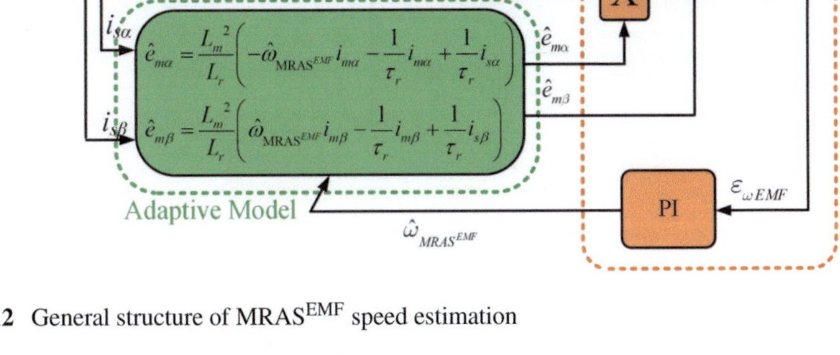

Fig. 12 General structure of MRAS^{EMF} speed estimation

$$e_{m\beta} = V_{s\beta} - R_s i_{s\beta} - \sigma L_s \frac{d}{dt} i_{s\beta} \tag{36}$$

$$\hat{e}_{m\alpha} = \frac{L_m^2}{L_r}\left(-\hat{\omega}_{MRAS^{EMF}} i_{m\alpha} - \frac{1}{\tau_r} i_{m\alpha} + \frac{1}{\tau_r} i_{s\alpha}\right) \tag{37}$$

$$\hat{e}_{m\beta} = \frac{L_m^2}{L_r}\left(\hat{\omega}_{MRAS^{EMF}} i_{m\beta} - \frac{1}{\tau_r} i_{m\beta} + \frac{1}{\tau_r} i_{s\beta}\right) \tag{38}$$

where e_m and $\hat{e}_m$ represent the estimated electromotive forces and the components of the magnetizing current ($i_{m\alpha}$, $i_{m\beta}$) are calculated by Eqs. 39 and 40 (Kumar et al. 2015; Marwali and Keyhani 1997).

$$\frac{di_{m\alpha}}{dt} = -\hat{\omega}_{MRAS^{EMF}} i_{m\beta} - \frac{1}{\tau_r} i_{m\alpha} + \frac{1}{\tau_r} i_{s\alpha} \tag{39}$$

$$\frac{di_{m\beta}}{dt} = \hat{\omega}_{MRAS^{EMF}} i_{m\alpha} - \frac{1}{\tau_r} i_{m\beta} + \frac{1}{\tau_r} i_{s\beta} \tag{40}$$

The error rate ($\varepsilon_{\omega EMF}$) is calculated with Eq. 41 according to the data acquired from the reference and adaptive models. The error rate obtained with the reference and predicted values of the electromotive force vectors is applied to the speed adaptation mechanism and the speed estimation ($\widehat{\omega}_{MRAS^{EMF}}$) is performed with Eq. 42 (Korzonek et al. 2019).

$$\varepsilon_{\omega EMF} = e_{m\beta}\hat{e}_{m\alpha} - e_{m\alpha}\hat{e}_{m\beta} \tag{41}$$

$$\hat{\omega}_{MRAS^{EMF}} = K_p \varepsilon_{\omega EMF} + K_i \int \varepsilon_{\omega EMF} dt \tag{42}$$

The reactive power-based MRAS^{RP} estimation method is the improved version of MRAS^{EMF}, and its general structure is shown in Fig. 13. The reactive power used in calculating the error rate is calculated by the electromotive force and the stator current of the motor and is expressed by Eqs. 43 and 44 (Korzonek et al. 2019).

$$q_m = e_{m\beta} i_{s\alpha} - e_{m\alpha} i_{s\beta} \tag{43}$$

$$\hat{q}_m = \hat{e}_{m\beta} i_{s\alpha} - \hat{e}_{m\alpha} i_{s\beta} \tag{44}$$

In the MRAS^{RP} method, the error rate ($\varepsilon_{\omega RP}$) depends on the reactive powers acquired from the reference and adaptive models and is obtained with Eq. 45. The error rate obtained is the input of the PI controller that creates the adaptation mechanism. The speed estimate of the induction motor ($\widehat{\omega}_{MRAS^{RP}}$) is acquired from the PI controller output that minimizes the error rate with Eq. 46 (Korzonek et al. 2019).

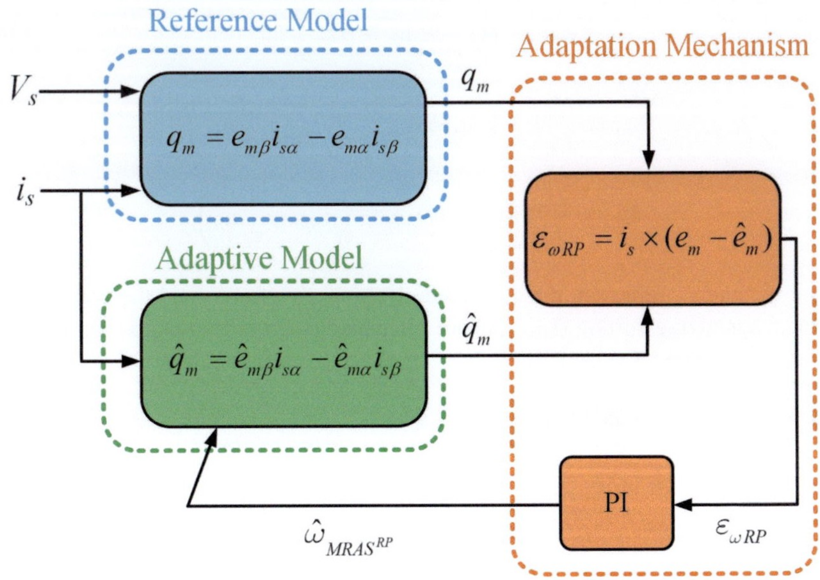

Fig. 13 General structure of MRAS^{RP} speed estimation

$$\varepsilon_{\omega RP} = i_s \times (e_m - \hat{e}_m) \tag{45}$$

$$\hat{\omega}_{MRAS^{RP}} = K_p \varepsilon_{\omega RP} + K_i \int \varepsilon_{\omega RP} dt \tag{46}$$

2.1.4 Active Power, Reactive Power, and Virtual Variable-Based Model Reference Adaptive Systems (P-, Q-, X-, Y-MRAS)

Active power (P-MRAS), reactive power (Q-MRAS), and virtual variable (X- and Y-MRAS)-based model reference adaptive speed estimation systems were proposed by Verma and his colleagues and their general structure is shown in Fig. 14 where $V^*_{sd,q}$ represents the reference stator voltage component and $i^*_{sd,q}$ represents the reference stator current component (Boyar 2024; Verma and Chakraborty 2014).

For Q-,P-,X-,Y-MRAS methods, the reference model is expressed by Eq. 47, while the adaptive model is stated by Eq. 48. The output of the reference model is expressed by V_{mr}, while the output of the adaptive model is expressed by V_{ma}. In Q-,P-,X-,Y-MRAS methods, the error rate (ε_ω) between the two models is calculated by Eq. 49 by performing the necessary calculations in the reference and adaptive models. The speed of the motor is obtained by applying the obtained error with the PI controller with Eq. 50 (Korzonek et al. 2019; Verma and Chakraborty 2014).

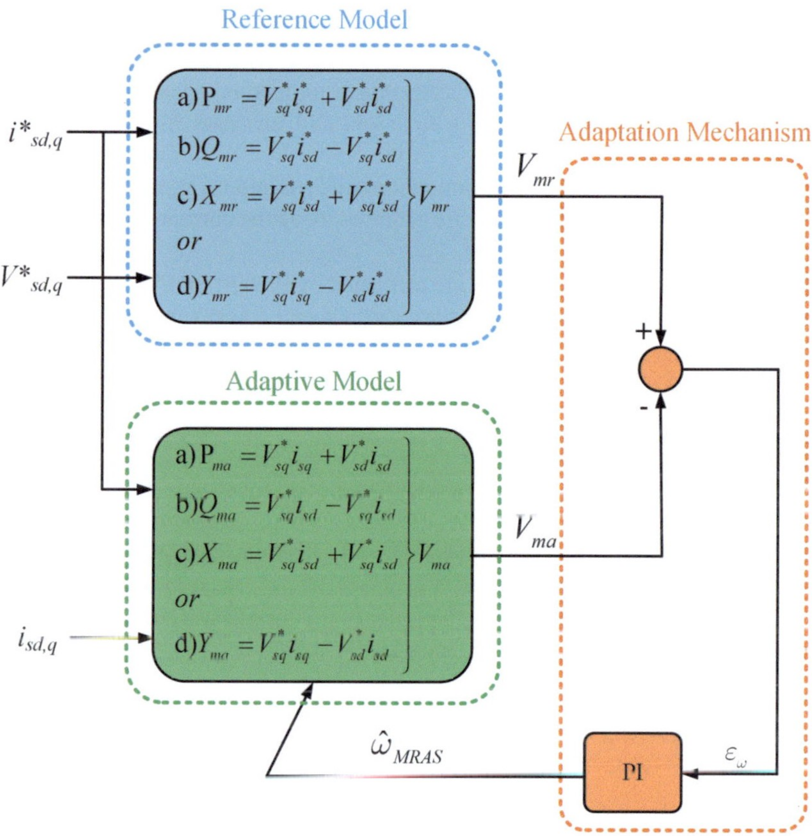

Fig. 14 General structure of Q-,P-,X-,Y-MRAS speed estimation

$$\left.\begin{aligned} a)\quad & P_{mr} = V^*_{sq} i^*_{sq} + V^*_{sd} i^*_{sd} \\ b)\quad & Q_{mr} = V^*_{sq} i^*_{sd} - V^*_{sq} i^*_{sd} \\ c)\quad & X_{mr} = V^*_{sq} i^*_{sd} + V^*_{sq} i^*_{sd} \\ d)\quad & Y_{mr} = V^*_{sq} i^*_{sq} - V^*_{sd} i^*_{sd} \end{aligned}\right\} V_{mr} \tag{47}$$

$$\left.\begin{aligned} a)\quad & P_{ma} = V^*_{sq} i_{sq} + V^*_{sd} i_{sd} \\ b)\quad & Q_{ma} = V^*_{sq} i_{sd} - V^*_{sq} i_{sd} \\ c)\quad & X_{ma} = V^*_{sq} i_{sd} + V^*_{sq} i_{sd} \\ d)\quad & Y_{ma} = V^*_{sq} i_{sq} - V^*_{sd} i_{sd} \end{aligned}\right\} V_{ma} \tag{48}$$

$$\varepsilon_{\omega} = V_{mr} - V_{ma} \tag{49}$$

$$\hat{\omega}_{MRAS} = K_p \varepsilon_{\omega} + K_i \int \varepsilon_{\omega} dt \tag{50}$$

2.1.5 Sliding Mode Observer (SMO)

The sliding mode observer (SMO) speed estimation method consists of the current model, current estimation, switching function, and low-pass filter (LPF) and its general structure is seen in Fig. 15. In the SMO speed estimation method, the rotor flux components of the induction motor are acquired using the current model, while the stator current components are acquired using the current estimation model (Dybkowski et al. 2013).

In the SMO method, $\widehat{\Psi}_{r\alpha}$ ve $\widehat{\Psi}_{r\beta}$ are estimated by Eqs. 51 and 52, where C is the constant coefficient and μ is the auxiliary variable (Fu 2005; Yan et al. 2000).

$$\frac{d}{dt}\hat{\psi}_{r\alpha} = -(1/\tau_r)\hat{\psi}_{r\alpha} - \hat{\omega}^e_{SMO}\hat{\psi}_{r\beta} + (L_m/\tau_r)i_{s\alpha} + C\hat{\psi}_{r\beta}\mu \quad (51)$$

$$\frac{d}{dt}\hat{\psi}_{r\beta} = -(1/\tau_r)\hat{\psi}_{r\beta} + \hat{\omega}^e_{SMO}\hat{\psi}_{r\alpha} + (L_m/\tau_r)i_{s\beta} - C\hat{\psi}_{r\alpha}\mu \quad (52)$$

where $\widehat{\omega}^e_{SMO}$ represents the speed estimate before the low-pass filter with the SMO method, and $\widehat{\omega}_{SMO}$ represents the speed estimate obtained with the SMO method.

In the SMO speed estimation method, the stator current is estimated by Eqs. 53 and 54 (Fu 2005; Yan et al. 2000).

$$\frac{d}{dt}\hat{i}_{s\alpha} = \frac{1}{\sigma L_s}\left(-R_s\hat{i}_{s\alpha} - \frac{L_m^2}{L_r\tau_r}\hat{i}_{s\alpha} + \frac{L_m}{L_r\tau_r}\hat{\psi}_{r\alpha} + \frac{L_m}{L_r}\hat{\omega}^e_{SMO}\hat{\psi}_{r\beta} + V_{s\alpha} - \frac{L_m}{L_r}\hat{\psi}_{r\alpha}\mu\right) \quad (53)$$

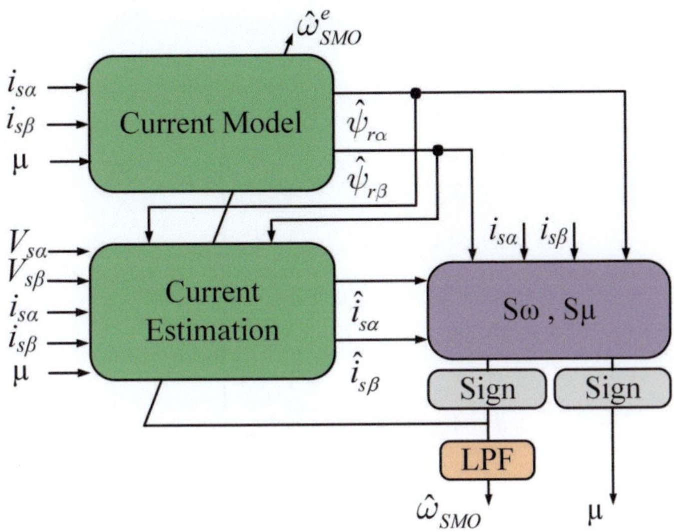

Fig. 15 General structure of SMO speed estimation

$$\frac{d}{dt}\hat{i}_{s\beta} = \frac{1}{\sigma L_s}\left(-R_s\hat{i}_{s\beta} - \frac{L_m^2}{L_r\tau_r}\hat{i}_{s\beta} + \frac{L_m}{L_r\tau_r}\hat{\psi}_{r\beta} - \frac{L_m}{L_r}\hat{\omega}^e_{SMO}\hat{\psi}_{r\alpha} + V_{s\beta} - \frac{L_m}{L_r}\hat{\psi}_{r\beta}\mu\right) \tag{54}$$

S_ω, S_μ speed and auxiliary variable switching functions are obtained with Eqs. 55 and 56 (Boyar 2024; Yan et al. 2000).

$$S_\omega = (\hat{i}_{s\beta} - i_{s\beta})\hat{\psi}_{r\alpha} - (\hat{i}_{s\alpha} - i_{s\alpha})\hat{\psi}_{r\beta} \tag{55}$$

$$S_\mu = (\hat{i}_{s\alpha} - i_{s\alpha})\hat{\psi}_{r\alpha} + (\hat{i}_{s\beta} - i_{s\beta})\hat{\psi}_{r\beta} \tag{56}$$

The speed information of the induction motor is predicted by the SMO method, Eq. 57, and μ is obtained by Eq. 58 where ω_0 and μ_0 represent constant coefficients (Fu 2005; Yan et al. 2000).

$$\hat{\omega}^e_{SMO} = \omega_0 sign(S_\omega) \tag{57}$$

$$\mu = \mu_0 sign(S_\mu) \tag{58}$$

The evaluation of the forecast performance can be performed by calculating the root mean square error (RMSE) given in Eq. 59 where $(\hat{\omega}_m)$ represents the predicted speed. RMSE is a commonly used calculation to evaluate the forecast performance. The difference between the actual values and the forecast values determines the RMSE (Lai et al. 2000).

$$\omega_{RMSE} = \sqrt{\frac{1}{n}\sum_{i=1}^{n}(\omega_m - \hat{\omega}_m)^2} \tag{59}$$

3 Sensorless Induction Motor Drive Application Examples

Boyar et al. (Boyar et al. 2024) developed a sensorless-based FCS-MPC method to control the induction motor with ANPC multi-level inverter. The proposed system consists of ANPC inverter, induction motor, FCS-MPC method, and $MRAS^{RF}$ and $MRAS^{CC}$ speed estimation units. Two different MRAS-based FCS-MPC methods were developed for torque and flux control of the induction motor. In this study, $MRAS^{RF}$ and $MRAS^{CC}$ speed estimation methods were designed separately to obtain the necessary speed information for FCS-MPC and their performance analyses were performed. Firstly, simulation studies of the proposed drives were performed and then analysis was performed with application studies.

3.1 Simulation Study

The simulation analysis of the designed sensorless induction motor drive includes the induction motor itself; an ANPC inverter; and the $MRAS^{RF}$, $MRAS^{CC}$, and FCS-MPC blocks. In this study, the rotor speed information required by the FCS-MPC controller was separately estimated using $MRAS^{RF}$ and $MRAS^{CC}$ methods. The performance of the drive system using these two different MRAS-based speed estimation techniques was comparatively evaluated. As illustrated in Fig. 16, both drive systems were tested under identical operating conditions (Boyar et al. 2024).

Firstly, the reference speed of the motor was set to 65 rad/s without any load. From the third second, the reference speed was reduced and stabilized at 50 rad/s. The load torque of 5 Nm was applied for 2 s at the fourth second. At the seventh second, the reference speed was increased back to 65 rad/s. When the speed reference was 50 rad/s and the motor was under a 5 Nm load, the actual motor speed was measured as 46.44 rad/s in the $MRAS^{RF}$-based controller and 49.7 rad/s in the $MRAS^{CC}$-based configuration. The $MRAS^{RF}$ method resulted in a speed estimation error of 7.12%, whereas the $MRAS^{CC}$ method provided a much closer match to the actual speed. Apart from the speed response, both sensorless drive approaches yielded similar performance in terms of output voltage, current, electromagnetic torque, and stator

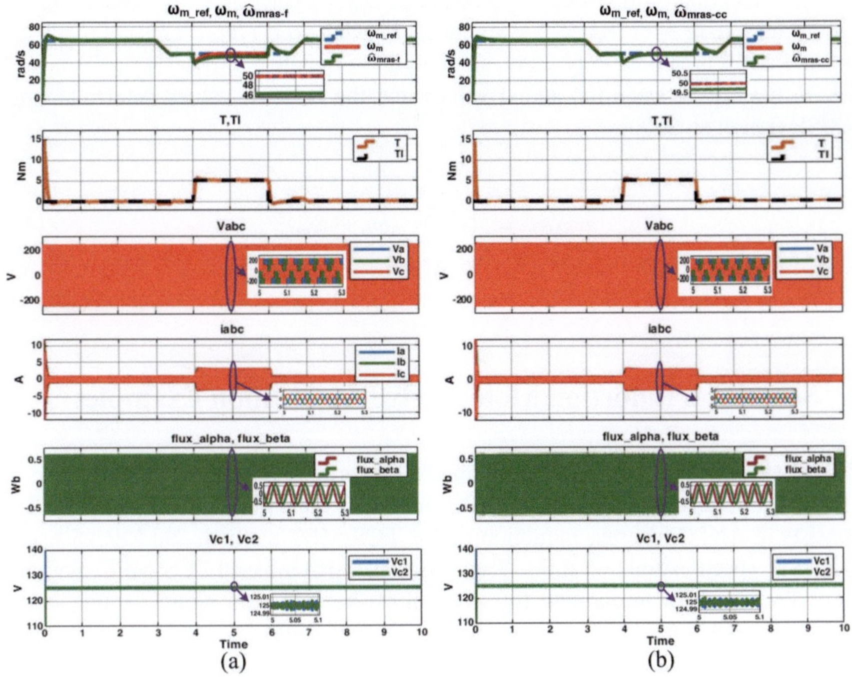

Fig. 16 Simulation results **a** $MRAS^{RF}$-based FCS-MPC **b** $MRAS^{CC}$-based FCS-MPC (Boyar et al. 2024)

flux. The ANPC inverter delivered a peak output voltage of 250 V. Under no-load conditions, the stator current was approximately 1 A, which increased to 3.17 A when a 5 Nm load was applied. The proposed FCS-MPC controller maintained the stator flux constant at its reference value of 0.6 Wb and ensured balanced input capacitor voltages of 125 V. According to these simulation outcomes, it was concluded that the $MRAS^{CC}$-based FCS-MPC configuration demonstrated superior speed estimation accuracy and overall performance for the sensorless induction motor drive (Boyar et al. 2024).

3.2 Implementation Study

The induction motor drives controlled by FCS-MPC based on $MRAS^{RF}$ and $MRAS^{CC}$ techniques were experimentally evaluated under identical speed and load conditions. The input voltage applied to the ANPC inverter was set to 250 V. The motor speed required for the FCS-MPC algorithm was estimated without the use of any physical sensor, utilizing $MRAS^{RF}$ and $MRAS^{CC}$ observer methods. Initially, the reference speed was established at 65 rad/s in a no-load scenario and maintained for 30 s, as illustrated in Fig. 17. Then the reference speed was reduced to 50 rad/s and a 5 Nm load was implemented. Eventually, the reference speed of induction motor was set to 65 rad/s. Thanks to the developed FCS-MPC, it was observed that the induction motor followed the determined reference speeds. Error values were acquired by taking the difference between the actual and predicted speeds for both systems. It was obvious that the speed error and fluctuation rate were higher in speed estimation with $MRAS^{RF}$ compared to $MRAS^{CC}$ (Boyar et al. 2024).

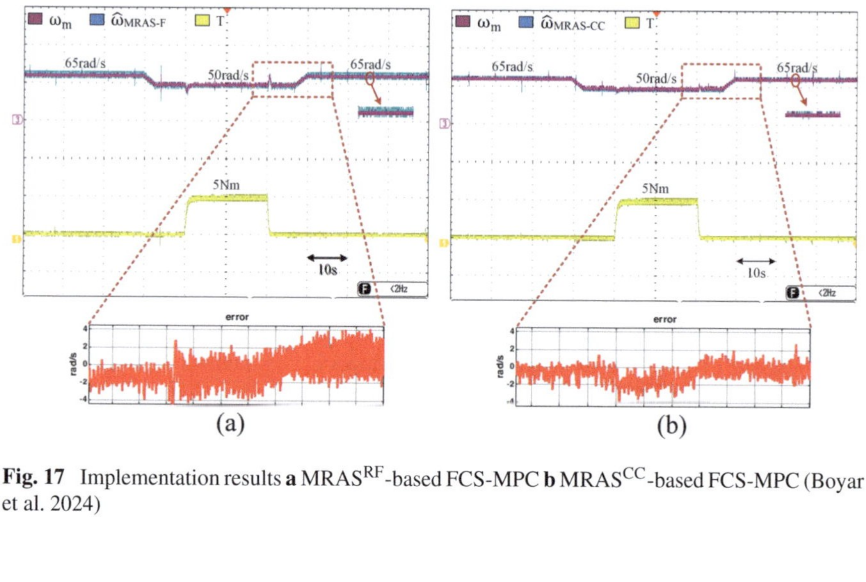

Fig. 17 Implementation results **a** $MRAS^{RF}$-based FCS-MPC **b** $MRAS^{CC}$-based FCS-MPC (Boyar et al. 2024)

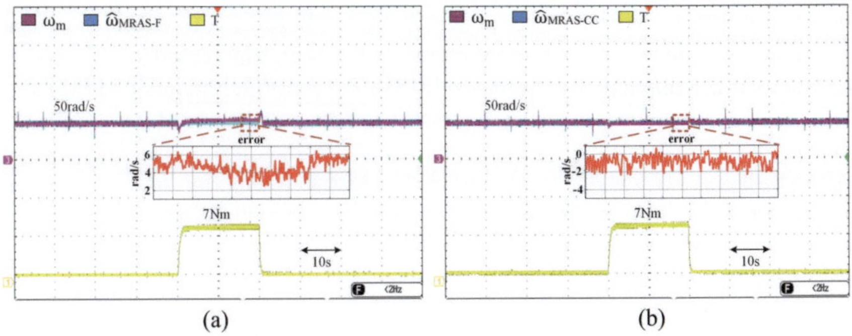

Fig. 18 Implementation results **a** $\mathrm{MRAS}^{\mathrm{RF}}$-based FCS-MPC **b** $\mathrm{MRAS}^{\mathrm{CC}}$-based FCS-MPC under 7 Nm load (Boyar et al. 2024)

A comprehensive analysis of the performance of induction motor drives controlled by $\mathrm{MRAS}^{\mathrm{RF}}$- and $\mathrm{MRAS}^{\mathrm{CC}}$-based FCS-MPC techniques under various load scenarios has been carried out. The reference speed has been determined as 50 rad/s and a load of 7 Nm has been applied to the induction motor. Figure 18.a and b indicates the error ratios between the actual speed and the predicted speed when the reference speed is 50 rad/s and the load is 7 Nm (Boyar et al. 2024).

When the RMSE ratios are assessed when the load is applied as 7 Nm, 2.23 RMSE values have been obtained for $\mathrm{MRAS}^{\mathrm{RF}}$ and 1.13 RMSE values have been obtained for $\mathrm{MRAS}^{\mathrm{RF}}$. It has been determined that the induction motor drive with $\mathrm{MRAS}^{\mathrm{RF}}$-based FCS-MPC controller has a higher RMSE ratio compared to the $\mathrm{MRAS}^{\mathrm{CC}}$-based system under the same conditions. With this comprehensive analysis study, it was found that $\mathrm{MRAS}^{\mathrm{CC}}$-based FCS-MPC performed better than $\mathrm{MRAS}^{\mathrm{RF}}$-based FCS-MPC, especially under changing load conditions. Reduced RMSE values in $\mathrm{MRAS}^{\mathrm{CC}}$ indicate that precise speed tracking is achieved with improved performance (Boyar et al. 2024).

4 Chapter Summary

This section, which covers basic and advanced methods for controlling electric motor drive systems, specifically emphasizes the importance of sensorless control techniques on power converters. It includes detailed explanations of classical and modern motor control methods such as scalar control, FOC, DTC, and MPC. These methods are evaluated in terms of their dynamic performance, application complexity, and suitability for industrial use. Following the explanations on control strategies, sensorless speed estimation techniques, which are important in terms of increasing the reliability and cost-effectiveness of motor drive systems, are discussed. In this chapter, the working principles, advantages, and application difficulties of methods such as MRAS and SMO are included. Advanced sensorless control approaches obtained by combining MRAS-based algorithms with motor control structures are also examined. Two different application examples are presented where $MRAS^{RF}$ and $MRAS^{CC}$ methods are integrated with MPC. In comparative analyses, the dynamic response, estimation accuracy, and durability of the proposed sensorless MPC methods under various operating conditions were evaluated.

References

Abu-Rub H, Iqbal A, Guzinski J (2021) Sensorless speed control of AC machines. In: High performance control of AC drives with MATLAB®/Simulink, John Wiley & Sons, Ltd, pp 433–467. https://doi.org/10.1002/9781119591313.ch8.

Alsofyani IM, Idris NRN (2013) A review on sensorless techniques for sustainable reliablity and efficient variable frequency drives of induction motors. Renew Sustain Energy Rev 24:111–121. https://doi.org/10.1016/j.rser.2013.03.051

Ammar A, Kheldoun A, Metidji B, Talbi B, Ameid T, Azzoug Y (2018) An experimental assessment of direct torque control and model predictive control methods for induction machine drive. In: 2018 international conference on electrical sciences and technologies in Maghreb (CISTEM), pp 1–6. https://doi.org/10.1109/CISTEM.2018.8613419.

Bao G, Hou Y (2017) Deadbeat direct torque control of induction motor based on MRAS. In: 2017 3rd IEEE international conference on control science and systems engineering (ICCSSE), pp 341–345. https://doi.org/10.1109/CCSSE.2017.8087953

Bednarz SA, Dybkowski M (2019) Estimation of the induction motor stator and rotor resistance using active and reactive power based model reference adaptive system estimator. App Sci 9(23), Art no 23. https://doi.org/10.3390/app9235145

Benlaloui I, Drid S, Chrifi-Alaoui L, Ouriagli M (2015) Implementation of a new MRAS speed Sensorless vector control of induction machine. IEEE Trans Energy Convers 30(2):588–595. https://doi.org/10.1109/TEC.2014.2366473

Boyar A, Kabalci E, Kabalci Y (2024) Sensorless speed controller of an induction motor with MRAS-based model predictive control. Comput Electr Eng 118:109350. https://doi.org/10.1016/j.compeleceng.2024.109350

Boyar A (2024) Improvement of IoT-based industrial motor drive including model predictive control and multilevel inverter, PhD Thesis, Nigde Omer Halisdemir University, Graduate School of Natural and Applied Sciences

Boyar A, Kabalci E (2024) Sensorless MPCC based PMSM drive with multilevel inverter for EVs. In: 2024 6th global power, energy and communication conference (GPECOM), pp 114–119. https://doi.org/10.1109/GPECOM61896.2024.10582756.

Boyar A, Kabalci E, Kabalci Y (2022) MRAS based model predictive torque control of induction motor drive for electric vehicles. In: 2022 4th global power, energy and communication conference (GPECOM), pp 51–56. https://doi.org/10.1109/GPECOM55404.2022.9815689.

Boyar A, Ersan K, Kabalcı Y (2023) Model predictive torque control-based induction motor drive with remote control and monitoring interface for electric vehicles. Electric Power Comp Syst 51(18):2159–2170. https://doi.org/10.1080/15325008.2023.2211581

Dybkowski M, Orlowska-Kowalska T, Tarchała G (2013) Sensorless traction drive system with sliding mode and MRASCC estimators using direct torque control. Automatika 54(3):329–336. https://doi.org/10.7305/automatika.54-3.169

El Merrassi W, Abounada A, Ramzi M (2021) Advanced speed sensorless control strategy for induction machine based on neuro-MRAS observer. Materials Today: Proceedings 45:7615–7621. https://doi.org/10.1016/j.matpr.2021.03.081

Fu T (2005) Speed-sensorless torque control of induction motors for hybrid electric vehicles, masters, Concordia University. Accessed 23 Jan 2024. https://spectrum.library.concordia.ca/id/eprint/8686/

Hannan MA, Ali JA, Mohamed A, Hussain A (2018) Optimization techniques to enhance the performance of induction motor drives: a review. Renew Sustain Energy Rev 81:1611–1626. https://doi.org/10.1016/j.rser.2017.05.240

Hughes A, Drury B (2013) Chapter seven–variable frequency operation of induction motors. In: Electric Motors and Drives (Fourth Edition), Hughes A and Drury B (eds) Boston: Newnes, pp 205–253. https://doi.org/10.1016/B978-0-08-098332-5.00007-3

Jannati M et al (2017) A review on variable speed control techniques for efficient control of single-phase induction motors: evolution, classification, comparison. Renew Sustain Energy Rev 75:1306–1319. https://doi.org/10.1016/j.rser.2016.11.115

Jun E-S, Nguyen MH, Kwak S-S (2020) Model predictive control method with NP voltage balance by offset voltage injection for three-phase three-level NPC inverter. IEEE Access 8:172175–172195. https://doi.org/10.1109/ACCESS.2020.3024634

Kabalcı E, Boyar A (2021) Chapter 7–Multilevel inverter applications for electric vehicle drives. In: Multilevel Inverters, Kabalcı E (ed), Academic Press, pp 185–208. https://doi.org/10.1016/B978-0-323-90217-5.00006-X.

Korzonek M, Tarchala G, Orlowska-Kowalska T (2019) A review on MRAS-type speed estimators for reliable and efficient induction motor drives. ISA Trans 93:1–13. https://doi.org/10.1016/j.isatra.2019.03.022

Korzonek M, Tarchala G, Orlowska-Kowalska T (2020) Simple stability enhancement method for stator current error-based MRAS-type speed estimator for induction motor. IEEE Trans Industr Electron 67(7):5854–5866. https://doi.org/10.1109/TIE.2019.2960726

Krušelj D, and Sumina D (2019) Modified predictive torque control method of induction machines for torque ripple reduction. Automatika 60(2): 227–238. https://doi.org/10.1080/00051144.2019.1614345.

Kumar R, Das S, Syam P, Chattopadhyay AK (2015) Review on model reference adaptive system for sensorless vector control of induction motor drives. IET Electr Power Appl 9(7):496–511. https://doi.org/10.1049/iet-epa.2014.0220

Kumar R, Das S, Chattopadhyay AK (2016) Comparative assessment of two different model reference adaptive system schemes for speed-sensorless control of induction motor drives. IET Electr Power Appl 10(2):141–154. https://doi.org/10.1049/iet-epa.2015.0121

Lai Y-S, Lin J-C, Wang JJ (2000) Direct torque control induction motor drives with self-commissioning based on Taguchi methodology. IEEE Trans Power Electron 15(6):1065–1071. https://doi.org/10.1109/63.892820

Manias SN (2017) 12–Introduction to motor drive systems. In: Power electronics and motor drive systems. In: Manias SN (ed) Academic Press, pp 843–967. https://doi.org/10.1016/B978-0-12-811798-9.00012-3.

Marwali MN, Keyhani A (1997) A comparative study of rotor flux based MRAS and back EMF based MRAS speed estimators for speed sensorless vector control of induction machines. In: IAS '97 conference record of the 1997 IEEE industry applications conference thirty-second IAS annual meeting, vol 1, pp 160–166. https://doi.org/10.1109/IAS.1997.643022.

Mossa MA, Echeikh H, Iqbal A, Duc Do T, Al-Sumaiti AS (2020) A novel Sensorless control for multiphase induction motor drives based on singularly perturbed sliding mode observer-experimental validation. Applied Sci 10(8) Art no 8. https://doi.org/10.3390/app10082776.

Orlowska-Kowalska T, Dybkowski M (2010) Stator-current-based MRAS estimator for a wide range speed-sensorless induction-motor drive. IEEE Trans Industr Electron 57(4):1296–1308. https://doi.org/10.1109/TIE.2009.2031134

Panchade VM, Chile RH, Patre BM (2013) A survey on sliding mode control strategies for induction motors. Annu Rev Control 37(2):289–307. https://doi.org/10.1016/j.arcontrol.2013.09.008

Peng FZ, Fukao T, Lai JS (1994) Low-speed performance of robust speed identification using instantaneous reactive power for tacholess vector control of induction motors. In: Proceedings

of 1994 IEEE industry applications society annual meeting, vol 1, pp 509–514. https://doi.org/10.1109/IAS.1994.345437
Peng F-Z, Fukao T (1994) Robust speed identification for speed-sensorless vector control of induction motors. IEEE Trans Ind Appl 30(5):1234–1240. https://doi.org/10.1109/28.315234
Poorfakhraei A, Narimani M, Emadi A (2021) A review of modulation and control techniques for multilevel inverters in traction applications. IEEE Access 9:24187–24204. https://doi.org/10.1109/ACCESS.2021.3056612
Ramamoorthy RT, Larimore B, Bhardwaj M (2021) Sensored field oriented control of 3-phase permanent magnet synchronous motors using TMS320F2837x. Accessed 18 Jun 2025. https://www.ti.com/lit/pdf/sprabz0
Reddy CU, Prabhakar KK, Singh AK, Kumar P (2020) Speed estimation technique using modified stator current error-based MRAS for direct torque controlled induction motor drives. IEEE J Emerg Sel Top Power Electron 8(2):1223–1235. https://doi.org/10.1109/JESTPE.2019.2901523
Rodriguez J et al (2022) Latest advances of model predictive control in electrical drives—Part I: basic concepts and advanced strategies. IEEE Trans Power Electron 37(4):3927–3942. https://doi.org/10.1109/TPEL.2021.3121532
Schauder C (1992) Adaptive speed identification for vector control of induction motors without rotational transducers. IEEE Trans Ind Appl 28(5):1054–1061. https://doi.org/10.1109/28.158829
Sutikno T, Idris NRN, Jidin A (2014) A review of direct torque control of induction motors for sustainable reliability and energy efficient drives. Renew Sustain Energy Rev 32:548–558. https://doi.org/10.1016/j.rser.2014.01.040
Tamai S, Sugimoto H, Yano M (1987) Speed sensorless vector control of induction motor with model reference adaptive system. In: Proceedings of IEEE industry application society (IAS) annual meeting, pp 189–195
Trzynadlowski AM (2001) 5–Scalar control methods. In: Control of Induction Motors, Trzynadlowski AM ed. In: Engineering, San Diego: Academic Press, pp 93–105. https://doi.org/10.1016/B978-012701510-1/50005-2.
Verma V, Chakraborty C (2014) New series of MRAS for speed estimation of vector controlled induction motor drive. In: IECON 2014 40th annual conference of the IEEE industrial electronics society, pp 755–761. https://doi.org/10.1109/IECON.2014.7048585.
Wang F, Chen Z, Stolze P, Stumper J-F, Rodríguez J, Kennel R (2014) Encoderless finite-state predictive torque control for induction machine with a compensated MRAS. IEEE Trans Industr Inf 10(2):1097–1106. https://doi.org/10.1109/TII.2013.2287395
Wang F, Zhang Z, Mei X, Rodríguez J, Kennel R (2018) Advanced control strategies of induction machine: field oriented control, direct torque control and model predictive control. Energie 11(1), art no 1. https://doi.org/10.3390/en11010120
Yan Z, Jin C, Utkin V (2000) Sensorless sliding-mode control of induction motors. IEEE Trans Industr Electron 47(6):1286–1297. https://doi.org/10.1109/41.887957
Yaramasu V, Wu B (2017) Modeling of power converters for model predictive control. In: Model predictive control of wind energy conversion systems, John Wiley & Sons, Ltd, pp 149–176. https://doi.org/10.1002/9781119082989.ch5.
Zhang Y, Jin J, Huang L (2021) Model-free predictive current control of PMSM drives based on extended state observer using Ultralocal model. IEEE Trans Industr Electron 68(2):993–1003. https://doi.org/10.1109/TIE.2020.2970660

Model Predictive Control (MPC) in Power Electronics

Doğan Çelik

Abstract Model predictive control (MPC) has become a cornerstone technique in power electronics, offering advanced capabilities in dynamic response, constraint handling and system optimization. By predicting future system behaviour over a finite horizon and adjusting control actions in real time, MPC enables superior regulation of voltage, current and power compared to traditional control strategies. This chapter examines the fundamental operating principles, modelling techniques and prediction enhancement methods for MPC, including discretization using the Forward Euler method and reference prediction through Lagrange interpolation. It also addresses the integration of prediction error compensation mechanisms to further improve control accuracy. Additionally, the chapter discusses state-space modelling and voltage reference generation for two-level voltage source inverters (2L-VSI). A comprehensive overview of MPC applications is presented, covering grid-connected inverters, active front-end (AFE) rectifiers, motor drives, uninterruptible power supplies (UPS), high-voltage direct current (HVDC) systems and renewable energy systems. Finally, emerging trends such as AI-assisted MPC, digital twin integration and hybrid MPC–machine learning approaches are introduced, highlighting MPC's critical role in the evolution of future smart grids, energy storage systems and sustainable energy infrastructures.

Keywords Model predictive control · Power electronic converters · Lagrange interpolation · Prediction error compensation

Nomenclature

Symbols	Description
v_I	Inverter-side voltage (V)
v_g	Grid voltage (V)
x_k	State vector at sampling instant *k*

D. Çelik (✉)
Department of Electrical and Electronics Engineering, Van Yuzuncu Yil University, Van, Turkey
e-mail: dogancelik@yyu.edu.tr

E. Kabalci (ed.), *Modern Control Methods for Power Electronics*, Energy Systems in Electrical Engineering, https://doi.org/10.1007/978-981-92-1974-2_10

u_k	Input (control) vector at sampling instant k
y_k	Output vector at sampling instant k
S_i	Discrete switching state
J	Objective function (cost)
$\in_s$	Prediction error
k_x	Weighting coefficient for error compensation
$\Delta T(k)$	Sampling time interval at k

1 Chapter Overview

Model predictive control (MPC) has emerged as a dominant and flexible control strategy in power electronics due to its superior performance in dynamic response, constraint handling and multi-objective optimization. Unlike traditional control methods such as proportional integral derivative (PID) and hysteresis controllers, MPC predicts future system behaviour over a finite horizon and optimally adjusts control actions in real time based on the minimization of a cost function. This predictive nature enables MPC to achieve high-precision control of voltage, current and power flow in complex and dynamic environments (Lam and Gong, 2025; Carnielutti et al. 2024; Young et al. 2016).

This chapter provides a structured discussion of MPC's operating principles, implementation techniques and applications in modern power electronic systems. First, the operating principle of MPC in power converters and electrical drives is presented, highlighting the basic prediction, cost evaluation and optimization procedures involved. Subsequently, advanced prediction techniques, including error correction methods based on Lagrange interpolation and discretization through the Forward Euler method, are discussed to enhance the predictive accuracy and real-time applicability of MPC. State-space modelling and the generation of voltage reference signals for two-level voltage source inverters (2L-VSI) are then introduced, demonstrating how predictive models can be used to regulate inverter output voltages and currents effectively. Furthermore, the chapter explores the broad range of MPC applications in power electronics, including grid-tied inverters, DC-DC converters, multi-level inverters and their integration into renewable energy systems. In the context of electrical drives, the MPC improves torque and flux control performance in induction motors and permanent magnet synchronous machines, offering faster transient responses, reduced switching losses and enhanced efficiency (Gao et al. 2024; Ran et al. 2024; An et al. 2018; Zarei et al. 2022; Hausberger et al. 2023; Harbi et al. 2023). Despite these advantages, the implementation of MPC presents challenges related to computational complexity and real-time feasibility. This has driven research into efficient numerical solvers, parallel computing and hardware acceleration techniques using field-programmable gate arrays (FPGAs) and digital signal processors (DSPs) (Purraji et al. 2025). Moreover, stability assurance and controller tuning under varying operating conditions remain critical research areas. The MPC is also increasingly applied in renewable energy systems, where it enhances the control of doubly fed induction generators (DFIGs) in wind energy applications by reducing reactive power losses and improving grid stability. In photovoltaic (PV) systems, the MPC improves maximum power point tracking efficiency and facilitates better grid integration through predictive inverter control strategies. These contributions position MPC as a key enabler for the modernization of energy systems, improving overall system efficiency, reliability and resilience (Bordons et al. n.d.; Model Predictive Control for Doubly-Fed Induction Generators and Three-Phase Power Converters, n.d.; Lu et al. 2024; Ullah et al. 2024; Garcés-Ruiz et al. 2024).

Finally, this chapter outlines emerging trends and future research directions, such as AI-assisted MPC for adaptive and intelligent control, digital twin integration for predictive maintenance and hybrid approaches combining MPC with machine learning techniques. The expanding role of MPC in smart grids, microgrids and energy storage management is also discussed, emphasizing its potential to enhance grid resilience, optimize load balancing and support the transition towards sustainable and intelligent energy systems.

2 MPC Methods for Power Electronics-Based Systems

The MPC has emerged as a powerful and versatile control strategy for power electronic systems, offering predictive capabilities, multivariable control and inherent constraint handling. Originally introduced in the 1960s for process industries, MPC gradually expanded into sectors such as petrochemical, automotive, aerospace and mining. Its application to power electronics started in the 1980s, though early adoption was limited by the computational demands exceeding the capabilities of processors at that time. Thanks to significant advancements in digital signal processors (DSPs), field-programmable gate arrays (FPGAs) and microcontrollers since the 2000s, MPC has gained renewed attention in power electronics and electric drives, enabling real-time implementations that were previously impractical. In modern power electronics, the MPC has been successfully applied in a broad range of systems including motor drives, active front ends, uninterruptible power supplies (UPSs), static synchronous compensators (STATCOMs), high-voltage direct current (HVDC) systems and flexible AC transmission systems (FACTS). The increasing demand for fast dynamic response, reduced switching losses, improved harmonic performance and operation under constraints has made MPC an attractive alternative to traditional linear controllers. The MPC methods for power electronic systems are generally categorized into indirect MPC and direct MPC, depending on whether a separate modulator is employed (Rodriguez et al. 2022a, 2022b).

Indirect MPC, also known as continuous control set MPC, generates continuous-valued control outputs, such as duty cycles or voltage references, which are then converted into switching signals by a dedicated modulator like pulse width modulation (PWM) or space vector modulation (SVM). This structure allows leveraging existing modulation schemes while embedding predictive features into the control loop. Quadratic programming (QP)-based MPC approaches can generally be divided into two main categories. The first category consists of controllers that employ explicit MPC, where the optimization problem—typically a mixed-integer or standard quadratic program (MIQP or QP)—is solved offline. In this case, all possible system states are considered in advance, and the corresponding control actions are precomputed and stored. During real-time operation, the appropriate control input is retrieved from a lookup table, allowing for extremely fast execution. Within the context of indirect MPC, both online and offline strategies are possible. While offline solutions significantly reduce the computational burden during operation, they require substantial memory to store the pre-evaluated control laws (Añon et al. 2024; Najafqolian et al. 2024).

Direct MPC, or finite control set MPC (FCS-MPC), integrates the control and modulation functions by directly selecting the optimal switching state from a finite set of available states. Unlike indirect MPC, no separate modulator is needed because the switching action is determined at every sampling interval based on predictions of the system's future behaviour. This approach offers faster dynamic response and inherently enforces switching constraints, but it increases computational burden due to the combinatorial nature of the optimization problem (Vazquez et al. 2014; Kouro et al. 2015). Several advanced strategies have been developed within both categories to address practical challenges. The FCS-MPC with long prediction horizons improves steady-state performance and mitigates issues like steady-state error and switching frequency variation, albeit with increased computational effort. The MPC with variable switching instants enhances control quality by optimizing not only the switching state but also the switching instant within a sampling period, enabling fixed switching frequency and reduced current ripple. Hybrid MPC approaches combine predictive control with classical modulation schemes or introduce hysteresis bands to balance computational complexity and performance. Explicit MPC precomputes control laws offline, facilitating fast execution in systems with tight real-time constraints but requiring significant memory resources. The MPC using programmed modulation patterns incorporates optimal or selective harmonic elimination patterns into the predictive framework to achieve improved harmonic performance with reduced switching losses. Beyond basic control, MPC frameworks have been expanded to address multi-objective control problems, fault-tolerant operation and nonlinear or non-ideal system modelling. Additionally, researchers have explored computationally efficient formulations (e.g. by reducing candidate switching sets or using approximate cost functions) and hardware acceleration using parallel processing on FPGAs and GPUs to enable real-time implementation for high-switching-frequency converters and high-speed drives (Al-Kaf et al. 2025; Iranian et al. 2024).

The main advantages of MPC in power electronics include its fast transient response, inherent multivariable and constraint handling capabilities, direct control of converter switching states and flexibility to include nonlinear models and objectives. However, challenges remain regarding computational complexity, variable switching frequency in direct MPC and sensitivity to parameter variations. Recent advances such as model-free MPC, self-tuning predictive controllers and learning-based predictive methods aim to further improve robustness, performance and adaptability. Figure 1 provides a comprehensive overview of the different categories of predictive control methods, highlighting their structural characteristics, key operational principles and main advantages such as fast dynamic response, inherent constraint handling and suitability for nonlinear and multivariable systems (Karamanakos et al. 2020). The MPC strategies in power electronics can be grouped into two main types, depending on whether a separate modulator is used. In indirect MPC, a modulation signal or duty cycle is calculated by the controller and then passed to a modulator to generate the switching signals, as illustrated in Fig. 2a. Here, the control output is a continuous-valued vector. In contrast, direct MPC combines control and modulation into a single step, removing the need for a separate modulator, as shown in Fig. 2b (Karamanakos et al. 2020; Rodriguez And Cortes 2012).

The MPC is a model-based control strategy that depends on an accurate system model to predict future behaviour, an essential requirement in power electronics, where control actions must be executed within microseconds. The system can be described by a state vector x (including key variables like inductor currents and capacitor voltages) and manipulated input variables u that influence system behaviour.

In direct MPC, the control input vector u represents discrete switching states of the power converter switches, expressed as follows:

$$u = \begin{bmatrix} u_a \; u_b \; u_c \end{bmatrix}^T \in U_\delta \subset \mathbb{Z}^3 \tag{1}$$

where U_δ denotes the set of all possible switching combinations in the converter.

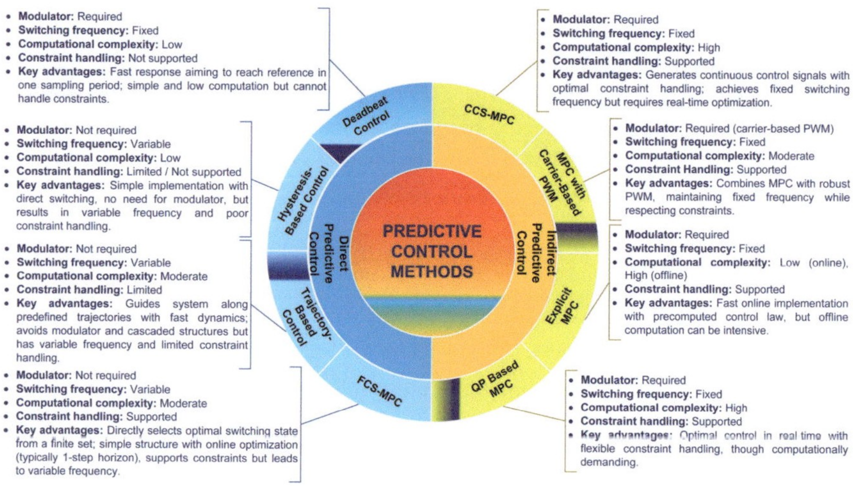

Fig. 1 Classification of predictive control methods

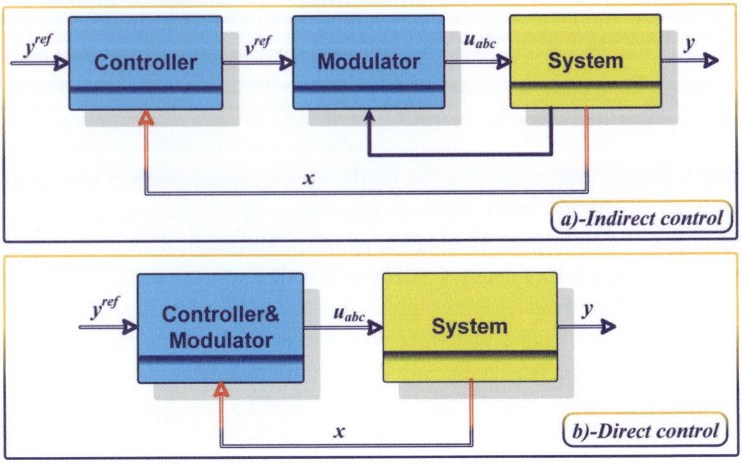

Fig. 2 Main structural configurations of predictive control methods

Conversely, in indirect MPC, the control input u corresponds to continuous modulation signals, typically normalized between -1 and 1:

$$u = \begin{bmatrix} u_a \ u_b \ u_c \end{bmatrix}^T \in U_\kappa = [-1, 1]^3 \subset \mathbb{R}^3 \tag{2}$$

The system dynamics are described by the continuous-time state-space equations:

$$\begin{aligned} \dot{x}(t) &= f_c(x(t), u(t)) \\ y(t) &= g_c(x(t)) \end{aligned} \tag{3}$$

where y is the output vector, selected based on the application. $x(t) \in \mathbb{R}^{n_x}$ represents the state vector (e.g. inductor currents, capacitor voltages), and $y(t) \in \mathbb{R}^{n_y}$ denotes the output vector, which is selected based on the control objectives. The functions $f_c(\cdot)$ and $g_c(\cdot)$ represent the state update and output equations, respectively. If the system is linear, these equations simplify to the standard linear state-space form:

$$\begin{aligned} \dot{x}(t) &= A_c x(t) + B_c u(t) \\ y(t) &= C_c x(t) \end{aligned} \tag{4}$$

where $A_c \in \mathbb{R}^{n_x \times n_x}$, $B_c \in \mathbb{R}^{n_x \times n_u}$ and $C_c \in \mathbb{R}^{n_y \times n_x}$ are the system, input and output matrices, respectively. For systems exhibiting bilinear dynamics, the state-update equation becomes:

$$f_c(x(t), u(t)) = [A_{c1} + A_{c2} u(t)] x(t) + B_c u(t) \tag{5}$$

Three-phase systems are often modelled in a two-dimensional reference frame, either rotating or stationary. This transformation, inspired by linear control methods to decouple control loops, converts three-phase variables into a simplified two-axis representation $\xi_{abc} = [\xi_a \xi_b \xi_c]^T$. In the dq plane:

$$\xi_{dq} = \mathcal{K}(\varphi)\xi_{abc} \tag{6}$$

where $\xi_{dq} = \left[\xi_d \xi_q\right]^T$. φ is the angular position of the rotating reference frame relative to phase-a axis. For a stationary reference frame $\left(\omega_{fr} = 0\right)$, the transformation simplifies to the $\alpha\beta$ plane:

$$\xi_{\alpha\beta} = \mathcal{K}(0)\xi_{abc} \tag{7}$$

where $\xi_{\alpha\beta} = \left[\xi_\alpha \xi_\beta\right]^T$.

3 Operating Principle of MPC in Power Electronics

MPC is an advanced control technique that determines the optimal control input by forecasting the system's future response over a finite horizon. It utilizes a dynamic model of the system to predict future behaviour and select control actions that minimize a predefined cost function, ensuring that the system output closely follows a desired reference trajectory [25, 26]. In the context of power electronic converters (PECs) or drive systems, the control objective typically involves selecting an appropriate switching signal $S(t)$ often corresponding to the gate signals of a PEC to regulate a system variable $x(t)$ as closely as possible to its reference value $x^*(t)$. To enable these predictions, a mathematical model of the system is required. This model is typically formulated in discrete-time state-space representation, which facilitates the prediction of future states based on current measurements and control inputs (Kouro et al. 2009; Vazquez et al. 2017; Zhou et al. 2025; Khalilzadeh et al. 2021). The general form of this model is expressed as follows (Khalilzadeh et al. 2021):

$$\begin{aligned} x(k+1) &= A_c x(k) + B_c u(k) \\ y(k) &= A_c x(k) + D_c u(k) \end{aligned} \tag{8}$$

where x, u and y represent the state, input and output vectors of the system, respectively. The matrices A_c, B_c, C_c and D_c correspond to the system dynamics, input influence, output mapping and feedforward (or direct transmission) paths. The indices k and $k+1$ denote the discrete-time steps, referring to the current and subsequent sampling instants, respectively (Kouro et al. 2009; Vazquez et al. 2017; Khalilzadeh et al. 2021).

In finite control set MPC (FCS-MPC), the control input $u(k)$ is selected from a limited set of discrete switching states S_i, where $i = 1, 2, \cdots, n$. At each sampling instant t_k, the controller evaluates every possible switching option S_i to forecast the system's subsequent state based on the dynamic model.

$$x_{pi}(t_{k+1}) = f_p(x(t_k), S_i \tag{9}$$

f_p denotes prediction function.

To evaluate system performance, an objective function J is formulated, typically to quantify tracking error or other control goals. In cases involving multiple performance criteria, the total cost function can be constructed by aggregating several sub-objectives, $j_1, j_2, \cdots, j_n$ each scaled by corresponding weighting factors $a, b, \cdots, n$, allowing flexible prioritization of control objectives (Khalilzadeh et al. 2021):

$$J = f((x(t), u(t)) = aj_1 + bj_2 + \cdots nj_n \tag{10}$$

A straightforward and widely used formulation for a sub-objective j_i involves calculating the absolute deviation between the predicted system output and the corresponding reference value:

$$j_g = \left| x^*(t_{k+1}) - x_{pi}(t_{k+1}) \right| \tag{11}$$

The controller assesses the objective function J_i for each candidate control action S_i and selects the optimal input corresponding to the action that yields the lowest overall cost:

$$\begin{aligned} J_{opt} &= \min(J_1, J_2, \cdots, J_n) \\ S_{opt} &= \arg\{J_i\}_{i=1,2,\cdots,n} \end{aligned} \tag{12}$$

Assuming that the reference value does not change significantly within a single sampling interval T_s, it is reasonable to approximate the future reference as $x^*(t_{k+1}) \approx x^*(t_k)$. The overall working principle is illustrated in Fig. 3a. In the ideal scenario shown in Fig. 3b, all measurements, computations and control decisions are executed instantaneously. The control input $S(t)$ corresponds to the discrete switching states of the converter, denoted as S_i, where $i = 0, \cdots, 7$, as depicted in Fig. 3c.

Figure 4 illustrates the overall structure of the MPC system. The control approach utilizes both the reference and the predicted current values at the $(k + 1)$ sampling instant to effectively counteract delays introduced by digital implementation.

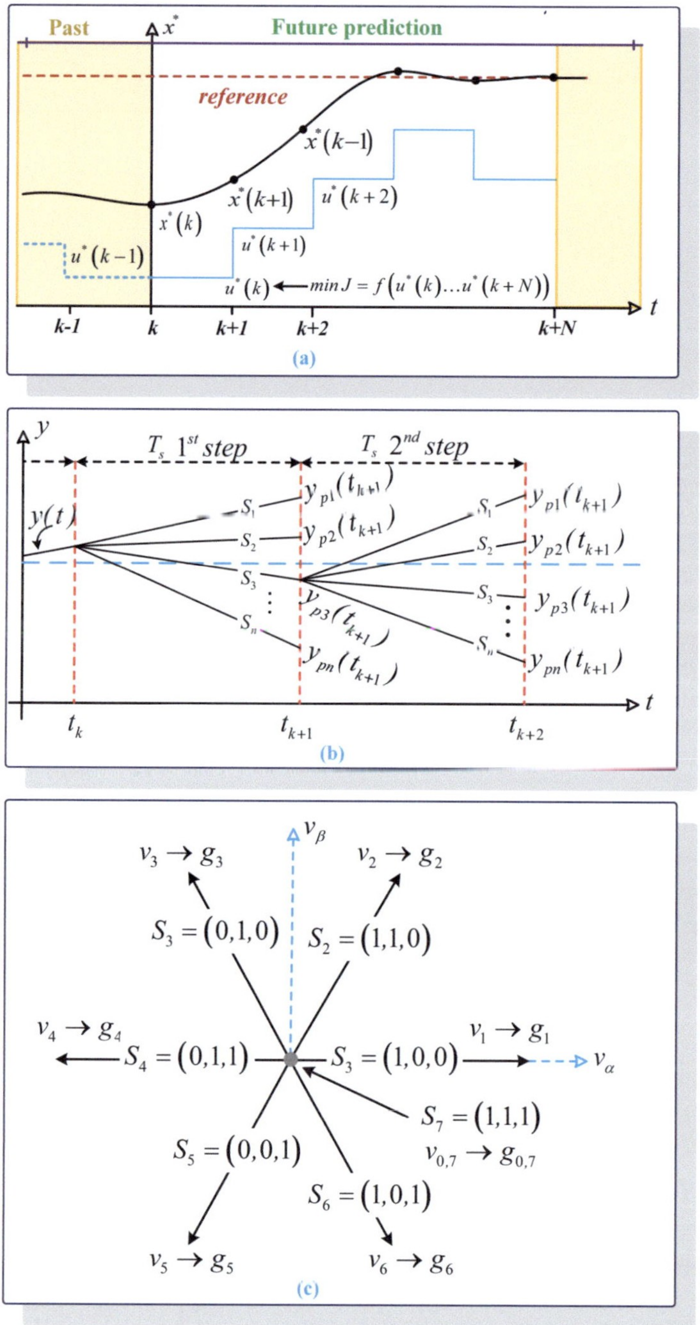

Fig. 3 **a** Principle of MPC; **b** idealized implementation with real-time measurement, computation and control execution and **c** relationship among voltage space vectors, inverter switching states and the associated cost function

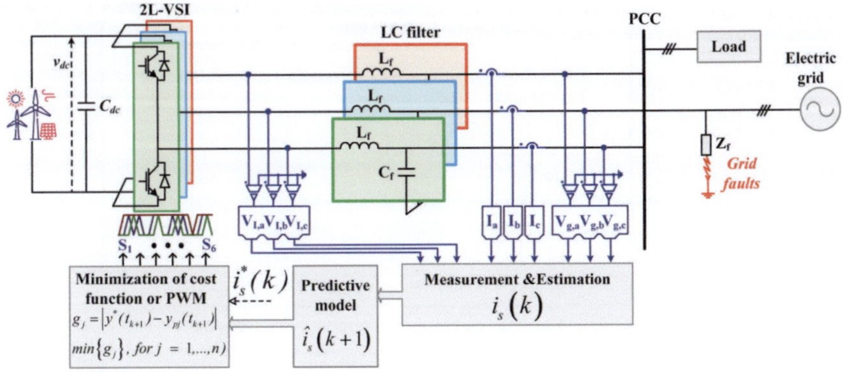

Fig. 4 MPC generic control diagram for PECs

4 Prediction Error Correction with Lagrange Interpolation and Forward Euler Approach

To develop a discrete-time model of a dynamic system, appropriate discretization techniques are essential. Two commonly used methods are the Forward Euler method and Lagrange interpolation, particularly effective for first-order dynamic systems (Kukrer, 1996; Bode et al. 2005).

4.1 Discretization via Forward Euler Method

In a grid-connected inverter system, the continuous-time derivative of a state variable can be approximated in discrete form as follows:

$$\frac{dx}{dt} \approx \frac{x(k+1) - x(k)}{\Delta T(k)} \tag{13}$$

where T_s is the sampling time interval.

4.2 4.1.1. Reference Prediction Using Lagrange Interpolation

In predictive control applications involving sinusoidal references or larger sampling intervals, tracking delays may occur. To overcome this, Lagrange interpolation is used to estimate future reference values based on current and past samples (Kukrer, 1996; Bode et al. 2005).

The general nth order Lagrange interpolation for predicting the reference current $\tilde{i}^*(k+1)$ is expressed as follows:

$$\tilde{i}^*(k+1) = \sum_{j=0}^{n} (-1)^{n-j} \begin{bmatrix} n+1 \\ j \end{bmatrix} i^*(k+j-n) \tag{14}$$

For sinusoidal references, $n = 2$ or higher is typically recommended. When $n = 2$, the interpolation simplifies to the following:

$$\tilde{i}^*(k+1) = 3i^*(k) - 3i^*(k-1) + i^*(k-2) \tag{15}$$

To estimate two steps ahead, the expression of (10.15) can be forward shifted:

$$\tilde{i}^*(k+2) = 3\tilde{i}^*(k+1) - 3i^*(k) + i^*(k-2) \tag{16}$$

Substituting (15) into (16), we derive:

$$\tilde{i}^*(k+2) = 6i^*(k) - 8i^*(k-1) + 3i^*(k-2) \tag{17}$$

4.2.1 Prediction Error Compensation

To further enhance the accuracy of predicted current values, a prediction error compensation mechanism is integrated. This correction accounts for system uncertainties, parameter variations and modelling inaccuracies (Zhou et al. 2025). The compensated predicted current is given by:

$$i_p^s(k+1)\Big|_u = i_{p0}^s(k+1)\Big|_u + k_x \varepsilon_s|_u \tag{18}$$

where u denotes voltage vector, ε_s is prediction error and represents a weight coefficient and $k_x \in [0, 1]$. At each sampling instant, some prediction error naturally occurs due to changes in system parameters over time (Zhou et al. 2025). This error can be defined as follows:

$$\varepsilon_s(k) = i_{p0}^s(k) - i_s(k) \tag{19}$$

Thus, integrating with the interpolation result, we have:

$$i_p^s(k+1)\Big|_u = \left[3i^*(k) - 3i^*(k-1) + i^*(k-2)\right] + k_x \varepsilon_s|_u \tag{20}$$

The MPC has gained widespread use in motor drives and PEC applications due to the complex and nonlinear nature of motor systems, where parameters are often hard to determine precisely and may vary over time. Although the overall objective remains consistent across different implementations, various prediction techniques have been developed within MPC frameworks. To provide a clear overview of MPC approaches, four categories are identified based on their model-free implementation strategies, as illustrated in Fig. 5a. Additionally, prediction errors are compensated for, and the corrected current values are used in the cost function to determine the most suitable voltage vector. An example of current prediction based on voltage vector selection is shown in Fig. 5b (Zhou et al. 2025).

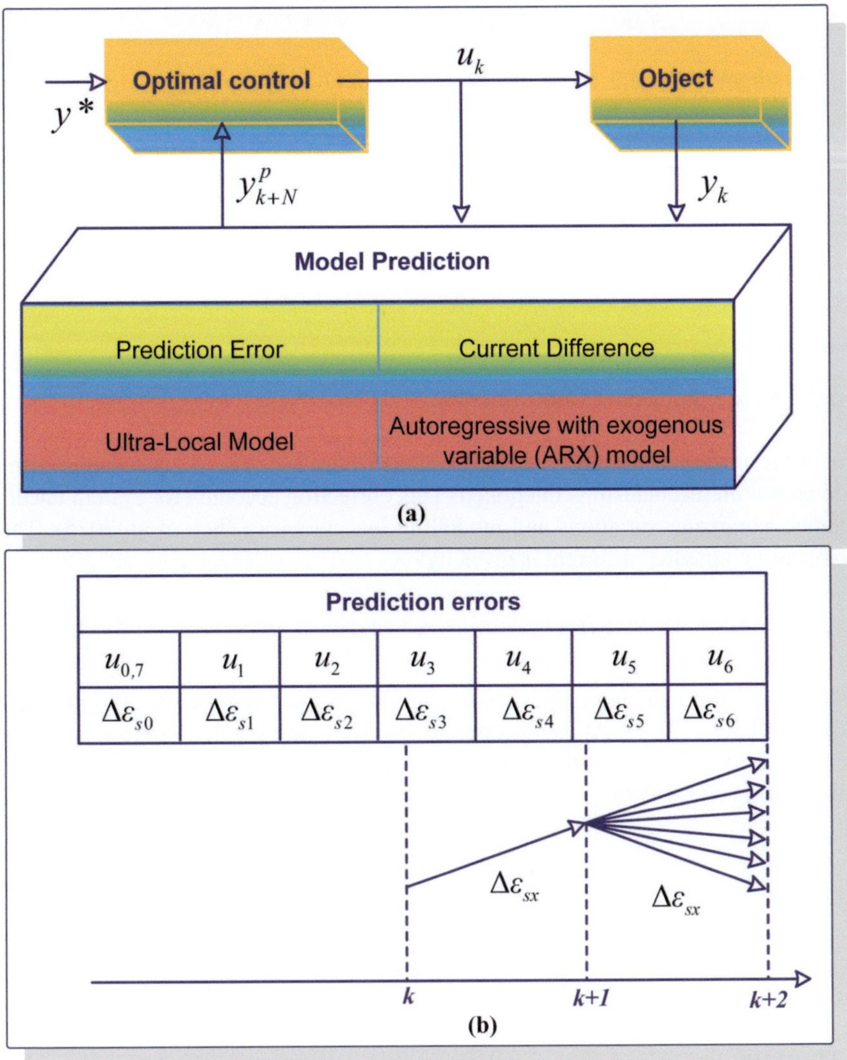

Fig. 5 **a** Model implementation techniques, and **b** approach for correcting prediction errors

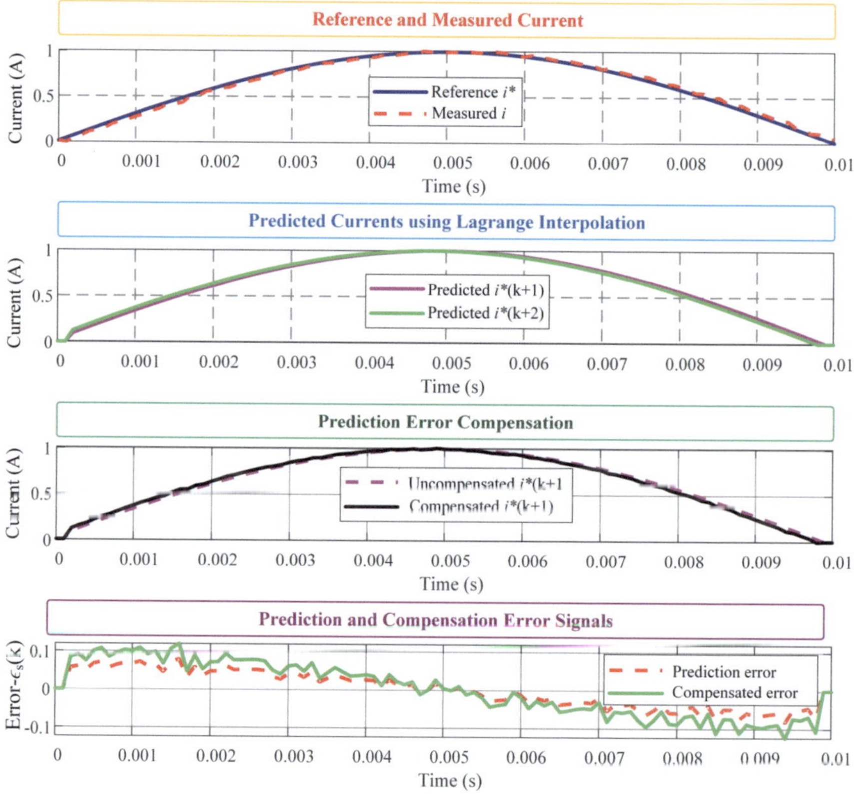

Fig. 6 MPC using Lagrange interpolation and error compensation

Figure 6 presents the predictive current control framework. It shows the difference between the reference and measured currents, the predicted currents using Lagrange interpolation at time steps $k+1$ and $k+2$, and the compensation of these predictions using a weighted error correction method. The variation of prediction and compensation errors over time highlights the role of interpolation and error correction in enhancing tracking accuracy in predictive control systems.

5 State-Space Modelling and Voltage Reference Generation of 2L-VSI

The inverter model in the stationary αβ-reference frame, derived using Kirchhoff's Voltage Law (KVL), is as follows:

$$v_I = v_g + L_f \frac{di}{dt} \tag{21}$$

Applying Clarke transformation, the inverter voltage relation becomes:

$$\vec{v}_{I,\alpha\beta} = \vec{v}_{g,\alpha\beta} + L_f \frac{di_{\alpha\beta}}{dt} \tag{22}$$

Discretizing this using Forward Euler yields:

$$\frac{L_f}{\Delta T(k)}\left(\vec{i}_{\alpha\beta}(k+1) - \vec{i}_{\alpha\beta}(k)\right) = \vec{v}_{I,\alpha\beta}(k+1) - \vec{v}_{g,\alpha\beta}(k+1) \tag{23}$$

Solving for the next-step current:

$$\vec{i}_{\alpha\beta}(k+1) = \vec{i}_{\alpha\beta}(k) + \frac{\Delta T(k)}{L_f}\left(\vec{v}_{I,\alpha\beta}(k+1) - \vec{v}_{g,\alpha\beta}(k+1)\right) \tag{24}$$

Let $\Delta\vec{v}_{I,\alpha\beta}(k+1) = \vec{v}_{I,\alpha\beta}(k+1) - \vec{v}_{g,\alpha\beta}(k+1)$, then:

$$\vec{i}_{\alpha\beta}(k+1) = \vec{i}_{\alpha\beta}(k) + \frac{\Delta T(k)}{L_f}\Delta\vec{v}_{I,\alpha\beta}(k+1) \tag{25}$$

Based on the reference current, the voltage increment can be derived:

$$\Delta\vec{v}_{I,\alpha\beta}(k+1) = \frac{L_f}{\Delta T(k)}\left(\vec{i}^{*}_{\alpha\beta}(k+1) - \vec{i}_{\alpha\beta}(k)\right) \tag{26}$$

The one-step predicted current is estimated as follows:

$$\vec{i}_{\alpha\beta}(k) = \vec{i}_{\alpha\beta}(k-1) + \frac{\Delta T(k)}{L_f}\Delta\vec{v}_{I,\alpha\beta}(k) \tag{27}$$

Substituting Eq. (27) into Eq. (26), we obtain the two-step predictive voltage increment:

$$\Delta\vec{v}_{I,\alpha\beta}(k+1) = \frac{L_f}{\Delta T(k)}\left(\vec{i}^{*}_{\alpha\beta}(k+1) - \vec{i}_{\alpha\beta}(k-1)\right) - \Delta\vec{v}_{I,\alpha\beta}(k) \tag{28}$$

Assuming the grid voltage vectors $\vec{v}_{g,\alpha\beta}(k+1)$ and $\vec{v}_{g,\alpha\beta}(k)$ are also predicted via Lagrange interpolation, the final reference inverter voltage vector becomes:

$$\begin{aligned}\vec{v}^{*}_{I,\alpha\beta}(k) = \frac{L_f}{\Delta T(k)}\left(\vec{i}^{*}_{\alpha\beta}(k+1) - \vec{i}_{\alpha\beta}(k-1)\right) - \left(\vec{v}_{I,\alpha\beta}(k) - \vec{v}_{g,\alpha\beta}(k)\right) \\ + \vec{v}_{g,\alpha\beta}(k+1)\end{aligned} \tag{29}$$

Figure 7 presents a flow diagram illustrating the various tasks carried out by the predictive controller. In this structure, the outer control loop operates at each sampling interval, while the inner loop evaluates all possible switching states to determine the optimal state to apply during the following sampling period. To further enhance the accuracy of the predicted current values, a prediction error compensation mechanism is incorporated into the control process. This approach ensures that the control objectives are continuously met by dynamically updating the switching actions based on real-time system conditions and correcting for any model inaccuracies.

The vector diagram in the $\alpha\beta$-reference frame, shown in Fig. 8, illustrates the dynamic performance of the current prediction model. In the Fig. 8, the blue trace represents the measured currents, depicting the actual trajectory of the current

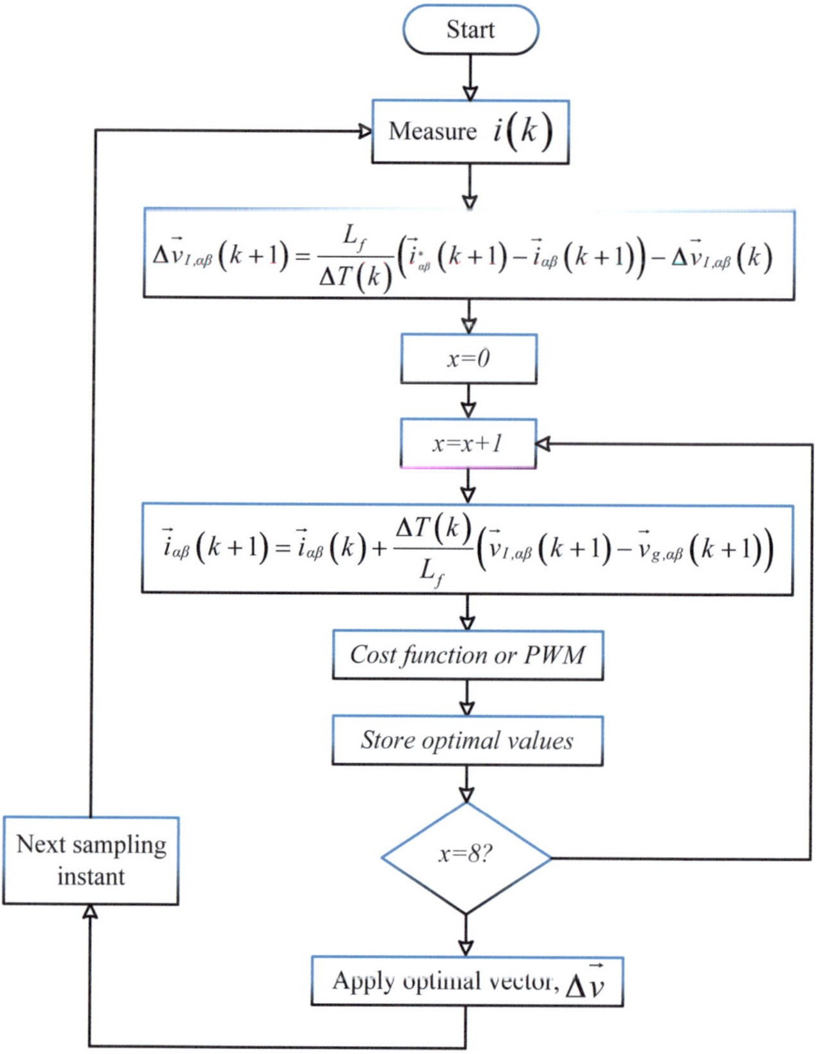

Fig. 7 Operational flow of the proposed predictive current control

vector over time in the stationary $\alpha\beta$ plane, while the red trace corresponds to the predicted currents generated by the model. A close alignment between the measured and predicted trajectories indicates that the model accurately captures the current behaviours in both magnitude and trend. The current tracking errors, represented by green vectors, highlight the instantaneous differences between predicted and measured values.

These errors are generally small, demonstrating the model's good overall performance. However, occasional deviations in the error vectors suggest opportunities for further refinement, particularly under specific operating conditions or transient events. Figure 8 provides a clear and effective visual validation of the current prediction model's ability to track real-world current dynamics in the $\alpha\beta$-reference frame.

The predicted currents under both balanced and unbalanced grid voltage conditions are illustrated in Fig. 9a and b for the α- and β-reference frames, respectively. To simulate an unbalanced condition, a single-phase fault was introduced on phase B, resulting in a 35% voltage drop. Despite this disturbance, the control system demonstrates a rapid adjustment in the amplitude of the load currents, accurately following the changes in the reference current. The predicted currents remain closely aligned

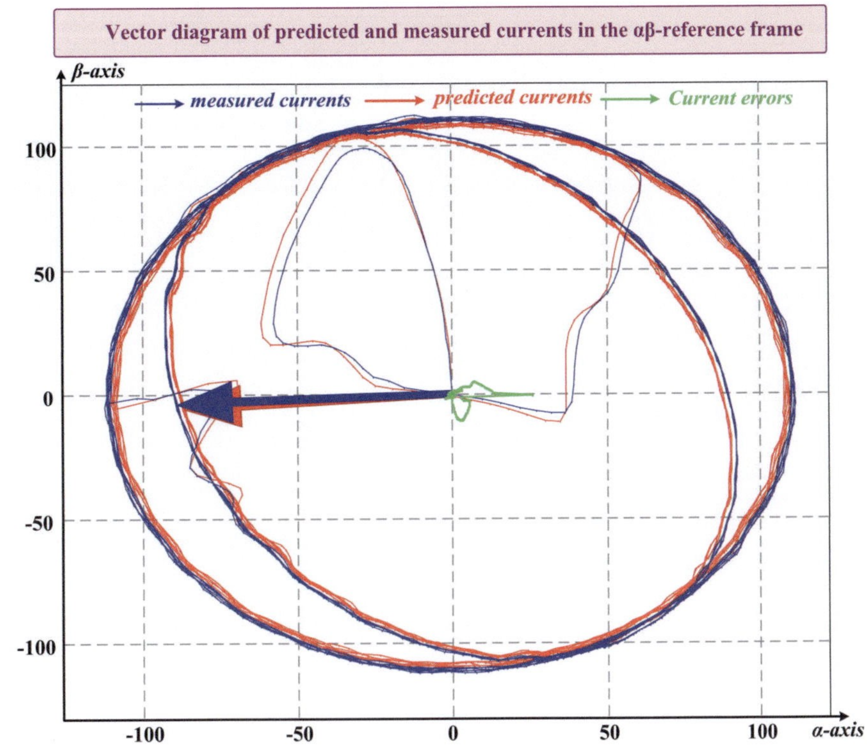

Fig. 8 Vector diagram of predicted and measured currents in the αβ-reference frame

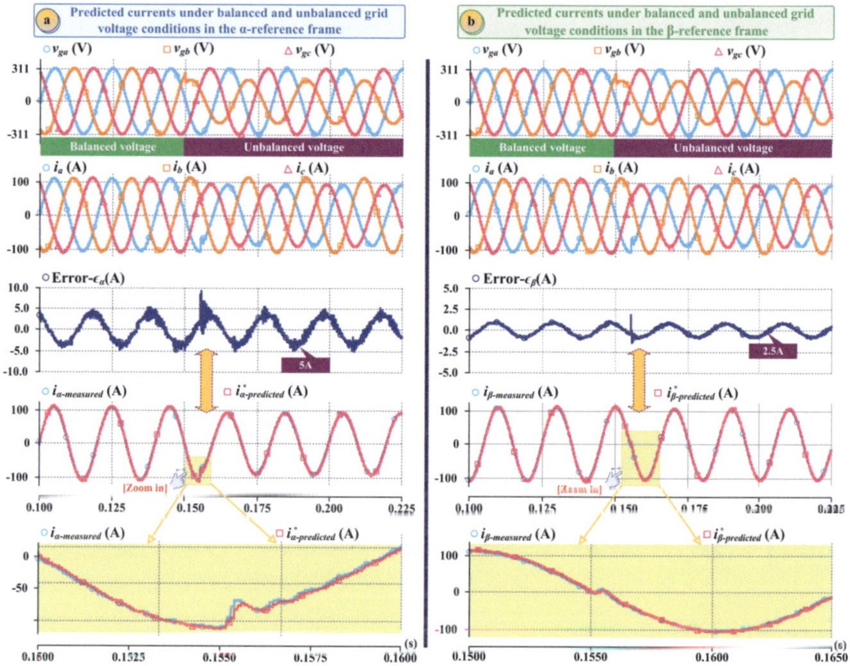

Fig. 9 Predicted currents under balanced and unbalanced grid voltage conditions in **a** α-and **b** β-reference frames

with the reference values, and the tracking error remains low throughout both the balanced and faulted conditions. This highlights the robustness and effectiveness of the predictive control strategy in maintaining stable current regulation even during asymmetrical grid events.

The dynamic response of the load currents and load voltage to a step change in the reference current amplitude is depicted in Fig. 10. In this case, the reference current amplitude is decreased from 110 to 70 A across all phases simultaneously. Figure 10 shows the reference current $\left(i^*_{\alpha\beta}\right)$, along with the predicted and measured currents, presented in both the abc and αβ-reference frames, as well as their corresponding tracking error signals. Following the change, the load currents adapt quickly, highlighting the controller's effectiveness in maintaining accurate current tracking. The tracking errors are minimal, remaining within 5 A under steady-state conditions and within 2.5 A during load transients.

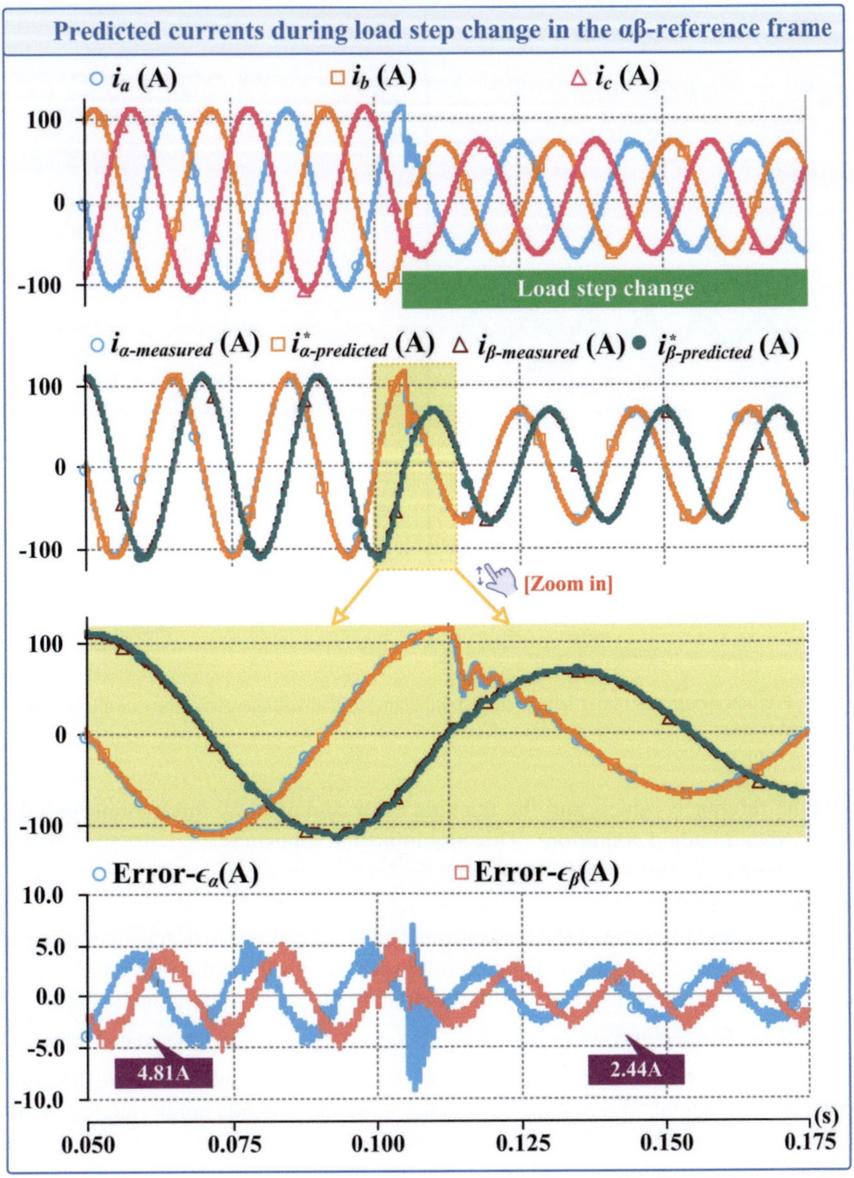

Fig. 10 Predicted and measured currents during load step change

6 Applications of MPC in Power Electronics

The MPC has received considerable attention in power electronics and electrical drives due to its capability to manage multivariable systems, handle operational constraints and optimize system performance over a predictive horizon (Karamanakos et al. 2020; Quevedo et al. n.d.). In power converters, MPC has been successfully applied for voltage regulation in DC-DC converters and for optimizing switching strategies in DC-AC and AC-DC converters, resulting in enhanced efficiency and dynamic response. Additionally, multi-level inverters used in high-power applications benefit from MPC's ability to improve voltage balancing and reduce harmonic distortion.

In the area of electrical drives, MPC facilitates predictive torque and current control of AC motors, enabling faster dynamic responses and higher operational efficiency in industrial applications. In renewable energy systems, MPC plays a crucial role in the control of wind energy systems, particularly those utilizing DFIGs, and in the predictive control of photovoltaic (PV) inverters for reliable grid integration. Moreover, MPC strategies have been employed for advanced energy storage management and the optimization of hybrid power systems, enhancing system reliability and economic viability.

These developments, as discussed extensively in recent literature (Vazquez et al. 2014, 2017; Zerdali et al. 2024; Elmorshedy et al. 2021), underscore the evolution of MPC into a robust and versatile control approach for a wide range of industrial and renewable energy applications. MPC has emerged as a powerful and flexible strategy for managing complex control tasks in PECs and electrical drives. Its predictive nature and ability to handle multivariable systems with constraints make it highly suitable for modern industrial and renewable energy applications. Figure 11 illustrates the most significant applications of MPC in this field, including a) current source converter–active front end (CSC-AFE), b) voltage source converter–uninterruptible power supply (VSC-UPS), c) STATCOM, d) motor drives and e) HVDC transmission systems. These applications highlight the versatility of MPC in improving dynamic performance, enhancing efficiency and ensuring robust operation under varying load and grid conditions.

7 Conclusion and Chapter Summary

This chapter has provided a comprehensive overview of MPC principles; implementation strategies and its applications in PECs, electrical drives and renewable energy systems. MPC's predictive capability, combined with its ability to handle multivariable systems and enforce operational constraints, makes it a highly effective and versatile control method for modern power electronics. Key topics addressed include the fundamental operating principles of MPC, enhancement of prediction accuracy through Forward Euler discretization and Lagrange interpolation, and the importance

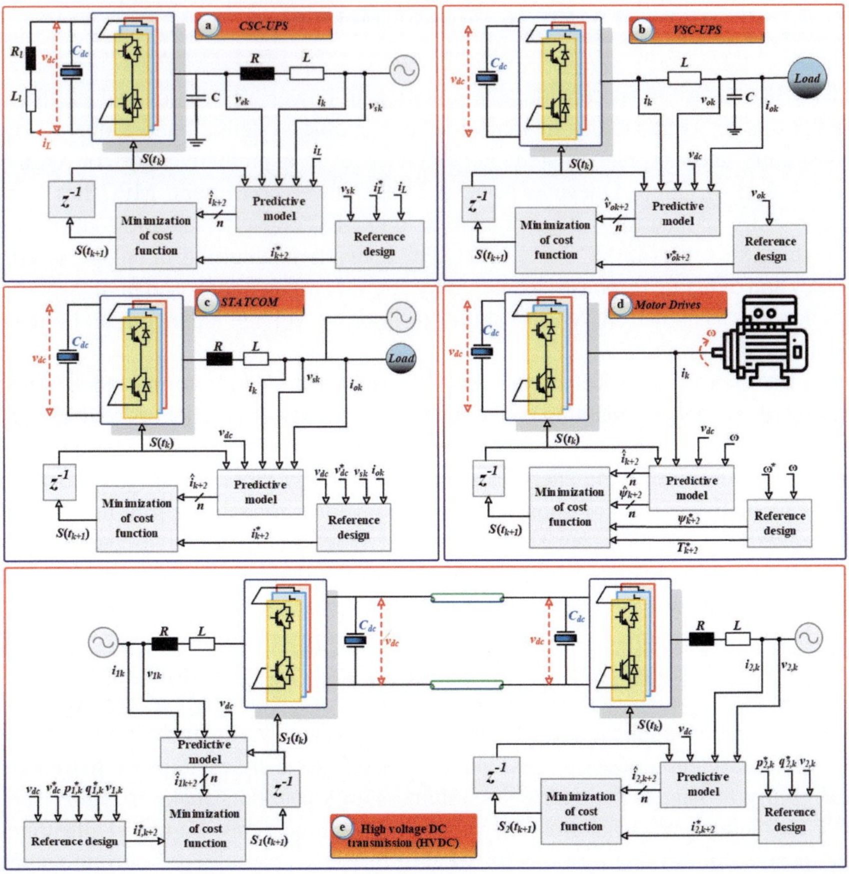

Fig. 11 The most relevant applications of MPC for PECs; **a** CSC-AFE, **b** VSC-UPS, **c** STATCOM, **d** motor drives and **e** high-voltage DC transmission

of state-space modelling for voltage reference generation in the 2L-VSI. Practical implementation of MPC has been demonstrated across DC-AC inverter systems, emphasizing its ability to improve system response, dynamic behaviour and overall efficiency.

The results validate the effectiveness of the proposed MPC approach. The vector diagram of predicted and measured currents in the αβ-reference frame confirms accurate current tracking, with minimal prediction errors. Under balanced and unbalanced grid voltage conditions, the predicted currents in both the α- and β-reference frames exhibit robust performance, maintaining stable behaviour even during a 35% voltage drop on one phase. Additionally, during a load step change, the predicted and measured currents showed fast and smooth transient responses, with tracking errors remaining within acceptable limits. These results highlight the strong tracking accuracy and disturbance rejection capabilities of the proposed predictive control strategy. Nevertheless, challenges such as computational complexity and the need for real-time processing capabilities necessitate the use of high-speed digital platforms such as FPGAs and DSPs. Stability assurance and robustness under variable operating conditions also remain critical areas of ongoing research.

For future works, emerging trends such as artificial intelligence (AI)-assisted MPC (Cotrufo et al. 2020) for adaptive control, digital twin integration for predictive maintenance and the application of MPC in smart grids and energy storage systems are expected to drive further advancements (Guevara et al. 2024). These developments will enhance the predictive control frameworks, contributing to the creation of smarter, more resilient and highly efficient next-generation power electronic systems.

References

Al Kaf HAG, Lee KB, Blaabjerg F (2025) Overview of DSP-based implementation of machine learning methods for power electronics and motor drives. J Power Electron 25:271–288. https://doi.org/10.1007/S43236-024-00975-2/FIGURES/18

An F, Song W, Yu B, Yang K (2018) Model Predictive Control With Power Self-Balancing of the Output Parallel DAB DC–DC Converters in Power Electronic Traction Transformer. IEEE J Emerg Sel Top Power Electron. 6:1806–1818. https://doi.org/10.1109/JESTPE.2018.2823364

Añon AM, Bae S, Saroya M, Isele, D (2024) Multi-profile quadratic programming (MPQP) for optimal gap selection and speed planning of autonomous driving. Proc IEEE Int Conf Robot Autom 12158–12164. https://doi.org/10.1109/ICRA57147.2024.10611173

Bode GH, Loh PC, Newman MJ, Holmes DG (2005) An improved robust predictive current regulation algorithm. IEEE Trans Ind Appl 41:1720–1733. https://doi.org/10.1109/TIA.2005.858324

Bordons C, Garcia-Torres F, Ridao MA (n.d.) Advances in industrial control model predictive control of microgrids

Carnielutti F, Aly M, Norambuena M, Hu J, Guerrero J, Rodriguez J (2024) Intelligent power electronic converters and control for microgrids: a master-slave model predictive approach. presented at the

Model predictive control for doubly-fed induction generators and three-phase power converters

Cotrufo N, Saloux E, Hardy JM, Candanedo JA, Platon R (2020) A practical artificial intelligence-based approach for predictive control in commercial and institutional buildings. Energy Build 206:109563. https://doi.org/10.1016/J.ENBUILD.2019.109563

Elmorshedy MF, Xu W, El-Sousy FFM, Islam MR, Ahmed AA (2021) Recent Achievements in Model Predictive Control Techniques for Industrial Motor: A Comprehensive State-of-the-Art. IEEE Access 9:58170–58191. https://doi.org/10.1109/ACCESS.2021.3073020

Gao X, Tian W, Yang Q, Chai N, Rodriguez J, Kennel R, Heldwein ML (2024) Model Predictive Control of a Modular Multilevel Converter Considering Control Input Constraints. IEEE Trans Power Electron 39:636–648. https://doi.org/10.1109/TPEL.2023.3318320

Garcés-Ruiz A, Gil-González W, Montoya OD (2024) Stability analysis for an ad-hoc model predictive control in DC/DC converters with a constant power load. Results in Engineering. 22:102262. https://doi.org/10.1016/j.rineng.2024.102262

Guevara BS, Recalde LF, Moya V, Varela-Aldas J, Gandolfo DC, Toibero JM (2024) Estimation of Unmodeled Dynamics: Nonlinear MPC and Adaptive Control Law With Momentum Observer Dynamic. IEEE Access 12:77121–77132. https://doi.org/10.1109/ACCESS.2024.3407684

Harbi I, Rodriguez J, Liegmann E, Makhamreh H, Heldwein ML, Novak M, Rossi M, Abdelrahem M, Trabelsi M, Ahmed M, Karamanakos P, Xu S, Dragičević T, Kennel R (2023) Model-Predictive Control of Multilevel Inverters: Challenges, Recent Advances, and Trends. IEEE Trans Power Electron 38:10845–10868. https://doi.org/10.1109/TPEL.2023.3288499

Hausberger T, Kugi A, Eder A, Kemmetmuller W (2023) Cooperative Model Predictive Control Concepts for Coupled AC/DC- and DC/DC-Power Converters. IEEE Trans Control Syst Technol 31:359–369. https://doi.org/10.1109/TCST.2022.3179583

Iranian ME, Zamiri E, de Castro A (2024) Optimal implementation of d-q frame finite control set model predictive control with LabVIEW. Electronics 2025, Vol 14, p 100. https://doi.org/10.3390/ELECTRONICS14010100

Karamanakos P, Liegmann E, Geyer T, Kennel R (2020) Model Predictive Control of Power Electronic Systems: Methods, Results, and Challenges. IEEE Open J Ind Appl 1:95–114. https://doi.org/10.1109/OJIA.2020.3020184

Khalilzadeh M, Vaez-Zadeh S, Rodriguez J, Heydari R (2021) Model-Free Predictive Control of Motor Drives and Power Converters: A Review. IEEE Access 9:105733–105747. https://doi.org/10.1109/ACCESS.2021.3098946

Kouro S, Cortés P, Vargas R, Ammann U, Rodríguez J (2009) Model predictive control-a simple and powerful method to control power converters. IEEE Trans Industr Electron 56:1826–1838. https://doi.org/10.1109/TIE.2008.2008349

Kouro S, Perez MA, Rodriguez J, Llor AM, Young HA (2015) Model Predictive Control: MPC's Role in the Evolution of Power Electronics. IEEE Ind Electron Mag 9:8–21. https://doi.org/10.1109/MIE.2015.2478920

Kukrer O (1996) Discrete-time current control of voltage-fed three-phase PWM inverters. IEEE Trans Power Electron 11:260–269. https://doi.org/10.1109/63.486174

Lam C-S, Gong C, Sou W-K (2025) Improved model predictive control for LC-HAPF. In: Advanced controllers design for hybrid active power filters. Springer Nature, Singapore, pp 39–60

Lu P, Zhang N, Ye L, Du E, Kang C (2024) Advances in model predictive control for large-scale wind power integration in power systems. Advances in Applied Energy. 14:100177. https://doi.org/10.1016/j.adapen.2024.100177

Najafqolian MA, Alipour K, Mousavifard R, Tarvirdizadeh B (2024) Control of Aerial Robots Using Convex QP LMPC and Learning-Based Explicit-MPC. IEEE Trans Industr Inform. https://doi.org/10.1109/TII.2024.3397353

Purraji M, Zamiri E, Sanchez A, de Castro A (2025) Rapid Prototyping for Design and Test of FPGA-Based Model Predictive Controllers for Power Converters. Journal of Electrical Engineering and Technology. 20:2419–2438. https://doi.org/10.1007/S42835-025-02173-X/FIGURES/11

Quevedo DE, Aguilera RP, Geyer T (n.d.) Model predictive control for power electronics applications

Ran X, Xu B, Liu K, Liu Y (2024) A Three-Vector Finite Control Set Model Predictive Control Variant With Low Power Fluctuations of AC/DC Converters. IEEE Journal of Emerging and Selected Topics in Industrial Electronics. 5:1338–1343. https://doi.org/10.1109/JESTIE.2024.3386179

Rodriguez J, Garcia C, Mora A, Davari SA, Rodas J, Valencia DF, Elmorshedy M, Wang F, Zuo K, Tarisciotti L, Flores-Bahamonde F, Xu W, Zhang Z, Zhang Y, Norambuena M, Emadi A, Geyer T, Kennel R, Dragicevic T, Khaburi DA, Zhang Z, Abdelrahem M, Mijatovic N (2022a) Latest Advances of Model Predictive Control in Electrical Drives - Part II: Applications and Benchmarking With Classical Control Methods. IEEE Trans Power Electron 37:5047–5061. https://doi.org/10.1109/TPEL.2021.3121589

Rodriguez J, Cortes P (2012) Model Predictive Control. In Predictive Control of Power Converters and Electrical Drives (eds J. Rodriguez and P. Cortes). https://doi.org/10.1002/9781119941446.ch3

Rodriguez J, Garcia C, Mora A, Flores-Bahamonde F, Acuna P, Novak M, Zhang Y, Tarisciotti L, Davari SA, Zhang Z, Wang F, Norambuena M, Dragicevic T, Blaabjerg F, Geyer T, Kennel R, Khaburi DA, Abdelrahem M, Zhang Z, Mijatovic N, Aguilera RP (2022b) Latest Advances of

Model Predictive Control in Electrical Drives - Part I: Basic Concepts and Advanced Strategies. IEEE Trans Power Electron 37:3927–3942. https://doi.org/10.1109/TPEL.2021.3121532
Rodriguez J, Cortes P (2012) Predictive Control of a three-phase inverter. In: Predictive control of power converters and electrical drives, pp 41–63. https://doi.org/10.1002/9781119941446.CH4
Ullah Q, Costa Resende E, Carlos Gomes Freitas L, Laaksonen H, Godoy Simões M (2024) Enhancing voltage stability of grid forming power converters based on model predictive controller. Int J Electr Power Energy Syst 163:110317. https://doi.org/10.1016/j.ijepes.2024.110317
Vazquez S, Leon JI, Franquelo LG, Rodriguez J, Young HA, Marquez A, Zanchetta P (2014) Model Predictive Control: A Review of Its Applications in Power Electronics. IEEE Ind Electron Mag 8:16–31. https://doi.org/10.1109/MIE.2013.2290138
Vazquez S, Rodriguez J, Rivera M, Franquelo LG, Norambuena M (2017) Model Predictive Control for Power Converters and Drives: Advances and Trends. IEEE Trans Industr Electron 64:935–947. https://doi.org/10.1109/TIE.2016.2625238
Young HA, Perez MA, Rodriguez J (2016) Analysis of Finite-Control-Set Model Predictive Current Control With Model Parameter Mismatch in a Three-Phase Inverter. IEEE Trans Industr Electron 63:3100–3107. https://doi.org/10.1109/TIE.2016.2515072
Zarei ME, Ramirez D, Prodanovic M, Venkataramanan G (2022) Multivector Model Predictive Power Control for Grid Connected Converters in Renewable Power Plants. IEEE J Emerg Sel Top Power Electron. 10:1466–1478. https://doi.org/10.1109/JESTPE.2021.3077953
Zerdali E, Rivera M, Wheeler P (2024) A Review on Weighting Factor Design of Finite Control Set Model Predictive Control Strategies for AC Electric Drives. IEEE Trans Power Electron 39:9967–9981. https://doi.org/10.1109/TPEL.2024.3370550
Zhou C, Zhang Y, Yang H (2025) Model-free predictive control of motor drives: a review. CES Trans Electr Mach Syst:76–90. https://doi.org/10.30941/CESTEMS.2025.00002

Internal Model-Based Control for Grid-Connected Inverters

Ersan Kabalci

Abstract Internal model control (IMC) method has emerged as a powerful approach for enhancing the performance and stability of grid-connected inverters in modern power systems. This chapter provides a comprehensive overview of IMC principles by focusing on their application to inverter control in renewable energy integration, microgrids, and smart grid systems. This chapter begins with the literature survey that is followed by mathematical foundation of internal model control, and its ability to handle system uncertainties and disturbances is effectively highlighted. The chapter then explores the design and implementation of IMC for grid-connected inverters, emphasizing techniques for improving power quality, dynamic response, and robustness against grid fluctuations. Additionally, the impact of grid disturbances, nonlinearities, and parameter variations on inverter performance is analysed along with strategies to mitigate these effects, and with the implementation of simulation environment. The simulation results demonstrate the effectiveness of IMC in ensuring stable operation, low Total Harmonic Distortion (THD) ratios, and compliance with grid codes. Finally, the chapter discusses emerging trends and future research directions in internal model-based control for next-generation grid-connected inverters, particularly in the context of increasing renewable energy penetration and the evolution of smart grids. This chapter serves as a valuable resource for researchers, engineers, and practitioners working on advanced control methodologies for power electronics and grid integration.

Keywords Internal model control (IMC) · Grid-connected inverter · Renewable energy integration · Power quality · Current control · Total harmonic distortion (THD) · Smart grid · Microgrid · Grid stability · Disturbance rejection

---

E. Kabalci (✉)
Electrical and Electronics Engineering, Faculty of Engineering and Architecture, Nevsehir Haci Bektas Veli University, Nevsehir, Turkey
e-mail: kabalci@nevsehir.edu.tr; ersan.kabalci@rub.de

Faculty of Electrical Engineering and Information Technology, Ruhr University Bochum, Bochum, Germany

E. Kabalci (ed.), *Modern Control Methods for Power Electronics*, Energy Systems in Electrical Engineering, https://doi.org/10.1007/978-981-92-1974-2_11

Nomenclature

Symbols	Description
C_b	Base capacitance, *F*
C_f	Filter capacitor, *F*
f_{sw}	Switching frequency, *Hz*
i_g	Grid current, *A*
i_{inv}	Inverter current, *A*
L_1	Inverter side inductor, *H*
L_2	Grid side inductor, *H*
L_b	Base inductance, *H*
L_{eq}	Minimum equivalent inductance
P_{ref}	Reference value for active power, *W*
Q_{ref}	Reference value for reactive power, *VAR*
r_d	Damping resistance, *Ω*
Ts	Sampling period, *s*
v_{aN}, v_{bN}, v_{cN}	Phase voltages relative to neutral, *V*
v_{cf}	Capacitor voltage, *V*
v_g	Grid voltage, *V*
V_{LL}	Line-to-line voltage, *V*
Z_b	The base impedance, *Ω*
α, β	Orthogonal axis of stationary reference frame
ΔI_{Lmax}	Maximum current ripple
θ	Phase angle of utility grid
θ^*	Estimated phase angle
ω_g	Angular frequency of grid, *rad/s*
ω_r	Resonant frequency

1 Introduction

The electricity grid is currently undergoing a major transformation with the increasing prominence and adoption of distributed energy resources (DER). This shift marks a transition from traditional and centralized power generation to a more decentralized, flexible, and sustainable energy infrastructure. The core of this transformation is the interface systems provided by inverters and power converters, which are crucial for connecting DERs to the utility grid. The DERs comprise various decentralized energy generation and storage technologies, including photovoltaic (PV) modules for solar power plants (SPPs), wind turbines, energy storage systems (ESS), and combined heat and power (CHP) plants. The broad deployment of these resources across the grid enhances system resilience, improves efficiency, and contributes to environmental sustainability. The integration of DERs into the electricity grid has introduced significant complexities and challenges in terms of control and management. As the shift from traditional centralized energy systems to decentralized frameworks accelerates, the need for advanced control strategies becomes increasingly evident (Liu et al. 2020; Kumar et al. 2020).

Inverters as the main power converters in these distributed energy circumstances, convert the variable output voltage and power delivered by DERs into a form that is compatible with the centralized utility grid. The inverters operate in two main modes: islanded operation, where they function independently of the grid, and normal operation, where they are grid-connected. In island mode, inverters regulate voltage and current for the loads they supply, while in normal mode, they must adhere to grid operating conditions that require more advanced control algorithms alongside voltage and current regulation (López-Alcolea et al. 2023; García-Triviño et al. 2023).

Proper controller design and stable control are critical for safe inverter operation, especially when switching between island and normal modes. The increasing penetration of renewable energy sources to the grid often leads to issues such as low inertia and fluctuating short circuit levels, particularly when the grid operates at a low level. These challenges make the voltage highly sensitive to power flow variations at the point of common coupling (PCC) of inverters. Various studies have highlighted that power quality degrades, the effectiveness of active damping diminishes, inverter control performance is compromised, and the stability of converter-grid connections deteriorates in weak grid conditions. The intermittency and unpredictability of renewable energy sources (RESs) pose significant risks to grid stability, particularly during disturbances such as faults, voltage sags, or imbalances between supply and demand. Grid-connected voltage source inverters (VSIs) mitigate these risks by offering features like fault ride-through capabilities, frequency regulation, and dynamic voltage support. Leveraging advanced control strategies, such as internal model-based controllers or model predictive control, VSIs can rapidly adapt to grid disturbances, ensuring continuous operation and resilience under dynamic conditions. In particular, the phase-locked loop (PLL) used for grid synchronization can introduce significant challenges under weak grid conditions characterized by low short-circuit ratios (SCRs). Specifically, the PLL can generate a negative resistance component in the output impedance of the inverter at low frequencies. This negative resistance destabilizes the inverter by magnifying small disturbances, causing oscillations or instability in the system. Such effects are intensified in weak grids due to their reduced capacity to absorb fluctuations (Chawda et al. 2024; Zhou et al. 2023; Xu et al. 2023; Zhang et al. 2025; Liu et al. 2023; Saranchimeg and Nair 2021).

Addressing these stability issues requires the adoption of advanced control techniques designed to enhance the performance of grid-connected inverters under weak grid conditions. The techniques such as virtual synchronous machines (VSMs) emulate the dynamic behaviour of synchronous generators, improving the system inertia and damping properties to counteract the destabilizing effects introduced by the PLL. Similarly, robust impedance shaping strategies focus on modifying the output impedance of inverter to achieve better compatibility with weak grids, thereby mitigating the negative resistance introduced by the PLL. The model predictive control (MPC) has emerged as a powerful alternative in addressing stability concerns in weak grid conditions. MPC leverages a system model to predict future system behaviour and optimize control actions over a finite prediction horizon. MPC provides superior handling of nonlinearity and system dynamics compared to conventional control methods by incorporating constraints directly into the optimization process. For grid-connected inverters, MPC can actively regulate output impedance and minimize harmonics while maintaining stability in weak grid environments. However, it relies on increased workload of the processor due to complex mathematical calculations. The internal model control (IMC) based controller is another promising approach which leverages mathematical models of the grid and inverter dynamics to design control strategies that ensure stability and robust performance. These controllers can compensate for PLL-induced negative resistance by adjusting the inverter response to grid conditions in real time. Additionally, adaptive control techniques that dynamically modify control parameters based on real-time grid measurements can further enhance the resilience of inverters in weak grids. These methods allow the inverter to adjust its operational behaviour to accommodate changing grid conditions, ensuring stable and efficient performance even in the presence of severe grid challenges. The integration of these advanced control strategies, such as VSMs, MPC, IMCs, and adaptive controls, is critical for overcoming the stability challenges posed by PLL-induced negative resistance. Their application not only mitigates instability but also enhances the overall reliability and efficiency of renewable energy systems connected to weak grids (Liu et al. 2020; Vigneysh et al. 2022; Komurcugil and Kukrer 2017; Zhou et al. 2022).

This chapter presents details of IMC method which is an alternative controller for power converter and inverters. The chapter also discusses developing a control method that enhances the robustness of grid-connected LCL-filtered inverters in the face of various disturbances and instability issues. To achieve this aim, a modelling study that defines several fundamental development targets related to model-based control has been presented in this chapter. The proposed IMC scheme is based on implementation with additional active damping of the LCL filter, and able to present various ancillary services in addition to its primary function of regulating power flows and injecting surplus power into the grid.

A robust IMC is developed to ensure stable and reliable operation of grid-connected inverters with enhanced adaptability to varying grid conditions, including weak and highly dynamic grids. The proposed control scheme enables independent regulation of active and reactive power by providing precise control over power flows and enhanced capability for voltage and frequency support. The passive damping techniques are incorporated into the design of the LCL filter by effectively suppressing resonance and ensuring stable operation without the need for additional control complexity or energy losses associated with active damping. On the other hand, the control strategy ensures high power quality by achieving low total harmonic distortion (THD) that meets the grid compliance requirements in the injected current and output voltage of the inverter.

These aspects present a comprehensive approach to improving the stability, efficiency, and power quality of grid-connected inverters that make them highly suitable for renewable energy and distributed generation applications. The literature review, fundamental features of IMC, evolving methods in IMCs, system description and IMC controller design are presented in the following subsections while the proposed control strategy is evaluated at the simulation setup and case studies are also represented along with the test scenarios.

2 Literature Review

The control strategies improved for grid-connected systems have advanced beyond traditional single-loop control towards a more sophisticated dual- or multiple-loop control system, which include both voltage and current control loops. The current control loop plays a crucial role in maintaining the quality of the power injected to the grid by regulating the output current. A common approach for current regulation in grid-connected inverters is the proportional-integral (PI) current controller, which helps the output current track a reference signal derived from the desired power transfer to the grid. While PI controllers are simple and effective under balanced grid conditions, they face significant limitations in complex and unbalanced environments. Their primary drawback is an inability to manage unbalanced grid conditions effectively. In real-world applications, the grid is not often perfectly balanced due to factors such as faults, phase differences, or fluctuating outputs from renewable energy sources. Under such conditions, PI controllers struggle to track sinusoidal reference currents without introducing steady-state errors. These errors occur because PI controllers are optimized for constant and predictable conditions, which are uncommon in grid-connected systems that are affected by renewable energy sources like solar and wind. Additionally, PI controllers are unable to dynamically adapt to variations in grid voltage and phase, leading to poor performance during disturbances. This issue is especially pronounced in grids with high penetration of DERs where voltage fluctuations and imbalances are frequent. Accordingly, power quality may deteriorate which leads to increased harmonic distortion and inefficiencies in the overall energy transfer process (Liu et al. 2020; Kumar et al. 2020).

Advanced control methods have been developed to overcome the limitations of conventional PI controllers in grid-connected systems. These methods provide more robust and adaptive solutions for maintaining power quality, especially in the presence of unbalanced grid conditions, harmonics, and fluctuating parameters. Unlike the fixed and linear approach of PI controllers, advanced control techniques can dynamically respond to changes in the grid, offering enhanced control over the current and voltage, as well as improved power injection into the grid. Addressing these issues to ensure stable and secure integration of renewable energy into the grid is recognized as a complex task, and it is widely accepted that inverters must be equipped with robust control algorithms to manage these challenges effectively (Liu et al. 2020; Kumar et al. 2020).

In addition to the control methods, LC and LCL filters are commonly used to reduce current harmonics in VSIs, which facilitate the transfer of energy from DERs to the grid. Recently, LCL filters have gained considerable attention due to their higher performance characteristics compared to simple L or LC filters. The LCL filters enable inverters to operate efficiently at lower switching frequencies, offer better harmonic attenuation, and contribute to lower grid current ripple. These advantages make LCL filters a highly effective solution for improving the overall quality of power injected into the grid. While LCL filters outperform in harmonic reduction, they introduce some complexities that require careful consideration. One of the primary drawbacks of LCL filters is the potential for unstable operation due to the additional two poles they introduce into the system stability region. This can lead to resonance, which may destabilize the grid connection unless properly mitigated. Additionally, the control algorithms required for LCL filter-based systems are fundamentally more complex than those for L or LC filters. These filters demand advanced control strategies to manage resonance issues, ensure system stability, and maintain optimal performance (Xu et al. 2023; Nam et al. 2023; Chowdhury and Singh 2022; Han et al. 2023; Li et al. 2020).

Designing control methods for inverters equipped with LCL filters involves managing the complexities introduced by the third-order filter dynamics that make it difficult to control the grid current directly, as they require regulating a filter with multiple poles. As a result, the control algorithm becomes more comprehensive and challenging to design. In response, various active and passive damping techniques have been proposed in the literature to mitigate the resonance issues caused by the filter. The passive damping can be achieved by incorporating resistors into the LCL filter, either in series or parallel, to reduce resonance and improve stability. However, passive damping can lead to power losses and can reduce overall efficiency. On the other hand, the active damping techniques utilize virtual resistors through advanced control algorithms to manage resonance without significant power loss. These methods provide greater flexibility and efficiency, but their implementation increases the complexity of the control system. In addition to damping methods, advanced control strategies such as Lyapunov-based controllers have been developed to address the challenges of LCL filters in grid-connected systems. Lyapunov-based control approaches optimize inverter operation by maintaining system parameters close to a stable operating point. These controllers are designed to adapt to the changing operating conditions and eliminate uncertainties while ensuring the stability and robustness in dynamic grid environments (Komurcugil and Kukrer 2017; Zhou et al. 2022; Komurcugil et al. 2018; Huang et al. 2022).

Several types of Lyapunov-based controllers have been applied in the control of LCL filter-based inverters, including the classical PI controllers, sliding-mode control, backstepping and passivity-based controllers, nonlinear Lyapunov stability algorithms, and MPCs. Given the increasing complexity of grid-connected systems with renewable energy integration, the development of advanced control algorithms remains a critical area of research. Efforts are being made to enhance the stability and responsiveness of LCL-filtered inverters, particularly through the refinement of MPC algorithms. The main goal is to create control systems that can handle dynamic grid conditions and provide optimal power quality, all while maintaining system stability and efficiency (Korada and Mishra 2023; Zhang et al. 2020).

In the MPC algorithm, the design of the basic network model is fixed around the parameters of the LCL filter with the goal of creating a straightforward effective controller. This algorithm follows a reference network voltage to produce a variable switching frequency while continuously monitoring the output current of the inverter to detect and mitigate any grid distortions. One of the key advantages of FCS-MPC is the flexible structure of *the cost function* used in the model predictive algorithm. This flexibility allows the integration of additional control objectives and features, which makes it a flexible tool for enhancing the performance of grid-connected inverters. The cost function acts as an output generator, selecting the optimal control action that minimizes errors and maintains system stability, while accounting for factors like harmonic distortion, grid current, and dynamic performance. However, despite its advantages, FCS-MPC has a significant drawback due to using the variable switching frequencies. This variability results in a broad harmonic distortion spectrum even at lower frequencies, which in turn increases THD rates in the system. Moreover, the computational complexity of the FCS-MPC algorithm is another limitation. The real time optimization processes required for MPC involve complex mathematical calculations due to the requirement of high-performance processors with substantial computational power. This complexity, particularly in generating the switching signals and controlling the power converter, makes FCS-MPC more demanding in terms of hardware requirements (Ahmed et al. 2017; Wu et al. 2022; Simonetti et al. 2023).

The core principle of the IMC method presented in this chapter is to combine the advantages of both feedforward and feedback control techniques. Although IMC features a basic design approach, it employs a single tuning parameter that effectively balances closed-loop performance with robustness by allowing it to compensate for model inaccuracies. The application of IMC which has been initially utilized in the fields of chemical and mechanical engineering, has expanded into electrical engineering. The first use of IMC in this domain was applied to model-based current control in AC machines in 1998 (Morari 1983; Rivera et al. 1986; Harnefors and Nee 1998). The subsequent literature reviews indicated that model-based control algorithms remain relatively underexplored in the context of grid-connected inverters, with only a limited number of studies employing IMC methods since their introduction. In 2014, an IMC algorithm was successfully utilized for current control in an RL filter-based VSI, while a similar method was later applied to voltage regulation in islanded operation of generators (Yazdanian and Mehrizi-Sani 2014). Moreover, three significant studies have specifically focused on examining the stability conditions of grid-connected inverters, two of which explore three-phase systems and one that addresses a single-phase system (Komurcugil et al. 2018; Gong et al. 2021; Leitner et al. 2018). In comparison to traditional feedback control strategies that utilize PI algorithms, model-based current controllers demonstrate considerable advantages. These controllers enhance the damping ratios of sensitive eigenvalues, exhibit improved resistance to parameter variations, and deliver faster response times. Furthermore, they effectively reduce transient overshoot and undershoot in both voltage and current, positioning them as a more robust solution for ensuring stable operation in grid-connected inverters.

The studies conducted to date indicate that the application of model-based control algorithms for inverter control is relatively sparse, particularly in the context of small-signal analysis. It is important to highlight that the model-based controller, referred to as IMC, differs significantly from traditional PI and proportional-integral-derivative (PID) feedback control algorithms. Specifically, the IMC algorithm requires a comprehensive analytical definition of the grid-connected inverter, encompassing the input DC voltage, the inverter model, output filters, and the grid filter, as well as the grid model. A model-based controller requires the analytical system model to be established in advance, rather than deriving it through iterative estimation processes unlike in MPC algorithm. The analytical system model will be introduced into the control microprocessor to facilitate effective operation. The primary motivation for researching a model-based controller originates from the fact that the desired stability and robustness have not been achieved with similar controllers in recent years. Various decision support methods including sliding mode control, synchronous reference frames, Lyapunov-based algorithms, and learning algorithms have been integrated into MPC algorithms to enhance their prediction and control accuracy (Korada and Mishra 2023; Ngo et al. 2019; Wang and Wai 2020; Kouro et al. 2009; Rodriguez et al. 2007; Rodriguez et al. 2013; Xue et al. 2021).

While these methods aim to address specific challenges and improve the robustness and adaptability of the MPC framework, they often come with the trade-off of increased mathematical complexity. The inclusion of these advanced techniques typically requires more intricate calculations, additional parameters to tune, and higher computational resources. This added complexity can result in slower response times and may necessitate more advanced hardware to implement in real-time applications. Despite these enhancements, the overall efficiency of the MPC algorithm may not see significant improvement. The marginal gains in prediction accuracy or control precision are often offset by the practical limitations introduced by the increased computational burden, making these approaches less appealing for systems where simplicity and real-time performance are critical.

While advanced decision support methods theoretically strengthen MPC algorithms, their practical utility is limited by the substantial increase in complexity relative to the modest gains in overall performance. This underlines the need for a balanced approach that considers both the benefits and the costs of such enhancements. Therefore, a model-based control algorithm has been dealt with improved and analyzed modelling studies to evaluate the impact of the approach on system stability and reliability of a grid-connected inverter in this chapter.

3 Theory of Internal Model Control Method

IMC is a powerful control strategy that has gained widespread use in process control, power electronics, and industrial automation. It is known for its robustness against system uncertainties, disturbances, and modelling errors. The IMC framework is based on the Internal Model Principle, which states that a controller must explicitly include a model of the system to effectively reject disturbances and track reference signals optimally. Unlike traditional feedback controllers, which rely solely on error correction, IMC leverages a process model to predict system behaviour and compute optimal control actions proactively. This predictive capability makes IMC particularly useful in applications where disturbances and parameter variations significantly impact system performance, such as grid-connected inverters in renewable energy systems.

The origins of IMC can be traced back to the work of Garcia and Morari (Morari 1983; Garcia and Morari 1982), who introduced the concept as a unified framework for process control applications. Since then, IMC has been widely applied in chemical process industries, power electronics, robotics, and aerospace systems, demonstrating its versatility in handling both linear and nonlinear systems. The key advantage of IMC lies in its ability to combine feedforward and feedback control mechanisms, allowing for proactive disturbance rejection while ensuring steady-state accuracy.

IMC is particularly well-suited for process control applications, where systems often exhibit slow and overdamped behaviour with significant time delays. Traditional controllers, such as PID control, struggles to accommodate these dynamics effectively. In contrast, IMC explicitly considers the dynamic model of the system and applies inverse-based compensation to ensure optimal performance while addressing modelling inaccuracies. The complexity of a controller is a crucial factor, as its structure and the influence of tuning parameters should remain transparent to the operator. However, this does not imply an outright preference for traditional PID controllers. The widespread use of three-mode PID controllers historically decreases hardware limitations, rather than an inherent superiority in control performance. The advanced dynamic compensation techniques can be effortlessly integrated with modern computer-based control by allowing for extensive flexibility in system design. In principle, one could strategically position poles and zeros across the complex plane to achieve desirable control characteristics.

The controller response to specific disturbances such as stochastic noise modelling or resonance frequency avoidance is of secondary importance, as any remaining performance issues can be mitigated through filtering within the overall control system. Extensive comparisons between optimal control methods and conventional controllers indicate that while optimal control improves tracking (servo) and regulation behaviour, it does not inherently address robustness. Instead, robustness must be achieved indirectly by tuning weighting matrices or noise models, which often requires an iterative approach validated through simulations. This is not unexpected, as the widely used quadratic performance index, despite being mathematically convenient, allows gain adjustment to approach an asymptotically "perfect" response. However, the critics of state-space representations, which serve as the foundation for most modern optimal control theories, have promoted the research on IMC method (Horowitz and Shaked 1975).

A straightforward open-loop control system as seen in Fig. 1a can be effective when both the process model and the controller are stable. In such cases, designing a controller is relatively simple. The fundamental concept behind open-loop control is

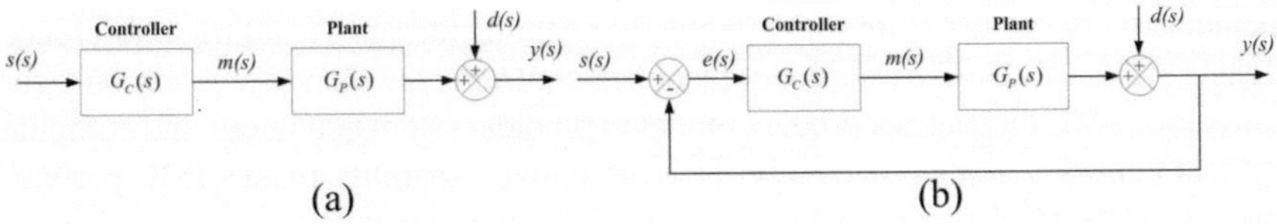

Fig. 1 The discrete-time control structures: **a** open-loop control **b** closed-loop feedback control

that the controller ($G_C(s)$) compensates for the process dynamics by using an inverse model, mathematically represented as:

$$G_C(s) = \frac{1}{G_M(s)} \tag{1}$$

where $G_M(s)$ is the process model that is assumed to be identical to the actual plant model $G_P(s)$. This approach theoretically allows ideal control by ensuring the desired output is achieved without external corrections. However, open-loop control systems suffer from critical limitations such as model-plant mismatch, inability to handle unmeasured disturbances and sensitivity to process changes. Any deviation between the assumed model $G_M(s)$ and the actual plant process $G_P(s)$ can lead to poor performance, instability, or inaccurate outputs. Open-loop systems do not have a mechanism to reject external disturbances $d(s)$ that may affect the process. If the process dynamics change over time due to nonlinearities, parameter variations, or aging effects, an open-loop system cannot adjust accordingly. A feedback control system can be introduced to overcome these drawbacks as depicted in Fig. 1b. The feedback loop helps to correct deviations in real time, thereby improving stability and robustness against disturbances. However, when model uncertainty exists, the controller needs to be re-tuned to maintain system stability and performance. A perfect control strategy can be theoretically achieved if the control architecture is designed using an exact process model.

This concept can be better understood by modifying the feedback path to incorporate the process model $G_M(s)$ without changing the control and output signals as illustrated in Fig. 2. A modified controller $Q_C(s)$ is obtained by introducing and subtracting the plant model in the feedback loop given as follows:

$$Q_C(s) = \frac{G_C(s)}{1 + G_M(s)G_C(s)}. \tag{2}$$

This equation allows the controller to compensate for model inaccuracies, leading to better disturbance rejection and improved robustness. IMC is a well-known approach that builds upon this concept by explicitly incorporating the process model into the control structure to achieve advanced performance even in the presence of uncertainties and disturbances.

The feedback control system consists of a process represented by the discrete transfer function $G_P(s)$, operating with a setpoint $s(s)$ and subject to an external

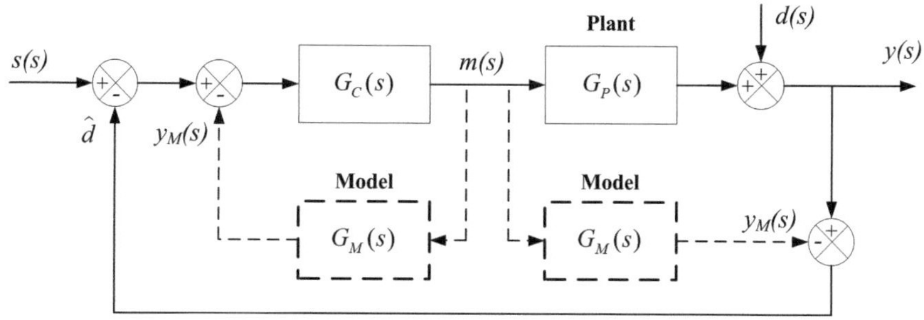

Fig. 2 The discrete-time feedback controller with the model modifications (Reprinted (adapted) with permission from Garcia and Morari (1982). Copyright {2025} American Chemical Society)

disturbance $d(s)$. By defining the model transfer function as $G_M(s)$, the effect of the input m on the measurement signal $y_M(s)$ can be both subtracted and added as shown by the dashed lines in Fig. 2 that results in an equivalent system representation.

By considering the control block $G_C(s)$ with model feedback as the new controller, we derive the fundamental IMC structure as illustrated in Fig. 3. Consequently, any traditional feedback controller can be reformulated into an IMC framework. Likewise, any IMC configuration can be transformed back into a conventional feedback control system by defining the appropriate equation as given in Eq. 3;

$$G_C(s) = \frac{Q(s)}{1 - G_M(s)Q(s)}. \tag{3}$$

The controller $G_C(s)$ is inherently simpler to design compared to $Q(s)$ as the IMC structure enables robustness to be explicitly incorporated as a design objective. This advantage arises from the unique formulation of the feedback signal $\hat{d}(s)$ which allows for better handling of uncertainties and disturbances.

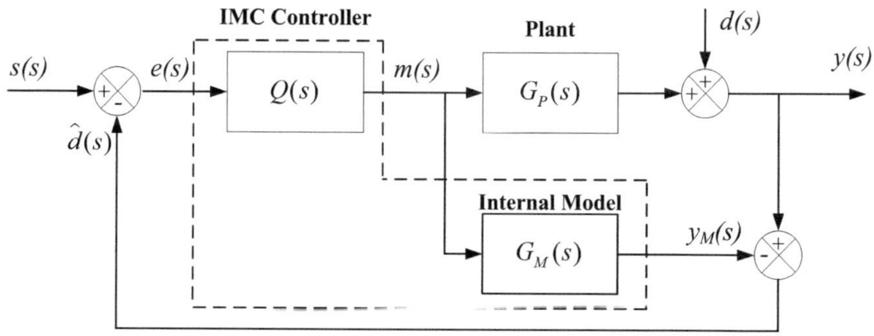

Fig. 3 Fundamental IMC structure

$$\hat{d}(s) = \frac{d(s)}{1 + (G_P(s) - G_M(s))G_M(s)}. \tag{4}$$

The control signal *m(s)* and the error signal *e(s) are* given by

$$m(s) = \frac{Q(s)(s(s) - d(s))}{1 + Q(s)(G_P(s) - G_M(s))} \tag{5}$$

$$e(s) = \frac{s(s) - d(s)}{1 + Q(s)(G_P(s) - G_M(s))}. \tag{6}$$

The output expression of IMC structure *y(s)* can be obtained as given in (7);

$$y(s) = \frac{G_P(s)Q(s)}{1 + Q(s)(G_P(s) - G_M(s))}s(s) + \frac{1 - G_M(s)Q(s)}{1 + Q(s)(G_P(s) - G_M(s))}d(s). \tag{7}$$

For stability, it is a necessary and sufficient condition that the roots of both following characteristic equations lie strictly to the left of the imaginary axis in the s-domain. All poles of the closed-loop system must have negative real parts. If the poles lie on or to the right of the imaginary axis, the system will either become unstable or experience marginal stability which leads to behaviours such as oscillations or an unbounded response.

$$\frac{1}{G_C(s)} + (G_P(s) - G_M(s)) = 0 \tag{8}$$

$$\frac{1}{G_P(s)G_C(s)} + \frac{1}{G_P(s)}(G_P(s) - G_M(s)) = 0. \tag{9}$$

If the process transfer function $G_P(s)$ is identical to the model transfer function $G_M(s)$, the system output *y(s)* closely follows the reference input *s(s)* by assuming no external disturbances *d(s)* exist in the system. This behaviour reflects that of an open-loop system while leveraging the advantages of feedback control. The IMC structure effectively integrates the simplicity of open-loop control with the robustness and adaptability of a feedback system to ensure reliable performance even in the presence of uncertainties. From the signal flow in the IMC framework, the following key characteristics defined with three properties can be observed (Morari 1983; Garcia and Morari 1982; Ranjan et al. 2023):

- **Dual Stability:** When the process model perfectly matches the actual system and no disturbances are present, the closed-loop system remains stable due to the property provided both $G_P(s)$ and the controller $G_C(s)$ are stable. This ensures predictable and reliable operation. Garcia and Morari (Garcia and Morari 1982) explain the proof of this property with the equality of $G_P(s) = G_M(s)$ which transforms the characteristic equations as follows

$$\frac{1}{G_C(s)} = 0 \tag{10}$$

$$\frac{1}{G_C(s)G_P(s)} = 0. \tag{11}$$

Dual stability property supports the idea that closed-loop stability can be automatically achieved for any stable control block $G_C(s)$ for a plant that is open-loop stable. However, this does not guarantee robustness. This holds true even if the control block is nonlinear, if it remains input–output stable. Therefore, any limitations on the control action such as constraints would not impact the overall stability of the system. In contrast, stability becomes a more complex issue when dealing with conventional feedback systems that involve nonlinear controllers. Furthermore, it is important to note that IMC cannot generally stabilize an open-loop unstable plant. Stabilization could only occur if the controller precisely cancels the unstable poles of the system, but this is practically unachievable. An open-loop unstable plant must first be stabilized through conventional feedback methods before IMC can be applied effectively.

Once the stability issue is addressed, the next consideration is determining the optimal choice for $G_C(s)$. Equation 6 suggests that "perfect" control could be achieved by selecting $G_C(s) = 1/G_M(s)$, where $G_M(s)$ represents the transfer function of plant internal model. However, it must not only ensure stability but also be feasible to implement in real systems for $G_C(s)$ to be a practical and viable control solution. This means that $G_C(s)$ needs to be physically realizable without violating constraints such as system dynamics, computational limitations, and hardware capabilities. In addition, the control law should be designed to avoid issues like excessive complexity or unfeasible precision, making sure the solution can be practically applied and operated in real-world conditions. To address these challenges, a factorization technique is introduced as follows:

$$G_M(s) = G_{M+}(s)G_{M-}(s) \tag{12}$$

where $G_{M+}(s)$ represents the part of the system that includes all time delays and zeros outside the unit circle. Inverting these components would require predictions to account for the delays, and attempting to invert the zeros outside the unit circle would result in an unstable controller. These characteristics of time delays and zeros outside the unit circle are known as "non-minimum phase" (NMP) in classical frequency response theory. Non-minimum phase systems are challenging to control because their dynamics introduce instability or undesirable behaviours when trying to achieve an ideal response. On the other hand, $G_{M-}(s)$ contains the minimum-phase elements of the system, which are invertible. These elements can be handled directly in the control design. In contrast, processes with time delays need to be approximated using techniques such as Taylor series expansions or Padé approximations, as these delays cannot be directly inverted in a practical control system.

In this step of the process, the IMC controller can be designed by taking the inverse of the invertible portion of the process model, which is represented by $G_{M-}(s)$. This allows the controller to effectively compensate for the controllable dynamics of the system, while the time delays and non-minimum phase elements are managed separately through approximations. In terms of minimizing the sum of the squared errors, the best performance that any control system can achieve is attained through the "perfect controller". This "perfect controller" would precisely match the plant dynamics by compensating for all variations and disturbances without introducing any error. In an ideal scenario, the perfect controller would be able to perfectly track the reference input and reject disturbances that result in zero error (Morari 1983; Garcia and Morari 1982; Ranjan et al. 2023).

- **Perfect Control:** Under the assumption that $G_P(s) = G_M(s)$ and that $G_P(s)$ is stable, the sum of the squared errors is minimized for both the regulator and the servo-controller when $G_C(s) = 1/ G_{P-}(s)$. This property can be derived from optimal control theory, which typically involves minimizing a cost function such as the sum of squared errors to find the optimal controller. It is required to focus on the consequences of $G_C(s) = 1/ G_{P-}(s)$ equality. For the perfect controller, Eq. (7) becomes

$$y(s) = G_{P+}(s)s(s) + (1 - G_{P+}(s))d(s). \tag{13}$$

It is noteworthy that $G_{P+}(s)$ represents the part of the process that limits the achievable control quality. These limitations are a fundamental part of the physical system and cannot be eliminated by any control system. If there were no time delays or non-minimum phase behaviour, achieving truly perfect control would be possible. However, it is important to recognize that when using the perfect controller, the system becomes highly sensitive to modelling errors. Small inaccuracies or deviations in the plant model can lead to significant performance issues, as the perfect controller relies on an exact match between the model and the actual system. This sensitivity to modelling errors makes the practical application of the perfect controller difficult in real-world systems, where precise modelling is often challenging (Morari 1983; Garcia and Morari 1982).

- **Zero Offset:** The controller satisfying $G_C(1) = 1/G_M(1)$ enables the IMC structure to guarantee zero offset performance. This means that the system will accurately track any constant reference input or external disturbance without experiencing any steady-state error. Thus, the IMC approach provides a powerful balance between control accuracy and robustness, which makes it highly effective for practical applications where both model uncertainties and external disturbances are common.

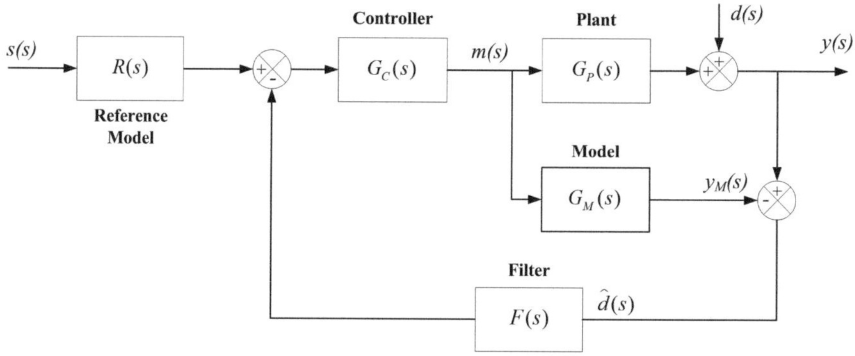

Fig. 4 Complete IMC structure with reference and filter (Reprinted (adapted) with permission from Garcia and Morari (1982). Copyright {2025} American Chemical Society)

One approach to account for model and process mismatches is to include a filter block *F(s)* in the feedback signal as shown in Fig. 4. This filter compensates for the inconsistencies between the estimated disturbance $\hat{d}(s)$ and the actual system dynamics. In the case of an exact model, this filter could also be used to counteract certain types of disturbance dynamics. As a result, the characteristic Eq. (8) is modified to

$$\frac{1}{G_C(s)} + F(s)(G_P(s) - G_M(s)) = 0. \tag{14}$$

This adjustment helps manage the mismatch between the model and the actual process by improving the system performance in the presence of disturbances and model inaccuracies. The filter *F(s)* effectively serves as a corrective mechanism that reduces the impact of modelling errors on system behaviour.

For a given model-plant mismatch, the filter *F(s)* can be selected such that Eq. (14) has all of its roots strictly to the left of the imaginary axis in the s-domain. This ensures that the closed-loop system remains stable. The characteristic equation can be modified to ensure that its poles are placed in the stable region of the s-plane by designing *F(s)* appropriately to account for the mismatch between the model and the actual plant. This guarantees the system stability, even when there are inconsistencies between the model and the plant dynamics.

4 System Modelling and Controller Design

4.1 System Description

The typical configuration of a grid-connected VSI with LCL filter is presented in Fig. 5 where the inverter-side inductance, filter capacitance, and grid-side inductance of LCL filter for each phase are denoted as L_1, C_f, and L_2, respectively. The single-phase equivalent circuit of this inverter model in the stationary $\alpha\beta$ coordinate frame is illustrated in Fig. 6. The VSI is analysed in a conventional two-level three-phase topology where the switching devices at each phase-leg are listed as $S_1 - S_4$, $S_3 - S_6$, $S_5 - S_2$ as seen in Fig. 5. The switching states of the VSI are determined by the gating signals S_a, S_b, and S_c as presented (Rodriguez et al. 2007; Rodriguez et al. 2013)

$$S_A = \begin{cases} 1, \ S_1 \ on \ \ and \ S_4 \ off \\ 0, \ S_1 \ off \ \ and \ S_4 \ on \end{cases} \tag{15}$$

$$S_B = \begin{cases} 1, \ S_3 \ on \ \ and \ S_6 \ off \\ 0, \ S_3 \ off \ \ and \ S_6 \ on \end{cases} \tag{16}$$

$$S_C = \begin{cases} 1, \ S_5 \ on \ \ and \ S_2 \ off \\ 0, \ S_5 \ off \ \ and \ S_2 \ on \end{cases} \tag{17}$$

The vectorial representation of switching states can be given as follows

$$S = \frac{2}{3}(S_a + aS_b + a^2 S_c) \tag{18}$$

where $a = e^{j2\pi/3}$.

Each switching state determines a specific set of phase voltages for the inverter relative to the negative busbar (v_{aN}, v_{bN}, and v_{cN}). The output voltage space vectors generated by the inverter are defined by Eq. (19)

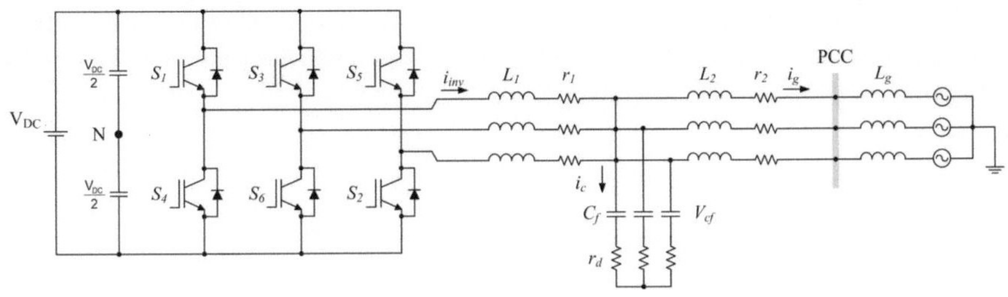

Fig. 5 The grid-connected VSI model with LCL filter

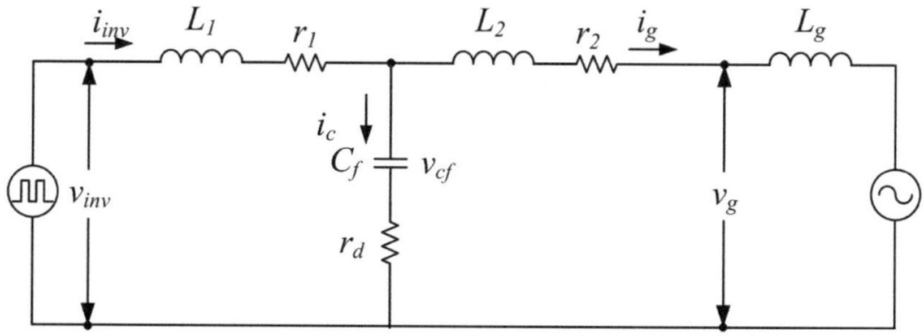

Fig. 6 Equivalent circuit of the grid-connected VSI with LCL filter in the stationary (αβ) frame

$$v = \frac{2}{3}(v_{aN} + av_{bN} + a^2 v_{cN}) \tag{19}$$

where v_{aN}, v_{bN}, and v_{cN} are the phase to neutral voltages of the inverter. In the analytical design studies, the discrete-time state-space model of the grid-connected VSI with LCL filter is developed to be used as the internal model for the proposed model-based controllers.

The state-space model of the system is derived from the differential equations governing its dynamics. For this purpose, the equivalent circuit of the system shown in Fig. 6 is used.

The VSI current i_{inv}, grid current i_g, capacitor voltage v_{cf}, and grid voltage v_g are defined as the state parameters. The inverter voltage v_{inv} is defined as a function of inverter side inductor L_1, capacitor voltage, and the damping resistance r_d as given in (20) while the grid voltage v_g is defined in (21);

$$v_{inv}(t) = L_1 \frac{di_{inv}(t)}{dt} + v_{cf}(t) + i_c(t) r_d \tag{20}$$

$$v_g(t) = -L_2 \frac{di_g(t)}{dt} + v_{cf}(t) + i_c(t) r_d \tag{21}$$

where L_2 is the grid-side inductor. The state variables are determined as follows referring to the inverter and grid voltage equations in (20) and (21):

$$\frac{di_{inv}}{dt} = \frac{1}{L_1}(v_{inv} - r_1 i_{inv} - v_c) \tag{22}$$

$$\frac{dv_{cf}}{dt} = \frac{i_{inv} - i_g}{C_f} \tag{23}$$

$$\frac{di_g}{dt} = \frac{1}{L_2}(v_{cf} - r_2 i_g - v_g). \tag{24}$$

The state-space model of grid-connected VSI in continuous time can be described as follows:

$$\frac{dx}{dt} = Ax(t) + Bu(t) \tag{25}$$

where the state vectors are $x(t) = [i_{inv}\ v_{cf}]^{\mathrm{T}}$ and $u(t) = [v_{inv}\ i_g]^{\mathrm{T}}$. The matrices A and B are given as follows (Li et al. 2020; Korada and Mishra 2023; Xue et al. 2021; Young et al. 2020; Reznik et al. 2014; Karamanakos et al. 2021):

$$A = \begin{bmatrix} -\frac{r_1}{L_1} & -\frac{1}{L_1} \\ \frac{1}{C_f} & 0 \end{bmatrix},\ B = \begin{bmatrix} \frac{1}{L_1} & 0 \\ 0 & -\frac{1}{C_f} \end{bmatrix}. \tag{26}$$

The discrete-time form of the voltage and current given in (20) for a sampling time T_s can be used to determine the future value of grid current with the voltage and measured current at the *k*th sampling instant. Approximating the derivative by the use of the forward-Euler method for the sampling period of *Ts* is as follows:

$$\frac{dv(t)}{dt} = \frac{v(k+1) - v(k)}{T_S} \tag{27}$$

$$\frac{di(t)}{dt} = \frac{i(k+1) - i(k)}{T_S} \tag{28}$$

The inverter voltage at time *(k)* and the inverter current at *(k + 1)* time can be extended to obtain inverter voltage at *(k)* time as follows:

$$v_{inv}(k) = \frac{L_1}{T_S}[i_{inv}(k+1) - i_{inv}(k)] + \frac{L_2}{T_S}[i_{inv}(k+1) - i_{inv}(k)] + v_g(k) \tag{29}$$

where inverter current at time *(k)* and *(k + 1)* are given as follows:

$$i_{inv}(k) = \frac{C_f}{T_S}\left[\frac{L_2}{T_S}[i_g(k) - i_g(k-1)] + v_g(k) - i_c(k)r_d - v_{Cf}(k-1)\right] + i_g(k) \tag{30}$$

$$\begin{aligned} i_{inv}(k+1) =& \frac{C_f}{T_S}\left[\frac{L_2}{T_S}[i_g(k+1) - i_g(k)] + v_g(k+1) - i_c(k+1)r_d - v_{Cf}(k)\right] \\ &+ i_g(k+1). \end{aligned} \tag{31}$$

4.2 LCL Filter Design and Dynamics

The power flow of an inverter can be precisely responded with proper control methods, ensuring stable and efficient grid interaction while meeting power quality criteria. The simplest form of output filter is comprised of a single inductor placed between the inverter and the grid. Its primary function is to smooth out the current waveform by reducing high-frequency switching harmonics that are generated by the inverter. The inductor achieves this by opposing rapid changes in current, thus acting as a first-order low-pass filter that provides 20 dB/decade harmonic attenuation. However, this basic filter design has limitations, particularly in its effectiveness at attenuating higher frequency harmonics. Large inductors which increase the physical size and cost of the system are often necessary to meet the harmonic requirements imposed by grid codes. The LC filter is an enhancement over the L filter which offers better harmonic reduction by combining an inductor with a capacitor. This arrangement creates a second-order filter, which provides more effective harmonic attenuation than the L filter alone. The inductor continues to reduce current ripple, while the capacitor bypasses high-frequency voltage harmonics, directing them to ground and smoothing the voltage waveform. This configuration results in a 40 dB/decade attenuation rate while striking a balance between performance and complexity which makes it suitable for systems requiring moderate harmonic suppression. However, it still requires a relatively large inductor, and the addition of a capacitor introduces resonance issues that must be carefully managed during design to avoid instability.

The LCL filter is the most advanced and effective filtering solution among the three filter types due to its third-order filtering structure, which provides better harmonic elimination with an attenuation rate of 60 dB/decade. The LCL filter effectively reduces both current and voltage harmonics, making it the most efficient filter for dealing with high-frequency switching harmonics. The LCL filter offers a higher degree of harmonic reduction with smaller inductor values, which reduces the size and cost of the inductors, and minimizes power losses compared to the L and LC filters. However, the increased complexity of the design requires more careful control strategies to avoid resonance, which can occur at certain frequencies and cause instability (Nam et al. 2023; Reznik et al. 2014; Guzman et al. 2018; Hollweg et al. 2023). The design procedure of the LCL filter used in the proposed modelling studies is performed referring to the cited publications and application notes in the literature. The following expressions and control parameters are improved regarding the parameters and network presented in Fig. 6. The filter capacitor C_f is determined according to the design base values that are defined as follows:

$$Z_b = \frac{V_{LL}^2}{P_n} \tag{32}$$

$$C_b = \frac{1}{\omega_g Z_b} \tag{33}$$

$$L_b = \frac{Z_b}{\omega_g} \tag{34}$$

where V_{LL} is the line-to-line voltage of the grid, Z_b is the base impedance, and $\omega_g = 2\pi f_g$ is the angular frequency of the grid. Filter capacitance is typically chosen as a percentage of the base capacitance value, C_b, which represents the nominal capacitance of the system. To ensure stability and effective filtering, the capacitance is adjusted based on the maximum allowable power variation seen by the grid, often limited to a range between 1 and 5%. By setting the maximum power variation at 5%, the filter capacitance can be calculated as $C_f = 0.05C_b$ to effectively smooth out voltage and current fluctuations which ensures that harmonic distortion remains within acceptable limits. At low frequencies, the LCL filter functions similarly to a single inductor with a total inductance, L_{dc}, which is the sum of the inductances of the inverter-side inductor L_1, and the grid-side inductor L_2 in Fig. 6. This total inductance is expressed as:

$$L_{dc} = L_1 + L_2. \tag{35}$$

In this low-frequency range, the current through the filter capacitor C_f is considered negligible in comparison to the current through the inductors, meaning the capacitors mainly filter high-frequency components, while the inductors dominate at lower frequencies. There is an optimal relationship between L_1 and L_2 that minimizes the voltage drop at the fundamental frequency while maximizing the filter capability to reduce higher-frequency harmonics. This relationship is expressed by representing L_1 and L_2 as percentages of the total inductance L_{dc} and L_2 using a variable α (where $\alpha \in [0, 1]$). The inductances are then given by:

$$L_1 = \alpha L_{dc}\ L_2 = (1 - \alpha)L_{dc}. \tag{36}$$

This α parameter determines how the total inductance is split between the inverter-side and grid-side inductors. By adjusting α, the filter performance can be optimized by balancing between minimizing losses (voltage drop at fundamental frequency) and maximizing harmonic attenuation. The LCL resonant circuit is comprised by the components L_1, L_2 and C_f all in parallel, which together determine the resonant frequency of the filter. The resonant frequency ω_r can be calculated as follows:

$$\omega_r = \sqrt{\frac{L_1 + L_2}{L_1 L_2 C_f}}. \tag{37}$$

To simplify the analysis, the equivalent inductance L_{eq} which sets the resonant frequency can be defined as the parallel combination of L_1 and L_2

$$L_{eq} = \frac{L_1 L_2}{L_1 + L_2} = \alpha(1 - \alpha)L_{dc}. \tag{38}$$

The minimum equivalent inductance L_{eq} occurs when $L_1 = L_2 = 0.5L_{2dc}$ meaning the two inductances are equal, which results in the most efficient harmonic attenuation at the resonant frequency. To design the filter, the total inductance L_{dc} is often chosen as 10% of the base inductance value L_b based on empirical data, and the individual inductances L_1 and L_2 can be determined as

$$L_1 = L_2 = 0.05L_b \tag{39}$$

referring to this data. This configuration balances the filtering performance and minimizes voltage drop across the inductors while ensuring effective attenuation of high-frequency harmonics (Reznik et al. 2014; Guzman et al. 2018).

A damping resistor can be introduced in the input filter as a passive damping method to reduce the high peak gain at the LCL resonant frequency as seen in Fig. 6. While passive damping is straightforward, it causes the drawback of additional losses due to the physical resistor. Alternatively, active damping addresses the issue by modifying the control system to emulate a virtual resistor, which achieves damping without introducing real power losses. Figure 3 illustrates a damped filter configuration, where a damping resistor is connected in series with the filter capacitor as shown in Fig. 6. This method is part of passive damping, which reduces resonance but introduces power losses due to the physical resistor. The s-domain block diagram in Fig. 7 further demonstrates that the same damping effect can be achieved within the control scheme of an ideal LCL filter by incorporating a virtual gain component which emulates the damping resistor "r_d" in Fig. 6. This approach, known as active damping, relies on modifying the control to simulate the resistor effect. For active damping, the filter capacitor currents need to be measured. This requires additional analogue inputs/outputs specifically, three measurements for a three-phase system or only two for balanced three-phase currents. The state equations that describe the filter behaviour in the s-domain are based on the configuration shown in Fig. 7a without damping resistor as follows:

$$v_{inv}(s) = sL_1 i_{inv}(s) + v_{cf}(s) \tag{40a}$$

$$v_g(s) = v_{cf}(s) - sL_2 i_g(s) \tag{40b}$$

$$v_{cf}(s) = \frac{i_c(s)}{sC_f} \tag{40c}$$

$$i_{inv}(s) = i_c(s) + i_g(s). \tag{40d}$$

The voltage and current transfer functions of the ideal LCL-filter in this configuration are

$$H_v(s) = \frac{v_g(s)}{v_{inv}(s)} = \frac{L_2 s}{L_1 L_2 C_f s^3 + (L_1 + L_2)s} \tag{41}$$

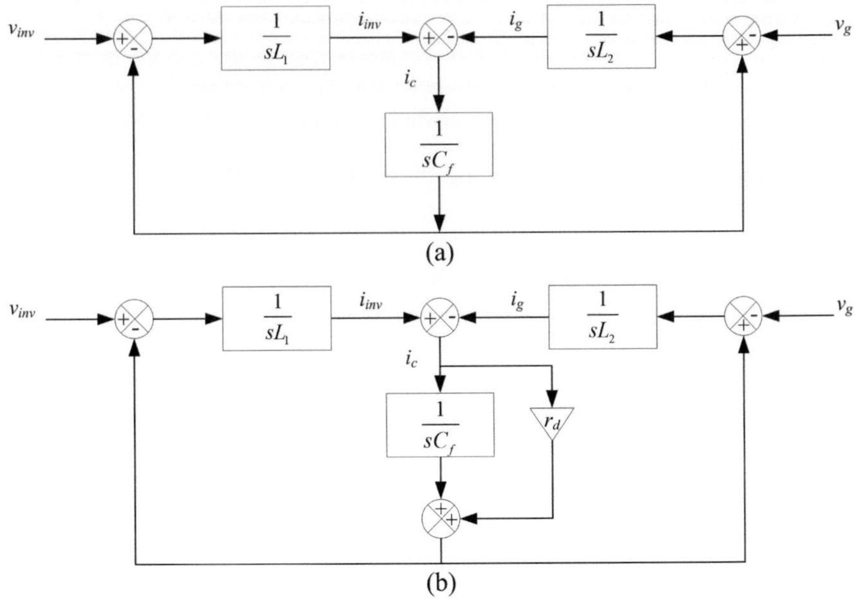

Fig. 7 The *s*-domain block diagram of the LCL filter at grid-connected inverter: **a** without damping resistor **b** with damping resistor

$$H_i(s) = \frac{i_g(s)}{i_{inv}(s)} = \frac{L_2}{L_1 L_2 C_f s^2 + (L_1 + L_2)}. \tag{42}$$

The voltage and current transfer functions of an LCL filter with a damping resistor shown in Fig. 3b considering the additional resistor placed in series with the capacitor, which is part of the passive damping approach. The damping resistor r_d helps suppress resonance at the LCL filter resonant frequency, but it also modifies the transfer functions of the system as given follows (Reznik et al. 2014; Guzman et al. 2018; Hollweg et al. 2023):

$$H_v(s) = \frac{v_g(s)}{v_{inv}(s)} = \frac{L_2 s}{L_1 L_2 C_f s^3 + r_d (L_1 + L_2) C_f s^2 + (L_1 + L_2) s} \tag{43}$$

$$H_i(s) = \frac{i_g(s)}{i_{inv}(s)} = \frac{L_2}{L_1 L_2 C_f s^2 + r_d (L_1 + L_2) C_f s + (L_1 + L_2)}. \tag{44}$$

The parameter definition of inductors along the LCL filter is dependent on the current ripple rates and related limits. The maximum current ripple ΔI_{Lmax} at the output of a VSI can be derived from the characteristics of the switching process and the filter inductance. For a typical Sinusoidal Pulse Width Modulation (SPWM) inverter, the ripple is influenced by the DC bus voltage, modulation index, and switching frequency among other factors. The maximum current ripple derived from the parameters is expressed follows:

$$\Delta I_{L\max} = \frac{2V_{DC}}{3L_1}\frac{(1-m)m}{f_{sw}} \tag{45}$$

where m is the modulation index of modulator, and f_{sw} is the switching frequency.

The peak-to-peak current ripple depends on the product of $m(1-m)$ which means that the ripple is minimized at $m = 0$ or $m = 1$ and maximized when $m = 0.5$:

$$\Delta I_{L\max} = \frac{V_{DC}}{6f_{sw}L_1}. \tag{46}$$

A 10% ripple of the rated current for the design parameters is given by $\Delta I_{Lmax} = 0.01I_{max}$ where the L_1 value is derived as in (33) (Reznik et al. 2014)

$$\mathrm{I}_{\max} = \frac{P_n\sqrt{2}}{3V_{LL}} \Rightarrow L_1 = \frac{V_{DC}}{6f_{sw}\Delta I_{L\max}} \tag{47}$$

Determining the value of the grid-side inductance (L_2) in an LCL filter of a grid-connected inverter involves balancing several factors including resonance frequency, harmonic attenuation, system stability, and power quality requirements. The process typically considers the overall inductance of the filter, the grid requirements, and the switching characteristics. Equations given in (36)–(39) are essential to determine the exact L_2 value after selecting the L_1. On the other hand, Reznik et al. describes the L_2 calculation as given in Reznik et al. (2014):

$$L_2 = \frac{\sqrt{\frac{1}{k_a^2}+1}}{C_f\omega_{sw}^2} \tag{48}$$

where k_a is the desired attenuation ratio. To avoid harmonic amplification near the switching frequency of the inverter, the resonant frequency f_{res} is generally chosen to be within a specific range

$$10f_g < f_{res} < 0.5f_{sw} \tag{49}$$

which ensures that the resonance occurs at a frequency high enough to avoid interaction with the grid frequency but low enough to avoid amplifying switching harmonics.

The damping resistor r_d placed in series with the filter capacitor attenuates part of the ripple around the switching frequency to avoid resonance. The value of the damping resistor is typically set as a fraction of the impedance of the capacitor at the resonant frequency. Specifically, r_d should be one-third of the capacitor impedance at the resonant frequency as expressed in

$$r_d = \frac{1}{3\omega_{res} C_f}. \tag{50}$$

The damping resistor r_d is a critical element in an LCL filter for a grid-connected inverter, as it reduces the impact of resonance without adding significant losses to the system. By carefully selecting r_d based on the resonant frequency and the capacitance, the system can achieve effective harmonic attenuation while maintaining stability. After determining the equations of grid-connected VSI and its LCL filter based on the specified principles, modelling studies have been carried out using MATLAB and Simulink to design the grid-connected inverter with internal model-based controller system.

4.3 Design Criteria of IMC and Control Objectives

It is stated that the proposed IMC controllers in the literature provide higher damping ratios, higher robustness against parameter changes and faster system response compared to classical PI and nonlinear control methods (Yazdanian and Mehrizi-Sani 2014; Leitner et al. 2018; Leitner et al. 2018). In the IMC method, mathematical models of the system are needed to design the controller. Model-based control of grid-connected inverters is designed to enhance performance, stability, and power quality by leveraging precise mathematical models of the inverter and its interactions with the grid. This approach allows the controller to predict system behaviour accurately and apply corrections proactively to ensure robust operation even under dynamic grid conditions. In comparison, MPC takes this predictive capability further by formulating control as an optimization problem solved over a finite prediction horizon. While model-based control methods often rely on linearized models and require careful parameter tuning, MPC adapts dynamically to changes in operating conditions. However, computational complexity of MPC is a trade-off since it requires real-time optimization solvers, which can be challenging for systems with high sampling rates such as grid-connected inverters.

The schematic of the grid-connected inverter controlled by the proposed IMC algorithm was developed in the research process and the block diagrams of the control algorithm are given in Fig. 8. The IMC algorithm is designed with two fundamental internal controllers responsible for power control relying on voltage and current controllers. Each of these controllers operates based on the model parameters generated from the state space equations of the inverter, filter, and grid. The simplified interconnection network of grid-connected inverter is obtained referring to the equivalent circuit given in Fig. 6. The filter branch can be expressed with a single inductor L which is the sum of the L_1 and L_2, while the resistance is comprised of grid-side resistance. The Kirchhoff's Voltage Law (KVL) is applied in the Clarke and Park transformations to the natural *abc* reference frame equations to obtain the two orthogonal axes of stationary reference frame (α, β), and then the *d–* and *q-axis*

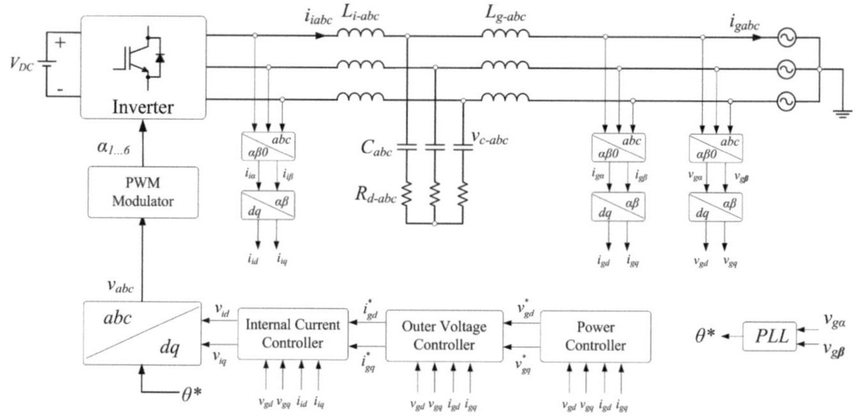

Fig. 8 The grid-connected LCL inverter with model-based controllers

rotating reference frame components, respectively. The transformation equations are described as follows:

$$V_{iabc} = L\frac{di_{iabc}}{dt} + Ri_{iabc} + V_{gabc} \tag{51}$$

$$V_{i\alpha\beta} = L\frac{di_{i\alpha\beta}}{dt} + Ri_{i\alpha\beta} + V_{g\alpha\beta} \tag{52}$$

$$V_{idq} = L\frac{di_{idq}}{dt} + j\omega Li_{idq} + Ri_{idq} + V_{gdq} \tag{53}$$

where V_{iabc} denotes the output voltage of the inverter in the natural reference frame, while stationary reference frame and rotating reference frame coefficients are presented as $V_{i\alpha\beta}$, and V_{idq} respectively. Similarly, i_{iabc} is the inverter current and V_{gabc} is the grid-side voltage. The stationary reference frame current of the inverter, $i_{i\alpha\beta}$, is transformed to the rotating reference frame $a_{\alpha\beta} = e^{j\theta}$ transformation. The *d–* and *q–axis* components of the inverter voltage in the rotating reference frame are expressed as shown in Eqs. (54a) and (54b), derived from the complex equation $a_{dq} = a_d + ja_q$

$$V_{id} = L\frac{di_{id}}{dt} - \omega Li_{iq} + Ri_{id} + V_{gd} \tag{54a}$$

$$V_{iq} = L\frac{di_{iq}}{dt} + \omega Li_{iq} + Ri_{iq} + V_{gq}. \tag{54b}$$

The block diagram of proposed cascaded IMC-based power regulator consisting of an internal current control algorithm is presented in Fig. 9. The cascaded voltage

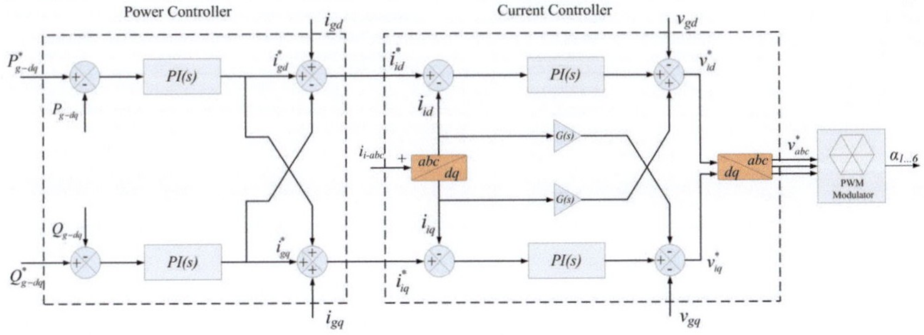

Fig. 9 Power and current control blocks in internal model-based controller

controller-based IMC model has been proposed in Bahrani et al. (2013). In the literature review, it is also seen that similar model-based control algorithms were developed by utilizing PI and PID controllers, whereas cascaded controller models have been considered (Leitner et al. 2018; Yazdanian and Mehrizi-Sani 2015). The tuning parameters of the PI controllers that are optimized for regulating the desired active and reactive power adjust the power values through the coefficients $K_{P\text{-}P}$, $K_{I\text{-}P}$ for active power and $K_{P\text{-}Q}$, $K_{I\text{-}Q}$ for reactive power, respectively. Similarly, the *d–* and *q–axis* components of the inverter current are controlled using the coefficients $K_{P\text{-}Cd}$, $K_{I\text{-}Cd}$ for the *d–axis* and $K_{P\text{-}Cq}$, $K_{I\text{-}Cq}$ for the *q–axis*, as illustrated in the PI controller blocks in Fig. 9. The integrator outputs in the general structure of the stepwise controller seen here are selected as the state variables of the system and thus the inputs are converted into derivatives of the state variables.

The state equations of the outer power controller and the internal current controller on the *d–* and *q-axis* are obtained as in (55) and (56), respectively.

$$\frac{d\vartheta_d}{dt} = v_{gd}^* - vgd \quad \frac{d\vartheta_q}{dt} = v_{gq}^* - vgq \tag{55}$$

$$\frac{d\varphi_d}{dt} = i_{id}^* - i_{id} \quad \frac{d\varphi_q}{dt} = i_{iq}^* - i_{iq} \tag{56}$$

where ϑ_d, ϑ_q, φ_d, and φ_q are the state variables.

The active (P) and reactive (Q) power transferred to the grid are defined based on the complex power equation $P + jQ = V_{gdq} i_{gdq}^*$, which relates the grid voltage and current components in the dq-rotating reference frame. In this equation, $V_{gdq} = V_{gd} + jV_{gq}$ represents the complex voltage of the grid, where V_{gd} and V_{gq} are the direct-axis and quadrature-axis components, respectively. On the other hand, $i_{gdq}^* = i_{gd} + ji_{gq}$ is the conjugate of the complex current, with i_{gd} and i_{gq} are being the respective *dq–axis* components of the inverter output current which is the same of current drawn by the grid. The real part of this equation, $P = \mathrm{Re}(V_{gdq} i_{gdq}^*)$ corresponds to the active power, while the imaginary part ($Q = \mathrm{Im}(V_{gdq} i_{gdq}^*)$) corresponds to the reactive power. These relationships emphasize that active power is primarily influenced by the alignment between the *d–axis* voltage (V_{gd}) and the *d–axis* current (i_{gd}). The reactive power depends on the interaction of the quadrature axis components (V_{gq} and i_{gq}) and their orthogonal relationship. Therefore, the active and reactive power control is based on the adjustment of i_{id} and -i_{gq} values, respectively. Thus, the outer power control loop and current controller loop presented in Fig. 9 are modelled referring to the Eqs. (40a) and (40b). In simulation modelling of proposed controller studies, the analytical designs are implemented regarding to the derived the state space equations of the inverter, filter and grid model. The analytical model is developed in real-time simulation environment using Matlab and Simulink software.

5 Simulation Setup and Case Studies

The simulation study is conducted using MATLAB/Simulink that is a widely used software environment for modelling and analysing power systems and control algorithms. The flexibility of this platform allows for accurate modelling of grid-connected VSI and its associated control schemes. The proposed internal model-based controller is implemented using discrete-time blocks to reflect its real-time operability in hardware. The simulation setup replicates real-world operational conditions while maintaining the adaptability needed to test diverse scenarios, ensuring a reliable assessment of the proposed system performance.

5.1 Simulation Environment

The simulation model represents a grid-connected VSI system that includes a full-bridge inverter, an LCL filter for harmonic attenuation, and a balanced three-phase grid complying with the block diagrams presented in Fig. 8. The modulator of VSI is modelled with sinusoidal pulse-width modulation (SPWM), which transforms the control signals into gate pulses for the switching devices. The LCL filter ensures that the switching harmonics produced by the inverter are mitigated, and the quality of the current injected into the grid is improved. The grid is modelled with nominal voltage and frequency with options to introduce disturbances such as voltage sags, harmonics, and frequency deviations for stress testing the system. The control structure comprises current controllers in the *dq*–reference frame, where separate PI controllers regulate the *d*– and *q*–axis currents. Decoupling terms are integrated into the control design to handle the cross-coupling effects between the axes for ensuring precise active and reactive power control.

The parameters of the simulation model are carefully selected to reflect realistic operational scenarios. The rated power of inverter, DC-link voltage, and switching frequency are aligned with typical values seen in medium-sized grid-tied systems. For the LCL filter, the inductance (L_f) and capacitance (C_f) are designed to achieve an optimal trade-off between harmonic attenuation and system stability. Resonance damping techniques are employed to address potential oscillations at the filter resonant frequency. The grid model operates at a nominal voltage and frequency of 50 Hz with provisions to introduce varying short-circuit ratios to mimic weak or strong grid conditions. These parameter settings presented in Table 1 ensure that the simulation is both representative of practical applications and robust enough to accommodate different operating scenarios.

Table 1 Parameters of the grid-connected VSI

Inverter parameters		
DC bus voltage	V_{dc}	800 V
Rated Power	S_n	100 kVA
Inverter side inductor	L_i	700 μH (0.15 pu)
Grid side inductor	L_g	15 μH (0.003 pu)
Filter capacitor	C_f	110 μF (0.05 pu)
Filter resistance	R_f	0.12 Ω (0.082 pu)
Nominal line voltage	V_{iabc}	380 V
System frequency	f_i	50 Hz
Base impedance	Z_b	1.444 Ω
Base voltage	V_b	310.26 V
Base current	I_b	151.94 A
Base current *(dq-frame)*	$I_{b\text{-}dq}$	214.87 A
Grid parameters		
Grid voltage	V_{gabc}	380 V
Angular frequency	ω_g	314 rad/s
Short circuit ratio	*SCR*	2.5–7
Reactance to resistance ratio	*X/R*	2–5
Controller parameters		
Switching frequency	f_{sw}	9 kHz
Sampling time	T_s	10 μs
Angular resonance frequency	ω_{res}	25,548 rad/s
Angular switching frequency	ω_{sw}	56,549 rad/s
Proportional gain of power controller	$K_{p\text{-}p}$	0.0018
Integral gain of power controller	$K_{i\text{-}p}$	0.1053
Proportional gain of current controller	$K_{p\text{-}c}$	3.4494
Integral gain of current controller	$K_{i\text{-}c}$	0.0049

5.1.1 LCL Filter Modelling

The LCL filter plays a crucial role in the designed grid-connected inverter system since it acts as a buffer to attenuate high-frequency switching harmonics and ensures the delivery of suitable current to the grid. Analysing the transfer function presents crucial insights into the frequency-domain characteristics of filter and the impact of damping on its overall performance. The fundamental transfer function of the LCL filter is defined as $H_{LCL} = i_g/v_i$ where i_g represents the grid-side current and v_i the inverter voltage. This analysis assumes that the grid voltage (v_g) acts as an ideal voltage source that is capable of absorbing harmonic frequencies. When operating under current-controlled conditions, where $v_g = 0$, the transfer function of the undamped LCL filter is expressed as (Reznik et al. 2014; Hollweg et al. 2023),

$$H_{LCL}(s) = \frac{1}{L_i C_f s^3 + (L_i + L_g)s}. \tag{57}$$

Here, L_i and L_g denote the inverter-side and grid-side inductances, respectively, while C_f represents the filter capacitance. The absence of damping leads to a sharp gain spike at the filter resonant frequency, which can compromise system stability if not mitigated. To suppress this resonance, a damping resistance (R_f) is introduced in series with the capacitor (C_f) that modifies the transfer function as follows:

$$H_{LCL}(s) = \frac{1 + C_f R_f s}{L_i L_g C_f s^3 + C_f (L_i + L_g) R_f s^2 + (L_i + L_g)s}. \tag{58}$$

The inclusion of the term C_f R_f s in the numerator introduces a damping factor that effectively reduces the gain spike. This modification not only smooths the response but also enhances the overall stability of the filter by mitigating the high-frequency resonant behaviour. An examination of the frequency response of the LCL filter through Bode plots presented in Fig. 10 shows significant improvements with damping. In the absence of damping, the filter displays a significant gain peak at its resonant frequency along with a phase shift nearing $-270°$ at higher frequencies. This behaviour is detrimental as it amplifies disturbances and limits the efficiency of the control loop. Conversely, the addition of damping eliminates the gain spike and results in a smoother response. At higher frequencies, the phase shift is limited to $-180°$, which enhances the stability margins of the system and ensures more reliable performance as seen in Fig. 10a where the effect of the change in grid-side inductor is analysed.

The Bode analysis also highlights the importance of limiting the closed-loop bandwidth of the inverter system. In order to maintain stability, the bandwidth should be limited to below 1000 Hz, where the phase shift of the LCL filter is approximately $-90°$. Exceeding this range may lead to excessive phase lag, which could compromise system stability and degrade the performance of current regulation.

The transfer function and Bode plots are used to design and to improve the stability of LCL filter with its damping components. The selected values of C_f, R_f, L_i and L_g

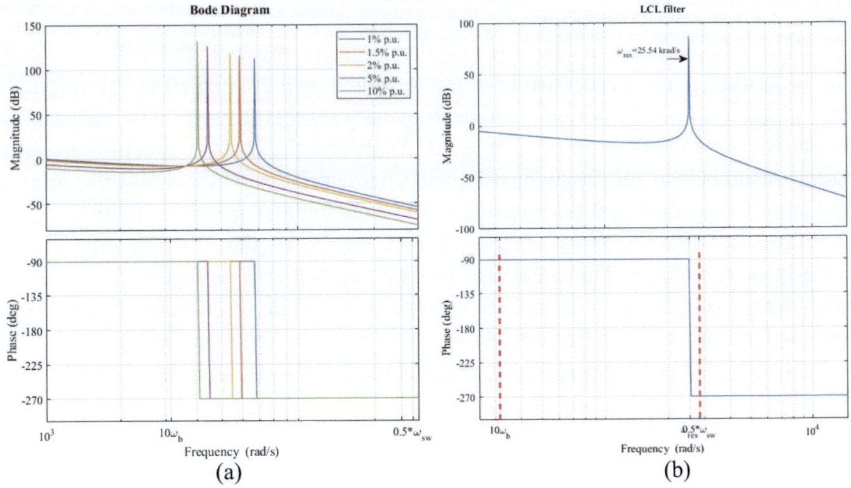

Fig. 10 Resonance frequency responses of LCL filter: **a** response to the change of grid side inductor **b** response of designed filter

as presented in Table 1 have been modelled in Simulink and stability analyses have been performed as shown in Fig. 10b to validate the reliability of the LCL filter. The effective resonance suppression obtained in the design ensures sufficient harmonic attenuation to meet the power quality standards and maintains compatibility with bandwidth requirements of the control loop. This balance ensures the filter enhances both the stability and performance of grid-connected inverter systems under a wide range of operating conditions.

5.1.2 Phase-Locked Loop (PLL) Modelling

The grid-connected inverter is controlled using the synchronous reference frame Phase-Locked Loop (SRF-PLL) that is a widely adopted method for three-phase systems. The block diagram of the implemented PLL is shown in Fig. 11, where the three-phase grid voltages (V_{abc}) are first transformed into a rotational reference frame. The rotational frame is further converted into a stationary frame through a dq transformation. The direct (d) and quadrature (q) components obtained from this transformation are used for active and reactive power control, respectively. This conversion process is analogous to the one used in Space Vector Modulation (SVM), where the abc components are first transformed into the $\alpha\beta$ stationary frame using the Clarke transformation, and then further into the dq frame using the Park transformation. The mathematical steps for these transformations are represented in Eqs. (45) and (46). Once the reference frame transformation is complete, the estimated phase

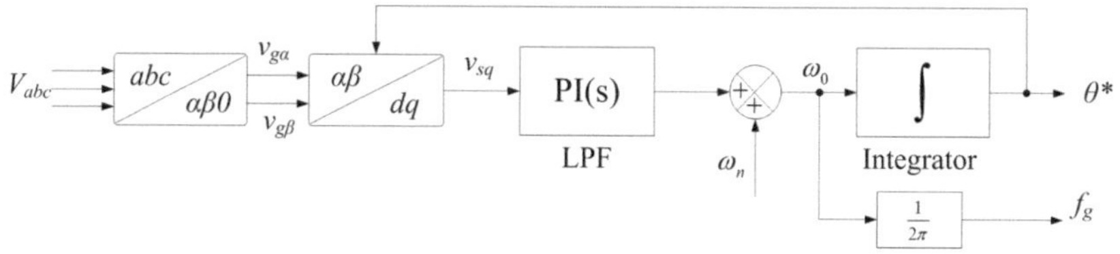

Fig. 11 Block diagram of SRF-PLL

angle of the utility grid (θ^*) is fed back to the Park transformation to ensure accurate compensation of the *dq* components (Karimi-Ghartemani 2014; Rodriguez et al. 2007; Kabalci 2021).

The Clark and Park transformation matrix are represented as follows respectively where θ^* denotes the estimated phase angle, and θ represents the phase angle of utility grid

$$\begin{bmatrix} V_\alpha \\ V_\beta \end{bmatrix} = \frac{2}{3}\begin{bmatrix} 1 & -\frac{1}{2} & -\frac{1}{2} \\ 0 & \frac{\sqrt{3}}{2} & -\frac{\sqrt{3}}{2} \end{bmatrix}\begin{bmatrix} V_a \\ V_b \\ V_c \end{bmatrix} \tag{59}$$

$$\begin{bmatrix} V_d \\ V_q \end{bmatrix} = \begin{bmatrix} \cos(\theta^*) & \sin(\theta^*) \\ -\sin(\theta^*) & \cos(\theta^*) \end{bmatrix}\begin{bmatrix} V_\alpha \\ V_\beta \end{bmatrix}. \tag{60}$$

The SRF-PLL is proposed to be operated in unbalanced grid voltages where the transformation components generate double the fundamental grid frequency that can be used in a phase detector. The Park transformation voltage coefficients represent direct and quadrature components as two-axis output, and the estimated phase angle is identical with the phase angle of the grid under balanced and undistorted grid conditions. In these operation conditions, V_d would be equal to the magnitude of the grid voltage while V_q represents zero as depicted in the following equations. SRF-PLL is designed to operate effectively under unbalanced grid voltage conditions. In such scenarios, the transformation components produce frequency components at twice the fundamental grid frequency, which can be utilized in the phase detection process. The Park transformation coefficients represent the direct (V_d) and quadrature (V_q) components as two-axis outputs. The estimated phase angle aligns precisely with the grid phase angle under balanced and undistorted grid conditions. In these ideal operating conditions, V_d corresponds to the magnitude of the grid voltage, while V_q equals zero, as illustrated in the following equations. This characteristic ensures accurate phase tracking and stability even under challenging grid conditions (Karimi-Ghartemani 2014; Kabalci 2021)

$$V_d = V_m \cos(\theta - \theta^*) \tag{61}$$

$$V_q = V_m \sin(\theta - \theta^*). \tag{62}$$

The loop filter in the SRF-PLL plays a critical role in ensuring precise dynamic performance for the estimated phase angle and a rapid locking response that is achieved through fine-tuning of the filter parameters. Under regular and balanced grid conditions, the filter provides both high accuracy and fast synchronization which allow the grid voltage and phase angle to be determined quickly and precisely. However, during grid disturbances or unbalanced operation, the bandwidth of loop filter must be reduced to maintain precision and suppress the influence of higher-order harmonic components and unstable conditions. To address this issue, a low-pass filter is incorporated into the PLL controller. This addition enhances system stability and improves the PLL tracking capability by ensuring reliable performance even under challenging grid conditions.

Figure 12 illustrates the operation of the SRF-PLL controller in synchronizing with the grid that is modelled in Simulink. It shows the key signals measured and processed during the operation of the SRF-PLL, which include the grid voltages, the derived phase angle (θ), and the voltage components in the stationary $\alpha\beta$-reference frame. The three-phase grid voltages (V_{abc}) are presented in their original sinusoidal form. These voltages are the input to the SRF-PLL and serve as the basis for phase synchronization. The grid voltages are transformed from the three-phase system to the stationary two-phase *αβ-reference* frame using the Clarke transformation. This step produces the orthogonal components V_α and V_β which are shown in the third axis of the figure as sinusoidal and cosine-like waveforms, respectively. These components serve as inputs to the PLL controller. The PLL extracts the phase angle (θ) of the grid voltage by processing the $V_{\alpha\beta}$ components. This derived phase angle is critical for synchronizing the inverter with the grid and ensuring proper alignment for current injection. The phase angle is depicted as a continuously increasing signal by representing the time-dependent position of the rotating reference frame. The figure also highlights the feedback nature of the SRF-PLL, where the derived phase angle (θ) is used to generate the dq-axis transformation matrix which ensures accurate phase tracking. The control loop maintains synchronization by minimizing the error between the projected V_d component and zero.

5.2 Case Studies

The modelled power system is subjected to various operating conditions to test its capabilities and to evaluate the performance of the proposed control scheme comprehensively. The first scenario examines steady-state operation under nominal grid conditions. In this case, the VSI operates with fixed active (P) and reactive (Q) power references that allow for the measurement of power tracking accuracy and steady-state errors. The metrics such as THD of the injected current are also evaluated to assess compliance with power quality standards. The second scenario involves dynamic power changes, where the active and reactive power setpoints are varied in steps to evaluate the transient response of the controller. The system's ability to track changes with minimal overshoot, fast settling time, and accurate tracking

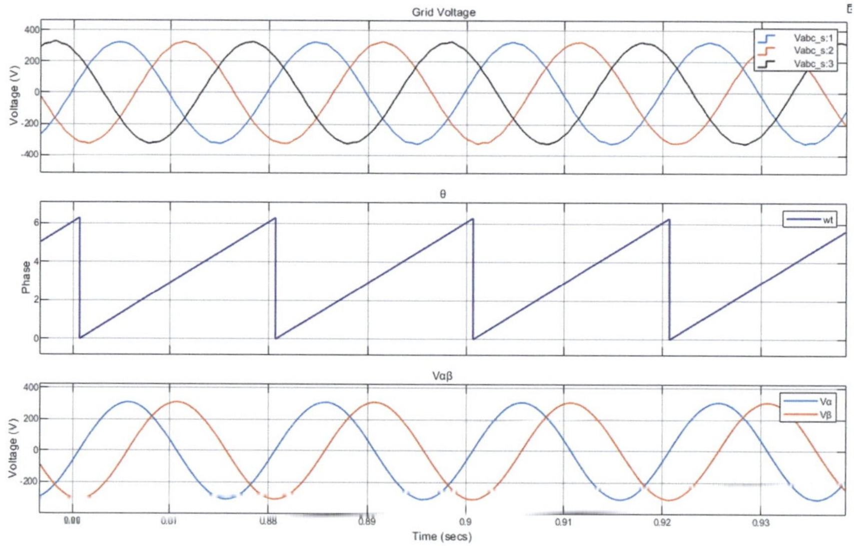

Fig. 12 Measurements of SRF-PLL: grid voltages, phase angle (θ), and stationary frame voltages ($V_{\alpha\beta}$)

is assessed. This scenario demonstrates the effectiveness of the control system in handling load variations or shifts in power demand.

In the third scenario, grid disturbances are introduced to test the robustness of the control system. The voltage sags, where the grid voltage is reduced to 80% of its nominal value, are simulated to analyse the ability of VSI to maintain the stability and regulate power under adverse conditions. Additionally, harmonic distortions in the grid voltage are modelled to evaluate the disturbance rejection capabilities of the controller and its impact on the current waveform quality. The fourth scenario relies on frequency deviations where the grid frequency ranging between 49 and 51 Hz is applied to test the robustness of the PLL and the synchronization mechanisms of the system.

- Scenario 1: Steady-state operation under nominal grid conditions.
- Scenario 2: Dynamic response to sudden load changes
- Scenario 3: Performance under grid disturbances (e.g., voltage sags/swells).
- Scenario 4: Efficiency in grid-following operation under frequency deviation.

5.2.1 Scenario 1: Steady-State Operation Under Nominal Grid Conditions

In this case study the system is operated with nominal 380 V grid line voltage and 50 Hz line frequency. The active power (P) and reactive power (Q) setpoints are fixed at predefined values as P = 50 kW, and Q = 10 kVAR. The inverter injects power

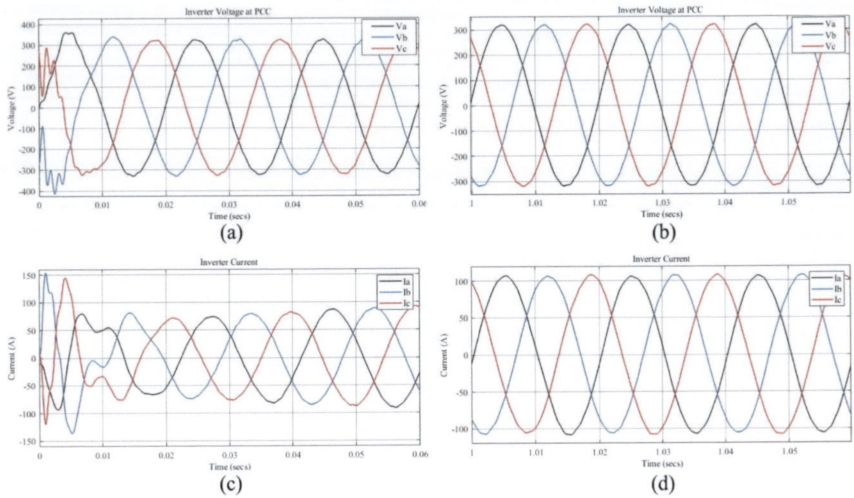

Fig. 13 Output voltage and current analyses of VSI at grid connection: **a** transient response of voltage, **b** voltage waveforms in steady-state, **c** transient response of current, **d** current waveforms in steady-state

into a balanced three-phase grid without significant variations in grid conditions or load demands. Figure 13 illustrates the voltage and current waveform analysis of VSI connection at PCC. The short transient intervals seen in Fig. 13a, c confirm the rapid response of proposed controller while the steady-state measurements denote the reliability and uniform structure of voltage and current waveforms.

The power quality performance of the proposed controller has been analysed with THD measurements that are run up to 200th order of harmonics to cover the switching frequency spectrum as shown in Fig. 14. The THD ratio of voltage during the grid connection is observed around 0.67% while the THD of line current is 1.28%. The comparison of THD ratios to the other methods proposed in the literature is presented in the following sections.

Figure 15 depicts the d- and q-axis components of the filter current (i_L) responses of the grid-connected inverter with IMC method. The i_d current stabilizes around 0.5 per unit (pu), indicating that the system is successfully delivering the desired amount of active power to the grid. This steady value reflects efficient control of active power flow. The i_q current stabilizes at approximately −0.1 pu, suggesting a controlled exchange of reactive power where the negative sign indicates that reactive power is injected to the grid depending on the system requirements. The i_d and i_q currents in this system demonstrate good transient performance, proper decoupling, and stable operation, ensuring effective active and reactive power control in the grid-connected inverter. Figure 16 illustrates the P and Q exchange over time in the grid-connected operation of IMC controlled inverter with the respective reference values (P_{ref} and Q_{ref}). The active power quickly reaches its steady-state value by matching the reference closely without any noticeable overshoot or oscillation sue

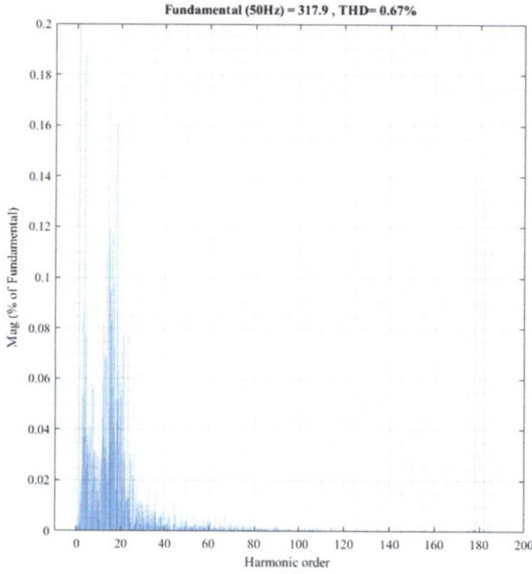

(a)

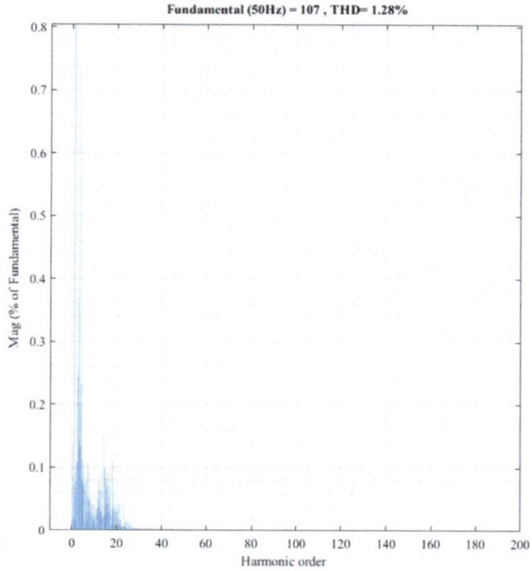

(b)

Fig. 14 Spectrum analyses of VSI: **a** voltage THD, **b** current THD

to its stable system response with fast dynamics. The green line (Q_{ref}) shows the reference for reactive power, while the brown line (Q) represents the actual reactive power delivered to the grid. The reactive power closely tracks the reference with minimal error and stabilizes quickly. Both P and Q measurements exhibit smooth and rapid transient responses, reaching steady-state values within approximately 100 ms. The detailed measurements of active and reactive power are shown in Fig. 16b where each one closely tracks the reference values with minimum ripple rates.

The steady-state operation under nominal grid conditions of the control system demonstrates minimal steady-state error, low THD in the output current and output

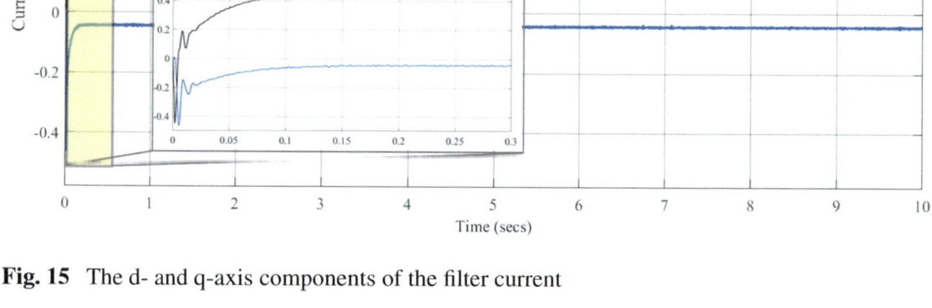

Fig. 15 The d- and q-axis components of the filter current

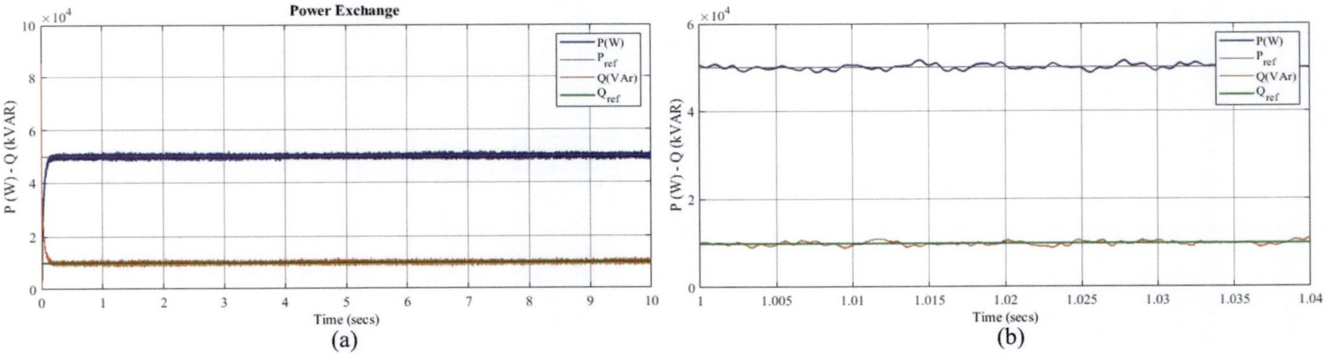

Fig. 16 Power exchange analysis of grid-connected VSI: **a** active and reactive power measurement **b** detailed view

voltages, and stable operation under nominal grid conditions. The presented analyses confirm the reliability of the proposed control to ensure the compliance of the controlled inverter with grid interconnection standards.

5.2.2 Scenario 2: System Performance Under Grid Disturbances

In the second case study, the system is operated with voltage sag by reducing the grid voltage to 80% of its nominal value for 2 s, and then a voltage swell is introduced by increasing the grid voltage to 120% of its nominal value for 2 s after the 3rd second of restoration. The active power (P) and reactive power (Q) setpoints are fixed at predefined values as P = 50 kW, and Q = 10 kVAR as in the first scenario. Figure 17a represents inverter voltages while Fig. 17b denotes the current waveforms measured in the grid disturbance scenario. The VSI is operated at normal grid conditions from [0 to 2 s] at the first interval, and the voltage waveforms are observed stable and sinusoidal with an amplitude of 1 pu during this period. The inverter operates under nominal grid conditions without disturbances that ensure balanced three-phase output voltages (V_a, V_b, V_c). At $t = 2$ s, a voltage sag is introduced by reducing the voltage magnitude to 0.8 pu to simulate a grid disturbance such as a fault or a sudden increase in load demand. The zoomed-in plot highlights the reduced amplitude of all three-phase voltages while maintaining their sinusoidal nature and synchronization. The internal model-based controller compensates for the disturbance by ensuring stable operation during the sag.

After the sag, the voltage amplitudes restored to 1 pu at $t = 4.05$ s that indicates return to nominal grid conditions. The zoomed-in plot during this interval shows smooth transitions back to sinusoidal voltages without overshoots or oscillations owing to the strong transient stability of the system. At $t = 7$ s, a voltage swell is introduced to the system by increasing the voltage magnitude to 1.2 pu. This disturbance mimics scenarios such as a sudden reduction in load or a grid switching event. The zoomed-in plot reveals an increase in amplitude across all three phases while maintaining sinusoidal waveforms.

The controller effectively mitigates the potential instability caused by the swell and ensures the inverter remains synchronized with the grid as shown in the last scope of Fig. 17a. After the swell, the voltage amplitude returns to 1 pu at $t = 9$ s which indicates the system's ability to restore nominal operation post-disturbance. The smooth recovery of waveforms shows the ability of the system to maintain power quality and stability. Across all scenarios, VSI maintains sinusoidal waveforms for the three-phase voltages, even during sag and swell conditions. The phase alignment between the three-phase voltages is maintained throughout the disturbance events, and it confirms that the internal model-based controller ensures grid synchronization. The controller also adapts the current amplitude in response to voltage disturbances by increasing it during sags and reducing it during swells. This adjustment highlights the robustness of the internal model-based controller in regulating power flow and maintaining system performance under varying grid conditions as shown in Fig. 17b.

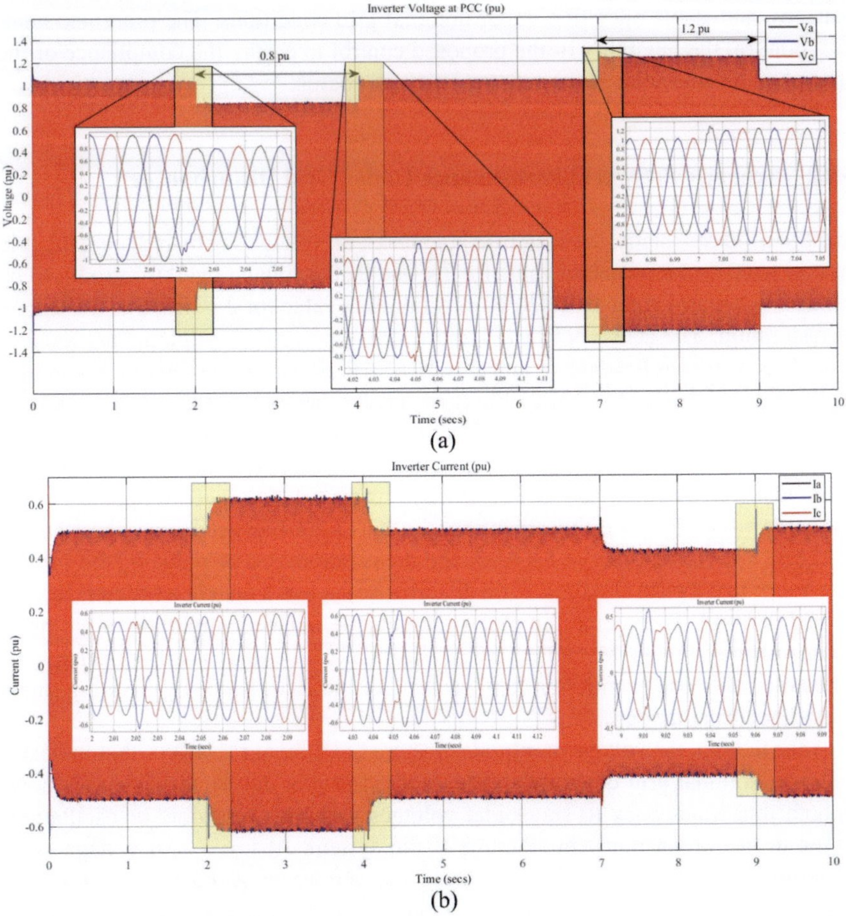

Fig. 17 Output voltage and current analyses of VSI at grid connection: **a** transient response of voltage, **b** voltage waveforms in steady state, **c** transient response of current, **d** current

The THD analyses of VSI inverter have been performed to observe the power quality of grid-connected operation under various disturbances. The voltage and current THD ratios have been shown in Fig. 18a, c during the voltage sag disturbances. The measured values ($THD_V = 0.80\%$ and $THD_I = 1.15\%$) validate that the proposed controller ensures stable and reliable operation under voltage sag conditions, and the measured harmonic rates comply with international standards of IEEE and IEC. On the other hand, the THD values of inverter voltage and currents under voltage swell disturbances are obtained at 0.63% and 1.62% as seen in Fig. 18b, d, respectively.

Figure 19 illustrates the P and Q exchange between the inverter and the grid during scenarios involving an 80% voltage sag and a 120% voltage swell. The analysis also compares the actual power (P, Q) with the reference power values (P_{ref} and Q_{ref}). During normal operation, the active power output tracks the reference value (P_{ref})

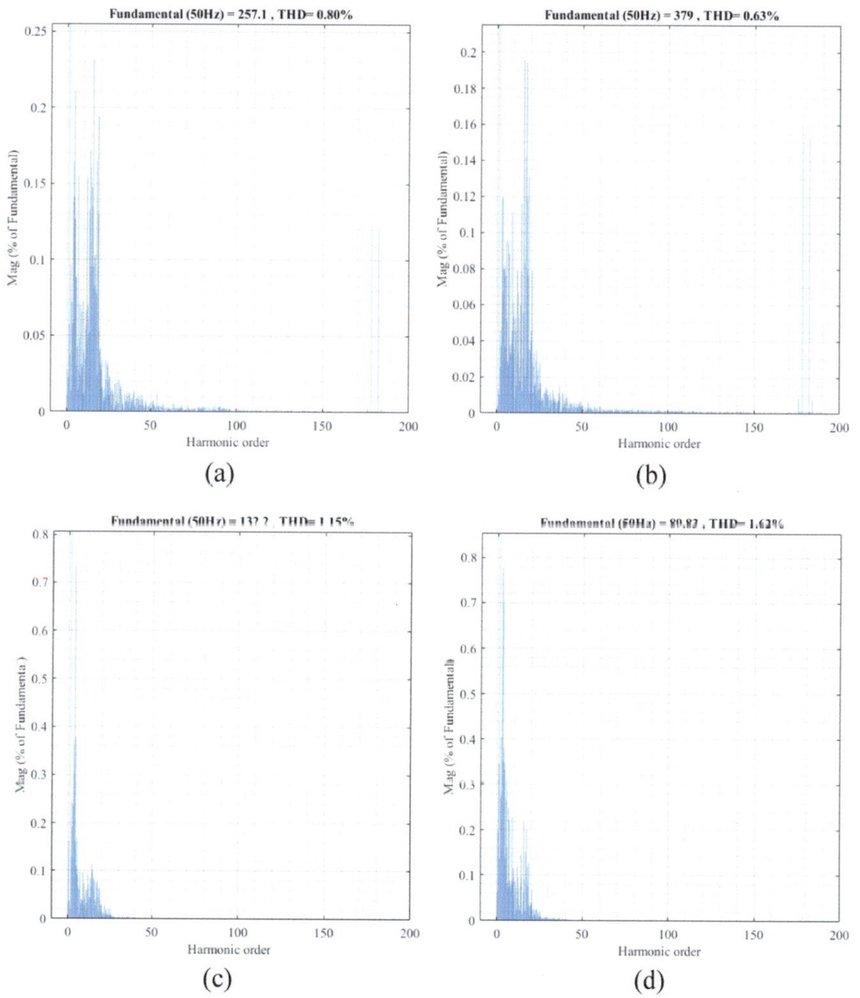

Fig. 18 Spectrum analyses of VSI: **a** THD_V during voltage sag, **b** THD_V during voltage swell, **c** THD_I during voltage sag, **d** THD_I during voltage swell

closely at approximately 0.5 pu. The reactive power output remains near 0.1 pu with zero shift from reference value. The steady and precise tracking of reference values confirms the efficiency of the proposed controller under nominal conditions. At $t =$ 2.02 s, the grid voltage drops significantly and causes a transient dip in active power. The zoomed-in section reveals that the active power output momentarily decreases due to the sudden voltage drop but quickly recovers to track P_{ref} around in 80 ms.

The reactive power output experiences a sharp transient spike which reflects the controller reactions to stabilize voltage and support the reactive power demand during the sag period. The spike decays rapidly and returns to Q_{ref} value as the system

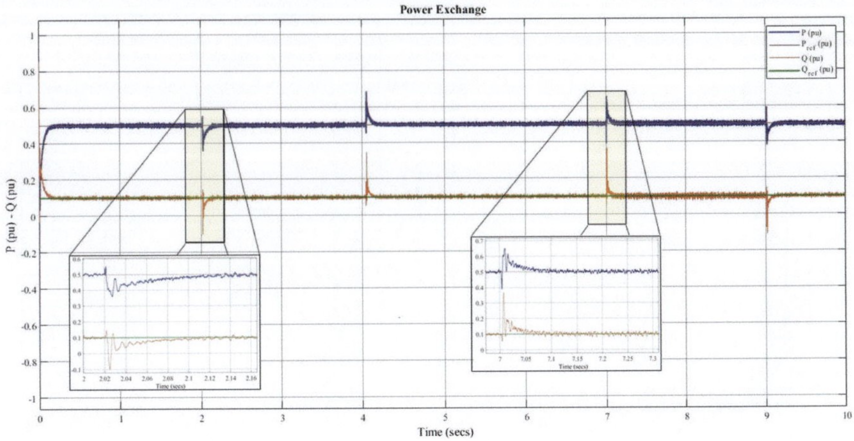

Fig. 19 Power exchange analysis of grid-connected VSI during voltage disturbances

stabilizes around in 80 ms likewise in active power control. Another screenshot seen at $t = 7$ s denotes the starting period of voltage swell. The grid voltage increases to 120% of nominal value which causes a transient overshoot in active power output. The zoomed-in section shows that the controller manages to suppress the overshoot, restoring the active power to its reference value within a short duration of 40 ms. Like the voltage sag, the swell induces a transient spike in reactive power that compensates for the overvoltage and quickly settles back to reference value.

The power response during both sag and swell demonstrates strong transient stability. The active power exhibits minimal overshoot or undershoot and closely tracks the reference values after disturbances. The reactive power spikes during disturbances indicate the ability of the inverter to provide dynamic voltage support. These transient responses play a critical role in stabilizing the grid under sag and swell conditions. The analysis demonstrates that the inverter controlled with its internal model-based controller successfully manages voltage disturbances while ensuring stable and reliable active and reactive power exchange with the grid. The system exhibits a remarkable ability to respond dynamically to both voltage sags and swells, as evidenced by the rapid recovery of power outputs to their reference values. This capability minimizes the impact of transients on the grid and ensures consistent operation under varying conditions.

5.2.3 Scenario 3: Dynamic Response to Sudden Load Changes

In the third case study scenario, the system is operated with varying load values in terms of both active and reactive power. The active power has been shifted to [60 kW 70 kW 80 kW 55 kW 45 kW] at [0 s, 2 s, 4 s, 6 s, 8 s] of the simulation while the reactive power was shifted [20 kVAR, 35 kVAR, 25 kVAR] at [0 s, 5 s, 6 s] intervals. Figure 20a represents inverter voltages while Fig. 20b denotes the current waveforms measured in the load change scenario. The VSI is operated at normal grid conditions with 60 kW active and 20 kVAR reactive power demand in [0 s, 2 s] at the first interval. The inverter operates under nominal grid conditions without disturbances that ensures balanced three-phase output voltages (V_a, V_b, V_c). At $t = 2$ s, an increment in active power is introduced by increasing P to 70 kW in load demand which is followed by 80 kW active power demand in $t = 4$ s. The first zoomed-in plot highlights that there is no remarkable variation in the inverter current as seen in Fig. 20a. The internal model-based controller compensates the sudden load changes by ensuring stable operation during the increasing power demand.

At $t = 5$ s, the reactive power demand is increased to 35 kVAR as seen in the voltage and current measurements of Fig. 20a, b respectively. The zoomed-in plot during this interval shows smooth increment in the inverter voltages without overshoots

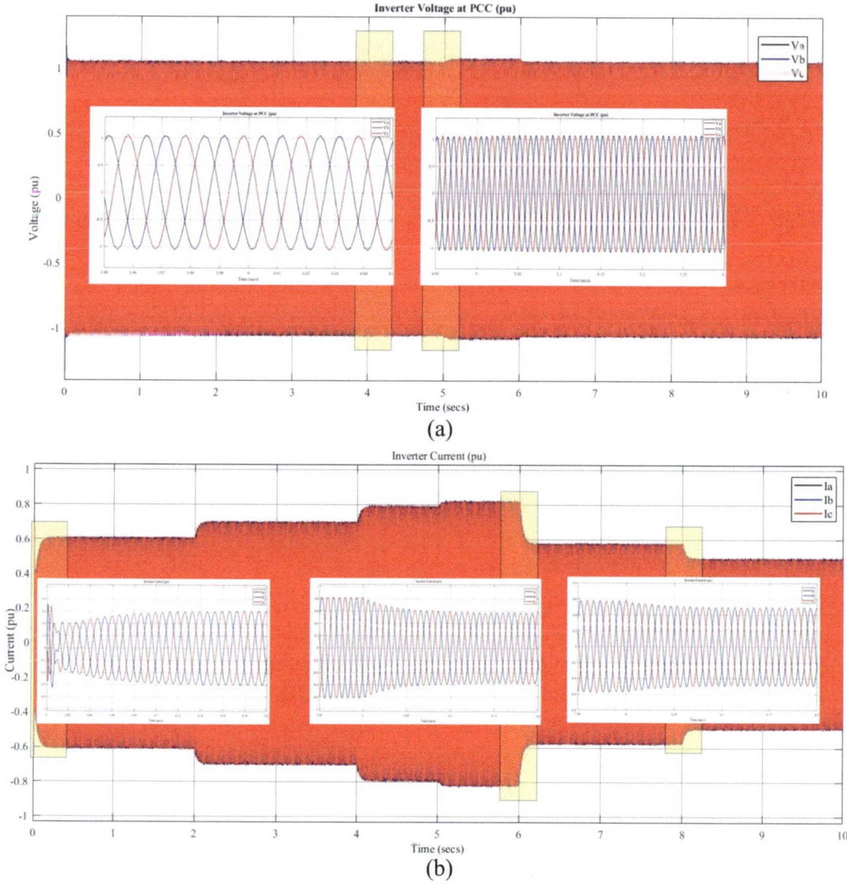

Fig. 20 Output voltage and current analyses of VSI at grid connection: **a** transient response of voltage **b** transient response of current

or oscillations owing to the strong transient stability of the system. At $t = 6$ s, a remarkable load disconnection is introduced to the system by decreasing the active power from 80 to 55 kW, and decreasing the reactive power from 35 to 25 kVAR. The zoomed-in plot of Fig. 20b reveals a decrease in current amplitude across all three phases while maintaining sinusoidal waveforms. Despite the abrupt load reduction, the inverter maintains synchronization with the grid as evidenced by the smooth waveforms and steady operation in the final segment of Fig. 20b. At $t = 8$ s, the active power demand drops further to 45 kW and the inverter responds swiftly and smoothly by ensuring stability and consistent output voltages.

The THD analyses of VSI inverter have been performed to observe the power quality of grid-connected operation under various load profiles. The voltage and current THD ratios have been shown in Fig. 21. There the two of voltage and current THD measurements shown in the figure regarding medium load and highest load intervals. The measured $THD_V = 0.76\%$ seen in Fig. 21a is obtained under 60 kW and 20 kVA load. When the P and Q are increased to 80 kW and 35 kVAR, the THD_V is measured at 0.94% as illustrated in Fig. 21b. The current THD ratios of 70 kW–20 kVAR and 80 kW–35 kVAR are measured as $THD_I = 0.94\%$ and $THD_I = 0.64\%$ respectively as shown in Fig. 21c, d. The measurements validate that the proposed controller ensures stable and reliable operation under varying load conditions, and the measured harmonic rates comply with international standards of IEEE and IEC.

Figure 22 illustrates a scenario involving sudden load changes that are reflected in the sharp transitions observed in both P and Q. These abrupt shifts represent the instantaneous change in load demand and operating conditions, which the inverter system must respond to effectively. The analysis also compares the actual power (P, Q) with the reference power values (P_{ref} and Q_{ref}). At specific time intervals (e.g., 2, 4, 6, and 8 s), there are noticeable steps in the active power output. Each sudden change corresponds to a new steady-state load level. The system rapidly adjusts the active power to match the new reference value (P_{ref}) with minimal overshoot or oscillations. The system starts in nominal grid conditions by supplying 60 kW of active power and 20 kVAR of reactive power to the load. The inverter delivers balanced three-phase output voltages (V_a, V_b, V_c) without observable disturbances or distortions, indicating robust initial performance. At $t = 2$ s, an increase in active power to 70 kW occurs which is followed by a further increase to 80 kW at $t = 4$ s.

The zoomed-in segment of Fig. 22 shows that the inverter faultlessly handles these load increments. The voltage waveforms remain sinusoidal, and no significant deviations in current amplitude are observed. This smooth transition is achieved by the internal model-based controller, which promptly compensates for the rising power demand, ensuring stability without introducing oscillations or overshoots. The quick stabilization after each transition indicates that the control system is highly responsive and capable of mitigating the effects of sudden load disturbances. The reactive power also exhibits sharp transitions at 5 and 6 s at similar points as the active power changes. At $t = 5$ s, the reactive power demand rises significantly from 20 to 35 kVAR. This adjustment is reflected in the inverter voltage and current waveforms (Fig. 20a, b). Both figures show a gradual and smooth transition without instability. The zoomed-in plot highlights that the voltage waveforms experience no overshoots

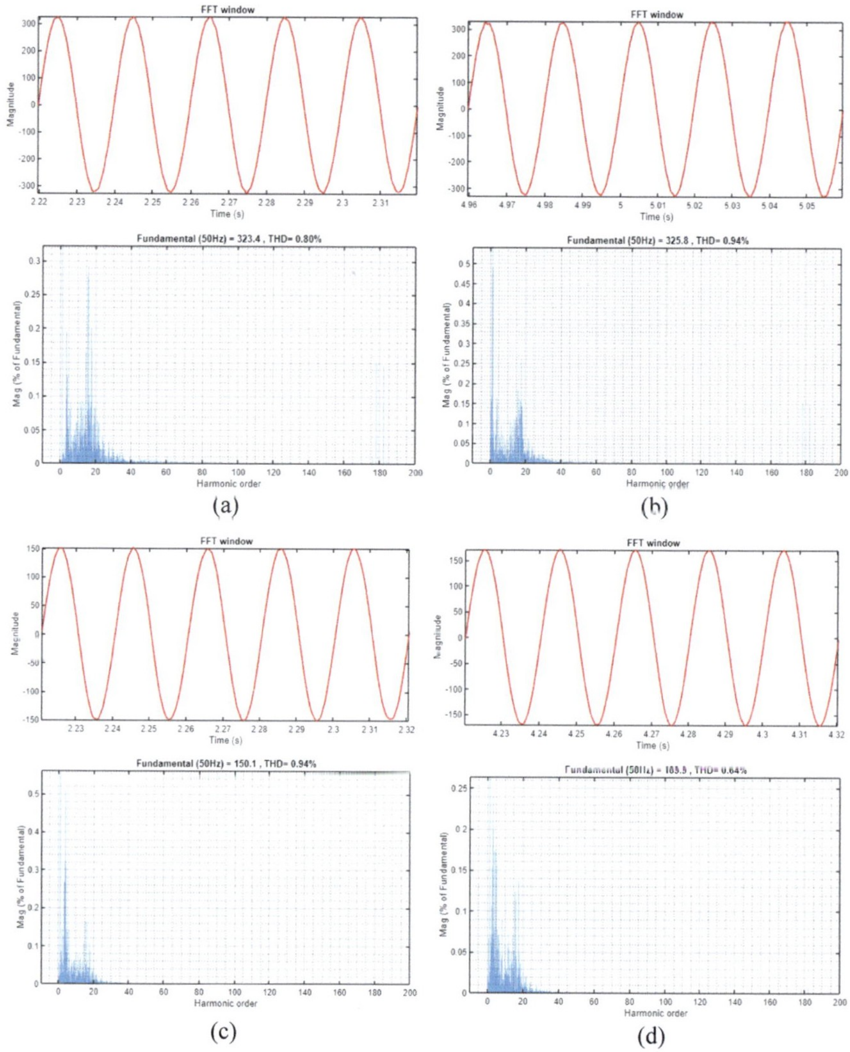

Fig. 21 Spectrum and waveforms of VSI: **a** THD_V during first case, **b** THD_V during highest P and Q demand, **c** THD_I during first case, **d** THD_I during highest P and Q demand

or oscillations during this interval. This behaviour underlines transient stability of the inverter and the controller's ability to manage dynamic reactive power demands.

At $t = 6$ s, a significant reduction in load occurs. The active power demand drops sharply from 80 to 55 kW, while the reactive power decreases from 35 to 25 kVAR as seen in Fig. 22. The zoomed-in section of Fig. 20b reveals a corresponding reduction in current amplitude across all three phases. Notably, the current waveforms maintain their sinusoidal shape that highlights effectiveness of the controller in preventing potential instability. Despite the abrupt load reduction, the inverter

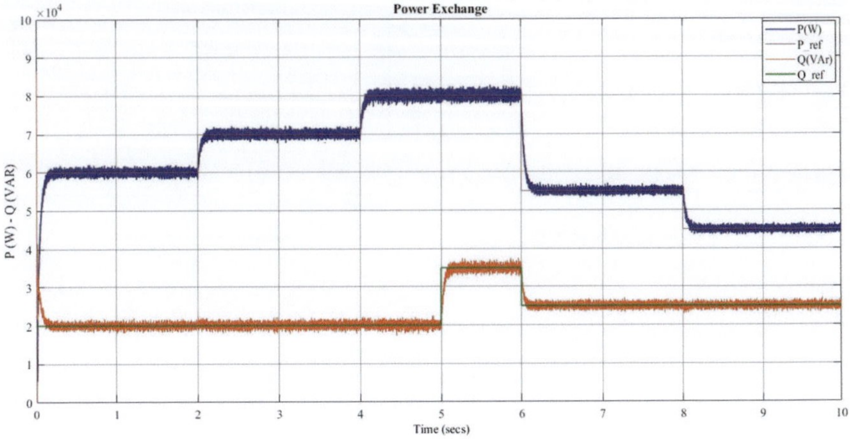

Fig. 22 Power exchange analysis of grid-connected VSI during sudden load changes

maintains synchronization with the grid as evidenced by the smooth waveforms and steady operation in the final segment. These shifts reflect the system response to maintain voltage stability and support reactive power demand under varying load conditions.

5.2.4 Scenario 4: Efficiency in Grid-Following Operation Under Frequency Deviation

In the fourth case study scenario, the system performance is evaluated under varying grid frequency conditions to test the reliability and stability of both the PLL and internal model-based control algorithms. The grid frequency is intentionally perturbed by decreasing it to 49 Hz and subsequently increasing it to 51 Hz, simulating realistic grid disturbances that can occur in practical operations. This dynamic frequency variation is introduced to analyse how effectively the PLL maintains synchronization and how well the IMC adapts to these changes without compromising the system stability or control accuracy.

The test deliberately introduces a controlled frequency deviation. The decrease to 49 Hz simulates a frequency sag that is often caused by load imbalances or supply constraints in the grid. The subsequent increase to 51 Hz represents a frequency overshoot, which can occur during transient grid recovery or load shedding. The smooth and continuous ramping of the phase angle ($\theta = \omega t$) observed in the results presented in Fig. 23 confirms the reliability of the PLL in tracking frequency changes. The absence of oscillations or phase-lock loss indicates a well-designed PLL with strong dynamic performance. The IMC control algorithm ensures that the inverter operates reliably despite the changing grid frequency. The smooth transition of $V_{\alpha\beta}$ voltages (stationary frame components) and the sinusoidal nature of the three-phase voltages (V_a, V_b, V_c) confirm that the IMC effectively compensates for frequency variations, maintaining stable output without introducing distortions. The system demonstrates grid support capability by successfully handling both frequency sag to 49 Hz as seen in Fig. 23a and overshoot to 51 Hz as in Fig. 23b.

This ensures that the inverter and its control systems can operate reliably under realistic and dynamic grid conditions. Stability is further demonstrated by the lack of oscillatory behaviour or transient deviations during the frequency changes. The test results show the output voltage and power generation capabilities of the system smoothly transition between steady-state conditions before, during, and

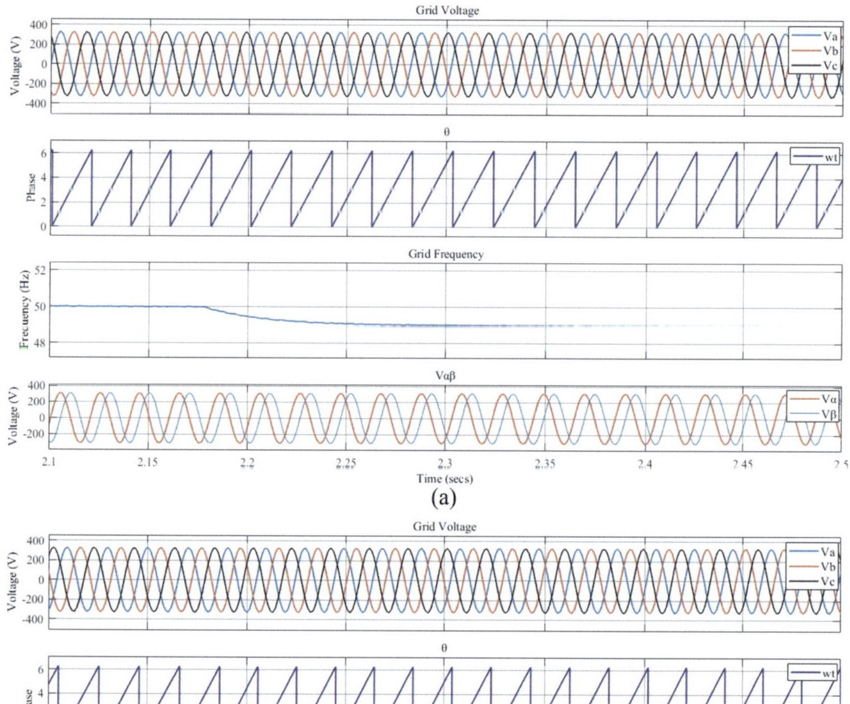

(a)

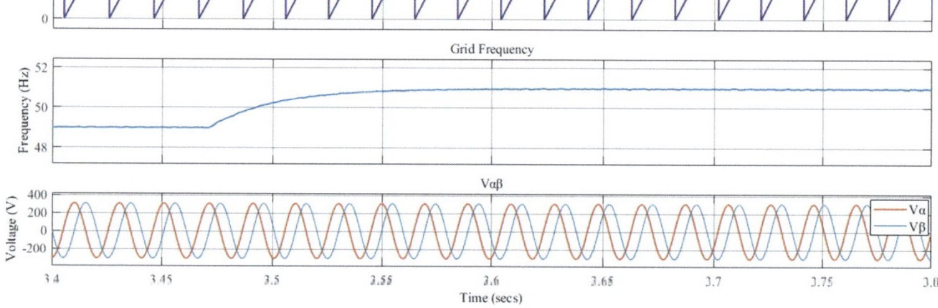

(b)

Fig. 23 PLL operation against frequency deviation tests: **a** frequency sag ($f_g = 49$ Hz), **b** frequency swell ($f_g = 51$ Hz) after frequency variations. This smooth transition is crucial in preventing disturbances that could propagate to other grid-connected systems. These results highlight the suitability of the PLL and IMC algorithms for real-world grid-connected applications, where maintaining stability under dynamic conditions is essential for safe and reliable operation. Further testing under more severe disturbances (e.g., abrupt frequency changes, harmonics, or voltage imbalances) in experimental studies provides additional insights into the limits of stability and performance.

6 Discussion

While the IMC approach has proven its effectiveness in the validation study, it is important to consider its advantages and limitations in comparison with other widely used control strategies such as PI control, PID control, MPC, and Lyapunov-based control methods. PI control is one of the simplest and most widely used controllers in power electronics, particularly for regulating voltage or current in grid-connected systems. It provides good performance in steady-state conditions and is highly efficient for controlling systems with relatively simple dynamics. However, PI controllers struggle with dynamic disturbances, nonlinearities, and time-varying conditions, which are common in real-world grids. While PI controllers are computationally efficient, they often fail to adapt quickly to changes in grid conditions such as voltage sags, swells, or sudden load variations. On the other hand, PID control is an extension of PI control that introduces a derivative term, which helps improve the transient response of the system. It is widely used in industrial applications due to its simplicity and effectiveness for many linear systems. While PID control can achieve good results for certain well-behaved systems, it suffers from similar limitations to PI control especially in the presence of nonlinearities and disturbances. It also has difficulty ensuring stability when applied to systems with high-order dynamics or those requiring robust performance under varying grid conditions. IMC integrates the dynamic model of the plant into the control design which enables it to react more effectively to these disturbances and provides superior performance under a wider range of operating conditions. IMC offers a more systematic way to handle complex, nonlinear, or time-varying performance by explicitly addressing these issues in the control design, leading to superior stability and performance under challenging conditions.

MPC is a modern control strategy that uses an optimization algorithm to predict future system behaviour and minimize a cost function, typically over a finite prediction horizon. In the context of grid-connected VSI systems, MPC offers a distinct advantage in handling multivariable systems with constraints such as voltage, current, and power. MPC can also account for system dynamics and disturbances in real-time which makes it highly suitable for applications where future system states need to be predicted and optimized. However, MPC requires significant computational resources, especially when applied to large-scale systems, as it involves solving an optimization problem at each control step. In contrast, IMC uses a simpler design based on inverse models and filter blocks, offering easier implementation with lower computational demands. The advantage of IMC lies in its simpler structure, which results in faster real-time performance and lower complexity compared to MPC.

Lyapunov-based control methods which are widely used alternative are rooted in stability theory, where a Lyapunov function is designed to prove the system stability. This method ensures global asymptotic stability by constructing a suitable Lyapunov function that proves the system will converge to its desired equilibrium. While Lyapunov-based control guarantees stability and is highly reliable for certain classes of systems, it often requires careful tuning and can be more complex to implement in systems with nonlinearities or high-dimensional dynamics, as is common in grid-connected inverters. Comparing to the Lyapunov-based control methods, IMC offers a more intuitive design with the ability to directly model and correct for plant dynamics and disturbances through its model inversion and filtering that makes it more practical for real-time applications in grid-connected systems.

In comparison with other advanced control methods like MPC and Lyapunov-based control, IMC offers a simpler, more computationally efficient solution that is still highly adaptable to a wide range of system dynamics and disturbances. This makes IMC an attractive option for real-time applications in renewable energy systems and smart grids, where robustness, adaptability, and efficiency are key requirements.

The future of IMC in grid-connected applications holds remarkable potential, particularly in terms of grids that are evolving to accommodate renewable energy sources and are becoming more decentralized. The role of advanced control strategies like IMC becomes increasingly critical in ensuring grid stability, reliability, and efficiency as energy systems transition to these more complex and dynamic architectures. One key area where IMC can have a significant impact is in dealing with the nonlinearities often found in modern grid conditions. Traditional control methods, such as PI or PID controllers, tend to struggle when faced with nonlinear dynamics, time-varying parameters, and disturbances such as voltage fluctuations, harmonic distortions, or the introduction of variable renewable energy sources like wind and solar power. IMC can be extended to better handle these nonlinearities by refining its internal models and incorporating more detailed system dynamics into the control process owing to its model-based design. This enhancement will allow IMC to better track desired outcomes and maintain stable grid operations even under challenging conditions, such as in weak grids where voltage drops, frequency instability, or poor power quality may otherwise pose serious risks.

In addition to addressing these nonlinear dynamics, there is great potential for integrating IMC with emerging technologies such as artificial intelligence (AI) and machine learning (ML). AI and ML algorithms can be used to continuously monitor and adapt the IMC controller parameters in real-time for enabling the system to respond dynamically to changes in grid conditions. For example, machine learning could be used to predict load demands, fluctuations in renewable energy generation, or grid disturbances, and adjust the control parameters of the IMC system accordingly. This real-time optimization capability would significantly enhance the system's ability to react quickly to unforeseen changes such as sudden power surges or drops and help to improve both efficiency and stability.

Furthermore, as the adoption of DERs and multi-inverter systems grows, IMC could evolve to become even more effective in coordinating power sharing and ensuring optimal operation across multiple and geographically dispersed energy sources. Distributed energy resources such as rooftop solar panels, battery storage systems, and microgrids require sophisticated coordination to efficiently manage energy generation, consumption, and storage while maintaining grid stability. IMC could be expanded to include coordination protocols that allow multiple inverters to work together, balancing the load across the system and improving the sharing of renewable energy. This would help mitigate issues like overloading, underutilization, or instability, which can arise when individual inverters operate without synchronization. IMC could ensure higher levels of reliability and resilience by continuously monitoring system performance and adjusting control strategies to compensate for faults.

7 Conclusions

In this chapter, an IMC approach integrated with an LCL filter design for grid-connected VSI has been thoroughly explored by focusing on its ability to enhance the performance and reliability of energy systems under varying grid conditions. A comprehensive literature review on IMC highlighted its theoretical foundation, which focuses on using a model of the system to predict and correct discrepancies between desired and actual system outputs. IMC is particularly valued for its simplicity, robustness, and ability to handle disturbances and uncertainties within the system. Additionally, the theory behind IMC was elaborated upon, explaining the design process of the controller, which involves the inversion of the system model, as well as the integration of a filter to compensate for model-plant mismatches.

The mathematical formulations and design procedures were detailed by focusing on how the IMC controller is structured to optimize the control process by minimizing errors and ensuring stability. The IMC approach is designed with an internal model that represents the plant dynamics, which provides the system with the ability to anticipate and adjust for disturbances before they impact the grid performance. The analytical background of IMC involved solving the characteristic equations for stability and using filtering techniques to address non-minimum phase dynamics and other system complexities such as time delays.

The validation of the IMC approach was conducted through extensive simulations across multiple case studies designed to test robustness, efficiency, and adaptability of the system. In the first case study, the system demonstrated stable and efficient performance during steady-state operation under nominal grid conditions, which achieved minimal harmonic distortion and ensured reliable power quality. The second case study examined dynamic response of the system to sudden load changes by showcasing its ability to rapidly adapt to fluctuating power demands while maintaining stable operation without compromising output quality.

The third case study emphasized resilience of the IMC controller to grid disturbances such as voltage sags and swells. The system effectively managed power delivery by maintaining voltage regulation even under challenging operating conditions. Additionally, high efficiency and precise grid-following behaviour of the modelled system were evident, especially when subjected to frequency deviations. Its capability to handle varying grid parameters and ensuring consistent performance in unstable conditions has been observed in case studies.

The results from these case studies that have been tested across four distinct scenarios reinforced the suitability of IMC-based controller for real-world applications, especially in renewable energy systems and smart grid integrations, where dynamic and unpredictable conditions are commonplace. The study also pointed to future research opportunities such as enhancing control strategies for nonlinear grid conditions and weak grids, developing advanced simulation models for multi-inverter systems, and investigating alternative filter topologies to further improve harmonic suppression.

The validated IMC approach provides a solid foundation for grid-connected inverters by establishing it as a valuable tool for modern energy systems and the evolving smart grid infrastructure. IMC has the potential to play a central role in improving grid resilience, optimizing energy usage, and enhancing the stability and reliability of future power grids with continuing advancements. IMC could become integral in managing the complexities of modern, decentralized, renewable energy-driven grids.

Acknowledgements This work was supported by the Scientific and Technological Research Council of Turkiye (TUBITAK) under Grant agreement no. 1059B192302145. Ersan Kabalcı acknowledges TUBITAK for the support to the research project.

References

Ahmed AA, Koh BK, Park HS, Lee K-B, Lee YI (2017) Finite-control set model predictive control method for torque control of induction motors using a state tracking cost index. IEEE Trans Ind Electron 64(3):1916–1928. https://doi.org/10.1109/TIE.2016.2631456

Bahrani B, Rufer A (2013) A cascade voltage controller for three-phase islanded microgrids. In: 2013 IEEE power & energy society general meeting, Vancouver, BC: IEEE, pp 1–6. https://doi.org/10.1109/PESMG.2013.6673060

Chawda GS, Shaik AG, Su W (2024) Efficient wind energy integration in weak AC Grid with a DLMF-based adaptive approach. Appl Energy 372:123779. https://doi.org/10.1016/j.apenergy.2024.123779

Chowdhury VR, Singh A (2022) Adaptive internal model based current control with embedded active damping of a three-phase grid-connected inverter with LCL filter for PV application. In: 2022 IEEE Energy Conversion Congress and Exposition (ECCE), Detroit, MI, USA: IEEE, p. 1–6. https://doi.org/10.1109/ECCE50734.2022.9947896

Garcia CE, Morari M (1982) Internal model control. A unifying review and some new results. Ind Eng Chem Process des Dev 21(2):308–323. https://doi.org/10.1021/i200017a016

García-Triviño P, Sarrias-Mena R, García-Vázquez CA, Leva S, Fernández-Ramírez LM (2023) Optimal online battery power control of grid-connected energy-stored quasi-impedance source inverter with PV system. Appl Energy 329:120286. https://doi.org/10.1016/j.apenergy.2022.120286

Gong H, Wang X, Harnefors L (2021) Rethinking current controller design for PLL-synchronized VSCs in weak grids. IEEE Trans Power Electron, pp 1–1. https://doi.org/10.1109/TPEL.2021.3105549

Guzman R, De Vicuna LG, Castilla M, Miret J, De La Hoz J (2018) Variable structure control for three-phase LCL-filtered inverters using a reduced converter model. IEEE Trans Ind Electron 65(1):5–15. https://doi.org/10.1109/TIE.2017.2716881

Han X, Xu B, Kang K, Zuo S (2023) Research on the control strategy of LCL grid-connected inverters based on improved auto disturbance rejection. Sci Prog 106(4):00368504231208520. https://doi.org/10.1177/00368504231208520

Harnefors L, Nee H-P (1998) Model-based current control of AC machines using the internal model control method. IEEE Trans Ind Appl 34(1):133–141. https://doi.org/10.1109/28.658735

Hollweg GV, Khan SA, Chaturvedi S, Fan Y, Wang M, Su W (2023) Grid-connected converters: a brief survey of topologies, output filters, current control, and weak grids operation. Energies 16(9):3611. https://doi.org/10.3390/en16093611

Horowitz I, Shaked U (1975) Superiority of transfer function over state-variable methods in linear time-invariant feedback system design. IEEE Trans Autom Control 20(1):84–97. https://doi.org/10.1109/TAC.1975.1100825

Huang M, Zhang Z, Wu W, Yao Z (2022) An improved three-level cascaded control for LCL-filtered grid-connected inverter in complex grid impedance condition. IEEE Access 10:65485–65495. https://doi.org/10.1109/ACCESS.2022.3183332

Kabalci E (2021) Multilevel inverters: control methods and advanced power electronic applications. London: Academic Press, 2021. Accessed: Aug. 02, 2022. [Online]. Available: https://search.ebscohost.com/login.aspx?direct=true&scope=site&db=nlebk&db=nlabk&AN=2712992

Karamanakos P, Nahalparvari M, Geyer T (2021) Fixed switching frequency direct model predictive control with continuous and discontinuous modulation for grid-tied converters with LCL filters. IEEE Trans Control Syst Technol 29(4):1503–1518. https://doi.org/10.1109/TCST.2020.3008030

Karimi-Ghartemani M (2014) Enhanced phase-locked loop structures for power and energy applications. John Wiley & Sons Inc., Hoboken, New Jersey

Komurcugil H, Bayhan S, Bagheri F, Kukrer O, Abu-Rub H (2018) Model-based current control for single-phase grid-tied quasi-Z-source inverters with virtual time constant. IEEE Trans Ind Electron 65(10):8277–8286. https://doi.org/10.1109/TIE.2018.2801778

Komurcugil H, Kukrer O (2017) Investigation of active damping methods for model-based-current-control scheme of single-phase grid-tied VSI with LCL filter. In: IECON 2017—43rd annual conference of the IEEE industrial electronics society, Beijing: IEEE, Oct. 2017, pp 1261–1266. https://doi.org/10.1109/IECON.2017.8216215

Korada DMR, Mishra MK (2023) Fixed switching frequency model predictive current control for grid-connected inverter with improved dynamic and steady state performance. IEEE Access 11:104094–104105. https://doi.org/10.1109/ACCESS.2023.3317537

Kouro S, Cortes P, Vargas R, Ammann U, Rodriguez J (2009) Model predictive control—a simple and powerful method to control power converters. IEEE Trans Ind Electron 56(6):1826–1838. https://doi.org/10.1109/TIE.2008.2008349

Kumar VN, Babu NP, Kiranmayi R, Siano P, Panda G (2020) Improved power quality in a solar PV plant integrated utility grid by employing a novel adaptive current regulator. IEEE Syst J 14(3):4308–4319. https://doi.org/10.1109/JSYST.2019.2958819

Leitner S, Yazdanian M, Mehrizi-Sani A, Muetze A (2018) Small-signal stability analysis of an inverter-based microgrid with internal model-based controllers. IEEE Trans Smart Grid 9(5):5393–5402. https://doi.org/10.1109/TSG.2017.2688481

Leitner S, Yazdanian M, Ziaeinejad S, Mehrizi-Sani A, Muetze A (2018) Internal model-based active resonance damping current control of a grid-connected voltage-sourced converter with an LCL filter. IEEE Trans Power Syst 33(6):6025–6036. https://doi.org/10.1109/TPWRS.2018.2828843

Li H, Wu W, Huang M, Shu-hung Chung H, Liserre M, Blaabjerg F (2020) Design of PWM-SMC controller using linearized model for grid-connected inverter with LCL filter. IEEE Trans Power Electron 35(12):12773–12786. https://doi.org/10.1109/TPEL.2020.2990496

Liu Q, Caldognetto T, Buso S (2020) Review and comparison of grid-tied inverter controllers in microgrids. IEEE Trans Power Electron 35(7):7624–7639. https://doi.org/10.1109/TPEL.2019.2957975

Liu J, Wu H, Huang H, Yang H (2023) Renewable energy design and optimization for a net-zero energy building integrating electric vehicles and battery storage considering grid flexibility. Energy Convers Manag 298:117768. https://doi.org/10.1016/j.enconman.2023.117768

López-Alcolea FJ, Molina-Martínez EJ, Parreño Torres A, Vázquez J, Roncero-Sánchez P (2023) 2DOF-based current controller for single-phase grid-connected voltage source inverter applications. Appl Energy 342:121179. https://doi.org/10.1016/j.apenergy.2023.121179

Morari M (1983) Internal model control—theory and applications. IFAC Proc 16(21):1–18. https://doi.org/10.1016/S1474-6670(17)64183-1

Nam NN, Nguyen N-D, Lee YI (2023) Model predictive control for a voltage sensorless grid-connected inverter with LCL filter using lumped disturbance observer. IEEE. J Emerg Sel Top Power Electron 11(3):3050–3063. https://doi.org/10.1109/JESTPE.2023.3240390

Ngo V-Q-B et al (2019) Lyapunov-induced model predictive power control for grid-tie three-level neutral-point-clamped inverter with dead-time compensation. IEEE Access 7:166869–166882. https://doi.org/10.1109/ACCESS.2019.2953784

Ranjan A, Mehta U, Saxena S (2023) A comprehensive review of modified Internal Model Control (IMC) structures and their filters for unstable processes. Annu Rev Control 56:100895. https://doi.org/10.1016/j.arcontrol.2023.04.006

Reznik A, Simoes MG, Al-Durra A, Muyeen SM (2014) LCL filter design and performance analysis for grid-interconnected systems. IEEE Trans Ind Appl 50(2):1225–1232. https://doi.org/10.1109/TIA.2013.2274612

Rivera DE, Morari M, Skogestad S (1986) Internal model control: PID controller design. Ind Eng Chem Process des Dev 25(1):252–265. https://doi.org/10.1021/i200032a041

Rodriguez P, Pou J, Bergas J, Candela JI, Burgos RP, Boroyevich D (2007) Decoupled double synchronous reference frame PLL for power converters control. IEEE Trans Power Electron 22(2):584–592. https://doi.org/10.1109/TPEL.2006.890000

Rodriguez J et al (2007) Predictive current control of a voltage source inverter. IEEE Trans Ind Electron 54(1):495–503. https://doi.org/10.1109/TIE.2006.888802

Rodriguez J et al (2013) State of the art of finite control set model predictive control in power electronics. IEEE Trans Ind Inform 9(2):1003–1016. https://doi.org/10.1109/TII.2012.2221469

Saranchimeg S, Nair NKC (2021) A novel framework for integration analysis of large-scale photovoltaic plants into weak grids. Appl Energy 282:116141. https://doi.org/10.1016/j.apenergy.2020.116141

Simonetti F, D'Innocenzo A, Cecati C (2023) Neural network model-predictive control for CHB converters with FPGA implementation. IEEE Trans. Ind. Inform. 19(9):9691–9702. https://doi.org/10.1109/TII.2023.3233973

Vigneysh T, Pachauri N, Suresh V (2022) Internal model-based cascaded control approach for multi-functional grid interactive converters. IEEE Access 10:19862–19874. https://doi.org/10.1109/ACCESS.2022.3151676

Wang Y, Wai R-J (2020) Design of discrete-time backstepping sliding- mode control for LCL-type grid-connected inverter. IEEE Access 8:95082–95098. https://doi.org/10.1109/ACCESS.2020.2995469

Wu X, Zhu Z-Q, Freire NMA (2022) High frequency signal injection sensorless control of finite-control-set model predictive control with deadbeat solution. IEEE Trans Ind Appl 58(3):3685–3695. https://doi.org/10.1109/TIA.2022.3162184

Xu F, Zhu M, Ye Y (2023) The Stability of LCL-Type grid-tied inverter based on repetitive control and grid voltage feed-forward. IEEE. J Emerg Sel Top Power Electron 11(2):1496–1506. https://doi.org/10.1109/JESTPE.2022.3213408

Xue C, Zhou D, Li Y (2021) Hybrid model predictive current and voltage control for LCL-filtered grid-connected inverter. IEEE. J Emerg Sel Top Power Electron 9(5):5747–5760. https://doi.org/10.1109/JESTPE.2020.3049083

Yazdanian M, Mehrizi-Sani A (2014) Internal model-based current control of the RL filter-based voltage-sourced converter. IEEE Trans Energy Convers 29(4):873–881. https://doi.org/10.1109/TEC.2014.2353035

Yazdanian M, Mehrizi-Sani A (2015) Case studies on cascade voltage control of islanded microgrids based on the internal model control. IFAC-Pap 48(30):578–582. https://doi.org/10.1016/j.ifacol.2015.12.442

Young HA, Marin VA, Pesce C, Rodriguez J (2020) Simple finite-control-set model predictive control of grid-forming inverters with LCL filters. IEEE Access 8:81246–81256. https://doi.org/10.1109/ACCESS.2020.2991396

Zhang J, Isobe T, Tadano H (2020) Model-based control for grid-tied inverters operated in discontinuous current mode with low harmonic current distortion. IEEE Trans Power Electron 35(10):11167–11180. https://doi.org/10.1109/TPEL.2020.2978871

Zhang D, Shafiullah GM, Das CK, Wong KW (2025) Optimal allocation of battery energy storage systems to improve system reliability and voltage and frequency stability in weak grids. Appl Energy 377:124541. https://doi.org/10.1016/j.apenergy.2024.124541

Zhou L, Eull M, Wang W, Preindl M (2022) Modular model-predictive control with regulated third-harmonic injection for zero-sequence stabilized inverter. IEEE Trans Ind Appl 58(6):7634–7647. https://doi.org/10.1109/TIA.2022.3193988

Zhou Z, Wang W, Ramasubramanian D, Farantatos E, Huang GM (2023) Small signal stability of phase locked loop based current-controlled inverter in 100% inverter-based system. IEEE Trans Sustain Energy 14(3):1612–1623. https://doi.org/10.1109/TSTE.2023.3241094

Artificial Intelligence Control in Power Electronics

Azadeh Kermansaravi, Hani Vahedi, and Mohamed Trabelsi

Abstract This chapter explores the transformative role of Artificial Intelligence (AI) in the control of power electronic systems, addressing the increasing complexity and dynamic behavior of modern energy infrastructures. It systematically introduces and compares major AI paradigms such as Symbolic AI, Computational Intelligence (CI), Machine Learning (ML), Deep Learning (DL), and Reinforcement Learning (RL), emphasizing their unique contributions to modeling, control, and optimization in nonlinear, uncertain, and real-time environments. The discussion includes a detailed evaluation of algorithmic structures, strengths, and limitations of each approach in the context of power converters control, with a focus on control robustness, adaptability, and computational feasibility. Moreover, the chapter presents a comprehensive case study on RL-based current control of a grid-connected multilevel inverter, highlighting its capacity for model-free learning and real-time adaptation under varying operating conditions. This chapter serves as both a theoretical primer and a practical guide, equipping readers with foundational knowledge and design insights necessary for leveraging AI to advance intelligent and resilient power electronics systems in renewable energy and smart grid applications.

Keywords Power electronics · Smart control · Artificial intelligence · Real-time · Reinforcement learning

---

A. Kermansaravi (✉)
Center of Expertise Cyber Security, The Hague University of Applied Sciences, The Hague, The Netherlands
e-mail: azadeh.kermansaravi@ieee.org

H. Vahedi
Delft University of Technology, Delft, Netherlands
e-mail: hani.vahedi@ieee.org

M. Trabelsi
Electronics and Communications Engineering Department, Kuwait College of Science and Technology, Kuwait, Kuwait
e-mail: m.trabelsi@kcst.edu.kw

E. Kabalci (ed.), *Modern Control Methods for Power Electronics*, Energy Systems in Electrical Engineering, https://doi.org/10.1007/978-981-92-1974-2_12

Nomenclature

Symbols	Description
C_i	Conditions (facts or logical expressions) in a rule-based system
$\mu_A(x)$	Degree of membership of input x in fuzzy set A
R_i	Fuzzy rule i in inference system
$f^i(x_1, x_2)$	Crisp functional output in Takagi–Sugeno-Kang (TSK) fuzzy inference
p_0^i, p_1^i, p_2^i	Coefficients of the linear output function in TSK model
$x_i \in R^n$	Candidate solution in n-dimensional space (for GA or PSO)
v_i^t	Velocity of particle i at time t in PSO
τ_{ij}	Pheromone level on edge (i,j) in ACO
ρ	Pheromone evaporation rate in ACO
$f(x; \theta)$	Mapping function learned in supervised learning
$(x^{(i)}, y^{(i)})$	Training data sample i in supervised learning
$\hat{y}$	Predicted output of the network
W_1, W_2	Weight matrices of the neural network
b_1, b_2	Bias vectors of the neural network
σ	Activation function (e.g., ReLU, tanh)
$\varnothing(x)$	Feature mapping (kernel transformation)
$T_k(x)$	Prediction of the k-th decision tree

1 Overview

Power electronics have become a foundational technology in modern energy systems, enabling efficient conversion, regulation, and integration of electrical energy across a wide range of applications. As the demand for higher energy efficiency, operational flexibility, and sustainability grows, power electronic converters play a pivotal role in supporting the energy transition. These converters bridge diverse and dynamically evolving components within modern power infrastructures (Qashqai et al. 2021, 2022a).

These interfaces have evolved significantly from basic, low-cost implementations to advanced, intelligent solutions capable of meeting the high-performance demands of challenging applications such as electric vehicles, smart grids, aerospace power systems, and high-voltage direct current (HVDC) transmission. This evolution has been driven by the following key factors:

- Improved reliability and fault tolerance capabilities (consider EV drivetrains and grid-tied renewable energy systems (Gros et al. 2023).
- Elaborate modulation capability and higher switching frequencies to achieve fast dynamic responses (Gao et al. 2023).

– Stronger performance in adverse conditions (consider supporting a fuel-cell-based electricity nano-grid that experiences fluctuations in source voltage)) (Kumar et al. 2020).

These converters are becoming increasingly complex, with multilevel architectures, such as Neutral Point Clamped (NPC), Modular Multilevel Converters (MMCs), Flying Capacitors Inverters (FCIs), and Packed U Cells (PUC) inverters, widely employed to enhance power density and waveform quality (Qashqai et al. 2021; Gao et al. 2022). While these topologies offer notable performance benefits, they also bring significant control challenges due to inherent nonlinearity, parameter sensitivity, and complex dynamics.

Traditional control strategies such as PI, Model Predictive Control (MPC), and sliding mode control (SMC) are being used frequently to regulate voltages and currents in power electronics systems. While these approaches may be simple and implemented successfully in industry, they have severe difficulties when involving model uncertainty, parameter variation, aging assumptions, and external disturbances (Qashqai et al. 2022a, 2019, 2023; Chen 2024; Bahrami and Khashroum 2310). Many of the aforementioned techniques rely on white-box modelling, which assumes detailed knowledge of system dynamics, signal flow, and internal parameters. However, this assumption often fails in practice, especially with commercial power electronic components treated as black boxes, where internal specifications are inaccessible or undocumented unless comprehensive datasheets are available (Qashqai et al. 2022a, 2019). Recent literature has recommended the need for black-box modelling where input-output data are only used. This leads to the possibility of data-driven and AI-enhanced control for these components and applications. This is especially feasible for commercial and industrial systems where a model does not exist or is highly difficult to successfully derive or accurately assess (Qashqai et al. 2022a, 2019, 2022b).

Moreover, the growing usage of power electronics in distributed energy resources and microgrids, whether DC, AC or hybrid topologies, also brings more challenges to control. Inverters interfacing with AC and DC buses will need to contend with grid unbalances, EMI filter coupling, load uncertainty, and dynamic mode changing (Zhang et al. 2023). In these cases, adaptive and intelligent control is even more needed.

Finally, interest continues to grow in inserting AI into power electronics control. Deep reinforcement learning (DRL), long short term memory (LSTM) networks, and adaptive neural network-based controllers have been studied as potential replacements for classical control to provide model-free self-learning that is resilient to disturbances (Gao et al. 2023; Qashqai et al. 2023; Chen et al. 2025; Zhao et al. 2020; Wan et al. 2024).

In conclusion, the non-linear, time-varying, and decentralized characteristics of power electronic systems require moving from static, model-based control to flexible, data-driven, and smart control.

This chapter provides an overview of AI-based control techniques in power electronic systems. These methods enhance further the potential of modern power converters and support the next generation of intelligent, efficient, and resilient control systems across diverse applications.

2 Fundamentals of AI Methods Used in Power Electronics

AI has rapidly emerged as a transformative force in power electronics, enabling intelligent control, adaptive optimization, and enhanced decision-making in power electronic systems that are non-linear, time-varying, and highly sensitive to uncertainties in modeling. AI applies data-driven strategies to improve upon previously established control and design methods while achieving real-time adaptation, fault diagnostics, and monitored system efficiency (Qashqai et al. 2023; Chen et al. 2025; Zhao et al. 2020). This section provides a comprehensive review of the established AI methods applied in power electronics and focus on the general explanatory principles, activity and classification, important architectures, and related implementation issues. AI methods can be categorized into three main groups (Gros et al. 2023; Gao et al. 2023; Zhao et al. 2020; Bose 2020):

1. Symbolic AI/ Knowledge-Based Systems
2. Computational Intelligence (CI)/Evolutionary AI
3. Machine Learning (ML)

The following section provides a detailed explanation of each category and examines their respective impacts on power electronic applications.

2.1 Symbolic AI Systems

These systems, often referred to as knowledge-based AI, represent some of the earliest forms of AI applied in power electronics. Unlike data-driven approaches such as neural networks, Symbolic AI relies on explicit domain knowledge encoded as rules or logical statements. Instead of learning from historical data, Symbolic AI methods make decisions based on predefined rules developed by domain experts (Huang et al. 2016; Simões and Bubshait 2019). In power electronics, these methods can be applied to fault diagnosis, condition monitoring, protection, and rule-based control methods for converters and inverters. There are two major symbolic AI methods: Expert Systems (ES) and Fuzzy Logic (FL) Controllers.

2.1.1 Expert Systems (ES)

These systems are rule-based and logic-driven systems. It means that they operate based on "if–then" logic statements rather than mathematical models or data training. The reasoning is formulized with logic-based notation, decision tables, or production rule syntax, evaluated by an inference engine using methods such as forward chaining or backward chaining. ES were popular in applications such as fault diagnosis and isolation, protection coordination in multilevel systems, rule-based control in DC-DC or AC-AC converters (Huang et al. 2016; Bose 2002; Feng et al. 2023). In propositional logic notation of ES, a rule can be expressed as:

$$IFC_1 \wedge C_2 \wedge \cdots \wedge C_n \Rightarrow A$$

where:

- $C_i$are conditions (facts or logical expressions)
- A is an action or conclusion

This rule is stored in a knowledge base and evaluated by an inference engine.

Illustrative Example 1. Rule-Based Control in Power Converters

This example illustrates how a basic Expert System (ES) can use rule-based reasoning to control a power converter and decide inverter output by modifying the switching duty cycle based on the state of the system. The intention is to demonstrate how human-expert knowledge can be represented as interpretable control rules that can be assessed in real-time.

Objective: To use linguistic and logical conditions to guide inverter switching behavior based on system measurements such as voltage deviation and switching frequency stability.

Rule Logic: A representative rule can be stated as follows:
IF the inverter output voltage deviation is high
AND the switching frequency remains stable
THEN issue a control signal to reduce the duty cycle.
This rule can be translated into logical form:

$$D_v = \text{"Voltage deviation is high"}$$

$$F_s = \text{"Switching frequency is stable"}$$

$$U = \text{"Control action : Reduce duty cycle"}$$

Then:

$$Dv\Lambda F_s \Rightarrow U$$

Table 1 presents a possible rule table format for this example.

Expected Outputs: In practice, these types of rule-based controllers could be simulated in order to assess how the controller would reverberate under specific opening

Table 1 ES rule table format

Rule #	Condition 1	Condition 2	Condition 3
R1	Voltage deviation high	Switching frequency stable	Reduce duty cycle

conditions. A possible validating step would involve measuring inverter output voltage across time and illustrating how the application of the rule diminishes the voltage deviation and stabilizes the output. A possible type of output includes a time-series plot of the output voltage across time before and after applying the rule. A comparing plot showing real-time duty cycle adjustment. A truth table or inference surface indicating the output space explored by the expert rules. While ES-based systems are not as prevalent today (~0.9% of AI application in power electronics (Zhao et al. 2020)), this example was intended as an important piece to allow us to describe interpretable control logic, particularly in systems where interpretability and explainability take precedence over adaptability.

2.1.2 Fuzzy Logic (FL)

FL on the other hand, provides a more flexible symbolic reasoning structure, extending traditional Boolean logic (true/false) into a continuous domain. That allows for multi-valued (as opposed to binary) logic. Unlike standard true/false statements, FL allows for truthfulness to a degree, between true and false, running 0 to 1, which is well-suited for complex, real-world systems. In power electronics, where systems are often non-linear, time-varying, and subject to uncertainties, FL provides a useful framework for designing effective controllers capable of dealing with a range of operating conditions without requiring accurate mathematical models (Simões and Bubshait 2019; Kececioglu et al. 2020; Cheok and Wang 2005).

FL is different from ES. ES relies on rigid rule-based logic, whereas FL uses linguistic variables, for example, "High Voltage" or "Slight Error" and represents them using membership functions. These functions map input values to a degree of belonging within a fuzzy set, allowing controllers to make decisions decision that simulate human reasoning. FL has seen several applications in power electronics, including load forecasting and demand estimation, control of power electronic converters, and regulation of system stability and voltage. It also plays a significant role in energy management for microgrids and smart grids, fault detection and classification, as well as speed and torque control of electric machines. Moreover, FL has been employed in the design of hybrid control strategies, such as Fuzzy-PID controllers and combined fuzzy logic with neural networks (Bose 2002; Mendonça et al. 2012; Çakmak et al. 2025).

There are two common fuzzy inference systems: the Mamdani and Takagi–Sugeno–Kang (TSK) styles. The Mamdani is more intuitive and is popular for control applications, while the TSK deterministic system is often considered more analytical and is best for optimization and system modelling. Both have been effectively applied to control voltage, current, and switching behavior in power electronic devices (Zhao et al. 2020; Bose 2020; Khandelwal and Kumar 2019). The general formulation of these methods is defined as follows:

- Mamdani Fuzzy Inference System (MFIS) is based on fuzzy rules with fuzzy outputs (Alajmi et al. 2010). A Mamdani-style FIS typically follows these 4 steps:

  1. Fuzzification of crisp inputs:

$$\mu_A(x) = \text{degree of membership of input } \times \text{ in fuzzy set } A$$

  2. Rule Evaluation using FL:

$$R_i = \mathit{If}\ x_1 \mathit{is}\, A_1^i \text{ and}\, x_2\ \mathit{is}\, A_2^i \rightarrow\ y\ \mathit{is}\ B^i$$

  where:

  - x_1, x_2 are input variables,
  - A_1^i, A_2^i are fuzzy sets (with membership functions) for inputs,
  - y is the output variable,
  - B^i is a fuzzy set for the output.

  3. Aggregation of outputs across all rules:

$$\mu_{agg}(y) = \max_i\left(\mu B^i(y)\right)$$

  4. Defuzzification to obtain a crisp output (often via centroid):

$$y^* = \frac{\int y.\mu_{agg}(y)dy}{\int \mu_{agg}(y)dy}$$

- Takagi–Sugeno–Kang (TSK) Fuzzy Inference is based on fuzzy rules with crisp functional outputs. It means that, the output is not a fuzzy set but a crisp function of the inputs. A TSK-style FIS typically follows these 4 steps (Kececioglu et al. 2020):

  1. Fuzzification of crisp inputs:

$$\mu_A(x) = \text{degree of membership of input } \times \text{ infuzzy set} A$$

  2. Rule Evaluation using FL:

$$R_i = \mathit{If}\ x_1 \mathit{is}\, A_1^i \text{ and}\, x_2\ \mathit{is}\, A_2^i \rightarrow\ y = f^i(x_1, x_2)$$

  where:

- x_1, x_2 are input variables,
- A_1^i, A_2^i are fuzzy sets (with membership functions) for inputs,
- y is the output variable,
- $f^i(x_1, x_2)$ is a crisp output function (often linear or constant), e.g.

$$f^i(x_1, x_2) = p_0^i + p_1^i x_1 + p_2^i x_2$$

3. Calculation of rule firing strengths(weights):

$$w_i = \mu A_1^i(x_1).\mu A_2^i(x_2)$$

4. Aggregation of rule outputs using a weighted average (no defuzzification needed):

$$y^* = \frac{\sum_{i=1}^{N} w_i.f^i(x_1, x_2)}{\sum_{i=1}^{N} w_i}$$

Illustrative Example 2. FL Control of a Buck Converter Duty Cycle

This example shows how fuzzy logic (FL) controllers, including both the Mamdani and Takagi–Sugeno–Kang (TSK) forms, can be used to control the duty cycle of a buck converter to improve and maintain the output voltage level to the desired voltage level. This example provides an overview on how FL controllers operate in power electronics, and highlights their interpretability and flexibility for controlling nonlinear systems.

Objective: Control the output voltage V_{out} of a buck converter to track a reference voltage V_{ref} by adjusting the duty cycle D of the converter's switching signal.

System Description: The controller operates on two inputs:

Voltage Error, $e(t)$: This is the difference between the desired output voltage (V_{ref}) and the actual output voltage (V_{out}):

$$e(t) = V_{\text{ref}} - V_{\text{out}}$$

- If $e(t)$ is positive, output voltage is too low and the duty cycle should increase.
- If $e(t)$ is negative, output voltage is too high and the duty cycle should decrease.

Change in Voltage Error, $\Delta e(t)$:

$$\Delta e(t) = e(t) - e(t-1)$$

- If $\Delta e(t)$ is positive, error is increasing (getting worse).
- If $\Delta e(t)$ is negative, error is decreasing (improving).

Table 2 Fuzzy rule base for buck converter (for both Mamdani and TSK)

Δ Error	Negative	Zero	Positive
Negative	Decrease	Decrease	No change
Zero	Decrease	No change	Increase
Positive	No change	Increase	Increase

Output: Control Signal: This is the duty cycle of the converter switch, controlling the amount of power transferred $\Delta D(t)$.

The fuzzy controller adjusts D based on $e(t)$ and $\Delta e(t)$ to stabilize the output voltage.

Solution Using Mamdani, TSK Methods:

Table 2 presents the fuzzy rule base for this example. In a Mamdani fuzzy inference system, the output is represented as a fuzzy label such as *Decrease*, *No Change*, or *Increase*. In contrast, a Takagi–Sugeno–Kang (TSK) inference system replaces each rule's output with a crisp functional expression. For instance, the rule *If error = Positive and Δerror = Positive* can be expressed in TSK form as $\Delta D = 0.5e + 0.2\Delta e$.

To illustrate how both methods operate, let's assume that the voltage error is defined as $e(t) = V_{ref} - V_{out} = 1.0V$, and the change in error is $\Delta e(t) = e(t) - e(t-1) = -0.2V$.

- Mamdani: In the Mamdani system, the fuzzy sets for the inputs are defined as *Negative*, *Zero*, and *Positive* for both error and change in error, and similarly, the output fuzzy set for the duty cycle adjustment includes the labels *Decrease*, *No Change*, and *Increase*. A possible rule base might be:

  1. If error is Positive and Δerror is Negative → Increase
  2. If error is Zero and Δerror is Zero → No Change
  3. If error is Negative and Δerror is Positive → Decrease

  Assuming the membership values for the inputs are $\mu Positive(e) = 0.7$ and $\mu Negative(\Delta e) = 0.5$. which means Rule 1 is firing with strength min(0.7,0.5) = 0.5. This value scales the output fuzzy set of increase to 0.5. During the defuzzification step, called the centroid method, the crisp control output will be calculated. Assume that the defuzzified adjustment of duty cycle suggests $\Delta D = +0.2$.

- TSK: For the TSK system, the rules are defined in terms of linear functions. The respective rules could be written as:

  1. If error is Positive and Δerror is Negative → $\Delta D = 0.5e + 0.2\Delta e$
  2. If error is Zero and Δerror is Zero → $\Delta D = 0$
  3. If error is Negative and Δerror is Positive → $\Delta D = -0.4e + 0.1\Delta e$

  The firing strength for Rule 1 could be calculated as $w_1 = 0.7.\ 0.5 = 0.35$, while the other rules do not fire because of zero membership. The final crisp output can be computed as a weighted average of the rule outputs. Substituting the variables yields

$$y^* = \frac{0.35.(0.5.1.0) + 0.2.(-0.2))}{0.35} = 0.5.1.0 - 0.04 = 0.46.$$

Therefore, the duty cycle change computed by the TSK method is ΔD = 0.46, providing a precise numerical result based on the input conditions.

Expected Outputs and Validation: To substantiate this example with simulation results readers should expect:

- A time-domain plot of V_{out} following V_{ref} before and after the control is applied.
- Control surface diagrams showing the mapping from inputs (e, Δe) to output ΔD.
- Membership functions for both error and Δerror.
- Comparison of Mamdani vs TSK outputs for the same inputs.

These validation outputs illustrate the interpretability (Mamdani) and precision (TSK) tradeoff and show that the fuzzy controller stabilized the converter output.

Summary and Comparison

Symbolic AI systems, including ES and FL, are interpretable and trustworthy systems that simulate human reasoning by explicitly encoding, via rule-based structures, symbols that reflect the rules of human reasoning (Table 3).

Despite these limitations, symbolic AI, particularly FL continues to play a valuable role in modern control strategies. When integrated into hybrid AI frameworks, such as Adaptive Neuro-Fuzzy Inference Systems (ANFIS) or fuzzy-ML hybrids, these systems combine the clarity of rule-based logic with the adaptability of ML, enabling transparent and intelligent decision-making in real-time power electronics control (Grabowski et al. 2002). As seen in Table 1, symbolic-based AI methods differ in their knowledge representation, their ability to handle uncertainty, and their potential to learn. Although ES are not popular anymore, FL is still relevant in new AI hybrid designs, including sophisticated power electronics applications, especially in hybrid

Table 3 Comparison of symbolic AI methods used in power electronics

Feature	Expert Systems (ES)	Fuzzy Logic (FL)
Knowledge representation	Rule-based (if–then rules)	Fuzzy rules and membership functions
Handling of uncertainty	Limited	Strong (via fuzzy sets and multi-valued logic)
Learning capability	None	None (unless hybridized, e.g., ANFIS)
Interpretability	High	High
Generalization across systems	Low	Moderate
Typical models	Rule Trees, Decision tables	Mamdani, Takagi–Sugeno–Kang (TSK)
Current usage trend	Declining (<1% of applications)	Still relevant in hybrid models

applications where both interpretability and adaptability are desired (Simões and Bubshait 2019; Alajmi et al. 2010).

2.2 *Computational Intelligence (CI)/Evolutionary AI*

Computational Intelligence, also known as Evolutionary AI, is a category of biologically inspired discovery-based optimization techniques that do not utilize gradient-based information or rely on mathematically sound models. CI optimization techniques mimic natural processes such as evolution, swarm behavior, and social foraging to optimize in a derivative-free environment. These methods are especially useful for the kinds of highly complicated, nonlinear and discontinuous problems often seen in power electronics systems (Gros et al. 2023; Kumar et al. 2020; Zhao et al. 2020).

CI methods are population-based, stochastic, and exploratory in nature. Therefore, the methods are well aligned to use case activities such as control tuning, multi-objective optimization, hardware configuration, and converter design. CI techniques optimize without requiring differentiability or convexity and can operate in ultra-large, high-dimensional, black-box search spaces in contrast to gradient based or model-driven approaches (Kumar et al. 2020; Divakar and Jacob 2019; Ali et al. 2021). Genetic Algorithms (GA), Particle Swarm Optimization (PSO), and Ant Colony Optimization (ACO) are some of the most common CI techniques explained as follows.

2.2.1 Genetic Algorithms (GA)

GAs are inspired by Darwinian evolution typically using the operators of selection, crossover, and mutation to evolve a population of candidate solutions. GA are also used to tune controllers (e.g., PID, fuzzy), optimize switching patterns, and even design structural conversion topologies (Kumar et al. 2020; Divakar and Jacob 2019; Ali et al. 2021). GA can be formulized as follows:

1. Initialize a population of solutions $x_i \in R^n$
2. Evaluate fitness $f(x_i)$
3. Selection: Choose parents based on fitness
4. Crossover: Combine parents to generate offspring
5. Mutation: Introduce random perturbations
6. Replace the old population and repeat

Fitness Objective Example:
Minimize Integral Time Absolute Error (ITAE) for PID tuning:

$$\min_{\theta} ITAE = \int_0^T t.|e(t)|dt$$

where θ includes controller gains (Kumar et al. 2020).

2.2.2 Particle Swarm Optimization (PSO)

PSO simulates the social behavior of bird flocking or fish schooling, as each particle modifies its movement based on its own best and the best global position in the swarm. The following update equations are going to be used:

$$v_i^{t+1} = wv_i^t + c_1 r_1 (p_i - x_i^t) + c_2 r_2 (g - x_i^t)$$

$$x_i^{t+1} = x_i^t + v_i^{t+1}$$

where x_i and v_i are the position and velocity, p_i is the personal best, and g is the global best. PSO has been shown to have good performance for fast convergence in the context of controller tuning and converter parameters optimization (Gros et al. 2023; Shi et al. 2015).

2.2.3 Ant Colony Optimization (ACO)

ACO models the seeking behavior of ants and utilizes pheromone trails to generate solutions in a probabilistic fashion over a graph. ACO is often used for discrete optimization problems, such as switching configurations, grid topologies, or routing problems (Zhao et al. 2020; Chen et al. 2019).

Pheromone update rule:

$$\tau_{ij} \leftarrow (1 - \rho)\tau_{ij} + \Delta\tau_{ij}$$

where ρ is the evaporation rate and $\Delta\tau_{ij}$ depends on the solution quality.

Illustrative Example3. Tuning a PI Controller for a Buck Converter

This examples illustrated three common metaheuristic methods (the Genetic Algorithm (GA), Particle Swarm Optimization (PSO), and Ant Colony Optimization (ACO)) to tune a PI controller for a DC-DC buck converter. The example serves to emphasize how bio-inspired optimization techniques can outperform manual tuning for control system design in power electronics.

Objective: Tune a PI controller for a DC-DC buck converter such that the output voltage $V_0(t)$ tracks a reference voltage $v_{ref} = 12V$ as quick and accurate as possible.

System Description: The system under consideration operates with an input voltage of $v_{in} = 24V$, while the output voltage $V_0(t)$ is measured across a resistive load. The control mechanism involves a proportional-integral (PI) controller characterized by two parameters: the proportional gain K_p, and the integral gain K_i. The control law is defined as:

$$u(t) = K_p e(t) + K_i \int_0^t e(\tau) d\tau$$

where $e(t) = V_{ref} - V_0(t)$, represents the instantaneous error between the desired and actual output voltage. To evaluate performance, the optimization problem is defined by the cost function known as the Integral of Time weighted Absolute Error (ITAE) computed over the time interval T:

$$J(K_p, K_i) = \int_0^T t.\left|V_{ref} - V_0(t)\right| dt$$

The optimization search space is defined as $K_p \in [0.1, 10], K_i \in [0.1, 5]$. The buck converter is simulated in a discretized model or circuit simulator. The goal is to minimize J for (K_p, K_i).

Solution using GA, PSO, ACO methods:

- GA: Using the GA method, a population of candidate (K_p, K_i) solutions is first initialized. Each individual's performance is assessed by evaluating the cost function J. The algorithm then performed selection, crossover, and mutation to guide the attention toward more optimal solutions. An example outcome of the GA method would be K_p= 4.2, K_i= 1.3, $J = 0.112$.
- PSO: With PSO, each candidate solution is represented by a particle, where a particle's position represents the (K_p, K_i) pair. Particles update their velocities and positions based on both personal and global best experiences, aiming to minimize J. As a method of optimization, PSO possesses a rapid convergence rate and produced one example result of K_p= 4.5, K_i= 1.1, $J = 0.098$.
- ACO: In solving a similar continuous optimization problem using the ACO, the continuous parameter space was discretized into graph nodes. The artificial ants traversed the graph nodes probabilistically based on pheromone levels and constructed solutions for the problem as a function of a pheromone level and their fitness value J that they returned to the starting node of the graph in response to. The ACO tends to work well for discrete optimization settings. An example result obtained via ACO is K_p= 4.0, K_i= 1.0, $J = 0.121$.

Expected Outputs and Validation: To enhance the example and emphasize its applied relevance, the following visual and quantitative outputs should be included in the final implementation or an appendix:

- Time-domain plots of $V_0(t)$ tracking V_{ref} for each method.
- Convergence curves showing the reduction of *ITAE J* over iterations.
- A surface plot of ITAE over the (K_p, K_i) search space, highlighting optimal points.
- A comparison table including rise time, settling time, overshoot, and steady-state error for each of the tuned controllers.
- Optional: time or number of iterations as a measure of computational efficiency.

These validation outputs will enable the reader not only to understand the characteristics of each method but also to evaluate their respective applicability to real-time tuning of controller parameters in power electronics applications.

3 Summary and Comparison

CI/Evolutionary AI methods such as GA, PSO, and ACO provide good options for optimization problems in power electronics where traditional methods or gradient-based methods may not be effective (Shi et al. 2015; Harrag and Messalti 2015). In the example of tuning a PI controller, PSO performed best, with the least error and fastest convergence to the optima, due in part to it being able to search continuously. GA performed quite well, it was robust and had a reasonable amount of flexibility in both discrete and continuous domains for search. ACO is more computationally expensive and is generally better suited for discrete search problems, such as topology selection or the best switching strategy for the given conditions.

The algorithms described provide a reasonable alternative to model-based methods, especially where the models are unavailable or unreliable. However, caution should be exercised, particularly in terms of their computational cost, as well as their lack of formal performance and stability guarantees that model-based methods may provide. This can become a major consideration when applied to real-time or safety–critical systems (Table 4).

Table 4 Comparison of evolutionary AI methods used in power electronics

Feature	GA	PSO	ACO
Inspiration	Natural selection and genetics	Swarm behavior of birds/fish	Foraging behavior of ants
Variable type support	Discrete/Continuous	Continuous	Primarily discrete
Tuning needs	Medium	Low	High
Convergence speed	Moderate	Fast	Slower
Best suited for	PID, fuzzy, MPC gain	Control weights, MPPT	Switching/topology design
Memory requirements	Medium	Low	Medium
Computational demand	Medium	Low	Medium–High

3.1 Machine Learning

ML is one of the most common types of AI utilized in power electronics. These data-driven techniques allow systems to learn from data, use self-adaptive techniques to respond to changing conditions, and accomplish automatable tasks like control, monitoring, diagnostics, and optimization without having to define explicit physical models (Gao et al. 2023; Bahrami and Khashroum 2310; Khan et al. 2023). ML methods can generally be divided into: supervised learning, unsupervised learning, and reinforcement learning. There are now also potential roles for self-supervised and semi-supervised learning (Bahrami and Khashroum 2310; Zhao et al. 2020). Each of these three categories confronts different challenges and has its own potential, depending on the availability of labelled data, the nature of the control task, and the system dynamics.

3.1.1 Supervised Learning (SL)

SL is the most common ML approach in power electronics and accounts for approximately 91% of ML applications in the field (Gros et al. 2023; Gao et al. 2023). SL methods rely on labelled datasets to learn mappings from the space of the input features to the output target for use in either classification or regression. SL approaches learn to fit a mapping $f(x;\theta)$ from input features $x \epsilon R^n$ using the labelled training data $\left\{(x^{(i)}, y^{(i)})\right\}_{i=1}^{N}$. In SL, the model is trained such that the prediction errors (as quantified by a loss function) are minimized

$$\min_{\theta} \sum_{i=1}^{N} \mathcal{L}(f(x_i;\theta), y_i)$$

where

- θ are model parameters
- $\mathcal{L}$ is a loss function (e.g., Mean Squared Error)

Popular types of SL include Feedforward Neural Networks (FFNNs), Support Vector Machines (SVMs), Random Forests (RF), Decision Trees, k-Nearest Neighbours (k-NN), and Gradient Boosting Models (Gros et al. 2023; Gao et al. 2023; Fujimoto et al. 2023). There are many applications of ML and AI techniques in power electronics. To illustrate, intelligent models can be used to detect and classify faults in converters and drives in real-time, allowing subsequent step to improve reliability of the converter or drive. In photovoltaic (PV) systems, AI techniques can be employed for maximum power point tracking (MPPT), enabling efficient energy harvesting under varying environmental conditions. AI can also be used to detect and mitigate harmonic distortion, thereby enhancing the power quality of electrical systems. These advancements not only improve system performance and reduce operational downtime but also support predictive maintenance by identifying potential failures before they occur (Zhao et al. 2020; Harrag and Messalti 2015; Yılmaz et al. 2023; Yang et al. 2023). These AI-driven improvements rely on advanced learning algorithms to analyze complex data patterns and make accurate predictions. Three widely used SL methods are FFNN, SVM, and RF.

- FFNN: are the most basic form of artificial neural networks where the connections between units do not form a cycle. They consist of layers of interconnected neurons, an input layer, one or more hidden layers, and an output layer (Gao et al. 2022; Yang et al. 2023). FFNN formulation for a network with one hidden layer is as follows:

$$h = \sigma(W_1 x + b_1)$$

$$\hat{y} = W_2 h + b_2$$

where x is an input feature vector (e.g., voltage, current), W_1, W_2 presents the weight matrices, b_1, b_2 are the bias vectors, σ is an activation function (e.g., Rectified linear unit (ReLU), tanh), and $\hat{y}$ is the network output (e.g., duty cycle or reference signal). The weights are learned by minimizing a loss function, typically MSE:

$$\mathcal{L} = \frac{1}{N}\sum_{i=1}^{N}\left(y^{(i)} - \hat{y}^{(i)}\right)^2$$

- SVM: are SL algorithms used for classification and regression tasks. The basic concept of SVM is to locate an optimal hyperplane that either separates data points into different classes (in classification task) or assumes a regression function (in regression cases) (Yang et al. 2023). In the classification case we are given a training sample (x_i, y_i) where $y_i \epsilon \{-1, +1\}$, the SVM aims to solve the following optimization problem:

$$\min_{w,b} \frac{1}{2}\|w\|^2 s.t. y_i\left(w^T x_i + b\right) \geq 1$$

This formulation ensures that the margin between the two classes is maximized while maintaining correct classification of the training data. In the case of regression, more commonly referred to Support Vector Regression (SVR), the purpose is altered only slightly by allowing a margin of tolerance around the regression function. The optimization problem becomes:

$$\min_{w,b} \frac{1}{2}\|w\|^2 + C\sum_{i=1}^{n}\left[\xi_i + \xi_i^*\right]$$

subject to the constraints:

$$y_i - w^T \phi(x_i) - b \leq \epsilon + \xi_i$$

$$w^T \phi(x_i) + b - y_i \leq \epsilon + \xi_i^*$$

where $\phi(x)$ is a kernel mapping (for example, radial basis function, or RBF), ϵ represents the margin of error or tolerance that the loss function applies no loss, C is a regularization parameter that controls the trade-off between model complexity and training error. The slack variables ξ_i, and ξ_i^* capture deviations beyond the ϵ-margin.

- RF is an ensemble of decision trees used for classification or regression tasks (Hannan et al. 2016). Each tree is built using a bootstrap sample of the data, and the final output is the majority vote (classification) or mean (regression) of all the trees. Let $T_1, T_2, \ldots, T_k$ be decision trees. For a regression problem RF is formulated as:

$$\hat{y} = \frac{1}{K} \sum_{k=1}^{K} T_k(x)$$

Each tree partitions the feature space recursively based on thresholds that minimize impurity (e.g., Gini index or variance).

Illustrative Example 4. PID Parameter Tuning in a DC-DC Buck Converter

This case study reviews how machine learning (ML) techniques are applied to the tuning of real-time PID controller parameters for a DC-DC buck converter in real time under varying operating conditions. The objective is to showcase that data-driven methods, such as feedforward neural networks (FFNN), support vector machines (SVM) and random forests (RF), can complement offline tuning and adjust the controller in dynamic conditions.

Objective: A buck converter requires real-time tuning of its PID controller to keep output voltage V_{out} at 12 V under variable load. The goal is to map operating conditions (input voltage, load resistance, inductor current, etc.) to optimal PID gains (K_p, K_i, K_d). The input features used to represent the operating state of the converter include the input voltage (V_{in}), the load resistance (R_{load}), the inductor current (I_L), and the voltage error $V_{error} = V_{ref} - V_{out}$. These inputs represent the electrical and control state of the system at any given time.

The output is the best PID controller gains: K_p, K_i, K_d. Since the PID controller gains require real-time tuning to maintain the same output voltage at different loads and input voltage so that the converter output voltage remains the output voltage with little steady-state error and overshoot.

Solution using FFNN, SVM, RF methods:

- FFNN: A neural network is trained on historical data that links system conditions to optimal PID gains. Each piece of training data has input $[V_{in}, R_{load}, I_L, V_{error}]$ and target output $[K_p, K_i, K_d]$. Following training, the neural network is able to infer the PID gains directly based on current system measurements, allowing for fast, adaptable control in real-time.

- SVM: Each of the PID gains is modeled separately using Support Vector Regression (SVR). SVR is especially effective in cases when there is less training (labeled) data available, as SVR models often have very good generalization performance. Each SVR model will attempt to learn to predict one of the gains based on the input features, making this modeling method appropriate for situations where it is expensive or impossible to collect large training datasets.
- RF: A random forest is trained on input features to estimate the PID gains from the same input variables. Random forests also have the added benefit for engineers of explainability; engineers can read through the decision trees, and evaluate the importance of the features so that they have a better understanding of how system variables are influencing the control parameters. Because of the interpretability of RF models, and where explanation is just as valuable as predictive capability RF models a valuable option for many applications.

Expected Outputs and Validation: To enhance clarity and demonstrate practical value, the following outputs should accompany the example:

- Tracking Performance Plots: $V_{out}(t) vs. V_{ref}$ for all three models under changing load or input voltage.
- Prediction accuracy metrics: Mean Absolute Error (MAE), Root Mean Square Error (RMSE), and R^2 score for K_p, K_i, K_d.
- Feature importance (RF model): Bar graph or table showing how input variables contribute to gain prediction.
- Comparative tuning behavior: Show time-domain response (rise time, overshoot) for FFNN, SVM, and RF-derived PID parameters.

3.1.2 Unsupervised Learning

Unsupervised Learning refers to the training of models through unlabeled data attempting to find its patterns, structure, or relationships without direct supervision. It represents approximately 9% of ML applications in power electronics (Gao et al. 2023, 2022) and is becoming popular in applications such as anomaly detection, feature extraction, and clustering of operating modes.

Given an **unlabeled** dataset consisting of input vectors $\left\{x^{(1)}, x^{(2)}, \ldots, x^{(N)}\right\}$, where each $x^{(i)} \epsilon R$, the goal is to determine the underlying structures or patterns in the data. Typical aims are to model clusters or latent groupings, uncover low-dimensional representations of high-dimensional data, and identify outliers or deviations from expected behavior. Learning is evaluated based solely on assumptions of the distributional or structural properties of the data, since there is no target variable.

Unsupervised learning algorithms can generally be categorized into clustering algorithms and dimensionality reduction algorithms. Clustering algorithms (e.g. k-means, DBSCAN) work to group data points together into clusters based on similarity. Dimensionality reduction algorithms (e.g. Principal Component Analysis (PCA), Autoencoders) work to map the data to a lower-dimensional space while preserving the meaningful structure in the data (Gao et al. 2023; Bahrami and Khashroum 2310; Zhao et al. 2020; Khandelwal and Kumar 2019).

Let's explore three widely used SL methods:

- K-Means Clustering: K-means is perhaps the most popularly used clustering algorithm. The aim of K-means is to partition the dataset into clusters such that the variance in each cluster is minimized. Given this intention, K-means mathematically solves the optimization problem:

$$\min_{\{c^{(i)}\},\{\mu_k\}} \sum_{i=1}^{N} \|x^{(i)} - \mu_c(i)\|^2$$

where μ_k is the centroid of the k th cluster, and $c^{(i)} \epsilon \{1, \ldots, K\}$ denotes the cluster assignment for point $x^{(i)}$. The algorithm begins by randomly initializing the K centroids $\mu_1, \ldots, \mu_k$, then in subsequent iterations it alternates between assigning a point to its nearest centroid and recomputing the centroids as means of the points assigned within the specific cluster. This continues until convergence (when cluster assignments stop changing).

- Principal Component Analysis (PCA): is a linear dimensionality reduction method that seeks to project high-dimensional data into a lower-dimensional subspace that maximizes the data variance. Given a centered data matrix $X \epsilon R^{Nxn}$, PCA finds a projection matrix $W \epsilon R^{nxk}$, such that the low-dimensional representation $Z = XW$, maximizes the variance. The optimization problem is:

$$\max_{W} Tr\left(W^T X^T XW\right), \quad s.t. \ W^T W = I$$

The solution is obtained by computing the top k eigenvectors of the covariance matrix $X^T X$, which form the columns of W. PCA is widely used for data compression, visualization, and noise reduction.

- Autoencoders (AE): An autoencoder is a neural network trained to produce its input at the output layer, making it useful in feature extraction and anomaly detection. It is comprised of two parts: an encoder $h = f_{enc}(x) = \sigma(W_1 x + b_1)$, which maps the input to a hidden layer representation, and decoder $h = f_{dec}(x) = \sigma(W_2 h + b_2)$, which tries to reconstruct the original input. The training objective involves minimizing the reconstruction error given by:

$$\min_{0} \frac{1}{N} \sum_{i=1}^{N} \|x^{(i)} - \hat{x}^{(i)}\|^2$$

where $\hat{x}$ is the reconstructed input and $\theta = \{W_1, b_1, W_2, b_2\}$ are the weights and biases we learn. Autoencoders also have different variations, including Sparse Autoencoders, Denoising Autoencoders, and Variational Autoencoders (VAEs), which were designed for various tasks such as noise robustness, or probabilistic encoding.

Illustrative Example 5. Operating Mode Detection in Grid-Tied Inverter

This example illustrates unsupervised learning methods that identify distinct operating modes of a grid-connected inverter without using labelled data. Students will learn to usekers, Principal Component Analysis (PCA), and autoencoders to real-world settings of multivariate time-series data in the area of power electronics and grid systems.

Objective: In a grid-connected inverter system, its operation can occur in up to four different modes: normal operation with balanced grid conditions and nominal power levels; low voltage ride through (LVRT) operation; faulted operation during different grid disturbances; and islanded operation when the grid is not connected to the inverter. Each mode represents a different behavior of the inverter system. Their respective behaviors might also be determined with several different time-series measurements. We collect data primarily on grid voltage waveforms, current waveforms, power output, inverter duty cycles, and harmonic content expressed as Total Harmonic Distortion (THD).

System Description: The system produces a multivariate time-series dataset. The features include: root mean square values of grid voltage (v_{rms}), and current (I_{rms}), active power output (P), harmonic distortion (THD), and inverter duty cycle (D). The task is to analyze this high-dimensional data to characterize the operation mode of an inverter in real time. The observation class (mode) may not have labeled data in many instances, so unsupervised learning algorithms would fit nicely into this context.

Solution Using K-Means, PCA, and Autoencoder:

- K-Means Clustering for Mode Detection: A straightforward and effective strategy is to apply K-means clustering to the feature vector $x = [v_{rms}, I_{rms}, P, THD, D]$, assuming K = 4 in order to capture the four known operational modes. K-means clusters the data by minimizing intra-cluster variance, and therefore groups together similar operating conditions. The assigned cluster labels can be viewed as the inferred mode of the inverter, and can be used for real-time unsupervised classification.
- PCA for reducing dimensionality and fault detection: PCA can help to reduce the dimensionality of sensor data, leading to improved visualization, and better methods of control. For example, sensor data from a system with 10 or more sensors could be projected onto 2 or 3 principal component subspaces which represent most of the variance of the system. In addition, PCA could help with fault detection: under normal operational conditions, the measurements from the system will be on or close to the (PC) subspace. If measurements deviate significantly from the PC subspace, this could indicate anomalous events, grid faults,

or a switch to islanded or LVRT mode. Reduce dimensionality for visualization or control (e.g., 10 sensors → 2 or 3 latent variables)

- An additional very effective method includes training an autoencoder neural network using data from the normal mode only. In this case, the autoencoder would learn to reconstruct a typical behavior of the system, capturing the intrinsic structure of healthy operation. When executing in runtime mode, the autoencoder computes a reconstruction error $E(x) = \|x - \hat{x}\|^2$, for each new data point, and if the error exceeds a threshold, the system would process the observation as abnormal. Thus, this approach is especially useful for detecting faults or slide deviations that were previously unseen, and does not require any fault labels.

Expected Outputs and Validation: To make the example pedagogical and assessable, include the following:

- Visualization of cluster assignments from K-means using PCA projections or t-SNE.
- Reconstruction error plots from autoencoder over time, with threshold crossings indicating anomalies.
- Confusion matrix or cluster interpretability table, mapping clusters to operational behaviors using known test scenarios.
- Explained variance plot from PCA to show how many principal components are needed.
- Real-time decision logic: pseudocode or block diagram illustrating how a control system would act based on mode detection.

3.1.3 Self-Supervised (SSL) and Semi-Supervised Learning (Semi-SL)

With more available data and fewer labeled samples, SSL, Semi-SL methods have started to be seen in the power electronics community. These methods utilize unlabeled data in novel ways to allow the model to learn useful representations without relying on large labeled data sets and improve generalization, especially in low-data systems (Gao et al. 2023; Zhao et al. 2020).

SSL: In SSL, the model is trained first to extract semantically meaningful representations from unlabeled data by solving a pretext task. This involves learning a mapping$f : X \rightarrow Z$, where Z is a latent representation of the input domain X. A downstream task model $g : Z \rightarrow Y$ is then trained with limited labels to perform classification, regression, or other supervised methods. Mathematically, SSL learning involves minimizing the pretext loss $\min_{\theta} \mathcal{L}_{pretext}(f(x; \theta))$ followed by optimizing the downstream task loss$\min_{\varnothing} \mathcal{L}_{task}(g(f(x; \varnothing), y)$.

SSL methods can be especially beneficial in cases where there is only limited labeled data. Contrastive learning methods, including SimCLR and MoCo, are the most commonly used SSL methods, and they learn invariant and discriminative features by learning that different augmented versions of the same input sample have similar latent encodings. The other major class includes autoencoders, such as denoising autoencoders, where the model will be learned to produce the input signal from corrupted or compressed representations. These methods learn the underlying structure of data without receiving any labels and can be useful for representation learning problems, anomaly detection problems, and modeling future signals in power electronic applications (Liu et al. 2021).

Semi-supervised learning (Semi-SL) techniques, on the other hand, uses both labeled and unlabeled data simultaneously. Typically, you will have a small labeled dataset (Fang et al. 2023).

Semi-supervised learning, on the other hand, uses both labeled and unlabeled data simultaneously. Typically, you will have a small labeled dataset $D_L = \{(x_i, y_i)\}$ and a large (or larger) dataset of unlabeled samples, $D_U = \{x_{ij}\}$. The learning objective consists of a supervised loss and unsupervised loss term given by:

$$\min_{\theta} \sum_{(x_i, y_i) \epsilon D_L} \mathcal{L}_{sup}(f(x_i), y_i) + \lambda \sum_{x_j \epsilon D_U} \mathcal{L}_{unsup}(f(x_i))$$

where $\mathcal{L}_{sup}$ is generally a classification or regression loss, and $\mathcal{L}_{unsup}$ enforces consistency, entropy minimization, or pseudo-labeling. It enforces a regularization on the unlabeled data. Typical formulations of the unsupervised term include pseudo-labeling, which creates provisional labels around unlabeled samples and retrains the model in an iterative way, consistency regularization, which penalizes the model for providing inconsistent outputs, such as when the same input is perturbed (which is used in FixMatch and Pi-Model); and entropy minimization, which pushes the model to provide confident (i.e., low entropy) predictions around the unlabeled input. The aim is to help refine decision boundaries.

SSL and Semi-SL provide strong solutions for building robust models in power electronics systems, especially where labeled data is highly valuable or inefficient to gather, meaning SSL and Semi-SL can utilize the structure in unlabeled data to help with generalization, improve performance, and provide intelligent and adaptable control and monitoring.

Illustrative Example 6. Fault Classification in Power Converters

This example presents contemporary machine learning techniques for fault detection in multilevel power converters. However, for this case study the labeled fault data is sparse. The goal for this example is for students to understand how to use self-supervised pretraining and semi-supervised learning to create a robust classifier for online fault diagnosis in industrial systems.

Objective: In the case of multilevel inverters, open-circuit conditions, short-circuit conditions and faults are challenging to detect, and datasets containing labeled fault cases are often limited due to the difficulties and costs associated with providing labels.

System Description: In this case, the goal becomes classifying such faults while only having a small amount of labeled data. The model inputs consist of time-series data containing the voltage and current waveforms, frequency-domain features computed using FFT, and inverter switching patterns.

Solution Using Self-Supervised and Semi-Supervised:

To address the labeling challenge, a self-supervised learning approach can be adopted to pretrain a feature encoder using only unlabeled waveform data. Contrastive learning is used to ensure the encoder learns representations that capture the underlying dynamics of the system. Following this, a semi-supervised learning strategy is employed to fine-tune the model using the limited labeled data. This may involve pseudo-labeling the unlabeled portion of the dataset and applying consistency regularization to improve generalization. Together, these techniques enable robust and data-efficient fault classification in power electronic converters, making them suitable for deployment in real-world industrial systems with constrained fault data availability (Gao et al. 2023, 2022; Liu et al. 2021; Fang et al. 2023).

Expected Outputs and Validation: To evaluate and demonstrate the effectiveness of the model and to support student learning, the following outputs should be included:

- Confusion matrix showing classification accuracy across fault types (open-circuit, short-circuit, normal)
- Visualization of learned feature embeddings using t-SNE or PCA (showing clusters before and after fine-tuning)
- Ablation study results (e.g., model accuracy with/without self-supervised pretraining)
- Training curves comparing supervised vs. semi-supervised approaches
- Sample waveform plots showing inputs for correctly and incorrectly classified examples

3.1.4 Reinforcement Learning (RL)

RL is a framework for decision-making in which an agent learns to take optimal actions through interaction with an environment and by receiving feedback through rewards. The agent, working to maximize its expected reward over time, learns an optimal policy for taking action. RL is well suited to environments that are complex, non-linear, and uncertain, making it an increasingly effective tool for adaptive control, decision-making in real-time, and online optimization within power electronics systems (Chen 2024; Wan et al. 2024; Adibi and Woude 2019; Bag et al. 2019; Belousov et al. 2021; Li et al. 2023).

RL problems are typically modeled as a Markov Decision Process (MDP), characterized by a tuple (S, A, P, r, γ), where S is the set of system states, A is the set of actions, P defines the state transition probabilities, r is the reward function, and γ is the discount factor. The agent iteratively refines its policy $\pi(a|s)$ (Belousov et al. 2021; Li et al. 2023), which determines the probability of taking action a in state s, in such a way that the expected return defined as $J(\pi) = E_\pi\left[\sum_{t=0}^{\infty} \gamma^t r(s_t, a_t)\right]$.

Reinforcement learning algorithms fall into three different families depending on how they accomplish the learning task (Chen 2024; Wan et al. 2024; Adibi and Woude 2019; Bag et al. 2019; Belousov et al. 2021; Li et al. 2023; Kermansaravi et al. 2024).

- Model-Free RL: Model-free methods do not require an explicit model of the environment's dynamics, and they learn from experience. The first of the model-free methods, value-based methods estimate a value function $Q(s, a)$, where the function predicts the expected return of taking an action a in state s. Classic implementations of value-based methods are Q-learning and Deep Q-Networks (DQN) which are now used extensively in a wide variety of control problems.
- Policy-based: The other model-free method is called policy-based learning, and here the algorithm learns the policy directly by maximizing the expected return based on various techniques including policy gradient methods. Prominent variations of policy methods are Proximal Policy Optimization (PPO), in which robust performance is the main observation for continuous control.
- Actor-Critic: A notable hybrid of these two methods is the actor-critic architecture. In this setting, the actor learns the policy, with the critic estimating the value function to give the actor improvement updates, providing value functions that have improved stability and sample efficiency in training. Actor-critic methods include the Deep Deterministic Policy Gradient (DDPG) and Soft Actor-Critic methods (SAC), which have shown to perform very well in continuous action space, which is important when applying RL in the domain of power electronics.
- Model-Based RL: In contrast, in model-based RL, a model of the environment's dynamics is learned, which can be used to simulate future states of the environment and evaluate various action sequences without executing them. While model-based RL methods often provide improved sample efficiency (in that they can create synthetic training data for the learning process) they are less common in power electronics problems due to the kwon challenges to model the high-frequency dynamics of switching devices, nonlinearities, and behaviors critical for stability in real-time power systems.

Overall, reinforcement learning represents a very attractive framework for constructing intelligent control strategies in power electronics. The relevant aspect of reinforcement learning is its ability to learn through interaction, and on the fly, being able to adapt to changing environments, and being able to operate under uncertainty makes it a significant piece of the future of intelligent autonomous energy efficient electronic systems.

Illustrative Example 7. RL for Inverter Control

This example demonstrates the use of reinforcement learning (RL) for the real-time control of a single-phase inverter, a common element in contemporary power electronics. The students learn to utilize Proximal Policy Optimization (PPO), a popular policy gradient method, and will train a control agent that is capable of tracking sinusoidal reference waveforms using only voltage, current, and integrated tracking error measurements.

Objective: Control a single-phase inverter to generate sinusoidal AC output from a DC source. The control objective is to control the output voltage and current to track reference waveforms so that the inverter produces high-quality power and operates stably.

System Description: The state of the system includes key electrical quantities that will characterize the inverter dynamics. The state characteristics include the AC-side current i_{ac}, output voltage v_{out}, reference current i_{ref}, reference voltage v_{ref}, and an accumulated integral error term that is useful for concentrating on steady-state tracking accuracy. The whole state vector can be written as:

$$s = \left[i_{ac}, v_{\text{out}}, i_{\text{ref}}, \text{integral error}\right].$$

The control actions of the inverter consist of discrete switching commands that each denote one of the three possible H-bridge switching states, $a \epsilon \{-1, 0, +1\}$, which rearranges the switching devices that internally modify the voltage applied to the load. This discrete action enables the reinforcement learning agent to select the switch state at each time step.

The reward function penalizes deviations from the reference signals, thus forcing the agent to lower the output errors over time. The reward function is a negative weighted sum of squared errors for current and voltage tracking:

$$r = -\left[w_1.(i_{ac} - i_{ref})^2 + w_2.(v_{out} - v_{ref})^2\right]$$

where w_1 and w_2 are tuning weights that balance how much focus is placed on either current or voltage accuracy as the control objective.

Solution using PPO for H-Bridge Inverter:

For a solution to this control problem, a Proximal Policy Optimization (PPO) utility (known as a policy gradient reinforcement learning algorithm) is selected as the training algorithm, which is known to be robust and popular for this type of learning. The training stage can be done using a simulated environment, either with tools like MATLAB/Simulink, or a fully open-source simulated environment that is built on some flavor of Python, such as OpenAI Gym or RLlib. The simulation environment works to replicate the electrical behavior of the inverter, including the nonlinear dynamics around both the voltage and current dynamics, and gives the agent simple control actions to learn effective switching parameters over time.

During training, the observations supplied to the agent needed to be normalized to help induce more stable learning and convergence times. The discrete action space, which emitted the three possible standardized switching states for the inverter, was incorporated into the policy structure within the PPO learning algorithm. Therefore, as the agent continued to interact with the simulated environment, it was able to improve on its control policy, issuing new control actions to maximize its total reward, which represented the fidelity of the sinusoidal waveform with only small deviations from the desired current or voltage output. Additionally, this example shows how to make use of reinforcement learning, specifically PPO to carry out real-time control in power electronics, thus allowing for model-free and adaptable methods of implementing control schemes that are intended to generalize to different operating conditions without needing complex control heuristics for specific controls or needing detailed system models.

Expected Outputs and Validation: The model will be generate the following outputs to assess and validate the RL controller's effectiveness:

- Time-domain waveforms showing reference and actual output signals (voltage and current) under the learned control policy
- Total harmonic distortion (THD) metrics before and after learning
- Learning curves showing cumulative reward vs. training episode
- Switching pattern visualizations to interpret learned strategies
- Comparison against PI or heuristic controllers under similar conditions

3.1.5 Deep Learning (DL)

DL refers to a specific category of ML that uses multiple-layered neural networks, also known as Deep Neural Networks (DNNs), to learn complicated hierarchical features from raw data. DL methods can also work with high-dimensional, non-linear, and sequential data. Since the majority of power electronics applications involve time-series signals, sensor measurements, and spatiotemporal features, we can see how DL is very appropriate for these applications (Qashqai et al. 2022a; Lucia et al. 2020; Hajihosseini et al. 2020).

A general deep learning model is tasked with learning functions of the type $\hat{y} = f(x; \theta)$, where x is the input (e.g., waveform, state vector), and y is the target (e.g., fault class, control signal), and θ represents the parameters/weights of the network that will be learned. During the training of the model, these parameters are standardized through the minimization of a loss function, typically defined as:

$$\mathcal{L}(y, \hat{y}) = \frac{1}{N} \sum_{i=1}^{N} \ell(f(x_i; \theta), y_i)$$

where ℓ is a cross-entropy for classification or MSE for regression.

The capacity to take this approach to learning enables a wide variety of learning paradigms including supervised, unsupervised, and reinforcement learning. Thus, resulting in various tasks within power electronics where this method will be used in control, diagnostics, and anomaly detection.

Several deep learning architectures have found widespread application in this field. For extracting spatial features models like convolutional neural networks (CNN), have been successful with image based fault detection in converter module and vibration-based diagnostics of power elements (Gros et al. 2023; Cheng et al. 2021). These models iteratively lower dimensionality by applying filters to input data before passing the data through layers where the data evolves through a series of activation functions such as ReLU, finally extracting spatial features at larger levels of abstraction.

$$x \rightarrow ReLU \rightarrow \cdots \rightarrow y$$

Recurrent neural networks (RNNs) use dynamic memory better suited for sequential data. LSTM and GRU approaches RNNs use memory cells coupled with gating to learn long-term holds within time-series data, thereby allowing for dynamic modeling of converter operation, as well as accurate forecasting of load and faults with high computational cost for each instance of prediction (Qashqai et al. 2021, 2022a; Khandelwal and Kumar 2019). Alternatively, GRU approaches are much lighter while still having comparable results in many sequence modeling tasks (Qashqai et al. 2021; Khandelwal and Kumar 2019).

In recent years, transformer-based models with self-attention mechanisms have experienced increased popularity and yielded better performance attributes when compared to other DL architectures because of their ability to model long-range dependencies in sequential data without recurrence. The efficiency and scalability of transformer-based models make them attractive options for applications like predictive analytics in smart grid systems and real-time monitoring of power converters (Zhao et al. 2020).

Transfer Learning (TL): Transfer learning (TL) has become another strategy that has proved valuable in the implementation of DL, where models trained on large data sets can be redeployed to perform new (but related) tasks. For example, a model for fault detection may have been trained on one converter topology and can be retuned for use on another converter topology. TL reduces the time for training the model while improving performance in new situations, especially when data that is labeled is limited (Khandelwal and Kumar 2019; Wang et al. 2020). In conclusion, deep learning offers a powerful and flexible framework for solving diverse problems in power electronics offered solutions for advanced signal classification, condition monitoring, control, and forecasting, because of its ability to leverage raw, complex, structured data.

Illustrative Example 8. Predictive Control of a 3-Level NPC Inverter

This example introduces the use of deep learning architectures–like CNNs, RNNs, transformers, and transfer learning (TL) for building predictive models to control the inverter switching behavior and predict the output current from historical sequences of current, voltage and switching actions. The students will investigate how supervised learning methods can be used to learn and generalize temporal patterns in power electronics.

Objective: The main goal is to create models that predict or control inverter switching behavior or output current based on historical input–output data (voltages, currents). These models will be able to learn from time-stamped sequences and can be used for real-time operation or control.

System Description: The training configuration for models like these will first create a data set that consists of $I(t)$, $V(t)$, $action(t)$ timestamped sequences. This data set is typically divided into three subsets for model training, validation, and test (e.g., 70% training, 15% validation, 15% test).

Having test and validation subsets is important to measure how well the model has learned generalizable patterns and perform adequately with unseen data. Depending on the prediction task, whether signal values or discrete switching actions, the appropriate loss function is selected: mean squared error (MSE) for regression problems or cross-entropy for classification. Models were trained using the Adam optimizer, using a batch size of 64 with a learning rate of 0.001.

Solution using CNN, LSTM, GRU, Transformer, TL:

Various deep learning architectures can be employed for this task. Convolutional neural networks (CNNs), while traditionally applied to spatial data, can be adapted for time-series input by reshaping sequences into two-dimensional representations or using windowed segments. For instance, a CNN model might take a fixed-size input window of previous samples such as

$$x = \begin{bmatrix} I(t-3) & I(t-2) & I(t-1) & I(t) \\ V(t-3) & V(t-2) & V(t-1) & V(t) \end{bmatrix}$$

which is reshaped into 2D Matrix (2,4). This data can then be processed using a 1D CNN with layers such as a convolutional filter (e.g., Conv1d), activation function (e.g., ReLU), and a fully connected (FC) layer to predict either the next current value $I(t+1)$ or switching decision or switching state. Outputs can vary depending the learning objective, such as MSE for a current prediction or cross-entropy for determining an appropriate expressed or inferred switching states.

Recurrent neural networks (RNNs), especially long short-term memory (LSTM) models, are also specifically designed for encoding temporal dependency for voltage and current sequences. An LSTM model may take an input sequence:

$$x_t = [I(t), V(t)], \ldots, [I(t-n), V(t-n)]]$$

where the sequence length is n, and this sequence can be processed through multiple LSTM layers. The final hidden state of the LSTM is often passed through a fully connected layer to generate its output that could be a control signal or a future current sequence value, respectively. LSTMs are well suited for modeling the dynamic behavior of inverter systems, though they can be slow to train and run computations.

Gated Recurrent Units (GRUs) are a more lightweight model compared to LSTMs. It follows the same input to output architecture, but has a simpler gating mechanism, and reduced number of parameters and time when training while still capturing the extenuating temporal loops. For example, a GRU model can be modeled with one hidden layer of 64 units and trained with the same voltage-current sequence inputs as an LSTM.

Transformers are a more recent development in modeling long-term dependencies in sequential data. Transformers can directly relate different time steps in a sequence using self-attention. Transformers use the self-attention mechanism to help them leverage long sequences more efficiently than RNN-based models.

For example, let us say a transformer encoder processes an input tensor of shape (sequence_length, batch_size, input_dim) (sequence_length,batch_size,input_dim), where input_dim consists of both current and voltage. The transformer model may contain several stacked transformer encoder layers (i.e. d model = 64 d model = 64 and 8 attention heads), plus a final embedding and a fully connected to provide the final output—either a control decision or a predicted signal value.

Transfer learning (TL) may also be useful in this context in a pre-trained model can be leveraged and reused across different but similar inverter topologies and datasets. For example, if an LSTM or transformer were trained with data collected from a Neutral-Point-Clamped inverter, you'd likely be able to transfer it to the H-bridge converter with just a small amount of additional data. When you transfer, you typically freeze the lower pre-trained layers and tune only the upper layers or output heads.

Expected Outputs and Validation: The model performance can be assessed through:

- Prediction error metrics (e.g., MSE for waveform prediction)
- Confusion matrices and accuracy for switching classification
- Visual comparison of predicted vs. actual signals
- Training/validation curves to evaluate generalization
- Cross-topology performance tests when using transfer learning

4 Summary and Comparison

In the past few years, ML methods have been increasingly employed in power electronics for modeling, control, fault detection, diagnostics, and optimization tasks, largely in applications where system dynamics are often nonlinear, time-varying, or ordinary analytical models are insufficient.

- SL can be used for tasks such as classification and regression (fault diagnosis, efficiency prediction, or power quality classification). It is well characterized by a huge requirement for labeled data and can be shown to guarantee lower error rates based on training datasets that cover wide variation.
- Unsupervised learning is usually for discovering anomalies or clustering operating conditions and not generally for learning distinct behaviors of the system without a labeled output. It has some utility where labeling data is costly or impractical, but lower performance can be expected.

- Semi-SL is somewhere in between, making use of labeled data, but the power of generalization is improved by adding additional, largely unlabeled data. This draws great interest in areas such as predictive maintenance, or where rare event detection is required.
- SSL may be relatively nascent in power electronics, but it has shown that reasonable features can be extracted without the need for an actual manual label which can be particularly helpful in the case of sensor time-series data or even power waveforms.
- RL seems appropriate for real-time control, planning, and management tasks such as inverter switching and control, MPPT in photovoltaic systems, fleet energy management in hybrid systems, and so on. RL learns optimal policies from online interaction with an environment but can present long training times and is not able to offer formal guarantees for safety–critical applications.
- DL was made to handle high-dimensional input data, for, e.g., waveform images, spectrograms, or raw sensor data. DL is a popular technique in fault classification, predictive modeling, and system identification applications. However, it often demands significant computational resources and large amounts of training data, making it resource-intensive.

ML techniques provide useful alternatives to traditional model-based or rule-based methods, particularly for uncertain, non-linear, or partially observable situations. However, explainability, data requirements, and real-time rollout cautiously need to be considered by users of these techniques, especially in contexts involving protection, control, or grid compliance (Table 5).

ML approaches are becoming a popular option to help solve the challenges associated with the complexities of power electronics systems where nonlinear dynamics, time-variant system-level behavior, and limited sensor information are potential issues. Each category of ML, including supervised, unsupervised, semi-supervised, self-supervised, RL, and DL, has its individual strengths which can be applied to modeling, control, and diagnostics tasks. SL is very accurate when sufficient labeled data is available; whereas unsupervised learning and semi-SL approaches can be helpful when the labeling can be very costly or limited. RL can be especially useful when optimal, model-free control can be executed in dynamic environments, and DL is useful for pattern recognition directly from raw data. SSL is a novel topic providing entirely new opportunities to take advantage of unlabeled data and extract useful features from it. The choice of ML approach will be context-dependent and relies heavily on the application situation that defines the data availability, the real-time aspect of the tasks, and the degree of complexity.

Best Use Cases in Power Electronics:

- SL: Fault detection and classification (e.g., IGBT or capacitor faults), efficiency estimation, voltage/current regulation.
- Unsupervised Learning: Anomaly detection in inverter outputs, clustering operational modes, sensor drift detection.
- Semi-SL: Predictive maintenance (e.g., remaining useful life estimation), classification of rare fault events or grid disturbances.

Table 5 Comparison of machine learning methods used in power electronics

Feature	Supervised learning	Unsupervised learning	Semi-supervised learning	Self-supervised learning	Reinforcement learning	Deep learning
Data requirement	High (labeled)	Low (unlabeled)	Moderated (some labeled)	Low (no labels)	Medium (interaction-based)	High (data-hungry)
Real-time suitability	Medium	High for simple models	Medium	Emerging	Yes (with training)	Low (unless optimized)
Explainability	Moderate	Low	Moderate	Low	Low	Low
Training complexity	Medium	Low	Medium	High	High	Very High
Adaptability	Limited (offline)	High	Medium	High	Very High	High
Model-free capability	No	Yes	Partially	Yes	Yes	Yes
Key challenge	Labeling effort	Interpretation	Balance of labeled/unlabeled	Maturity in Power Electronics	Stability and safety	Overfitting and computing demand

- SSL: Feature extraction from power waveforms, signal reconstruction, and pre-training models for diagnosis tasks.
- RL: Real-time control of power converters, MPPT in PV systems, grid-tied inverter current control, and voltage regulation in DC microgrids.
- DL: Harmonic distortion detection, raw signal-based fault classification, converter behavior modeling, virtual sensor development.

5 Applications of AI in Power Electronics Converters Control

AI techniques have become important for managing robust, adaptive, and efficient control as power electronic systems continue to increase in complexity. As discussed in the previous section, AI methods can be categorized into three principal categories: Symbolic AI, CI, and ML that primarily includes traditional ML techniques but also advanced learning methodologies such as DL and RL.

Each group addresses different challenges in power electronics, such as rule-based decision-making, optimization techniques, data-driven approaches, real-time control, etc. These approaches vary in their limitations or strengths depending on the task, control objective, and system constraints. This section provides a systematic classification of AI techniques and demonstrate their applications in the control of power electronic converters.

5.1 Symbolic AI

Symbolic AI systems provide solutions that are typically highly interpretable and transparent for the operation being performed. Symbolic AI systems are often quite useful for operations in safety–critical systems, particularly when validation is required. However, symbolic AI systems tend to be rather inflexible, and as such, do not adapt to changes in environment nor can they adapt to the changing conditions under which they may operate. Symbolic AI systems also tend to have low computational load which makes them suitable for embedded systems with limited resources. Symbolic AI is most effective in static or dominated by rule scenarios such as diagnostics, protection logic, and supervisory control (Huang et al. 2016; Simões and Bubshait 2019; Bose 2002; Feng et al. 2023; Kececioglu et al. 2020; Cheok and Wang 2005; Hannan et al. 2020) (Table 6).

Table 6 Use cases for applied symbolic AI techniques in control power electronics

Application area	AI method	Example/Description	References
Controller parameter tuning	ES	Adaptive Controller Parameter Optimization Using Rule-Based ES	Huang et al. (2016)
	FL	FL Controller for Dynamic Droop Adaptation in Decentralized Power Systems	Simões and Bubshait (2019)
Maximum Power point tracking (MPPT) in Renewable Energy systems	FL	Rule-based Maximum Power Point Tracking (MPPT)	Kececioglu et al. (2020)
Flux & rotor position estimation	ES	Intelligent Control in Power Electronics: Applications of ES, FL, and Neural Networks	Bose (2002)
	FL	Intelligent Flux and Rotor Position Estimation Using Fuzzy Inference Techniques	Cheok and Wang (2005)
Fault detection & diagnostics	ES	Intelligent Fault-Tolerant Control via Rule-Based ES in Distributed Environments	Feng et al. (2023)
	FL	Model-Based Fuzzy Predictive Control for Fault-Tolerant System Operation	Mendonça et al. (2012)
Modulation scheme optimization	FL	Intelligent Fuzzy Control Optimization for Real-Time System Performance: A QLSA-Based Approach	Hannan et al. (2020)

5.2 *CI*

CI includes biologically inspired techniques including evolutionary algorithms, FL, and swarm intelligence. Their usage is based on learning and optimization rather than exact models to handle uncertainty and nonlinearities. They are relevant if the interpretability, adaptability, and computational needs are between low and high. CI techniques are applicable for multi-objective optimization, tuning, and control in parameter uncertainty cases (Shi et al. 2015; Chen et al. 2019; Harrag and Messalti 2015; Hemachandu et al. 2017; Taghizadeh and Hagh 2010) (Table 7).

5.3 *Ml*

ML includes regular ML methods (e.g., supervised, unsupervised), DL, and RL to learn control among other things in the following ways. In the past, ML techniques

Table 7 Use cases for applied CI techniques in control power electronics

Application area	AI method	Example/Description	References
Controller parameter tuning	GA	Optimal parameter tuning for PI controllers	Kumar et al. (2020), Divakar and Jacob (2019), Ali et al. (2021)
	PSO	Application of PSO for Enhanced Voltage Control in 15-Level Inverters	Hemachandu et al. (2017)
	ACO	Application of Improved ACO in Fuzzy PID-Based Multi-Area Load Frequency Control	Chen et al. (2019)
Maximum power point tracking (MPPT) in renewable energy systems	GA	Genetic Algorithm Enhanced P&O MPPT with Adaptive Duty Cycle Control for Boost Converter Application	Harrag and Messalti (2015)
	PSO	Optimized Particle Swarm and Conductance-Based MPPT for PV Arrays Under Dynamic Shading Conditions	Shi et al. (2015)
	ACO-ANN	ACO-Tuned Neural MPPT Controller for PV Systems with Power Quality Enhancement via Shunt APF	Babes et al. (2022)
Motor speed estimation and control	PSO	Adaptive Neural Control of a Stand-Alone Induction Generator Using RBFN and IPSO Techniques	Lin et al. (2008)
	ACO	Modified ACO for Fast and Accurate MPPT in Standalone PV Systems	Phanden et al. (2021)
Modulation scheme optimization	PSO	Harmonic Elimination in Cascaded Multilevel Inverters with Unequal DC Sources Using PSO	Taghizadeh and Hagh (2010)

relied heavily on hand-coded features (usually constrained to engineering principles). Supporting data was often limited to the algorithm developer's practical experience or other less reliable data. By processing large data sets along with actions in an environment, policy representations (representation-falsifiability) can be learned that allow a system to predict future behavior or classify patterns, improving control of uncertain situations. The traditional ML and newer DL and RL methods have great adaptability, and are especially capable for real-time, dynamic control of uncertainty. However, they typically have lower interpretability making certification and safety validation a considerable challenge, and they have a high or even extremely high computing cost (especially in training) (Gao et al. 2022; Cheng et al. 2021; Hua et al. 2021; Hosseinzadehtaher et al. 2021; Kim 2023) (Table 8).

Table 8 Use cases for applied machine learning techniques in control power electronics

Application area	AI method	Example/Description	References
Fault tolerant control	SL -ANN	Enhances fault tolerance in MMCs using ANN-based ripple reduction	Gao et al. (2022)
	CNN	The CNN-based intelligent fault-tolerant control system	Cheng et al. (2021)
	RL	Control Strategy Optimization under multiplicative faults	Hua et al. (2021)
Controller parameter tuning	SVM	Data-Driven Model Predictive Control	Yang et al. (2023)
	ML (ANN)	tuning PI controller gains for transient response	Khan et al. (2023)
	RL	Reinforcement Learning for Intelligent Control of Power Electronic Systems	Chen (2024), Kermansaravi et al. (2024), Alquennah et al. (2023)
	DL	DC-DC Converter control (Boost Topology)	Hosseinzadehtaher et al. (2021), Alquennah et al. (2024) Kim (2023), Mohammadzadeh et al. (2022)
Modulation scheme optimization	DL	MPC Replacement via Policy Learning	Lucia et al. (2020)
	RL	Power conversion control with maximum power extraction	Srungaram and Yadlapati (2024)

Among many AI techniques, RL has received more attention than the others because it allows for continuous, real-time, adaptive control of power electronic systems. SL relies on substantial amounts of labelled data, and unsupervised learning focuses on discovering hidden patterns, but RL takes a different approach in that the agent learns through interactions with the environment. The agent is able to learn optimal control policies by continual interactions with the environment and it receives feedback about actions as rewards or penalties. By receiving environmental feedback and reinforcement of rewards (depending on the task setting), RL improves over time. Therefore, RL can help in making direct improvements without an explicit model (i.e., starting from scratch) (Kermansaravi et al. 2024; Alquennah et al. 2023, 2024).

SL approaches aim to address problems in power electronics such as fault detection, parameter estimation, and predictive diagnostics, and rarely break away from offline and static/mirror classes of datasets which typically go hand-in-hand, but do require initial/static signals/streams. RL addresses the control aspect directly and can take care of the control task simply by developing a control policy from real-time feedback from the system. RL has the additional strength of developing a control policy that can develop well in to the system where the dynamics are complex, uncertain, and/or changing with time, an area where traditional model-based or rule-based controllers may fail, or have limitations due to a lack of performance under new conditions (Yılmaz et al. 2023; Fang et al. 2023; Belousov et al. 2021).

RL's model-free characteristic can decrease reliance on extensive system models, which often take a lot of time or are not possible because detailed models for nonlinear, high-frequency switching systems such as power converters may not exist. Moreover, using a DL paradigm with RL allows for high-dimensional state and action spaces, and demonstrates RL's scalability for various converter architectures and operating conditions.

As the power grid shifts toward decentralized, more variable, and digitally controlled processes, the need for intelligent control strategies capable of autogenous adaptation to that variability is increased. The agent-based framework of RL provides a potential step toward this ambition, wherein converters could learn, adapt, and optimize autonomously informed from the "world" feedback of real operating conditions (Chen 2024; Adibi and Woude 2019; Belousov et al. 2021; Alquennah et al. 2023, 2024).

The next section provides an application example to show how RL-based control has the potential to be applied to a grid-connected power converter, using both the control architecture, and demonstrating the performance benefits of an agent and real-time learning and decision-making.

6 Application Examples: Reinforcement Learning-Based Control of Grid-Connected Power Converters

This section demonstrates a practical example of intelligent control in power electronics, specifically the current control of a grid-connected multilevel inverter based on RL (Kermansaravi et al. 2024). The example illustrates how a data-driven, model-free control strategy can take advantage of RL to improve the real-time performance of a power converter in a smart grid environment.

Context: Grid-connected inverters are a fundamental part of DC microgrids, renewable energy interfaces, and EV charging infrastructure. Control techniques such as proportional-integral (PI), model predictive control (MPC), or sliding-mode control (SMC) all depend upon good models and are sensitive to parameter changes. In this application, an RL-based controller have developed to control a 5-level Packed U-Cell (PUC5) inverter.

System Overview: The PUC5 inverter is a single source, a five-level topology that produces quasi-sinusoidal output with low harmonic distortion through the use of an auxiliary capacitor that generates more voltage levels. The RL controller will be responsible for the regulation of the magnitude of the current from the inverter that is

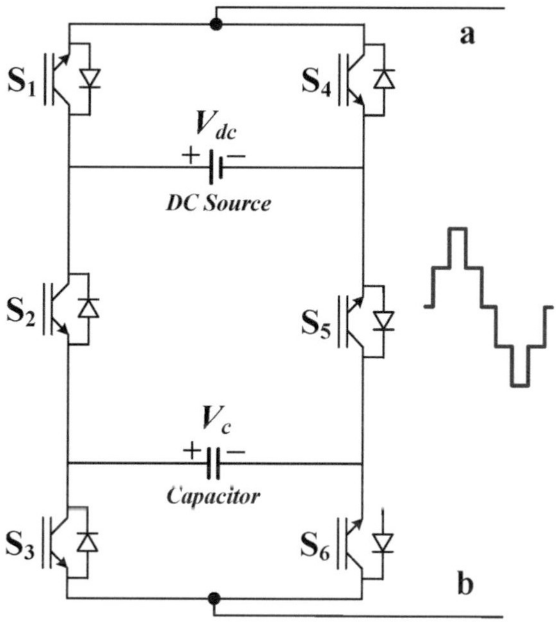

Fig. 1 PUC5 inverter topology

injected into the grid while the capacitor voltage balancing is handled by switching to logic (Fig. 1).

RL Methodology:

- Environment and Agent Design: The RL environment was created in MATLAB/Simulink to model the inverter's dynamic under grid-connected conditions while the agent employs PPO using an actor-critic architecture.

  The observation space contains:

- The actual AC current (I_{ac})
- The reference current waveform I^*
- The maximum current amplitude (I^*_{max})

The action space consists of five discrete output levels representing the states of the inverter voltage (+2E, +E, 0, −E, −2E), which makes it a simpler way of problem control.

- Reward Function: Design The reward function is a custom PI-based one that is designed to guide the agent: It penalizes the proportional error $|I_* - I_{ac}|$. The integral error term is included so that the agent can track the signal accurately over the long term. It is designed to penalize overshooting and control effort. It has a small positive bias that encourages exploring. The structure created to keep learning stable and effective, but without knowing any part of the actual overall systems' model for all the states (Fig. 2 and Table 9).

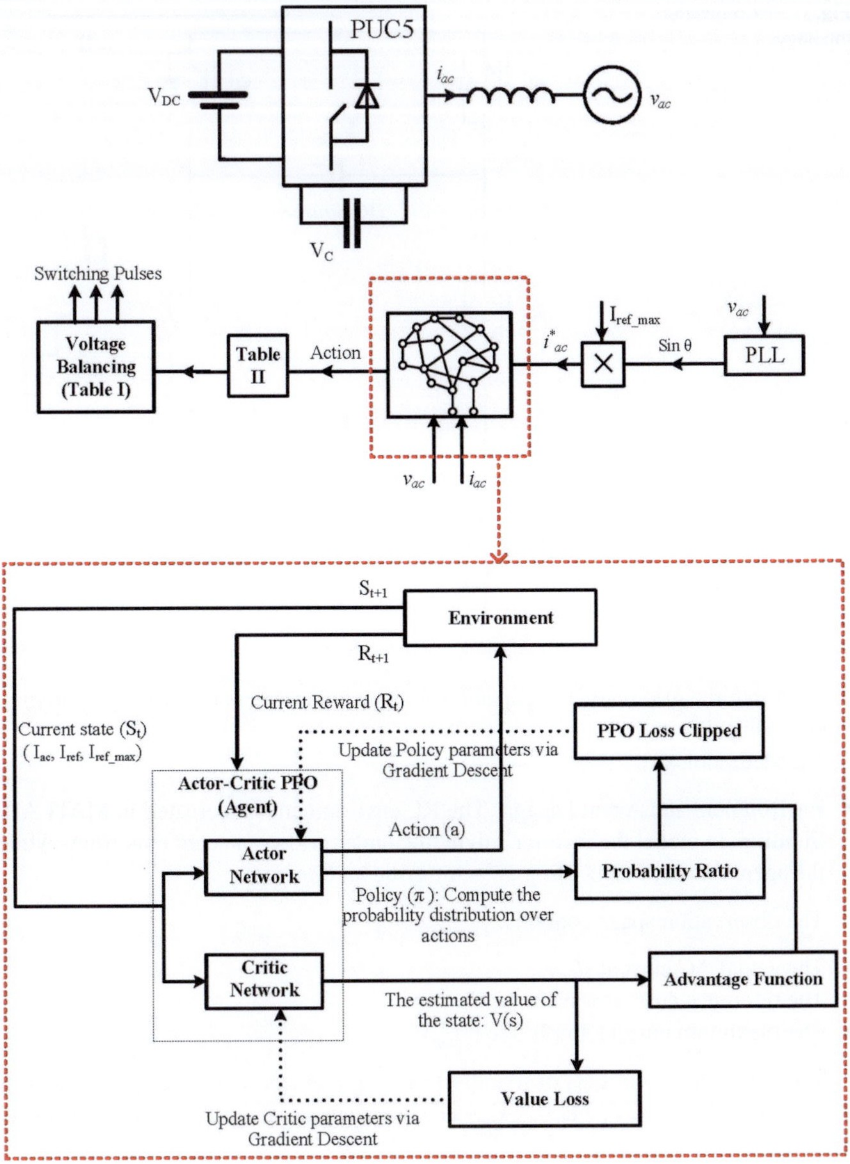

Fig. 2 Grid-Connected PUC5 inverter with RL Controller

- Training Process: The training of the RL controller was done using a MATLAB/Simulink environment, which simulated the grid-connected PUC5 inverter system in a wide range of operating conditions. The RL agent was trained using the PPO algorithm in an actor–critic architecture, which facilitates continual improvement of the policy when interacting with the environment. The training was run in

Table 9 Action space Of RL control or PUC5 inverter

Action	V_{out}
A1	+2E
A2	+E
A3	0
A4	−E
A5	−2E

episodes with each run being limited to one complete sine wave, which was approximately 833 simulation steps. The training was run up to 10,000 episodes. The training used a discrete action space of five inverter output voltage states and the observation space normalized values of the grid current, the reference, and its amplitude. The target training outcomes were directed by a custom-designed reward function which offered penalties for current tracking errors, overshooting, and control effort. To ensure the RL agent learning remained stable and efficient, the training used an experience horizon of 1000 steps, a mini-batch size of 512, and a discount factor of 0.99, and the policy update was clipped to maintain stability in policy updates. During training, the RL agent learned switching actions which optimized performance in both steady-state and transient conditions; the RL agent demonstrated a good ability to generalize over the training data, including varying current and amplitude, power factors, and voltage disturbances.

Results and Performance Evaluation: The performance of the RL-based controller was assessed in three different situations that proved their robustness and ability to adapt. In the equilibrium condition, the performance of the RL controller had relatively stable output, with a clean 5-level voltage waveform (with an injected current THD of only 2.5%). The auxiliary capacitor voltage was tracked and achieved the target, 100 V, which would keep the inverter functioning appropriately (Fig. 3).

For dynamic tracking, the system was subjected to a step change in the reference current amplitude from 5 to 10 A. The controller responded instantly and accurately, tracking the reference waveform without overshoot or delay (Fig. 4) The disturbance rejection capability were evaluated based on two tests. The first consisted of changing the power factor from 1.0 to 0.86, which engaged the inverter in the exchange of reactive power with the grid. In this situation the RL controller quickly adapted to the new operating point while still providing quality of performance (Fig. 5).

The second test was implemented by applying a drop to the grid voltage which decreased from 120 V RMS to 100 V RMS. The controller maintained an appropriately regulated current even through a voltage sag, indicating that it handled the voltage disturbance well (Fig. 6).

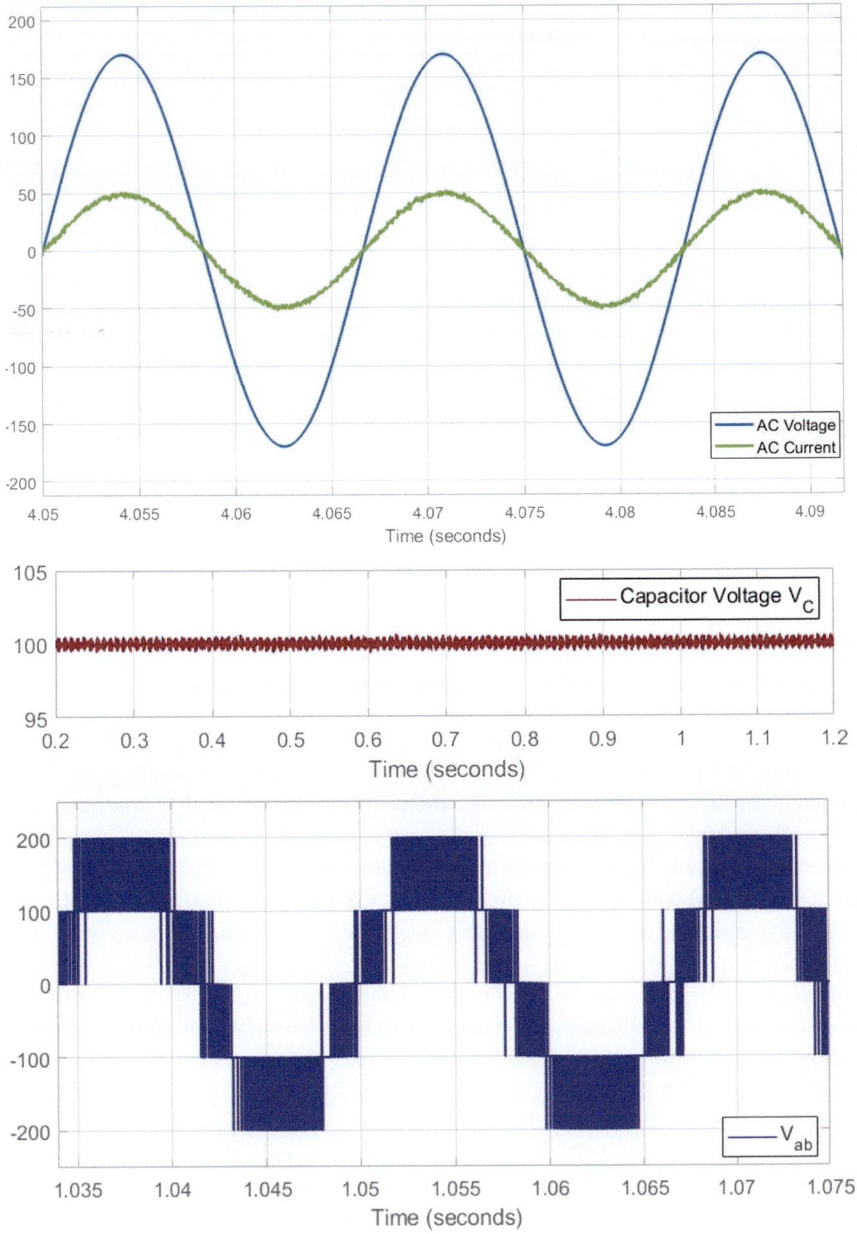

Fig. 3 Steady state results pf PUC5 grid-connected inverter with RL-based current controller

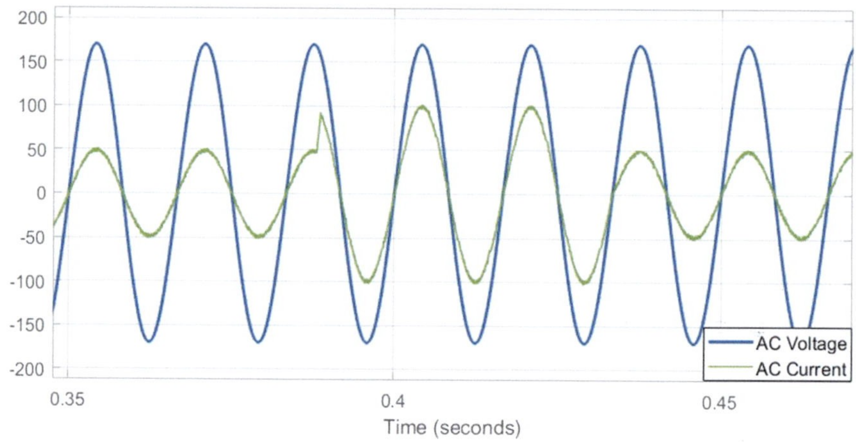

Fig. 4 Results during a change in current amplitude from 5 to 10A (AC Current magnified by 10x)

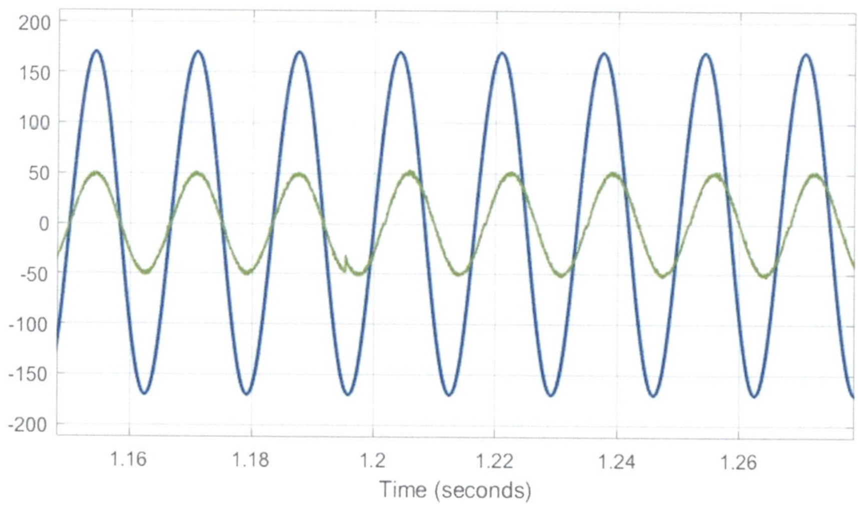

Fig. 5 Results during a change in power factor from 1 to 0.86

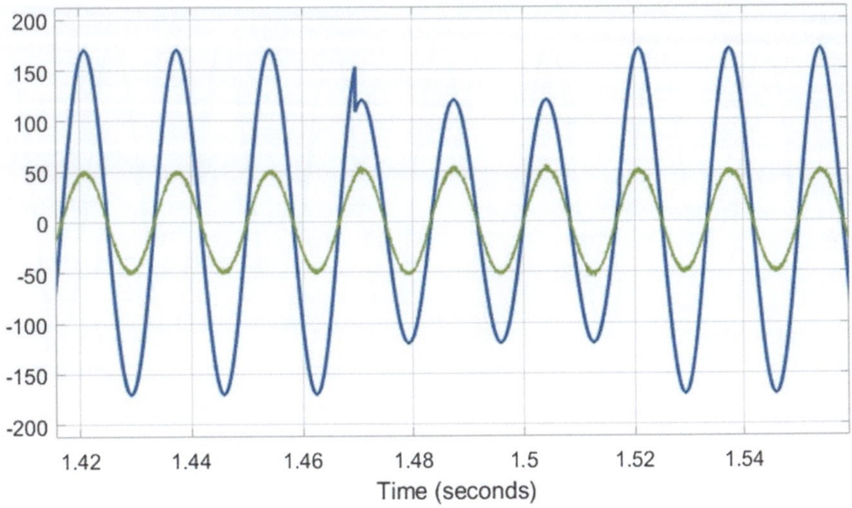

Fig. 6 Results during a change in AC voltage variation from 120 V RMS to 100 V RMS intelligent power systems

7 Conclusion

The chapter presents an overview of how ML and AI will provide more features with power electronics systems. A range of learning paradigms were presented, and examples of each were provided, including both applications of supervised and unsupervised learning bases, semi-supervised learning, self-supervised learning, DL methods and RL, in terms of modeling, fault detection, diagnostics, and control.

Each ML method has its pros/cons. Supervised learning method performs best when there is sufficiently labeled data available which can demonstrate the extent of accuracy; Unsupervised and semi-supervised learning methods are useful for settings where data labels need to be quantified; RL is applied on more complex domains of "adaptive control" where real-time behaviors need to be considered, DL methods can work with high-dimensional input data, from which ML can extract features from raw signals. Independently of specific ML methodologies and approaches to AI, symbolic AI and CI have been useful for interpretable and flexible means of control and for optimization problems in the field of power electronics converters.

Even with the advancements in ML and AI methodology and platform development, there continues to be challenges in using ML and AI/ML processes in the context of power appliance systems. For example, large datasets are a must, confidently building "explainability" capabilities, large computational power required for real-time modelling applications, and difficulty with operational behavior generalization in the situational space of complex systems. In the future, as these power electronics systems operate in more autonomous, learning from data environments, ideally research moving forward will include improved robustness, trust, transparency, real-time processes, and greater flexibility by way of compliance with grids codes and safety measures.

Overall, ML and AI represent a promising set of tools to advance the intelligence, adaptability, and efficiency of modern power electronic systems.

References

Adibi M, van der Woude J (2019) A reinforcement learning approach for frequency control of inverted-based microgrids. IFAC-PapersOnLine 52(4):111–116

Alajmi BN, Ahmed KH, Finney SJ, Williams BW (2010) Fuzzy-logic-control approach of a modified hill-climbing method for maximum power point in microgrid standalone photovoltaic system. IEEE Trans Power Electron 26(4):1022–1030

Ali M et al (2021) Robust ANN-based control of modified PUC-5 inverter for solar PV applications. IEEE Trans Ind Appl 57(4):3863–3876

Alquennah AN, Chida M, Zamzam T, Trabelsi M (2023) Reinforcement learning based controller for grid-connected PUC PV inverter. In: IECON 2023–49th annual conference of the IEEE industrial electronics society, IEEE, pp 1–6

Alquennah AN, Chida M, Zamzam T, Trabelsi M, Ghrayeb A, Khatri S (2024) Double deep q networks reinforcement learning based controller for grid-connected 7 level PUC inverter. In: 2024 4th international conference on smart grid and renewable energy (SGRE), IEEE, pp 1–6

Babes B, Boutaghane A, Hamouda N (2022) A novel nature-inspired maximum power point tracking (MPPT) controller based on ACO-ANN algorithm for photovoltaic (PV) system fed arc welding machines. Neural Comput Appl 34(1):299–317

Bag A, Subudhi B, Ray PK (2019) A combined reinforcement learning and sliding mode control scheme for grid integration of a PV system. CSEE J Power Energy Syst 5(4):498–506

Bahrami M, Khashroum Z (2023) Review of machine learning techniques for power electronics control and optimization. arXiv preprint arXiv:2310.04699

Belousov B, Abdulsamad H, Klink P, Parisi S, Peters J (2021) Reinforcement learning algorithms: analysis and applications

Bose BK (2002) Expert system, fuzzy logic, and neural network applications in power electronics and motion control. Proc IEEE 82(8):1303–1323

Bose BK (2020) Artificial intelligence techniques: How can it solve problems in power electronics?: An advancing frontier. IEEE Power Electron Mag 7(4):19–27

Çakmak R, Bayrak G, Koç M (2025) A fuzzy logic-based energy management approach for fuel cell and photovoltaic powered electric vehicle charging station in DC microgrid operations. IEEE Access

Chen G, Li Z, Zhang Z, Li S (2019) An improved ACO algorithm optimized fuzzy PID controller for load frequency control in multi area interconnected power systems. IEEE Access 8:6429–6447

Chen M et al (2025) Power for AI and AI for power: the Infinite entanglement between artificial intelligence and power electronics systems. IEEE Power Electron Mag 12(1):37–43

Chen P et al (2024) A review on the applications of reinforcement learning control for power electronic converters. IEEE Trans Indus Appl

Cheng D, Liu L, Yu Z (2021) CNN-based intelligent fault-tolerant control design for turbofan engines with actuator faults. IEEE Access 9:28122–28139

Cheok AD, Wang Z (2005) Fuzzy logic rotor position estimation based switched reluctance motor DSP drive with accuracy enhancement. IEEE Trans Power Electron 20(4):908–921

Divakar A, Jacob J (2019) Genetic algorithm based tuning of nonfragile and robust PI controller for PSFB DC-DC converter. In: 2019 international conference on communication and electronics systems (ICCES), IEEE, pp 1846–1851

Fang B, Li X, Han G, He J (2023) Rethinking pseudo-labeling for semi-supervised facial expression recognition with contrastive self-supervised learning. IEEE Access 11:45547–45558

Feng Z, Zhou Z, Yang R, Ban X, Hu C (2023) Fault-tolerant control based on belief rule base expert system for multiple sensors concurrent failure in liquid launch vehicle. Nonlinear Dyn 111(5):4357–4373

Fujimoto Y, Kaneko A, Iino Y, Ishii H, Hayashi Y (2023) Challenges in smartizing operational management of functionally-smart inverters for distributed energy resources: a review on machine learning aspects. Energies 16(3):1330

Gao Y et al (2022) Inverse application of artificial intelligence for the control of power converters. IEEE Trans Power Electron 38(2):1535–1548

Gao Y, Wang S, Dragicevic T, Wheeler P, Zanchetta P (2023) Artificial intelligence techniques for enhancing the performance of controllers in power converter-based systems—an overview. IEEE Open J Indus Appl 4:366–375

Grabowski PZ, Kazmierkowski MP, Bose BK, Blaabjerg F (2002) A simple direct-torque neuro-fuzzy control of PWM-inverter-fed induction motor drive. IEEE Trans Industr Electron 47(4):863–870

Gros I-C, Lü X, Oprea C, Lu C, Pintilie L (2023) Artificial intelligence (AI)-based optimization of power electronic converters for improved power system stability and performance. In: 2023 IEEE 14th international symposium on diagnostics for electrical machines, power electronics and drives (SDEMPED), IEEE, pp 204–210

Hajihosseini M, Andalibi M, Gheisarnejad M, Farsizadeh H, Khooban M-H (2020) DC/DC power converter control-based deep machine learning techniques: real-time implementation. IEEE Trans Power Electron 35(10):9971–9977

Hannan M et al (2020) Role of optimization algorithms based fuzzy controller in achieving induction motor performance enhancement. Nat Commun 11(1):3792

Hannan M, Abd Ali J, Mohamed A, Uddin MN (2016) A random forest regression based space vector PWM inverter controller for the induction motor drive. IEEE Trans Indus Electron 64(4):2689–2699

Harrag A, Messalti S (2015) Variable step size modified P&O MPPT algorithm using GA-based hybrid offline/online PID controller. Renew Sustain Energy Rev 49:1247–1260

Hemachandu P, Veera Reddy V, Kusuma N, Mohan Reddy D, Divya Prasad P (2017) A unique tuning of PID controller using particle swarm optimizer for modern multilevel inverter fed micro-grid system. In: Emerging trends in electrical, communications and information technologies: proceedings of ICECIT-2015, Springer, pp 415–424

Hosseinzadehtaher M, Fard AY, Shadmand MB, Fajri P (2021) Artificial intelligence inspired model predictive control for frequency regulation in power electronics dominated grids. In: 2021 IEEE 12th international symposium on power electronics for distributed generation systems (PEDG), IEEE, pp 1–6

Hua C, Li L, Ding SX (2021) Reinforcement learning-aided performance-driven fault-tolerant control of feedback control systems. IEEE Trans Autom Control 67(6):3013–3020

Huang H, Crouch DL, Liu M, Sawicki GS, Wang D (2016) A cyber expert system for auto-tuning powered prosthesis impedance control parameters. Ann Biomed Eng 44:1613–1624

Kececioglu OF, Gani A, Sekkeli M (2020) Design and hardware implementation based on hybrid structure for MPPT of PV system using an interval type-2 TSK fuzzy logic controller. Energies 13(7):1842

Kermansaravi A et al (2024) Reinforcement learning based control of grid-connected PUC5 inverter. In: IECON 2024–50th annual conference of the IEEE industrial electronics society, IEEE, pp 1–6

Khan MA, Yousaf MZ, Luo M, Wenhuan W, Mirsaeidi S, Muttaqi KM (2023) AI-driven optimization approach for enhanced performance of power converters through efficient controller tuning. In: 2023 IEEE International conference on energy technologies for future grids (ETFG), IEEE, pp 1–5

Khandelwal A, Kumar J (2019) Applications of AI for power electronics and drives systems: a review. 2019 Innovations in power and advanced computing technologies (i-PACT), vol 1, pp 1–6
Kim DG (2023) Artificial intelligence and agent based modeling for power system engineering. J Mach Comput 3(3):340–350
Kumar S, Krishnasamy V, Neeli S, Kaur R (2020) Artificial intelligence power controller of fuel cell based DC nanogrid. Renew Energy Focus 34:120–128
Li J, Lewis FL, Fan J (2023) Reinforcement learning: optimal feedback control with industrial applications. Springer Nature
Lin F-J, Teng L-T, Yu M-H (2008) Radial basis function network control withimproved particle swarm optimizationfor induction generator system. IEEE Trans Power Electron 23(4):2157–2169
Liu X et al (2021) Self-supervised learning: generative or contrastive. IEEE Trans Knowl Data Eng 35(1):857–876
Lucia S, Navarro D, Karg B, Sarnago H, Lucia O (2020) Deep learning-based model predictive control for resonant power converters. IEEE Trans Industr Inf 17(1):409–420
Mendonça LF, Sousa J, Da Costa JS (2012) Fault tolerant control using a fuzzy predictive approach. Expert Syst Appl 39(12):10630–10638
Mohammadzadeh M, Akbari E, Salameh AA, Ghadamyari M, Pirouzi S, Senjyu T (2022) Application of mixture of experts in machine learning-based controlling of DC-DC power electronics converter. IEEE Access 10:117157–117169
Phanden RK, Sharma L, Chhabra J, Demir Hİ (2021) A novel modified ant colony optimization based maximum power point tracking controller for photovoltaic systems. Mater Today Proc 38:89–93
Qashqai P, Babaie M, Zgheib R, Al-Haddad K (2023) A model-free switching and control method for three-level neutral point clamped converter using deep reinforcement learning. IEEE Access 11:105394–105409
Qashqai P, Vahedi H, Al-Haddad K (2019) Applications of artifical intelligence in power electronics. In: 2019 IEEE 28th international symposium on industrial electronics (ISIE), IEEE, pp 764–769
Qashqai P, Zgheib R, Al-Haddad K (2021) GRU and LSTM comparison for black-box modeling of power electronic converters. In: IECON 2021–47th annual conference of the IEEE industrial electronics society, IEEE, pp 1–5
Qashqai P, Al-Haddad K, Zgheib R (2022) Deep neural network-based black-box modeling of power electronic converters using transfer learning. In: 2022 IEEE energy conversion congress and exposition (ECCE), IEEE, pp 1–6
Qashqai P, Zgheib R, Al-Haddad K (2022) A programmatical method for real-time simulation of black-box lstm-based models of power electronic converters in hypersim. In: 2022 IEEE 1st industrial electronics society annual on-line conference (ONCON), IEEE, pp 1–5
Shi J, Zhang W, Zhang Y, Xue F, Yang T (2015) MPPT for PV systems based on a dormant PSO algorithm. Electric Power Syst Res 123:100–107
Simões MG, Bubshait A (2019) Frequency support of smart grid using fuzzy logic-based controller for wind energy systems. Energies 12(8):1550
Srungaram RT, Yadlapati K (2024) Machine-learning based control of bi-modular multilevel PWM inverter for high power applications. PLoS ONE 19(7):e0305759
Taghizadeh H, Hagh MT (2010) Harmonic elimination of cascade multilevel inverters with nonequal DC sources using particle swarm optimization. IEEE Trans Industr Electron 57(11):3678–3684
Wan Y, Xu Q, Dragičević T (2024) Reinforcement learning-based predictive control for power electronic converters. IEEE Trans Indus Electron
Wang S, Dragicevic T, Gontijo GF, Chaudhary SK, Teodorescu R (2020) Machine learning emulation of model predictive control for modular multilevel converters. IEEE Trans Industr Electron 68(11):11628–11634
Yang Z, Zeng J, De La Cruz A (2023) Learning and outperforming the model predictive control with a linear ε-support vector machine for power converter. IEEE Trans Industr Electron 71(1):5–13

Yılmaz M, Kaleli A, Çorapsız MF (2023) Machine learning based dynamic super twisting sliding mode controller for increase speed and accuracy of MPPT using real-time data under PSCs. Renew Energy 219:119470

Zhang X, He J, Ma H, Ma Z, Ge X (2023) Stability enhancement methods of inverters based on lyapunov function, predictive control, and reinforcement learning. Springer

Zhao S, Blaabjerg F, Wang H (2020) An overview of artificial intelligence applications for power electronics. IEEE Trans Power Electron 36(4):4633–4658

Machine Learning Control in Power Electronics

Özgür Çelik, Necdet Sinan Özbek, and Ahmet Teke

Abstract Power electronic devices are fundamental components of modern technological developments due to play a critical role in management, conversion, and control of electrical energy. In the context of cutting-edge technology, these devices are crucial in enabling a wide range of applications. To accomplish such a pivotal role, a paradigm shift is also required in the control of power electronic devices. In this manner, advanced control of power electronic devices has a major role for achieving high dynamic performance, minimizing energy losses, enhancing system robustness, facilitating the integration of renewable energy sources, and smart grid technologies. Therefore, machine learning (ML) control based strategies have prominent features in control of power electronic systems. The main objective of this chapter is to provide a thorough and in-depth analysis of ML methods that applicable to the control of power electronic systems. Through detailed analysis, the significant potential of these approaches in advancing the future development of control and optimization in power electronic systems is explained.

Keywords Power electronics · Control of power converters · Machine learning · Adaptive control

Ö. Çelik (✉) · N. S. Özbek
Department of Electrical and Electronic Engineering, Adana Alparslan Türkeş Science and Technology University, Adana, Türkiye
e-mail: ozgurcelik@atu.edu.tr

N. S. Özbek
e-mail: nozbek@atu.edu.tr

A. Teke
Department of Electrical-Electronics Engineering, Çukurova University, Adana, Türkiye
e-mail: ahmetteke@cu.edu.tr

E. Kabalci (ed.), *Modern Control Methods for Power Electronics*, Energy Systems in Electrical Engineering, https://doi.org/10.1007/978-981-92-1974-2_13

1 Chapter Overview

This chapter investigates the use of machine learning based control mechanisms in power electronic systems, with an emphasizes on supervised and reinforcement learning techniques. The potential of machine learning based control to enhance system reliability and efficiency by adapting to operate under variable operating conditions, optimizing control parameters, and predicting failures are highlighted.

2 Introduction

Electrical energy is primarily managed and transformed by power electronic devices that use solid-state components. These technologies are used in many different fields, such as advanced manufacturing systems, energy storage management, sustainable transportation, smart grid technologies, and industrial automation. Between sources and consumers, power electronics systems serve as the vital interface in the form of power conditioning units. These cutting-edge systems handle electrical energy conversion, control, and management. In an effort to ensure high levels of reliability and efficiency, power electronic converters have been implemented across low-, medium-, and high-power applications (Lopez-Santos et al. 2021). In this way, power electronic converters and their control systems have been continuously improving due to the evolving energy landscape. Power circuit topologies and control strategies can be considered as two major area to investigate the primary developments in power electronic converters (Zhang et al. 2018). In Fig. 1, each type, namely DC–DC, AC–DC, DC–AC, and AC–AC converters, is depicted (Maroti et al. 2022).

In recent advancements, many power electronic converter topologies, which play a significant role in achieving efficient, reliable, and flexible operation, have been designed. In parallel with these developments, control methods are also gaining great importance. In the current literature, many conventional control techniques for power electronic converters are designed to ensure stable and efficient operation.

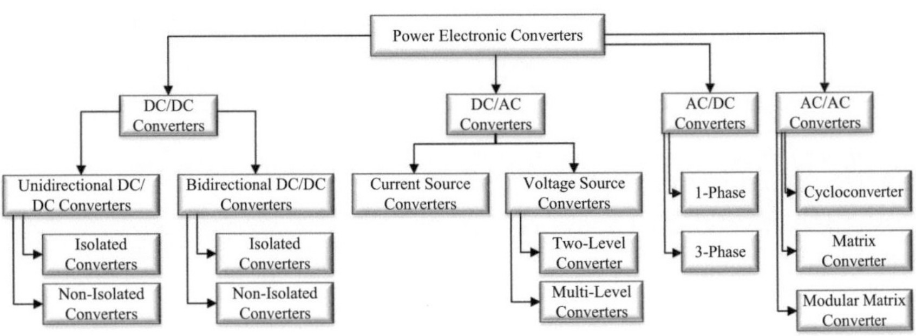

Fig. 1 Main categories for power electronic converters

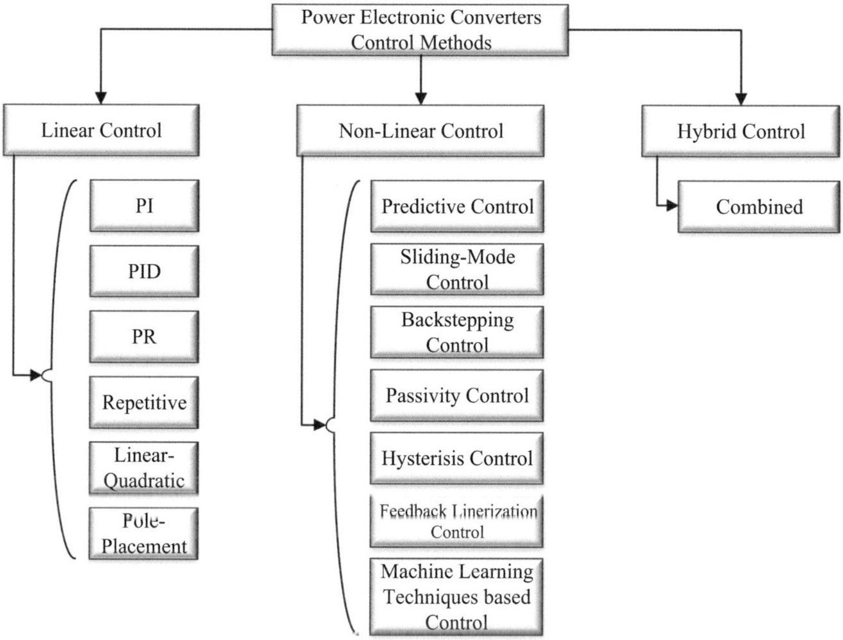

Fig. 2 Control methods for power electronic converters

These control techniques encompass a range of techniques such as linear control and nonlinear control. The power electronic converters control methods can be mainly classified as illustrated in Fig. 2 (Aghdam et al. 2020).

Linear control methods are grounded in classical feedback control theory and widely used in the regulation of power electronic converters due to their simplicity, robustness, and ease of implementation. These methods, mainly including proportional–integral (PI), proportional–integral–derivative (PID), and proportional-resonant (PR), are developed according to the dynamics and characteristics of linear systems (Boscaino et al. 2024). These methods are commonly utilized in commercial and industrial applications. To design these controller structures, the system dynamics are assumed to be properly represented by linear time-invariant (LTI) models which allow for analytical design and stability analysis using well-established techniques such as bode plot or root locus (Chao and Blaabjerg 2021). The tasks, consisting of current and voltage regulation, can be achieved sufficiently by PI controllers owing to simplicity and effectiveness in regulating output under steady-state conditions (Çelik 2025). The transfer function of the PI controller is given as

$$G_{pi}(s) = k_p + \frac{k_i}{s} \tag{1}$$

where k_p and k_i are the proportional and integral gains, respectively.

PR controllers are especially effective in tracking sinusoidal references and are primarily preferred in AC systems. The transfer function of the ideal PR controller can be obtained through shifting integrator of PI controller in positive and negative fundamental frequency as shown in Chao and Blaabjerg (2021)

$$G_{PR}(s) = k_p + \frac{2k_i s}{s^2 + w^2} \tag{2}$$

where k_p, k_i are the proportional and integral gains, respectively, and w is fundamental frequency.

To obtain high-precision tracking of periodic signals and disturbance rejection capability, repetitive control is utilized for periodic signals such as current, voltage, or power waveforms. A repetitive controller typically uses a delay-line filter to store the periodic signal characteristics. The general repetitive controller in the continuous domain can be represented as (İnci and Çelik 2025)

$$G_{RC}(s) = \frac{Q(s)e^{-sT_o}}{1 - Q(s)e^{-sT_o}} \tag{3}$$

where T_o represents delay element and $Q(s)$ is a filter.

On the other hand, linear quadratic regulator (LQR) and pole-placement control are among the widely used linear control methods. LQR is a linear optimal control method derived from state-space representations. It involves formulating a linear system model and defining a quadratic cost function to evaluate control performance. Basically, used for stabilizing a time-invariant linear system to the origin. By minimizing this cost function, an optimal feedback gain matrix is obtained (Maghfiroh et al. 2022). In pole-placement control, the system considered is completely measurable and controllable. This control approach is based on feedback control theory and allows the designer to assign the closed-loop poles of a system to desired locations in the complex plane. In this manner, the steady-state and dynamic performance criteria such as settling time, over/undershoot, or stability margins can be adjusted (Zhang et al. 2023). The linear methods bearing advantages such as simplicity, low computational requirements, and ease of implementations. However, limited performance in nonlinear systems, sensitivity to model parameters, fixed gain limitations, and low performance under disturbances make these control methods disadvantageous.

Compared to linear control methods, nonlinear methods offer better performance of power electronic converters across a wide range of operating conditions and disturbances. Among the nonlinear control methods, Model Predictive Control (MPC), despite being more complex, offers superior performance by predicting future system behavior and optimizing control actions over a finite horizon (Hu et al. 2021). Another strong technique that is well-known for its ability to withstand changes in parameters and outside disruptions is Sliding Mode Control (SMC), which offers superior dynamic response in highly nonlinear systems. Despite their differences in performance and complexity, these control strategies serve as the cornerstone for managing power, voltage, and current flow in a broad range of power electronic applications (Mousavi et al. 2022). On the other hand, passivity control proposes better performance in terms of coping with the instability challenge, external disturbances, and parameter variations. Passivity implies that the real component of the output impedance or admittance of a voltage source converter (VSC) is greater than or equal to zero in the frequency domain (Chopra et al. 2022). Backstepping control is a recursive, Lyapunov-based nonlinear control technique particularly suited for nonlinear systems. The primary strengths of backstepping control are its capability for tackling with unmatched disturbances and uncertainties through the integration of adaptive or robust components. The main consideration for backstepping control lies behind selecting a proper Lyapunov function in line with the control targets (Li and Tong 2024). In the feedback linearization method, the nonlinear system is transferred to a fully or partially decoupled linear system (Meng et al. 2022). The hysteresis control is simple in comparison with the aforementioned nonlinear control methods. However, this method is sensitive to low power levels and it leads to high switching losses and resonance problems (Cheng et al. 2021). Although these methods provide promising performance for power electronic converters, design of these methods is strongly based on precise mathematical models, accurate parameter knowledge, structured disturbance characteristics, and defined operating limits. An accurate mathematical model requirement due to model dependency of these methods, dynamic response under parameter variations and disturbances still remains as a significant issue. In this context, machine learning-based control methods that can adaptively learn and respond to complex system dynamics, nonlinearities, and uncertainties without requiring exact mathematical formulations are proposed in the literature.

Machine learning-based methods, such as supervised, unsupervised, or reinforcement learning are highly capable of handling high-dimensional state spaces and variations in system operating parameters more flexible when compared to aforementioned controllers. They can also optimize performance over time through adaptive learning, making them well-suited for applications with disturbances and nonstationary behaviors for systems including industrial automation, sustainable transport, smart grid technologies, energy storage management, and advanced manufacturing systems. The promising aspects of the machine learning-based methods that provide a data-driven, model-free, or model-light approach are widely used to form control structure robust to external disturbances and uncertainties caused by parameters. Therefore, many machine learning-based control approaches are proposed in the literature to optimize or form the control structures.

3 Machine Learning Basics for Control Applications

Machine learning (ML) has revolutionized control engineering by enabling adaptive mechanism, data-based decision-making for complex dynamical systems. Traditional control techniques such as PID control and optimal control are based on the premise of having adequately formulated mathematical models. However, real-world systems of interest tend to be nonlinear, uncertain, and time-varying, making the traditional approaches inapplicable. Compared to traditional control approaches that utilize heavily analytical models derived from physical principles, ML provides data-driven approaches that possess the ability to handle unmodeled dynamics, uncertainty, high-dimensional systems, and nonlinearity. In this context, ML techniques provide a stage for directly learning the control policies from data with less need for explicit modeling and increased robustness and flexibility.

From control system point of view, ML provides techniques to define how information is abstracted, represented, and utilized. Rather than relying solely on traditional classification of learning schemes, control based applications demand a critical understanding of how each learning mode contributes to the design and adaptation of controllers under uncertainty.

This section introduces the theoretical foundation of ML for control applications with the focus on unsupervised and supervised learning, reinforcement learning, and neural networks.

3.1 Function Approximation and Universal Representability

Function approximation lies at the heart of ML-based control. Consider a nonlinear control-affine system:

$$\dot{x}(t) = f(x(t)) + g(x(t))u(t), x \in \mathbb{R}^n, u \in \mathbb{R}^m \tag{4}$$

In many practical settings, f and g are partially known or unknown. ML models—particularly artificial neural networks (ANNs), radial basis functions (RBF), and Gaussian processes (GPs)—are employed to approximate these functions from data. For parametric models $\hat{f}_\theta(x)$, the learning objectives becomes

$$min_\theta \sum_{i=1}^{N} \|\dot{x}_i - \hat{f}_\theta(x_i)\|^2 \tag{5}$$

This leads to optimization landscapes that are typically nonconvex, and gradient-based solvers, such as stochastic gradient descent (SGD) or Adam, are used for convergence.

3.2 Stability and Safety Under Learning-Based Control

Closed-loop stability of the ML-based power electronic systems is yet another challenge. The feedback interconnection of a learned policy π_θ with a physical plant can induce instabilities due to approximation errors and distributional shift. Control strategies to address this include:

Lyapunov-constrained training, where policies are trained to satisfy

$$\dot{V}(x) = \nabla V(x)^T (f(x) + g(x)\pi_\theta(x)) < 0 \tag{6}$$

Barrier functions, wherein learning-based control laws are regularized using control barrier functions to guarantee constraint satisfaction. Safe Reinforcement Learning (RL): Algorithms are designed to explore without violating safety constraints, often using reachability analysis or conservative value iteration. Recent works also explore contraction metrics, dissipativity, and incremental stability as alternative tools to assess ML-controlled systems.

3.3 Overview of Supervised, Unsupervised, and Reinforcement Learning

A unifying framework is to view supervised, unsupervised, and reinforcement learning in terms of constructing approximators, observers, and policies. The classification of ML algorithms and their applications is presented in Fig. 3 (Qashqai et al. 2023a).

Supervised Learning: Formally, given a dataset $D = \{(x_i, y_i)\}_{i=1}^N$, supervised learning aims to find a function $f : \mathbb{R}^n \to \mathbb{R}^m$ that minimizes a risk functional $R(f) = \mathrm{E}_{(x,y)\backsim D}[\ell(f(x), y]$, where ℓ is a the suitable loss. In control applications, this formulation is widely employed for system identification, inverse dynamics modeling, and policy imitation.

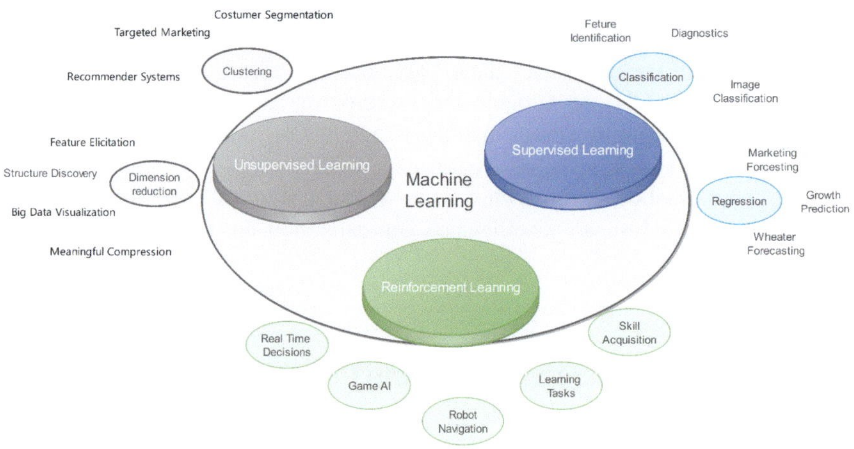

Fig. 3 Machine learning algorithms and their application

Unsupervised Learning: This paradigm seeks latent representations or structures in unlabeled data $\{x_i\}_{i=1}^{N}$. Methods such as Principal Component Analysis (PCA), clustering, and autoencoders are used in control for dimensionality reduction, sensor fusion, and anomaly detection.

On the other hand, RL involves an agent interacting with an environment (S, A, P, R, γ) aiming to find a policy $\pi : S \rightarrow A$ that maximizes the expected discounted return $E[\sum_{t=0}^{\infty}\gamma^t R(s_t, a_t)]$. RL is suited to deal with optimal design of feedback and adaptive control in stochastic and unknown environments. The core principle of RL techniques is to achieve a desired result by organizing the reward or punishment of an event. The reward values determined for each combination are tabulated in the table created for the possible state vector and control signal values, and the control signal that will produce the desired behavior is chosen from the states with the highest reward value (Özbek 2019). This learning strategy is applicable both model-based and model-free case (Jang et al. 2019).

3.4 *Key Algorithms Relevant to Control (e.g., Decision Trees, Neural Networks, Q-Learning, Deep RL)*

It is important to emphasize that the ML algorithms available for control applications have different advantages and disadvantages. To mention a few, decision trees and their ensemble versions, namely random forests and gradient boosted trees, are interpretable and computationally inexpensive, making them suitable for fast decision-making problems. However, their efficiency is reduced for high-dimensional continuous control issues. Neural networks, on the other hand, can represent complex control policies thanks to their complex nonlinear dynamic structure, but they require large amounts of training data as well as careful regularization to avoid overfitting. Gaussian processes provide a probabilistic method that is particularly useful in safety–critical applications as they can represent predictive uncertainty, but they scale poorly with the amount of data. Value-based methods like Q-learning and deep Q-network (DQN) are suitable for discrete-time and discrete-action control systems, while policy optimization methods like PPO, DDPG, and SAC excel in high-dimensional continuous control environments by combining policy optimization with function approximation. The selection of a particular algorithm involves trade-offs between interpretability, sample efficiency, generalization, and computation expense (Table 1).

3.5 *Data Requirements and Training Considerations*

The successful realization of ML for control tasks depends on high-quality attention to data preparation, algorithm training protocols, and operating restrictions. The

Table 1 Comparison of the several ML algorithms

Algorithm	Description	Strengths	Limitations
Decision Trees and Ensembles	Tree-based models including Random Forests and Gradient Boosted Trees	Interpretable, efficient on tabular data	Poor performance in high-dimensional continuous control
Artificial Neural Networks (ANNs)	Deep networks for nonlinear function approximation	Flexible, handles high-dimensional input, expressive	Requires large datasets, sensitive to overfitting
Gaussian Processes (GPs)	Bayesian regression model with uncertainty quantification	Probabilistic predictions, confidence intervals	Computationally expensive for large-scale data
Q-Learning and Deep Q-Networks (DQN)	Value-based RL for discrete control	Stable for low-dimensional, discrete systems	Limited to discrete action spaces
Deep RL (PPO, DDPG, SAC)	Policy optimization with function approximation for continuous control	Effective in complex, high-dimensional control environments	High sample complexity, challenging hyperparameter tuning

training data set must first deliver rich and diverse coverage across the relevant state-action space so as to avoid errors of extrapolation as well as ensure generalizability. Imperfect representation of states of interest may lead to unpredictable or unsafe closed-loop behavior. The exploration–exploitation dilemma is a central theme in reinforcement learning: agents must balance best learning of novel information with optimization of learned control policies to avoid converging to suboptimal policies. Overfitting is also a risk, particularly with the use of overparameterized neural models.

Computational resources, training procedures, and model complexity must all be consistent with the underlying control goals of an efficient ML infrastructure for control. From data-driven modeling to adaptive decision-making, machine learning provides strong tools for control applications. Assuring stability, robustness, and real-time performance are important challenges. Formal safety guarantees and hybrid model-based/data-driven control are potential avenues for future research.

4 Integration of Machine Learning into Power Electronics Control

Power electronics systems are redefined by the incorporation of ML in terms of data analysis, control, and adaptation of the power delivery mechanism, as well as the ability to monitor the systems and make the required adjustments in real-time performance to guarantee increased efficiency. The highly efficient nature of ML-based control systems solves the processes of tracking data, learning from data, pattern recognition, and gaining actionable insights through deep learning and RL. According to Hemavathi et al. (2023), ML has the potential to transform several industries by promoting sustainability goals, making better resource utilization, and improving decision-making processes.

ML-integrated systems rely on complex algorithms to determine the most efficient power electronics applications, optimize the energy in machines, and thereby increase the efficiency of system (Tajjour and Shyam 2023). Furthermore, predictive models can be used to manage solar and wind energy applications, allowing for high-energy output at specified periods. These models make energy purchase planning easier, ensuring that energy sources are used appropriately. Furthermore, deep learning strategies improve the operational performance, system management, and optimization of power system, which contributes to the development of smart grids and makes possible collaborative energy-sharing initiatives (Brunton and Kutz 2019).

ML control based power electronics systems play a critical part in smart grid functioning. Moreover, the utilization of power electronic converters in distributed generation, electrification, and integration into the grid is also growing with the development of renewable energy installations. The grid reliability may come down due to the uncertainties and disturbances (Özbek and Çelik 2022). Thus, overall power processing must be improved in terms of reliability and efficiency. One approach to accomplish this is to increase health by smart monitoring and making the controller fault resistant.

4.1 ML-Based Control Architectures: Model-Based versus Model-Free

Learning based control is classified as model-free and model-based approaches for achieving optimal system performance in the presence of uncertainties and disturbances. Among them, model-free methods directly learn control policies or value functions through interaction with the environment, without requiring a system model. Prominent examples include Q-learning, Deep Q-Networks (DQN), and actor–critic methods. These algorithms are attractive due to their general applicability and independence from prior knowledge; however, they typically require extensive sampling, leading to high data and computational demands.

4.2 Adaptive Control Using Supervised Learning

Supervised learning offers useful tools for developing adaptive controllers through offering real-time system identification and function approximation. For adaptive control, parameters or dynamics of the system can change over time due to environmental changes, aging of actuators, or modeling errors. Adaptation of control laws based on such variations can be achieved through approximating them with supervised learning algorithms using measured input–output data. For instance, kernel regression models or neural networks trained on labeled trajectory can learn to predict system dynamics or inverse dynamics, which are subsequently incorporated in control policies. Such integration ensures that the control policy adjusts according to the altered model, thereby providing robustness with respect to uncertainty and improved tracking performance. These adaptive plans are optimal for incremental and online learning variants, which permit ongoing model estimate updates without discarding everything in retraining.

4.3 Reinforcement Learning for Real-Time Decision-Making

RL approaches assign a reward or punishment value based on whether the amount of error produced is less or greater than the amount of error made in the previous phase. According to this determined reward or penalty score, system learning continues, and the parameters that will offer the optimal control are attempted to be attained (Stavrev and Ginchev 2024). As a consequence of determining these characteristics, a state-action chart will be generated, which will be utilized for control. Reinforcement learning methods are divided into model-based and model-free learning algorithms. For example, the Value Iteration model falls within the category of model-based algorithms. Ref (Jang et al. 2019) indicate that the Q-Learning, Sarsa, and Sarsa (λ) algorithms do not require a system model to function. These paradigms capture different levels of observability and structural assumptions, with varying convergence guarantees and computational constraints (Jang et al. 2019).

4.4 Hybrid Approaches

Yet another hybrid ML paradigm called Physics-Informed Neural Networks (PINNs) incorporates physical concepts directly into the learning process. These concepts are frequently expressed as partial differential equations (PDEs). PINNs improve model generalizability, data efficiency, and interpretability by restricting neural network predictions to follow physical relations, in contrast to data-driven models. When physical equations like Kirchhoff's laws, differential switching dynamics, or electromagnetic relationships are known but may be impacted by modelling errors. In this context, PINNs are very useful for power electronics applications in control systems and modeling situations. For instance, PINNs can be utilized to model grid-connected systems, inverters, or switching converters by integrating the governing circuit equations directly into the loss function.

Convolutional neural networks (CNNs) and recurrent neural networks (RNNs) can be utilized for predictive maintenance, including the detection of system faults prior to their occurrence. Meanwhile, deep learning (DL) technologies improve the working efficacy of power electronic systems, facilitating the deployment of smart grids and increasing the possibility for collaborative energy-sharing programs. DL models can analyze longer histories of input data, producing more accurate results than previous approaches. However, they require a considerable amount of data for training, which can be time-consuming to do in practice. Because of advances in deep learning and increased processing capability, RL-based optimization techniques have emerged as viable tools. Table 2 describes the ML algorithms and their applications in power electronics.

Table 3 evaluates the ML-based control approaches utilized in power electronics.

5 Case Studies and Applications

The integration of machine learning techniques into power electronic converters control systems represents a paradigm shift in the design and operation of modern power converters (Çelik et al. 2022). The primary objective of these methods is defined as handling fixed models and predefined parameters, which can significantly limit performance under dynamic or uncertain conditions. The parameter optimization or evolving control scheme is possible by using the machine learning methods (Yorat et al. 2023). The general block diagram for machine learning-based power electronics control is given in Fig. 4.

There are many studies conducted in the current literature. In Ref. (Marahatta et al. 2022), the authors proposed a nonlinear model predictive control by utilizing reinforcement learning for DC/DC boost converter topology. The PWM duty cycle control is performed based on load impedance and integral compensators are used. The main contributions can be summarized as reduced transient oscillations and better settling time that verified experimentally. In Ref. (Mahesh et al. 2023), support vector regression-based MPPT method is employed to generate duty cycle for controlling DC/DC boost converter. Better maximum power tracking efficiency and stability are obtained under variable operating conditions. In Ref. (Bindi et al. 2022), multilayer neural network with multi-valued neurons and support vector machine techniques are utilized for monitoring the health status of passive components of the zeta converter. Classification rate higher than 90% is obtained for converter health monitoring. In Ref. (Mohammadzadeh et al. 2022), hybrid control combining feedback control and model predictive control using mixture of experts with neural networks are applied experimentally to a DC/DC boost converter. The experimental results validate the reduced steady-state error and settling time. In Ref. (Harison et al.

Table 2 Evaluation of ML algorithms with their power electronics applications

Algorithm	Application in power electronics	Strengths	Limitations
Probabilistic Inference for Learning Control (PILCO)	Learning-based inverter control and MPPT	Uncertainty quantification, efficient learning from few samples	Computationally intensive, limited scalability
Probabilistic Ensembles with Trajectory Sampling (PETS)	Power converter switching strategy planning	Robustness via ensemble dynamics, trajectory-aware planning	Requires careful tuning, sensitive to ensemble diversity
Model-Based Policy Optimization (MBPO)	DC–DC converter regulation, real-time control optimization	Combines sample efficiency with robust learning	Model training instability, higher algorithmic complexity
Dyna: A hybrid RL algorithm that integrates model learning with simulated experience	Fault-tolerant inverter control	Uses simulated experience for faster learning	Quality of learned model crucial for effectiveness
MPC with NN	Grid-connected converter control	Predictive and adaptive under nonlinear conditions	Dependent on accuracy of NN model, real-time feasibility concerns
Proximal Policy Optimization (PPO): A policy-gradient reinforcement learning algorithm designed for reliable policy updates	Switching control in multilevel converters	Stable training, suitable for continuous control tasks	High sample demand, policy tuning required
DDPG: Deep Deterministic Policy Gradient DDPG	Active power balancing in microgrids	Suitable for high-dimensional control problems	Exploration safety and training instability
Safe RL	Grid-forming inverter control under uncertainty	Ensures safe exploration, robustness to disturbances	Conservative policies, lower exploration efficiency
Model Predictive Path Integral (MPPI): A stochastic model predictive control method used in real-time continuous control	Real-time optimal control of power converters	Continuous-time path optimization, handles stochastic noise	Requires access to system model, tuning of noise parameters
Online NN	Adaptive output voltage control in DC–DC converters	Continuous learning and adaptation, low memory requirements	Risk of drift without periodic reset, sensitivity to noise

(continued)

Table 2 (continued)

Algorithm	Application in power electronics	Strengths	Limitations
Kernel Regression	Load forecasting and adaptive impedance estimation	Non-parametric learning, good for small data	Computationally expensive at inference, less suitable for high-frequency control
PINNs	Converter and inverter modeling, thermal prediction, grid interaction	Physically consistent learning, improved generalization, low data requirements	Sensitive to PDE complexity, training instability, hyperparameter tuning

Table 3 Evaluation of ML-based control approaches

Approach	Core concept	Advantages	Limitations	Examples
Model-Free	Learn policy or value function from interaction	Simple, model-agnostic, flexible	High sample complexity, less interpretability	Q-learning, DQN, PPO, SAC
Model-Based	Learn a system model	Sample efficient, allows for planning and prediction	Sensitive to model errors, requires accurate dynamics	PILCO, PETS, MPC with NN model
Adaptive	Use supervised learning to track and adapt to changing dynamics	Robust to parametric variation, improves tracking	Requires labeled data, sensitive to noise	Online NN, Kernel regression
RL	Reinforcement learning with low-latency deployment	Enables fast, autonomous decision-making	Risky exploration, safety–critical adaptations needed	PPO, DDPG, Safe RL, MPPI
Hybrid	Combine learned models with direct policy/ value learning	Data-efficient, balances accuracy and generalization	Increased algorithmic complexity, requires careful design	Dyna, MBPO, PETS

2024), a buck converter is controlled through a scientific machine learning technique combining physics-based modeling with long short-term Memory (LSTM). The main consideration is reducing the training time and data size for training. The experimental and simulation results verify the improved prediction accuracy along with reduced training time. In Ref. (Hajihosseini et al. 2020), deep reinforcement learning using proximal policy optimization is applied to ultra-local model control with sliding mode observer for adaptive tuning of control parameters. The proposed control structure is applied to a DC/DC buck–boost converter. Enhanced voltage

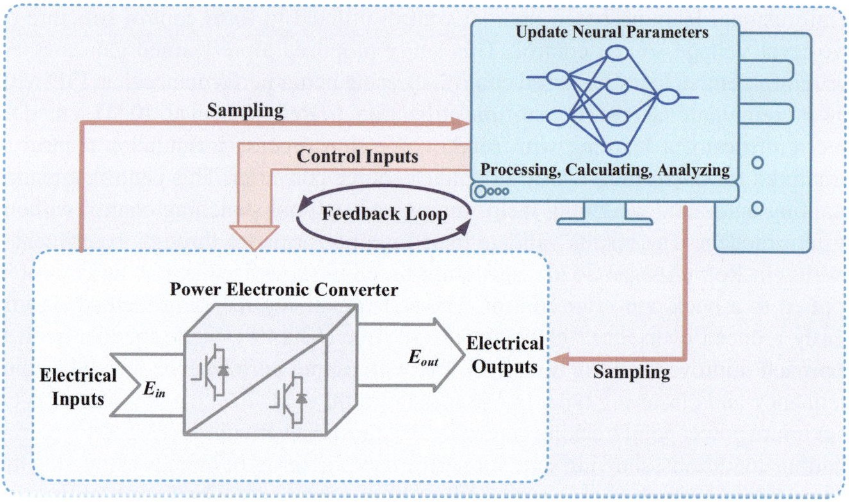

Fig. 4 The general block diagram for machine learning-based power electronics control

stabilization across the reference voltage variations is obtained in real-time performance. In Ref. (Mazaheri et al. 2024), a widely known deep reinforcement learning method, namely proximal policy optimization, is applied to control a buck converter. The obtained experimental and simulation results clearly indicate that the proposed controller shows high adaptability for variable operating conditions, low steady-state error, and enhanced dynamic response. In addition, the real-time applicability of the control structure is illustrated. In Ref. (Li et al. 2025), a bidirectional interleaved DC/DC converter is controlled through a linear quadratic regulator based on Youla parameterized recurrent equilibrium network. The performance of hybrid machine learning neural network model for buck and boost modes based on experimental setup is verified. In Ref. (Srungaram and Yadlapati 2024), a multilayer artificial neural network (ANN) based on reinforcement learning is utilized as a pattern recognition machine learning approach to estimate the maximum power point and forecast hourly photovoltaic (PV) power output. The estimated parameter is used to control bi-modular multilevel inverter. The simulation results show that the proposed intelligent control system, based on machine learning, provides efficient power conversion while ensuring maximum power extraction and supervisory optimal load demand management. In Ref. (Wang et al. 2021), a control structure based on model predictive control emulated by neural network is designed for controlling the modular multilevel converter. The dynamic response and steady-state performance of the system is improved. In addition, computational cost is remarkably reduced. The performance improvements are proved by experimental results. In Ref. (Liao et al. 2023), feedforward neural network-based power flow control is applied for a six-port multi-active-bridge converter. Experimental validation focused on evaluating the machine learning-based controller's performance in both trapezoidal and quasi-resonant operation modes is performed. In Ref. (Alfred et al. 2024), discrete-time reinforcement learning-based optimal control utilized to form control structure of two-level voltage source control. This study proposes a pre-learned gain method for reinforcement learning-based control, offering better performance than PID with lower computational cost for real-time utilization. In Ref. (Wan et al. 2024), a model-free reinforcement learning with Markov decision process formulation control is structured for controlling two-level voltage source converter. This control structure enabling automatic weighting factor tuning and optimal switching control without system models. The results validate their high performance through experimental results. In Ref. (Abegaz 2020), clustering-based unsupervised switching control is applied to a buck converter control. The self-organizing mapping method significantly reduced overshoot (99.77%) and fall time (63.24%), while the hierarchical approach improved settling time by 3.9%, with output performance linked to input frequency and clustering type. In Ref. (Padmavathi et al. 2021), a regression-based maximum power point tracking controller for boost converter is developed for partial shading conditions, showing superior performance in terms of quickly estimating the optimal duty cycle through a strong PV voltage confirmed in both simulations and experiments. In Ref. (Omotoso et al. 2023), a machine learning-based droop control with regression trees that improves voltage and frequency regulation and lowers harmonics in islanded microgrids. In Ref. (Aslay and Ting 2022), linear regression, random forest, and artificial neural network model are developed to predict output current ripple in a PFC AC/DC boost converter for electric vehicle chargers. Artificial neural network-based control scheme outperforming linear regression and random forest methods with high accuracy ($R^2 = 0.9995$), enabling faster, reliable predictions that support safer charging and efficient charger design. In Ref. (Qashqai et al. 2023b), a model-free deep reinforcement learning method for three-level neutral point clamped is proposed. Robust voltage and current control are proved through simulations and experiments. The proposed structure outperforms model predictive control in noisy and uncertain conditions despite long training time. In Ref. (Mahesh et al. 2022), decision tree-based maximum power point tracking control scheme is designed for boost converter. The presented control structure demonstrates strong performance under varying conditions. In Ref. (Zengin 2022), DC bus voltage of dual active bridge converter is regulated by using model-free Q-learning-based reinforcement learning control structure. Better dynamic performance, low-THD response, and better load handling are achieved; however, it requires high memory. In Ref. (Çelik and Teke 2017), a hybrid maximum power point tracking method using artificial neural network is applied for maximum power point tracking control. A hybrid artificial neural network-based maximum power point tracking method is used to control boost converter for a two-stage grid-tied PV system. The obtained results reveal the improvement in tracking speed, efficiency, and power quality under shading and dynamic conditions. In Ref. (Asif et al. 2023), neuro-fuzzy control for adjusting reference operation voltage for controlling a non-inverting DC/DC buck–boost converter. The obtained results show that the presented method achieves fast and accurate tracking performance under dynamic conditions. In Ref. (Rajamallaiah et al. 2024), a deep reinforcement learning-based modified twin delayed deep deterministic policy-gradient algorithm is applied for voltage regulation in buck converters. The performance of the proposed method is validated through real-time implementation. The developed control structure achieves smoother and more stable performance under uncertainties with lower training complexity than deep Q-network and deep deterministic policy-gradient methods. In Ref. (Ghidewon-Abay and Mehrizi-Sani 2023), a long short-term memory-based current control method is developed for three-phase voltage source converter. The simulation results reveal the superior performance and robustness compared to conventional current control methods for grid-connected voltage source converters. In Ref. (Çelik 2024), data-driven methods including three-layer neural network, Gaussian process regression, support vector regression, decision tree regression, and bagging ensemble decision tree-based maximum power tracking control structures are presented for vehicular fuel cell using zeta converter. The findings show optimized power-harvesting capability, increased efficiency, adapting to variable operating conditions, responding to load variations, and ensuring robust and reliable operation of vehicular fuel cells.

The case studies and applications conducted in the current literature are rigorously tabulated in Table 4.

6 Implementation Considerations and Challenges

The integration of machine learning methods for controlling power electronic converters offers significant advantages such as improved transient response, reduced steady-state error, optimal operation, and predictive capabilities; however, it raises several technical issues. The implementation of machine learning control in power electronics involves several critical considerations in terms of real-time performance constraints, data availability and quality, model generalization and robustness, explainability and interpretability, hardware limitations, and cybersecurity risks (Marahatta et al. 2022; Mahesh et al. 2023, 2022; Bindi et al. 2022; Mohammadzadeh et al. 2022; Harison et al. 2024; Hajihosseini et al. 2020; Mazaheri et al. 2024; Li et al. 2025; Srungaram and Yadlapati 2024; Wang et al. 2021; Liao et al. 2023; Alfred et al. 2024; Wan et al. 2024; Abegaz 2020; Padmavathi et al. 2021; Omotoso et al. 2023; Aslay and Ting 2022; Qashqai et al. 2023b; Zengin 2022; Çelik and Teke 2017; Asif et al. 2023; Rajamallaiah et al. 2024; Ghidewon-Abay and Mehrizi-Sani 2023; Çelik 2024). It should be also noted that computational cost, architectural accordance, and reliability aspects are significant aspects that need to be discussed during the design stage. Selecting an appropriate supervised, unsupervised, or reinforcement learning model such as deep neural networks, support vector machine regression, long short-term memory, convolutional neural networks, or reinforcement learning agents must be compatible with the control task, whether it be real-time electrical parameter regulating, dynamic load balancing, fault prediction, or maximum power point tracking. The challenges for implementation of machine learning control of power electronic converters can be explained as follows:

– *Real-Time Performance Constraints*

Table 4 The case studies and applications conducted in the current literature

Ref	Converter topology	ML/optimization methods	Case studies and applications	
			Simulation	Experimental
Marahatta et al. (2022)	Boost converter	Reinforcement Learning	✓	✓
Mahesh et al. (2023)	Boost converter	Support Vector Machine Regression	✓	×
Bindi et al. (2022)	Zeta converter	Multilayer Neural Network with Multi-valued Neurons, Support Vector Machine	✓	×
Mohammadzadeh et al. (2022)	Boost converter	Mixture of Experts with Neural Networks for feedback controller and model predictive controller	✓	✓
Harison et al. (2024)	Buck converter	Scientific Machine Learning combining Physics-Based Modeling with Long short-term memory	✓	✓
Hajihosseini et al. (2020)	Buck–boost converter	Deep Reinforcement Learning using Proximal Policy Optimization	✓	✓
Mazaheri et al. (2024)	Buck converter	Proximal Policy Optimization; actor–critic neural network architecture	✓	✓
Li et al. (2025)	Bidirectional interleaved DC/DC converter (IBC)	Youla REN (hybrid machine learning neural network model)	✓	✓
Srungaram and Yadlapati (2024)	Bi-modular multilevel inverter (BMI)	Multilayer Reinforcement learning	✓	✓
Wang et al. (2021)	Modular multilevel converter (MMC)	Machine Learning MPC	✓	✓
Liao et al. (2023)	A six-port multi-active-bridge (MAB) converter	Feedforward neural network	✓	✓
Alfred et al. (2024)	Buck converter (used for validation)	Reinforcement Learning (actor–critic, value iteration, policy iteration)	✓	×
Wan et al. (2024)	Two-level voltage source converter	Reinforcement Learning with Markov Decision Process formulation	✓	✓

(continued)

Table 4 (continued)

Ref	Converter topology	ML/optimization methods	Case studies and applications	
			Simulation	Experimental
Abegaz (2020)	Buck converter	Unsupervised ML: Self-organizing mapping, K-means, Gaussian mixture model, Hierarchical Clustering	✓	✓
Padmavathi et al. (2021)	Boost converter	Linear Regression with Gradient Descent	✓	✓
Omotoso, et al. (2023)	Grid-forming inverters	Regression Tree	✓	×
Aslay and Ting (2022)	Interleaved boost converter	Linear Regression, Random Forest, Artificial Neural Network	✓	✓
Qashqai et al. (2023b)	Three-level neutral point clamped (NPC) converter	Deep Reinforcement Learning	✓	✓
Mahesh et al. (2022)	Boost converter	Decision Tree and Cuckoo search	✓	×
Zengın (2022)	Dual active bridge (DAB) converter	Model-free Q-learning-based Reinforcement Learning compared to PI controller	✓	×
Çelik and Teke (2017)	Interleaved Boost Converter	Hybrid MPPT method based on Perturb & Observe and Artificial Neural Network	✓	×
Asif et al. (2023)	A non-inverting buck–boost converter	Neuro-fuzzy network for reference voltage estimation	✓	✓
Rajamallaiah et al. (2024)	Buck converter	A Deep Reinforcement Learning-based modified Twin Delayed Deep Deterministic Policy Gradient algorithm	✓	✓
Ghidewon-Abay and Mehrizi-Sani (2023)	Three phase Voltage source converter	Long Short-Term Memory	✓	×
Çelik (2024)	Zeta converter	Data-driven methods	✓	×

Given the stringent timing requirements of power electronic systems due to operating at high switching frequencies (kHz–MHz), where control loops often operate within microseconds, the computational latency of machine learning models must be minimized. Moreover, computational intensive under strict timing constraints requires specialized hardware. This demands the use of lightweight architectures or hardware acceleration platforms such as field-programmable gate arrays (FPGAs), system-on-chips (SoCs), or tensor processing units (TPUs).

– *Data quality*

To cover all system dynamics accurately under various operating conditions, creating a sufficient real dataset is significant. The dataset must be relevant, balanced, diverse, and labeled.

– *Generalization and Robustness*

To ensure model robustness, training on diverse, high-fidelity datasets that handle a wide range of operating conditions, including noise, disturbances, nonlinearities, and fault events, is required. This can lead to safe and stable operation.

– *Hardware Limitations*

The processing and memory limits can highly affect the performance of the controller. Utilization of complex machine learning methods may require specialized upgraded microcontrollers.

– *Interpretability*

The lack of interpretability often makes these models hard to be applied to many practical applications. Owing to being "black box" models, mathematical proofs, especially in terms of stabilization, are relatively difficult.

– *Reliability and Safety*

Applying machine learning techniques to critical systems including grid-connected applications, aerospace, electric vehicles, and microgrids requires safe operation criteria such as thermal stress testing, accelerated aging tests, fault tolerance, and cybersecurity for operational trust.

– *Cybersecurity*

These control methods that are especially used in grid-connected systems and microgrids are susceptible to cyber-attacks. There should be a robust defense mechanism to protect the safe operation of these systems.

There are continued advances in embedded machine learning platforms and edge computing; the practical deployment of intelligent control in power electronics is becoming increasingly viable and impactful. If the aforementioned challenges and implementation considerations are addressed properly, the promising aspects of the machine learning-based control methods for application areas including industrial automation, sustainable transport, smart grid technologies, energy storage management, and advanced manufacturing systems can provide the paradigm shift in terms of the control mechanism.

6.1 Conclusion and Future Directions

Machine learning based control of power electronic converters has emerged as an advanced approach that offering superior features in terms of improved dynamic response, high adaptability, reduced steady-state error, and high robustness. These control approaches are capable of predictive capabilities and intelligent decision making beyond the limitations of traditional control methods. Therefore, they are able to surpass the capabilities of traditional model-based techniques in terms of achieving high dynamic performance, minimizing energy losses and enhancing system robustness. In this manner, the incorporation of machine learning techniques based control strategies has become critically important. Machine learning-based control methods can adaptively learn and respond to complex system dynamics, nonlinearities, and uncertainties without requiring exact mathematical models. In this context, machine learning-based control methods have demonstrated significant potential in power electronic converters. However, practical utilization still remains limited by challenges including real-time computational constraints, data availability and quality, model generalization, model explainability, hardware limitations, and cybersecurity risks. To promote the application of machine learning algorithms in control of power electronic converters, future research prioritizes the development of general, interpretable, hardware-efficient machine learning models that suitable for embedded platforms, as well as hybrid control architectures that interactively combine classical and data-driven methods. Furthermore, reduced computational cost and secure communication protocols will foster the development of intelligent, resilient, and autonomous systems tailored for the demands of next-generation power electronic converters.

References

Abegaz BW (2020) Dynamic switching control of buck converters using unsupervised machine learning methods. J Eng 2020(12):1155–1164. https://doi.org/10.1049/joe.2020.0086

Aghdam MM, Li L, Zhu J (2020) Comprehensive study of finite control set model predictive control algorithms for power converter control in microgrids. IET Smart Grid 3(1):1–10

Alfred D, Czarkowski D, Teng J (2024) Reinforcement learning-based control of a power electronic converter. Mathematics 12(5):1–22. https://doi.org/10.3390/math12050671

Asif WA, Qureshi MB, Khan MM, Fayyaz MAB, Nawaz R (2023) Optimizing large-scale PV systems with machine learning: a neuro-fuzzy MPPT control for PSCs with uncertainties. Electron 12(7). https://doi.org/10.3390/electronics12071720

Aslay F, Ting NS (2022) Machine learning-based estimation of output current ripple in PFC-IBC used in battery charger of electrical vehicles: a comparison of LR, RF and ANN techniques. IEEE Access 10:50078–50086. https://doi.org/10.1109/ACCESS.2022.3174100

Bindi M et al (2022) Machine learning-based monitoring of DC-DC converters in photovoltaic applications. Algorithms 15(3). https://doi.org/10.3390/a15030074

Boscaino V, Ditta V, Marsala G, Panzavecchia N, Tinè G, Cosentino V, Cataliotti A, Di Cara D (2024) Grid-connected photovoltaic inverters: grid codes, topologies and control techniques. Renew Sust Energ 189:113903

Brunton SL, Kutz JN (2019) Data-driven science and engineering. Cambridge University Press. https://doi.org/10.1017/9781108380690

Çelik Ö (2024) Data-driven MPPT techniques for optimizing vehicular fuel cell performance in hybrid DC microgrid. Int J Hydrogen Energy 79:715–727

Çelik Ö (2025) An efficient control scheme for operational performance enhancement of vehicular fuel cell integrated power system. J Power Sources 625:235660

Çelik Ö, Teke A (2017) A Hybrid MPPT method for grid connected photovoltaic systems under rapidly changing atmospheric conditions. Electr Power Syst Res 152:194–210. https://doi.org/10.1016/j.epsr.2017.07.011

Çelik Ö, Zor K, Tan A, Teke A (2022) A novel gene expression programming-based MPPT technique for PV micro-inverter applications under fast-changing atmospheric conditions. Sol Energy 239:268–282

Chao W, Blaabjerg F (2021) Advanced control of power electronic systems—an overview of methods. Control of power electronic converters and systems. Academic Press, ch 1, pp 1–33

Cheng L, Jianhui H, Jing S (2021) Dual-vector predictive current control of open-end winding PMSM with zero-sequence current hysteresis control. IEEE J Emerg Sel Top Power Electron 10(1):184–195

Chopra N, Fujita M, Ortega R, Spong MW (2022) Passivity-based control of robots: theory and examples from the literature. IEEE Contr Syst Mag 42(2):63–73

Ghidewon-Abay S, Mehrizi-Sani A (2023) Control of a three-phase grid-connected voltage-sourced converter using long short-term memory networks. Energies 16(1). https://doi.org/10.3390/en16010453

Hajihosseini M, Andalibi M, Gheisarnejad M, Farsizadeh H, Khooban MH (2020) DC/DC power converter control-based deep machine learning techniques: real-time implementation. IEEE Trans Power Electron 35(10):9971–9977. https://doi.org/10.1109/TPEL.2020.2977765

Harison C, Sun W, El Mezyani T (2024) Applying scientific machine learning techniques to power electronics parameter prediction. In: 2024 IEEE Texas power energy conf. TPEC 2024, pp 1–6.https://doi.org/10.1109/TPEC60005.2024.10472282

Hemavathi B et al (2024) Machine learning in the era of smart automation for renewable energy materials. e-Prime Adv Electr Eng 7:100458. https://doi.org/10.1016/j.prime.2024.100458

Hu J, Shan Y, Guerrero JM, Ioinovici A, Chan KW, Rodriguez J (2021) Model predictive control of microgrids–an overview. Renew Sust Energ 136:110422

İnci M, Çelik Ö (2025) Repetitive control strategy for grid integration of vehicular fuel cells to enhance system stability and robustness. J Power Sources 642:236992

Jang B, Kim M, Harerimana G, Kim JW (2019) Q-learning algorithms: a comprehensive classification and applications. IEEE Access 7:133653–133667

Li Y, Tong S (2024) Adaptive backstepping control for uncertain nonlinear strict-feedback systems with full state triggering. Automatica 163:111574

Li R et al (2025) A stable machine learning based control for bidirectional interleaved DC/DC converter in battery systems. IET Power Electron 18(1):1–10. https://doi.org/10.1049/pel2.12847

Liao M, Li H, Wang P, Sen T, Chen Y, Chen M (2023) Machine learning methods for feedforward power flow control of multi-active-bridge converters. IEEE Trans Power Electron 38(2):1692–1707. https://doi.org/10.1109/TPEL.2022.3215459

Lopez-Santos O, Dantonio DS, Flores-Bahamonde F, Torres-Pinzon CA (2021) Hysteresis control methods, multilevel inverters: control methods and advanced power electronic applications. Academic Pressch, ch 2, pp 35–60

Maghfiroh H, Nizam M, Anwar M, Ma'Arif A (2022) Improved LQR control using PSO optimization and Kalman filter estimator. IEEE Access 10:18330–18337

Mahesh PV, Meyyappan S, Alla RKR (2022) Maximum power point tracking using decision-tree machine-learning algorithm for photovoltaic systems. Clean Energy 6(5):762–775. https://doi.org/10.1093/ce/zkac057

Mahesh PV, Meyyappan S, Alia RKR (2023) Support vector regression machine learning based maximum power point tracking for solar photovoltaic systems. Int J Electr Comput Eng Syst 14(1):100–108. https://doi.org/10.32985/ijeces.14.1.11

Marahatta A, Rajbhandari Y, Shrestha A, Phuyal S, Thapa A, Korba P (2022) Model predictive control of DC/DC boost converter with reinforcement learning. Heliyon 8(11):e11416. https://doi.org/10.1016/j.heliyon.2022.e11416

Maroti PK, Padmanaban S, Bhaskar MS, Ramachandaramurthy VK, Blaabjerg F (2022) The state-of-the-art of power electronics converters configurations in electric vehicle technologies. Power Electron Dev Comp 1:100001

Mazaheri N, Santamargarita D, Bueno E, Pizarro D, Cobreces S (2024) A deep reinforcement learning approach to DC-DC power electronic converter control with practical considerations. Energies 17(14). https://doi.org/10.3390/en17143578

Meng X, Yu H, Zhang J, Xu T, Wu H, Yan K (2022) Disturbance observer-based feedback linearization control for a quadruple-tank liquid level system. ISA Trans 122:146–162

Mohammadzadeh M, Akbari E, Salameh AA, Ghadamyari M, Pirouzi S, Senjyu T (2022) Application of mixture of experts in machine learning-based controlling of DC-DC power electronics converter. IEEE Access 10(October):117157–117169. https://doi.org/10.1109/ACCESS.2022.3218667

Mousavi Y, Bevan G, Kucukdemiral IB, Fekih A (2022) Sliding mode control of wind energy conversion systems: trends and applications. Renew Sust Energ 167:112734

Omotoso HO et al (2023) Machine learning supervisory control of grid-forming inverters in Islanded mode. Sustain 15(10). https://doi.org/10.3390/su15108018

Özbek NS (2019) An evaluation of model-free control strategies for quadrotor type unmanned aerial vehicles. In: International conference on applied automation and industrial diagnostics (ICAAID), pp 1–6

Özbek NS, Çelik Ö (2022) Design and analysis of a novel adaptive learning control scheme for performance promotion of grid-connected PV systems. Sustain Energy Technol Assess 52:102045

Padmavathi N, Chilambuchelvan A, Shanker NR (2021) Maximum power point tracking during partial shading effect in pv system using machine learning regression controller. J Electr Eng Technol 16(2):737–748. https://doi.org/10.1007/s42835-020-00621-4

Qashqai P, Babaie M, Zgheib R, Al-Haddad K (2023a) A model-free switching and control method for three-level neutral point clamped converter using deep reinforcement learning. IEEE Access 11:105394–105409

Qashqai P, Babaie M, Zgheib R, Al-Haddad K (2023) A model-free switching and control method for three-level neutral point clamped converter using deep reinforcement learning. IEEE Access 11:105394–105409.https://doi.org/10.1109/ACCESS.2023.3318264

Rajamallaiah A, Karri SPK, Shankar YR (2024) Deep reinforcement learning based control strategy for voltage regulation of DC-DC buck converter feeding CPLs in DC microgrid. IEEE Access 12(January):17419–17430. https://doi.org/10.1109/ACCESS.2024.3358412

Srungaram RT, Yadlapati K (2024) Machine-learning based control of bi-modular multilevel PWM inverter for high power applications. PLoS One 19(7):1–23. https://doi.org/10.1371/journal.pone.0305759

Stavrev S, Ginchev D (2024) Reinforcement learning techniques in optimizing energy systems. Electronics (Switzerland) 13(8):1–17. https://doi.org/10.3390/electronics13081459

Tajjour S, Shyam SC (2023) A comprehensive review on sustainable energy management systems for optimal operation of future-generation of solar microgrids. Sustain Energy Technol Assess 58:103377

Wan Y, Xu Q, Dragicevic T (2024) Reinforcement learning-based predictive control for power electronic converters. IEEE Trans Ind Electron 72(5):5353–5364. https://doi.org/10.1109/TIE.2024.3472299

Wang S, Dragicevic T, Gontijo GF, Chaudhary SK, Teodorescu R (2021) Machine learning emulation of model predictive control for modular multilevel converters. IEEE Trans Ind Electron 68(11):11628–11634. https://doi.org/10.1109/TIE.2020.3038064

Yorat E et al (2023) Day-ahead electricity price forecasting using artificial intelligence-based algorithms. In: International conference on innovation and intelligence for informatics, computing, and technologies (3ICT). IEEE, pp 121–126. https://doi.org/10.1109/3ICT60104.2023.10391547

Zengin S (2022) Reinforcement learning-based control of improved hybrid current modulated dual active bridge AC/DC converter. Neural Comput Appl 34(7):5417–5430. https://doi.org/10.1007/s00521-021-06698-w

Zhang G, Li Z, Zhang B, Halang WA (2018) Power electronics converters: past, present and future. Renew Sust Energ 81:2028–2044

Zhang J, Huang Y, Peng F, Yao Y (2023) Pole placement-based current control method for CSI-fed PMSM drive with eliminating capacitor voltage sampling. IEEE Trans Power Electron 38(8):9409–9420

MIX
Papier aus verantwortungsvollen Quellen
Paper from responsible sources
FSC® C105338

If you have any concerns about our products,
you can contact us on
ProductSafety@springernature.com

In case Publisher is established outside the EU,
the EU authorized representative is:
Springer Nature Customer Service Center GmbH
Europaplatz 3, 69115 Heidelberg, Germany

Printed by Libri Plureos GmbH
in Hamburg, Germany